Absolute-Value Inequalities

$|x - a| < b$ is equivalent to $-b < x - a < b$

$|x - a| > b$ is equivalent to $x - a > b$ and $x - a < -b$

The Number e

$e = 2.718281828459\ldots$

Definitions of the Number e

$$e = \lim_{x \to \infty}\left(1 + \frac{1}{x}\right)^x$$

$$e = \lim_{h \to 0}(1 + h)^{1/h}$$

Definition of a Logarithm

$y = \log_b x$ is equivalent to $x = b^y$

Natural Logarithm

$\log_e x = \ln x$

$y = \ln x$ is equivalent to $x = e^y$

Laws of Logarithms

$\log_b MN = \log_b M + \log_b N$

$\log_b \dfrac{M}{N} = \log_b M - \log_b N$

$\log_b M^c = c\log_b M$

Properties of Logarithms

$\log_b b = 1$

$\log_b 1 = 0$

$\log_b b^x = x$

$b^{\log_b x} = x$

Changing From Base b to Base e

$$\log_b x = \frac{\ln x}{\ln b}$$

Review of Graphs

To Find Intercepts

y-intercepts: Set $x = 0$ in an equation and solve for y

x-intercepts: Set $y = 0$ in an equation and solve for x

Vertex (h, k) of a Parabola

Complete the square in x for $f(x) = ax^2 + bx + c$ to obtain $f(x) = a(x - h)^2 + k$. Alternatively, compute the coordinates

$$\left(-\frac{b}{2a}, f\left(-\frac{b}{2a}\right)\right).$$

Even and Odd Functions

Even: $f(-x) = f(x)$;   Symmetry of graph: y-axis

Odd: $f(-x) = -f(x)$;   Symmetry of graph: origin

Rigid Transformations

Graph of $y = f(x)$ for $c > 0$:

$y = f(x) + c$, shifted up c units

$y = f(x) - c$, shifted down c units

$y = f(x + c)$, shifted left c units

$y = f(x - c)$, shifted right c units

$y = f(-x)$, reflection in y-axis

$y = -f(x)$, reflection in x-axis

Asymptotes

If the polynomial functions P and Q have no common factors, then the graph of a rational function

$$f(x) = \frac{P(x)}{Q(x)} = \frac{a_n x^n + \cdots + a_1 x + a_0}{b_m x^m + \cdots + b_1 x + b_0}$$

has a vertical asymptote where $Q(x) = 0$. The graph has the horizontal asymptote

$$y = a_n/b_m \text{ when } n = m,$$

and the horizontal asymptote

$$y = 0 \text{ when } n < m.$$

The graph has no horizontal asymptote when $n > m$. The graph has a slant asymptote when $n = m + 1$.

Essentials of

PRECALCULUS

with Calculus Previews

SIXTH EDITION

The Jones & Bartlett Learning Series in Mathematics

For more information on this series and its titles, please visit us online at http://www.jblearning.com. Qualified instructors, contact your Publisher's Representative at 1-800-832-0034 or info@jblearning.com to request review copies for course consideration.

The Jones & Bartlett Learning International Series in Mathematics

Essentials of Mathematical Statistics
Albright (978-1-4496-8534-8) ©2014

A Transition to Mathematics with Proofs
Cullinane (978-1-4496-2778-2) © 2013

Linear Algebra: Theory and Applications, Second Edition, International Version
Cheney/Kincaid (978-1-4496-2731-7) © 2012

Multivariable Calculus
Damiano/Freije (978-0-7637-8247-4) © 2012

Complex Analysis for Mathematics and Engineering, Sixth Edition, International Version
Mathews/Howell (978-1-4496-2870-3) © 2012

A Journey into Partial Differential Equations
Bray (978-0-7637-7256-7) © 2012

Association Schemes of Matrices
Wang/Huo/Ma (978-0-7637-8505-5) © 2011

Advanced Engineering Mathematics, Fifth Edition, International Version
Zill/Wright (978-1-4496-8980-3) © 2014
Canada/UK (978-1-4496-9302-2)

Calculus: Early Transcendentals, Fourth Edition, International Version
Zill/Wright (978-0-7637-8652-6) © 2011

Real Analysis
Denlinger (979-0-7637-7947-4) © 2011

Mathematical Modeling for the Scientific Method
Pravica/Spurr (978-0-7637-7946-7) © 2011

Mathematical Modeling with Excel®
Albright (978-0-7637-6566-8) © 2010

An Introduction to Analysis, Second Edition
Bilodeau/Thie/Keough (978-0-7637-7492-9) © 2010

Basic Real Analysis
Howland (978-0-7637-7318-2) © 2010

For more information on this series and its titles, please visit us online at http://www.jblearning.com. Qualified instructors, contact your Publisher's Representative at 1-800-832-0034 or info@jblearning.com to request review copies for course consideration.

Essentials of
PRECALCULUS
with Calculus Previews

SIXTH EDITION

Dennis G. Zill
Loyola Marymount University

Jacqueline M. Dewar
Loyola Marymount University

JONES & BARTLETT
LEARNING

World Headquarters
Jones & Bartlett Learning
5 Wall Street
Burlington, MA 01803
978-443-5000
info@jblearning.com
www.jblearning.com

Jones & Bartlett Learning books and products are available through most bookstores and online
booksellers. To contact Jones & Bartlett Learning directly, call 800-832-0034, fax 978-443-8000, or visit
our website, www.jblearning.com.

Production Credits
Publisher: Cathy L. Esperti
Acquisitions Editor: Laura Pagluica
Editorial Assistant: Taylor Ferracane
Senior Production Editor: Tiffany Sliter
Associate Production Editor: Sara Kelly
Associate Marketing Manager: Cassandra Peterson
VP, Manufacturing and Inventory Control: Therese Connell
Composition: Aptara®, Inc.
Cover Design: Kristin E. Parker
Rights and Photo Research Coordinator: Ashley Dos Santos
Cover, title, half title, and back cover image: © ESO
Printing and Binding: LSC Communications
Cover Printing: LSC Communications

Library of Congress Cataloging-in-Publication Data
Zill, Dennis G.
    Essentials of Precalculus with Calculus Previews / Dennis G. Zill, Loyola Marymount University,
    Jacqueline M. Dewar, Loyola Marymount University.
    —Sixth edition.
        pages ; cm
Includes bibliographical references and index.
ISBN 978-1-284-05632-7 (casebound)
1. Precalculus—Textbooks. I. Dewar, Jacqueline M., author. II. Title.
QA39.3.Z55 2016
512'.1—dc23
                                                                            2014031642

6048
Printed in the United States of America
21 20 19 18 17 10 9 8 7 6 5 4 3

Contents

© Brian A Jackson/ShutterStock, Inc.

© gualtiero boffi/ShutterStock, Inc.

Contents

Preface

To the Instructor

☐ About This Text

As the first word in the title implies, *Essentials of Precalculus with Calculus Previews* 6e is the shorter of our two precalculus texts published by Jones & Bartlett Learning. In this text we have deliberately kept the coverage to a reasonable number of topics to allow time for instructors to work with their students to strengthen their algebraic, logarithmic, and trigonometric skills and yet cover most of the material in a one-term course. Instructors wishing a wider variety of topics such as systems of equations, sequences, series, mathematical induction, and principles of counting are referred to our *Precalculus with Calculus Previews* 5e.

Emphases   The focus of this text is squarely on basic mathematics encountered in a typical calculus course. Many times we have seen students in a calculus class perform an operation such as differentiation of a function flawlessly, but fail to complete the problem successively because of difficulties in simplifying the resulting expression or in solving a related equation. So in this new edition we continue to emphasize the function concept while making an additional effort to reinforce algebraic and trigonometric skills. Through the many examples and numerous and varied exercises, we provide opportunities for students to practice operations such as factoring, expanding a power of a binomial, completing the square, synthetic and long division, rationalization, and solving inequalities and equations in situations similar to those they will encounter in calculus. Throughout we stress the importance of being familiar with key formulas from algebra, the laws of exponents, the laws of logarithms, and fundamental trigonometric definitions and identities. Marginal side notes and in-text annotations fill in the details of solutions of examples and convey additional information to the reader.

Notes from the Classroom   Selected sections of this text conclude with remarks called *Notes from the Classroom*. These remarks are aimed directly at the student and address a variety of student/textbook/classroom/calculus issues such as alternative terminology, reinforcement of important concepts, what material is or is not recommended for memorization, misinterpretations, common errors, solution procedures, calculators, and advice on the importance of neatness and organization.

Calculus Previews   Each of the six chapters in this text concludes with a section subtitled *Calculus Previews*. Each of these special sections is devoted to a single calculus concept (for example, *The Tangent Line Problem*). The discussion of these topics is kept at a level easily within the reach of a precalculus student. The emphasis in these previews is *not* on the calculus; the calculus topic provides a framework and motivation for the precalculus mathematics used in the discussion. The focus in these sections is on the algebraic, logarithmic, and trigonometric manipulations that are necessary for the successful completion of typical calculus problems related to the *Calculus Preview* topic. We wish to emphasize that the limit concept covered in the *Calculus Preview* for Chapter 1 does not impinge (except for a few words and the arrow notation) on other sections in the text. We also realize that the introduction to limits in Section 1.5 may be a bit ambitious for some classes. However, we feel that the algebra discussed in that section (factoring, binomial expansions, rationalization of a denominator, and so on) should be covered, but the discussion of the limit concept could be postponed and taught, perhaps in conjunction with Section 2.10, at a more appropriate time in the course. Thus, in Exercises 1.5 most of the problems are given in (a) and (b) format, where the part (a) of the problem involves algebra and part (b) is optional because it involves a limit. Of course, all the *Calculus Previews* could be delayed to the end of the course providing a bridge to a subsequent course in calculus. If time does not allow, these topics could be covered lightly or skipped entirely.

Gradual Use of Calculus Terminology Calculus-related words such as "continuous function" are used where appropriate. The idea is to give the student a good intuitive sense of what these words mean prior to their exposure to their formal definitions in calculus.

Building Functions from Words As teachers we know that the related rate and applied max-min, or optimization, problems can be a discouraging experience for some students of calculus. Correctly interpreting the words of such a problem in order to set up an equation or a function is a challenge for many students. It follows then that it is appropriate to emphasize such material in a precalculus course. In Section 2.9, entitled *Building a Function from Words*, we begin by illustrating how to translate a verbal description into a symbolic representation of a function. We then present actual problems taken from *Calculus* 4e by Dennis G. Zill and Warren S. Wright (Jones & Bartlett Learning, 2011), and demonstrate how to analyze the statement of the problem and transform those words into an objective function. We discuss the importance of drawing pictures, using variables to describe pertinent quantities, identifying a constraint between the variables, using the constraint to eliminate an extra variable, and observing that the domain of the objective function may not be the same as its implicit domain. To ensure that the focus is squarely on the process of fashioning a symbolic function from the words, we do not discuss how such optimization problems are actually solved.

Exercises The exercise sets contain a wide variety of different kinds of problems. In addition to the usual drill and applied problems, many of the exercise sets conclude with conceptual problems that are labeled *For Discussion*. We hope that instructors will utilize these problems, which are primarily conceptual in nature, and their expertise to engage in a classroom exchange of ideas with the students on how these problems can be solved. These problems could also be the basis for assigned writing projects. To encourage original thought we purposely have not included answers to these problems.

Final Examination Following the six chapters of the text we present a list of 70 questions called the *Final Examination*. This "test" consists of fill in the blank questions, true/false questions, and exercises that review topics in the six chapters of the text. It was not our intention to emulate an actual final examination in a precalculus course, rather our thought was to offer a vehicle for an informal wrap-up of the entire course. We suggest that a part of a class period be devoted to a discussion of these questions to help students prepare for their actual final examination and their subsequent transition to calculus. To facilitate the students' review, the answers of the *Final Examination* are given both in the *Student Resource Manual* as well as in the instructor's *Complete Solutions Manual*. Of course, the instructor is free to utilize this material in whatever manner he/she chooses (including ignoring it completely).

New to the Sixth Edition

Here are the changes made to the previous edition.

- A new section on implicitly defined functions has been added to Chapter 2 (Section 2.7). Although this topic was touched on in the fifth edition, we felt that it deserved a special consideration given its importance in calculus and in the definition of an inverse function.
- Many sections have been partially rewritten to improve their clarity.
- By request, additional examples and figures were inserted to improve the readability of the text.
- A discussion of the area of a circular sector has been added to Chapter 4 (Section 4.1).
- By request, problems on verification of trigonometric identities have been added to Chapter 4 (Exercises 4.5).
- A new section on the Product-to-Sum and Sum-to-Product trigonometric identities has been added to Chapter 4 (Section 4.6). This material was added because these formulas are arguably useful in solving certain kinds of trigonometric equations and in calculus when techniques of integration are studied.
- To reinforce the definitions of the inverse trigonometric functions, the discussion of trigonometric equations (Section 4.8) has been placed after the coverage of these inverse functions (Section 4.7).
- The discussion of applications of right triangles has been expanded and given its own section (Section 4.11).
- We felt that covering both Laws of Sines and the Law of Cosines in one section was asking a lot of the student. So these topics now are discussed in separate sections (Sections 4.12 and 4.13).
- The Chapter Review Exercises have been partitioned into three parts: A (fill in the blanks), B (true/false), and C (exercises).
- We did not think that there was a sufficient emphasis of solving equations involving exponential and logarithmic functions in the previous edition. As a consequence, exponential and logarithmic equations are now covered in a separate section (Section 5.3).
- Most of the exercise sets have been updated and many new problems have been added. We also feel that the problems added to the sections on triangle trigonometry and to the section on exponential and logarithmic models may be of particular interest (Exercises 4.11, Chapter 4 Review Exercises, Exercises 5.4).
- The *Final Examination* at the end of the text has been expanded.

Supplements for the Instructor

The following materials are available online at

go.jblearning.com/precalc6e

Complete Solutions Manual (CSM) prepared by Warren S. Wright provides worked-out solutions for every problem in the text as well as answers to all questions in the *Final Examination*.

Computerized Testing System or both Mac OS® and Windows® computer operating systems. This testing system allows instructors to create customized quizzes and tests. The questions and answers are sorted by chapter and can be easily installed on a computer. Publisher-supplied .rtf files can also be uploaded to the instructor's learning management system.

Image Bank in PowerPoint format features all labeled figures as they appear in the text. This useful tool allows instructors to easily display and discuss figures and problems that appear in the textbook.

WebAssign® developed by instructors for instructors is a premier independent online teaching and learning environment, guiding several million students through their academic careers since 1997. With WebAssign, instructors can create and distribute assignments using selected questions specific to this textbook. Instructors can also grade, record, and analyze student responses and performance instantly; offer more practice exercises, quizzes, and homework; and upload additional resources to share and communicate with their students seamlessly, such as the image bank in PowerPoint format and the test items supplied by Jones & Bartlett Learning's Computerized Testing System.

eBook Format offers the complete textbook in eBook format for instructors and students through WebAssign.

CourseSmart® is a new way for instructors and students to access this textbook in digital format, anytime from anywhere. Jones & Bartlett Learning has partnered with CourseSmart to make many of our leading mathematics textbooks available in the CourseSmart eTextbook store. For more information on CourseSmart Editions please visit

www.jblearning.com/elearning/econtent/coursesmart/

Please contact your Jones & Bartlett Learning Account Specialist for information on, and access to, online demonstrations of the supplements and services described above.

☐ Supplements for the Student

Student Resource Manual (SRM) prepared by Warren S. Wright. Unlike the traditional student solutions manual, where a selected subset of the problems are worked out, the *SRM* is divided into five parts: *Algebra Topics, Use of a Calculator, Basic Skills, Selected Solutions,* and *Answers to the Final Examination.* In *Algebra Topics,* selected topics from algebra (such as multiplication of an inequality by an unknown, implicit conditions in a word problem, Pascal's triangle, factoring techniques, binomial expansions, rationalizations of numerators and denominators, adding symbolic fractions, long division of polynomials, synthetic division of polynomials, factorial notation, and so on) are reviewed because of their relevance to calculus. Because we do not discuss how to use technology within the text proper, we have devoted the section *Use of a Calculator* to the review of graphing calculator essentials. In *Selected Solutions,* a detailed solution of every third problem in the exercise sets is given. *Answers to the Final Examination* is a list of answers for all the questions in the *Final Examination.*

Available in both print and online formats, this student manual can be purchased separately or ordered bundled with the textbook at a substantial savings.

Exploring Mathematics: Solving Problems with the TI-84 Plus Graphing Calculator by Jeffery M. Gervasi, Ed.D., Cuesta College, San Luis Obispo, CA is a graphing calculator manual that can be ordered either through the bookstore or online directly from Jones & Bartlett Learning.

WebAssign Access Card can be bundled with this text or purchased separately by the student online at

www.webassign.com

eBook with Course Access Card can be bundled with this text or purchased separately by the student online at

www.webassign.com

CourseSmart is a new way for students to access college textbooks in digital format. Jones & Bartlett Learning has partnered with CourseSmart to make this textbook available in the CourseSmart eTextbook store. The CourseSmart Edition has many features designed to make studying more efficient such as highlighting, online search, note taking, and print capabilities. For more information on purchasing this CourseSmart Edition please visit

www.jblearning.com/elearning/econtent/coursesmart/

To the Student

After teaching collegiate mathematics for many years, we have seen almost every type of student, from a budding genius who invented his own calculus, to students who struggled to master the most rudimentary mechanics of the subject. Frequently the source of difficulty in calculus can be traced to weak algebra skills or an inadequate background in trigonometry. Calculus builds immediately on your prior knowledge and skills because there is much new ground to be covered. Consequently there is very little time to review precalculus mathematics in the calculus classroom. So those who teach calculus must assume that you can factor, simplify and solve equations, solve inequalities, handle absolute-value equations and inequalities, use a calculator, correctly apply the laws of exponents, find equations of lines, plot points, sketch basic graphs, and apply important trigonometric and logarithmic identities. The ability to do algebra and trigonometry, work with exponentials and logarithms, and sketch *by hand* basic graphs quickly and accurately, are keys to success in a calculus course.

In this text we have tried to give you as much help as possible within the confines of the printed page using such features as marginal annotations, arrow annotations within examples, notes of caution, *Notes From the Classroom*, and the *Final Examination*. The many marginal and in-text annotations provide additional information or further explanation of the steps in the solution of an example. The *Student Resource Manual* (described above) was written just for you. It contains review material not found in the text, extra examples, information on calculators, solutions of problems, and answers to the *Final Examination*.

Those of us who teach and write mathematics texts strive to communicate clearly *how* to do mathematics. This text reflects our philosophy that a mathematics text for the beginning college/university level should be readable, straightforward, and loaded with motivation. The principal reason for studying precalculus is to become well-prepared for calculus. To show you how the material covered in this text is essential for success in calculus, we end each chapter with a section called *Calculus Preview*. In each of these previews a calculus problem provides a framework and motivation for precalculus mathematics and shows you how this mathematics plays a vital role in solving the problem.

Finally, we caution you that *learning* mathematics is not like learning how to ride a bicycle, that once learned, the ability sticks for a lifetime. Mathematics is more like learning another language or learning to play a musical instrument; it requires time and effort to memorize basic formulas and to understand when and how to apply them, and most importantly, it requires a lot of practice to develop and maintain proficiency. Even experienced musicians practice the fundamental scales before playing their instrument. So ultimately, you the student can learn mathematics (that is, make it stick) only through the hard work of doing mathematics.

We wish you the best of luck in this preparatory course and in your subsequent study of calculus.

Acknowledgments

We would like to take this opportunity to extend our heartfelt thanks to everyone at Jones & Bartlett Learning who were involved in the production of this text. But a special appreciation goes to our editor Laura Pagluica, and to Tiffany Sliter who, in addition to her real jobs as Senior Production Editor and endless-meetings attendee, once again guided the manuscript through the shoals of production with skill and unflagging patience.

We would also like to acknowledge the students and instructors for their feedback on the previous editions as well as the following reviewers who were generous enough to take time out from their busy schedules to offer suggestions for improvement:

Amal Mostafa, New Mexico State Univeristy

Marwan Abu-Sawwa, Florida State College at Jacksonville

We end with our usual apology for any mathematical or typographical errors that you may find in the text. These are the sole fault of the authors. In order to correct errors expeditiously, please send them directly to our editor at:

LPagluica@jblearning.com

Dennis G. Zill

Jacqueline M. Dewar

1 Inequalities, Equations, and Graphs

Chapter Contents

1.1 The Real Line

INTRODUCTION In calculus you will study quantities described by real numbers. Therefore, we begin with a review of the set of real numbers using the terminology and notation you will encounter in calculus.

☐ **Real Number System** Recall that the set R of **real numbers** consists of numbers that are either **rational** or **irrational**. Rational numbers are numbers of the form a/b, where a and $b \neq 0$ are integers. For example, -3, $-\frac{1}{2}$, $\frac{2}{3}$, 5, and $\frac{127}{4}$ are rational numbers. Irrational numbers are numbers that are not rational, that is, they are numbers that cannot be expressed as a quotient of integers. For example, $\sqrt{2}$ and π are irrational numbers. Every real number can also be written as a decimal. A rational number can be expressed either as a *terminating decimal*, such as $\frac{1}{8} = 0.125$, or a *nonterminating and repeating decimal*, such as $\frac{1}{3} = 0.333\ldots$. Repeating decimals, such as $0.666\ldots$ and $8.545454\ldots$, are often written as $0.\overline{6}$ and $8.\overline{54}$, respectively, where the bar indicates the digit or block of digits that repeat. An irrational number is always a *nonterminating and nonrepeating decimal* such as $\sqrt{2} = 1.41421\ldots$ or $\pi = 3.14159\ldots$. The following chart summarizes the relationship between the principal sets of real numbers.

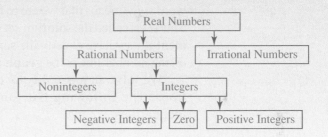

☐ **The Number Line** The set R of real numbers can be put into a one-to-one correspondence with the set of points on a line. As a consequence, we can visualize or represent real numbers as points on a *horizontal line* called the **number line** or **coordinate line**. The point chosen to represent the number 0 is called the **origin**. The direction to the right of 0 is said to be the **positive direction** on the number line; the direction to the left of 0 is the **negative direction**. Real numbers corresponding to points to the right of 0 are called **positive numbers** and numbers corresponding to points to the left of 0 are **negative numbers**. As indicated in FIGURE 1.1.1, the number 0 is considered to be neither positive nor negative. From here on, we will not distinguish between a point on the number line and the number that corresponds to this point.

FIGURE 1.1.1 The real number line

☐ **Inequalities** The number line is useful in demonstrating order relations between two real numbers a and b. As shown in FIGURE 1.1.2, we say that the number a is **less than** the number b, and write $a < b$, whenever the number a lies to the left of the number b on the number line. Equivalently, because the number b lies to the right of a on the number line we say that b is **greater than** a and write $b > a$. For example, $4 < 9$ is the same as $9 > 4$. We also use the notation $a \leq b$ if the number a is either **less than**

FIGURE 1.1.2 a is less than b

or equal to the number b. Similarly, $b \geq a$ means b is **greater than or equal** to a. For example, $2 \leq 5$ since $2 < 5$. Also, $4 \geq 4$ because $4 = 4$. For any two real numbers a and b, exactly *one* of the following is true:

A *less than* inequality $a < b$ can always be written as an equivalent *greater than* inequality $b > a$.

$$a < b, \qquad a = b, \qquad \text{or} \qquad a > b.$$

The symbols $<$, $>$, $\geq$, and $\leq$ are called **inequality symbols** and expressions such as $a < b$ or $b \geq a$ are called **inequalities**. The inequality $a > 0$ means the number a lies to the right of the number 0 on the number line, and so a is **positive**. We signify that a number a is **negative** by the inequality $a < 0$. Because the inequality $a \geq 0$ means a is either greater than 0 (positive) or equal to 0 (which is neither positive nor negative), we say that a is **nonnegative**.

☐ **Solving Inequalities** We are interested in solving various kinds of inequalities containing a variable. If a real number a is substituted for the variable x in an inequality such as

$$8x + 4 < 16 + 5x, \tag{1}$$

and if the result is a true statement, then a is said to be a **solution** of the inequality. For example, -2 is a solution of (1) because if x is replaced by -2, then the resulting inequality $8(-2) + 4 < 16 + 5(-2)$ simplifies to the true statement $-12 < 6$. The word *solve* means that we are to find the set of *all* solutions of an inequality such as (1). This set is called the **solution set** of the inequality. Two inequalities are said to be **equivalent** if they have exactly the same solution set. The representation of the solution set on the number line is the **graph** of the inequality.

We solve an inequality by finding an equivalent inequality with obvious solutions. The following list summarizes three operations that yield equivalent inequalities.

THEOREM 1.1.1   Properties of Inequalities

Suppose a and b are real numbers and c is a nonzero real number. Then the inequality $a < b$ is equivalent to:

   (*i*) $a + c < b + c$
  (*ii*) $ac < bc$, for $c > 0$
 (*iii*) $ac > bc$, for $c < 0$

Property (*iii*) of Theorem 1.1.1 is frequently forgotten. In words, (*iii*) states that:

> *If an inequality is multiplied by a negative number, then the direction of the resulting inequality is reversed.*

For example, if we multiply the inequality $-2 < 5$ by -3 then the *less than* symbol is changed to a *greater than* symbol:

$$-2(-3) > 5(-3) \qquad \text{or} \qquad 6 > -15.$$

EXAMPLE 1   Solving the Inequality (1)

Solve $8x + 4 < 16 + 5x$.

Solution We solve the inequality by using the properties of inequalities to obtain a sequence of equivalent inequalities:

$$8x + 4 < 16 + 5x$$
$$8x + 4 - 4 < 16 + 5x - 4 \quad \leftarrow \text{by } (i) \text{ of Theorem 1.1.1}$$
$$8x < 12 + 5x$$
$$8x - 5x < 12 + 5x - 5x \quad \leftarrow \text{by } (i) \text{ of Theorem 1.1.1}$$
$$3x < 12$$
$$\left(\tfrac{1}{3}\right)3x < \left(\tfrac{1}{3}\right)12 \quad \leftarrow \text{by } (ii) \text{ of Theorem 1.1.1}$$
$$x < 4.$$

Using set-builder notation, the solution set is $\{x \mid x \text{ real and } x < 4\}$. ∎

FIGURE 1.1.3 Solution set in Example 1 in interval notation is $(-\infty, 4)$

☐ **Interval Notation** The solution set in Example 1 is graphed on the number line in FIGURE 1.1.3 as a colored arrow over the line pointing to the left. In the figure, the right parenthesis at 4 indicates that the number 4 is *not* included in the solution set. Because the solution set extends indefinitely to the left—the negative direction—the inequality $x < 4$ can also be written as $-\infty < x < 4$, where ∞ is the **infinity symbol**. In other words, the solution set of the inequality $x < 4$ is

$$\{x \mid x \text{ real and } x < 4\} = \{x \mid -\infty < x < 4\}.$$

Using **interval notation** this set of real numbers is written $(-\infty, 4)$ and is an example of an **unbounded interval**. Table 1.1.1 summarizes various inequalities and their solution sets, as well as interval notations, names, and graphs. In each of the first four entries

TABLE 1.1.1   Inequalities and Intervals

Inequality	Solution Set	Interval Notation	Name	Graph
$a < x < b$	$\{x \mid a < x < b\}$	(a, b)	Open interval	
$a \leq x \leq b$	$\{x \mid a \leq x \leq b\}$	$[a, b]$	Closed interval	
$a < x \leq b$	$\{x \mid a < x \leq b\}$	$(a, b]$	Half-open interval	
$a \leq x < b$	$\{x \mid a \leq x < b\}$	$[a, b)$	Half-open interval	
$a < x$	$\{x \mid a < x < \infty\}$	(a, ∞)		
$a \leq x$	$\{x \mid a \leq x < \infty\}$	$[a, \infty)$		
$x < b$	$\{x \mid -\infty < x < b\}$	$(-\infty, b)$	Unbounded intervals	
$x \leq b$	$\{x \mid -\infty < x \leq b\}$	$(-\infty, b]$		
$-\infty < x < \infty$	$\{x \mid -\infty < x < \infty\}$	$(-\infty, \infty)$		

of the table, the numbers a and b are called the **endpoints** of the interval. As a set, the **open interval**

$$(a, b) = \{x \mid a < x < b\}$$

does not include either endpoint, whereas the **closed interval**

$$[a, b] = \{x \mid a \leq x \leq b\}$$

includes both endpoints. Note, too, that the graph of the last interval in Table 1.1.1, which extends indefinitely both to the left and to the right, is the entire real number line. In calculus the interval notation $(-\infty, \infty)$ is generally used to represent the set R of real numbers.

A word of caution is in order as you peruse Table 1.1.1. The **infinity symbols** $-\infty$ ("minus infinity") and ∞ ("infinity") do not represent real numbers and should *never* be manipulated arithmetically like a number. The infinity symbols are merely notational devices: $-\infty$ and ∞ are used to indicate unboundedness in the negative direction and in the positive direction, respectively. Thus when using interval notation, the symbols $-\infty$ and ∞ can never appear next to a square bracket. For example, the expression $(2, \infty]$ is meaningless.

An inequality of the form $a < x < b$ is sometimes referred to as a **simultaneous inequality** because the number x is *between* the numbers a and b. In other words, $x > a$ *and* simultaneously $x < b$. For example, the real numbers that satisfy $2 < x < 5$ is the intersection of the intervals defined by the inequalities $2 < x$ and $x < 5$. Recall that the **intersection** of two sets A and B, written $A \cap B$, is the set of elements that are in A *and* in B—in other words, the elements that are common to both sets. As illustrated in FIGURE 1.1.4 by the overlapping arrows extending indefinitely to the right and to the left, the solution set of the inequality $2 < x < 5$ can be written as the intersection $(2, \infty) \cap (-\infty, 5) = (2, 5)$.

FIGURE 1.1.4 The numbers in (2, 5) are the numbers common to both (2, ∞) and (−∞, 5)

▌EXAMPLE 2 Solving a Simultaneous Inequality

Solve $-2 \leq 1 - 2x < 3$.

Solution As previously discussed, one way of proceeding is to solve two inequalities:

$$-2 \leq 1 - 2x \qquad \text{and} \qquad 1 - 2x < 3$$

and then take the intersection of the two solution sets. A faster method is to solve both of the inequalities simultaneously in the following manner:

$$-2 \leq 1 - 2x < 3$$
$$-1 - 2 \leq -1 + 1 - 2x < -1 + 3 \qquad \leftarrow \text{by } (i) \text{ of Theorem 1.1.1}$$
$$-3 \leq -2x < 2.$$

We isolate the variable x in the middle of the last simultaneous inequality by multiplying by $-\frac{1}{2}$:

$$\left(-\tfrac{1}{2}\right)(-3) \geq \left(-\tfrac{1}{2}\right)(-2x) > \left(-\tfrac{1}{2}\right)2 \qquad \leftarrow \text{by } (iii) \text{ of Theorem 1.1.1}$$
$$\tfrac{3}{2} \geq x > -1,$$

where we note that multiplication by the negative number has reversed the direction of the inequalities. To express this inequality in interval notation, we first rewrite it with the leftmost number on the number line on the left side of the inequality: $-1 < x \leq \frac{3}{2}$. The solution set of the last inequality is the half-open interval $\left(-1, \frac{3}{2}\right]$; the square bracket on the right signifies that $\frac{3}{2}$ is included in the solution set. The graph of this interval is given in FIGURE 1.1.5. ▬

FIGURE 1.1.5 Solution set in Example 2

☐ **Sign-Chart Method** In Examples 1 and 2 we solved **linear inequalities** in one variable x, that is, inequalities that can be put into one of the forms

$$ax + b < 0, \quad ax + b \leq 0, \quad ax + b > 0, \quad ax + b \geq 0,$$

where $a \neq 0$. In the next several examples we illustrate the **sign-chart method** used in calculus for solving **nonlinear inequalities**. A nonlinear inequality in one variable x is simply an inequality that is not linear. For example, $x^2 \geq -2x + 15$ is a nonlinear inequality because of the presence of the x^2 term. The two properties of real numbers given next are fundamental to constructing a sign chart of an inequality.

THEOREM 1.1.2   Sign Properties of Products

(*i*) The product of two real numbers is **positive** if and only if the numbers have the same signs, that is, either $(+)(+)$ or $(-)(-)$.

(*ii*) The product of two real numbers is **negative** if and only if the numbers have opposite signs, that is, $(+)(-)$ or $(-)(+)$.

Here are some of the basic steps of the sign-chart method illustrated in the next example.

Guidelines for the Sign-Chart Method

- Use the properties of inequalities to recast the given inequality into a form where all variables and nonzero constants are on the same side of the inequality symbol and the number 0 on the other side.
- Then, if possible, factor the expression involving the variables and constants into linear factors $ax + b$.
- Mark the number line at the points where the factors are zero. These points divide the number line into intervals.
- In each of these intervals, determine the sign of each factor and the corresponding sign of the product using (*i*) and (*ii*) of Theorem 1.1.2.

EXAMPLE 3   **Solving a Nonlinear Inequality**

Solve $x^2 \geq -2x + 15$.

Solution We begin by rewriting the inequality with all terms to the left of the inequality symbol and 0 to the right. By (*i*) of Theorem 1.1.1,

$$x^2 \geq -2x + 15 \qquad \text{is equivalent to} \qquad x^2 + 2x - 15 \geq 0.$$

Factoring, the last expression is the same as $(x + 5)(x - 3) \geq 0$.

Then we indicate on the number line where each factor is 0—in this case, $x = -5$ and $x = 3$. As shown in **FIGURE 1.1.6**, this divides the number line into three disjoint, or nonintersecting, intervals: $(-\infty, -5)$, $(-5, 3)$, and $(3, \infty)$. Note, too, that since the given inequality requires the product to be nonnegative, that is, "greater than or *equal to* 0," the numbers -5 and 3 are two solutions. Next, we must determine the signs of the factors $x + 5$ and $x - 3$ on each of the three intervals. We are looking for those intervals on which the two factors are either both positive or both negative, for then their product will be positive. Since the linear factors $x + 5$ and $x - 3$ cannot change signs within these intervals, it suffices to obtain the sign of each factor at just *one* test value

$(-\infty, -5)$ $(-5, 3)$ $(3, \infty)$

−5   0   3

FIGURE 1.1.6 Three disjoint intervals

See (*i*) of Theorem 1.1.2 ▶

chosen from inside each interval. For example, on the interval $(-\infty, -5)$, if we use $x = -10$, then

Interval	$(-\infty, -5)$	
Sign of $x + 5$	$-$	$\leftarrow$ at $x = -10, x + 5 = -10 + 5 < 0$
Sign of $x - 3$	$-$	$\leftarrow$ at $x = -10, x - 3 = -10 - 3 < 0$
Sign of $(x + 5)(x - 3)$	$+$	$\leftarrow (-)(-)$ is $(+)$

Continuing in this manner for the remaining two intervals we get the sign chart in FIGURE 1.1.7. As can be seen from the third line of this figure, the product $(x + 5)(x - 3)$ is nonnegative on either of the unbounded intervals $(-\infty, -5]$ or $[3, \infty)$.

FIGURE 1.1.7 Sign chart for Example 3

Because the solution set in Example 3 consists of two nonintersecting, or **disjoint**, intervals it cannot be expressed as a single interval. The best we can do is to write the solution set as the union of the two intervals. Recall that the **union** of two sets A and B, written $A \cup B$, is the set of elements that are in either A or in B, or in both. Thus the solution set in Example 3 can be written $(-\infty, -5] \cup [3, \infty)$.

■ EXAMPLE 4　　Solving a Nonlinear Inequality

Solve $(x - 4)^2(x + 8)^3 > 0$.

Solution Since the given inequality already has the form appropriate for the sign-chart method (a factored expression to the left of the inequality symbol and 0 to the right), we begin by finding the numbers where each factor is 0, in this case, $x = 4$ and $x = -8$. We place these numbers on the number line and determine three intervals. Then in each interval we consider the signs of the powers of each linear factor. Because of the even power, we see that $(x - 4)^2$ is never negative. However, because of the odd power, $(x + 8)^3$ has the same sign as the factor $x + 8$. Observe that the numbers $x = 4$ and $x = -8$ are not solutions of the inequality because of the "greater than" symbol. Therefore, as we see in FIGURE 1.1.8, the solution set is $(-8, 4) \cup (4, \infty)$.

$$
\begin{array}{lcccccc}
(x-4)^2 & +\ + & + & +\ + & 0 & +\ + \\
(x+8)^3 & -\ - & 0 & +\ + & + & +\ + \\
(x-4)^2\,(x+8)^3 & -\ - & 0 & +\ + & 0 & +\ + \\
\end{array}
$$

$$\xleftarrow{\qquad\quad(\ \underset{-8}{}\qquad\quad \underset{4}{\times}\qquad\quad}\longrightarrow$$

FIGURE 1.1.8 Sign chart for Example 4

CHAPTER 1 INEQUALITIES, EQUATIONS, AND GRAPHS

Solve $x \le 3 - \dfrac{6}{x + 2}$.

Solution We begin by rewriting the inequality with all variables and nonzero constants to the left and 0 to the right of the inequality sign,

$$x - 3 + \frac{6}{x + 2} \le 0.$$

Next we put the terms over a common denominator,

$$\frac{(x - 3)(x + 2) + 6}{x + 2} \le 0 \quad \text{and simplify to} \quad \frac{x(x - 1)}{x + 2} \le 0. \tag{2}$$

Now the numbers that make the three linear factors in the last expression equal to 0 are $-2, 0,$ and 1. On the number line these three numbers determine four intervals. As a result of the "less than or *equal to* 0," we see that 0 and 1 are members of the solution set. However, -2 is excluded from the solution set since substituting this value into the fractional expression results in a zero denominator (making the fraction undefined). As we can see from the sign chart in FIGURE 1.1.9, the solution set is the union of intervals $(-\infty, -2) \cup [0, 1]$.

One thing we *don't do* is clear the denominator by multiplying the inequality by the variable expression $x + 2$. See Problem 70 in Exercises 1.1.

FIGURE 1.1.9 Sign chart for Example 5

NOTES FROM THE CLASSROOM

(*i*) Terminology used in mathematics often varies from teacher to teacher and from textbook to textbook. For example, inequalities using the symbols $<$ or $>$ are sometimes called *strict* inequalities, whereas inequalities using $\le$ or $\ge$ are called *nonstrict*. As another example, the *positive integers* 1, 2, 3, . . . are often referred to as the *natural numbers*.

(*ii*) Suppose the solution set of an inequality consists of the numbers such that $x < -1$ *or* $x > 3$. An answer seen very often on homework, quizzes, and tests is $3 < x < -1$. This is a misunderstanding of the notion of *simultaneity*. The statement $3 < x < -1$ means that $x > 3$ *and* at the same time $x < -1$. If you sketch this on the number line you will see that it is impossible for the same x to satisfy both inequalities. The best we can do in rewriting "$x < -1$ or $x > 3$" is to use the union of intervals $(-\infty, -1) \cup (3, \infty)$.

(*iii*) Here is another frequent error: The notation $a < x > b$ is meaningless. If, say, we have $x > -2$ *and* $x > 6$, then only the numbers $x > 6$ satisfy *both* conditions.

(*iv*) In the classroom we frequently hear the response "positive" when in reality the student means "nonnegative." Question: x under the square root sign $\sqrt{x}$ must be positive, right? Raise your hand if you agree. Invariably, lots of hands go up. Correct answer: x must be nonnegative, that is, $x \ge 0$. Don't forget that $\sqrt{0} = 0$.

Exercises 1.1   Answers to selected odd-numbered problems begin on page ANS–1.

In Problems 1–6, write the given statement as an inequality.

1. $a + 2$ is positive

2. $4y$ is negative

3. $a + b$ is nonnegative

4. a is less than -3

5. $2b + 4$ is greater than or equal to 100

6. $c - 1$ is less than or equal to 5

In Problems 7–14, write the given inequality using interval notation and then graph the interval.

7. $x < 0$

8. $0 < x < 5$

9. $x \geq 5$

10. $-1 \leq x$

11. $8 < x \leq 10$

12. $-5 < x \leq -3$

13. $-2 \leq x \leq 4$

14. $x > -7$

In Problems 15–18, write the given interval as an inequality.

15. $[-7, 9]$

16. $[1, 15)$

17. $(-\infty, 2)$

18. $[-5, \infty)$

In Problems 19–34, solve the given linear inequality. Write the solution set using interval notation. Graph the solution set.

19. $x + 3 > -2$

20. $3x - 9 < 6$

21. $\frac{3}{2}x + 4 \leq 10$

22. $5 - \frac{5}{4}x \geq -4$

23. $\frac{3}{2} - x > x$

24. $-(1 - x) \geq 2x - 1$

25. $2 + x \geq 3(x - 1)$

26. $-7x + 3 \leq 4 - x$

27. $-\frac{20}{3} < \frac{2}{3}x < 4$

28. $-3 \leq -x < 2$

29. $-7 < x - 2 < 1$

30. $3 < x + 4 \leq 10$

31. $7 < 3 - \frac{1}{2}x \leq 8$

32. $100 + x \leq 41 - 6x \leq 121 + x$

33. $-1 \leq \dfrac{x - 4}{4} < \frac{1}{2}$

34. $2 \leq \dfrac{4x + 2}{-3} \leq 10$

In Problems 35–58, solve the given nonlinear inequality. Write the solution set using interval notation. Graph the solution set.

35. $x^2 - 9 < 0$

36. $x^2 \geq 16$

37. $x(x - 5) \geq 0$

38. $4x^2 + 7x < 0$

39. $x^2 - 8x + 12 < 0$

40. $(3x + 2)(x - 1) \leq 0$

41. $9x \geq 2x^2 - 18$

42. $4x^2 > 9x + 9$

43. $(x + 1)(x - 2)(x - 4) < 0$

44. $(1 - x)\left(x + \frac{1}{2}\right)(x - 3) \leq 0$

45. $(x^2 - 1)(x^2 - 4) \leq 0$

46. $(x - 1)^2(x + 3)(x - 5) \geq 0$

47. $\dfrac{5}{x + 8} < 0$

48. $\dfrac{10}{x^2 + 2} > 0$

49. $\dfrac{5}{x} \geq -1$

50. $\dfrac{x - 3}{x + 2} < 0$

51. $\dfrac{x + 1}{x - 1} + 2 > 0$

52. $\dfrac{x - 2}{x + 3} \leq 1$

53. $\dfrac{x(x - 1)}{x + 5} \geq 0$

54. $\dfrac{(1 + x)(1 - x)}{x} \leq 0$

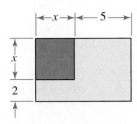

FIGURE 1.1.10 Rectangle in Problem 60

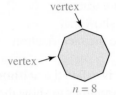

FIGURE 1.1.11 Octagon in Problem 61

FIGURE 1.1.12 Triangular arrays of dots in Problem 62

55. $\dfrac{x^2 - 2x + 3}{x + 1} \leq 1$

56. $\dfrac{x}{x^2 - 16} > 0$

57. $\dfrac{2}{x + 3} - \dfrac{1}{x + 1} < 0$

58. $\dfrac{4x + 5}{x^2} \geq \dfrac{4}{x + 5}$

59. If 7 times a number is decreased by 6, the result is less than 50. What can be determined about the number?

60. The sides of a square are extended to form a rectangle. As shown in FIGURE 1.1.10, one side is extended 2 inches and the other side is extended 5 inches. If the area of the resulting rectangle is less than 130 in.², what are the possible lengths of a side of the original square?

61. A **polygon** is a closed figure made by joining line segments. For example, a *triangle* is a three-sided polygon. Shown in FIGURE 1.1.11 is an eight-sided polygon called an *octagon*. A *diagonal* of a polygon is defined to be a line segment that joins any two nonadjacent vertices. The number of diagonals d in a polygon with n sides is given by $d = \frac{1}{2}(n - 1)n - n$. For what polygons will the number of diagonals exceed 35?

62. The total number N of dots in a triangular array with n rows is given by the formula $N = \frac{1}{2}n(n + 1)$. See FIGURE 1.1.12. How many rows can the array have if the total number of dots is to be less than 5050?

Applications

63. **Flower Garden** A rectangular flower bed is to be twice as long as it is wide. If the area enclosed must be greater than 98 m², what can you conclude about the width of the flower bed?

64. **Fever** The relationship between degrees Celsius T_C and degrees Fahrenheit T_F is given by $T_F = \frac{9}{5}T_C + 32$. A person is considered to have a fever if he or she has an oral temperature greater than 98.6°F. What temperatures on the Celsius scale indicate a fever?

Oral thermometer

65. **Parallel Resistors** A 5-ohm resistor and a variable resistor are placed in parallel. The resulting resistance is given by $R_T = \dfrac{5R}{5 + R}$. Determine the values of the variable resistor R for which the resulting resistance R_T will be greater than 2 ohms.

66. **What Goes Up . . .** With the aid of calculus it is easy to show that the height s of a projectile launched straight upward from an initial height s_0 with an initial velocity v_0 is given by $s = -\frac{1}{2}gt^2 + v_0 t + s_0$, where t is in seconds and $g = 32$ ft/s². If a toy rocket is shot straight upward from ground level, then $s_0 = 0$. If its initial velocity is 72 ft/s, during what time interval will the rocket be more than 80 ft above the ground?

67. **Linear Depreciation** The value V of a new car, which cost $50,000 initially, when depreciated linearly over 20 years is given by $V = 50,000(1 - x/20)$, where x represents years. Determine the values of x such that $0 < V < 20,000$.

68. **Pulse Rate** The pulse rate of a healthy person while engaged in aerobic exercises can vary widely. To obtain the maximum beneficial effect from the exercises, the pulse rate P_R should be maintained in a certain interval $[a, b]$. For jogging, one mathematical model determines the endpoints a and b of that interval subtracting the jogger's age from 220 and multiplying the result by 0.70 and 0.85, respectively. Write the desired interval for the pulse rate of a 40-year old jogger as a simultaneous inequality.

69. Discuss how you might determine the set of numbers for which the given expression is a real number.

    (a) $\sqrt{2x - 3}$ (b) $\sqrt{4 - 10x}$ (c) $\sqrt{x(x - 5)}$ (d) $\dfrac{1}{\sqrt{x + 2}}$

    Carry out your ideas.

70. In Example 5, explain why one should not multiply the last expression in (2) by $x + 2$.

71. (a) If $0 < a < b$, then use the properties of inequalities to show that $a^2 < b^2$.
    (b) If $a < b$, then explain why, in general, $a^2 < b^2$ is *not* true.

72. If $0 < a < b$, then use the properties of inequalities to show that $0 < \sqrt{a} < \sqrt{b}$. [*Hint*: One way of proceeding is to use the factorization $b - a = \left(\sqrt{b} + \sqrt{a}\right)\left(\sqrt{b} - \sqrt{a}\right) > 0$.]

73. If a and b are real numbers, then the number $(a + b)/2$ is called the **arithmetic mean**, or **average**, of a and b. Use the properties of inequalities to show that if $a < b$, then $a < \dfrac{a + b}{2} < b$.

74. If a and b are positive real numbers, then the number $\sqrt{ab}$ is called the **geometric mean** of a and b. Use the properties of inequalities to show that if $0 < a < b$, then $a < \sqrt{ab} < b$.

75. If $0 < a < b$, then show that the geometric mean of a and b is less than the arithmetic mean of a and b, that is, $\sqrt{ab} < \dfrac{a + b}{2}$. See Problems 73 and 74.

76. Using the definition in Problem 74 as a model, how would you define the geometric mean of three positive numbers a, b, and c? Of n positive numbers?

77. Do a little bit of research and find an application where the geometric mean of positive real numbers is used rather than the arithmetic mean of the numbers. You might have to use Problem 76.

1.2 Absolute Value

INTRODUCTION We can use the number line to picture distance. As shown in FIGURE 1.2.1, the distance between the number 0 and the number 3 is 3, and the distance between -3 and 0 is also 3. In general, for any *positive* real number x, the distance between x and 0 is x. If x represents a *negative* number, then the distance between x and 0 is $-x$. The concept of distance from a number on the number line to the number 0 is described by the **absolute value** of that number.

FIGURE 1.2.1 Distance is 3 units

DEFINITION 1.2.1   Absolute Value

For any real number x, the **absolute value** of x, denoted by $|x|$, is

$$|x| = \begin{cases} x, & \text{if } x \geq 0 \\ -x, & \text{if } x < 0 \end{cases} \tag{1}$$

Be careful. It is a common mistake to think that the symbol $-x$ represents a negative quantity simply because of the presence of the minus sign. If a symbol x represents a negative number (that is, $x < 0$), then $-x$ is a positive number. For example, if $x = -10 < 0$, then $|x| = -x = -(-10) = 10$.

As our first example shows, the symbol x in (1) is a placeholder. Other quantities can be placed inside the absolute value symbols $|\ |$.

EXAMPLE 1   Absolute Value

Write $|x - 5|$ without absolute value symbols.

Solution Wherever the symbol x appears in (1) we replace it by $x - 5$:

$$|x - 5| = \begin{cases} x - 5, & \text{if } x - 5 \geq 0 \\ -(x - 5), & \text{if } x - 5 < 0. \end{cases}$$

Let's consider each part of the foregoing definition separately. First, the inequality $x - 5 \geq 0$ means that $x \geq 5$. Therefore,

$$|x - 5| = x - 5 \quad \text{if} \quad x \geq 5.$$

Check this result (that is, $x - 5$ is nonnegative) by substituting numbers such as 5, 8, and 10. Next, $x - 5 < 0$ means that $x < 5$. In this case,

$$
\begin{array}{cc}
 & \text{distributive law} \\
 & \downarrow \qquad\qquad \downarrow \\
|x - 5| = & -(x - 5) = -x + 5 \quad \text{if} \quad x < 5.
\end{array}
$$

Again, you should convince yourself that this is correct (that is, $-x + 5$ is positive) by substituting a few numbers, such as 2 and -3. ∎

As illustrated in Figure 1.2.1, for any real number x and its negative $-x$, the distance to 0 is the same. That is, $|x| = |-x|$. This is one property in a list of properties of the absolute value that is given next.

THEOREM 1.2.1   Properties of Absolute Values

   (i) $|a| = |-a|$
  (ii) $|a| = 0$ if and only if $a = 0$
 (iii) $|ab| = |a||b|$
 (iv) $\left|\dfrac{a}{b}\right| = \dfrac{|a|}{|b|}, \quad b \neq 0$
  (v) $|a + b| \leq |a| + |b|$ (**Triangle inequality**)

For example, by virtue of property (iii) of Theorem 1.2.1 we can rewrite the expression $|-2x|$ as $|-2||x| = 2|x|$.

□ **Distance** If we wish to find the **distance** between any two numbers on the real number line, then all we have to do is subtract the leftmost number from the rightmost number. For example, the distance between 10 and -2 is

$$
\begin{array}{cc}
\text{rightmost number} & \quad \text{leftmost number} \\
\downarrow & \qquad \downarrow
\end{array}
$$

$$10 - (-2) = 12.$$

As we saw in the introduction, the distance between -3 and 0 is $0 - (-3) = 3$. If an absolute value is used to define the distance, then we do not have to worry about the order of subtraction.

DEFINITION 1.2.2 Distance Between Two Numbers

If a and b are any two numbers on the number line, the **distance** between a and b is

$$d(a, b) = |b - a| \tag{2}$$

Using the properties of absolute values,

by property (iii) of Theorem 1.2.1

$$|b - a| = |(-1)(a - b)| = |-1| \, |a - b| = |a - b|,$$

and so we have $d(a, b) = d(b, a)$. For example, the distance between $\sqrt{2}$ and 3 is

$$d(\sqrt{2}, 3) = |3 - \sqrt{2}| = 3 - \sqrt{2}$$

because $3 > \sqrt{2}$ or $3 - \sqrt{2} > 0$, or

$$d(3, \sqrt{2}) = |\sqrt{2} - 3| = -(\sqrt{2} - 3) = 3 - \sqrt{2}$$

because $\sqrt{2} < 3$ or $\sqrt{2} - 3 < 0$.

☐ **Midpoint** Suppose a and b represent two distinct numbers on the number line such that $a < b$. The **midpoint** m of the line segment between the numbers a and b is given by the average of the two endpoints of the interval $[a, b]$, that is

$$m = \frac{a + b}{2}. \tag{3}$$

$d(a, m) = d(m, b)$

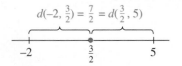

FIGURE 1.2.2 Midpoint m between a and b

As shown in FIGURE 1.2.2, (3) is easy to verify by using (2) to show that $d(a, m) = d(m, b)$.

◼ EXAMPLE 2   **Midpoint**

From (3), the midpoint of the line segment joining the numbers -2 and 5 is

$$\frac{(-2) + 5}{2} = \frac{3}{2}.$$

$d(-2, \frac{3}{2}) = \frac{7}{2} = d(\frac{3}{2}, 5)$

-2 $\frac{3}{2}$ 5

FIGURE 1.2.3 Midpoint in Example 2

See FIGURE 1.2.3.   ◼

☐ **Equations** Since (*i*) of Theorem 1.2.1 implies that $|-6| = |6| = 6$, we can conclude that the simple equation $|x| = 6$ has two solutions, either $x = -6$ or $x = 6$. In general, if a is a positive real number, then

$$|x| = a \quad \text{if and only if} \quad x = a \text{ or } x = -a. \tag{4}$$

◼ EXAMPLE 3   **An Absolute-Value Equation**

Solve **(a)** $|5x - 3| = 8$ **(b)** $|x - 4| = -3$.

Solution **(a)** In (4) the symbol x is a placeholder for any quantity. By replacing x by $5x - 3$, the given equation is equivalent to two equations

$$5x - 3 = 8 \quad \text{or} \quad 5x - 3 = -8.$$

We solve each of these. From $5x - 3 = 8$, we obtain

$$5x = 11 \qquad \text{which implies} \qquad x = \frac{11}{5}.$$

From $5x - 3 = -8$, we have

$$5x = -5 \qquad \text{which implies} \qquad x = -1.$$

Therefore, the solutions are $\frac{11}{5}$ and -1.

(b) Since the absolute value of a real number is always nonnegative, there is no solution to an equation such as $|x - 4| = -3$.

$|x| < a$

(a) The distance between x and 0 is *less* than a

$|x| > a$

(b) The distance between x and 0 is *greater* than a

FIGURE 1.2.4 Graphical interpretation of (i) and (ii) of Theorem 1.2.2

☐ **Absolute-Value Inequalities** Many important applications of inequalities involve absolute values. We have just seen that $|x|$ represents the distance along the number line between the number x and the number 0. Thus the inequality $|x| < a$, where $a > 0$, means that the distance between x and 0 is less than a. We can see in FIGURE 1.2.4(a) that this is the set of real numbers x such that $-a < x < a$. On the other hand, $|x| > a$ means that the distance between x and 0 is greater than a. In Figure 1.2.4(b), we see that these are the numbers that satisfy either $x > a$ or $x < -a$. These graphical observations suggest two additional properties of absolute value.

THEOREM 1.2.2 Absolute-Value Inequalities

Let a be a positive real number.

  (i) $|x| < a$ if and only if $-a < x < a$
  (ii) $|x| > a$ if and only if $x > a$ or $x < -a$

Properties (i) and (ii) of Theorem 1.2.2 also hold with the inequality symbols $<$ and $>$ are replaced by $\leq$ and $\geq$, respectively.

■ EXAMPLE 4 **Two Absolute-Value Inequalities**

(a) From (i) of Theorem 1.2.2, the absolute-value inequality $|x| < 1$ is equivalent to the simultaneous inequality $-1 < x < 1$.

(b) From (ii) of Theorem 1.2.2, the absolute-value inequality $|x| \geq 5$ is equivalent to two inequalities: $x \geq 5$ or $x \leq -5$.

■ EXAMPLE 5 **Two Absolute-Value Inequalities**

Solve **(a)** $|3x - 7| < 1$ **(b)** $|2x - 5| \leq 0$.

Solution (a) As in Example 3, the symbol x in the inequality $|x| < a$ is simply a placeholder for other quantities. If we replace x by $3x - 7$ and a by the number 1, then (i) of Theorem 1.2.2 yields the simultaneous inequality

$$-1 < 3x - 7 < 1$$

FIGURE 1.2.5 Solution set in Example 5

which we solve in the usual manner (see Example 2 in Section 1.1):

$$-1 + 7 < 3x - 7 + 7 < 1 + 7$$
$$6 < 3x < 8$$
$$\left(\tfrac{1}{3}\right)6 < \left(\tfrac{1}{3}\right)3x < \left(\tfrac{1}{3}\right)8$$
$$2 < x < \tfrac{8}{3}.$$

The solution set is the open interval $\left(2, \tfrac{8}{3}\right)$ shown in FIGURE 1.2.5.

(b) Since the absolute value of any expression is never negative, the values of x that satisfy the inequality $\leq$ are those for which $|2x - 5| = 0$. By (ii) of Theorem 1.2.1 we conclude that $2x - 5 = 0$. Hence the only solution is $\tfrac{5}{2}$. ∎

FIGURE 1.2.6 The distance between x and b is *less* than a

▢ **Distance Again** An absolute-value inequality such as $|x - b| < a$ can also be interpreted in terms of distance along the number line. Since $|x - b|$ is distance between x and b, the inequality $|x - b| < a$ is satisfied by all real numbers x whose distance between x and b is less than a. This interval is shown in FIGURE 1.2.6. Note that when $b = 0$ we get (i) of Theorem 1.2.2. Similarly, the set of numbers satisfying $|x - b| > a$ are the numbers x whose distance between x and b is greater than a.

▨ **EXAMPLE 6**   **An Absolute-Value Inequality**

Solve $\left|4 - \tfrac{1}{2}x\right| \geq 7$.

Solution If we replace x and a in $|x| \geq a$ by $4 - \tfrac{1}{2}x$ and 7, respectively, then we see from (ii) of Theorem 1.2.2 that $\left|4 - \tfrac{1}{2}x\right| \geq 7$ is equivalent to the two different inequalities

$$4 - \tfrac{1}{2}x \geq 7 \qquad \text{or} \qquad 4 - \tfrac{1}{2}x \leq -7.$$

We solve each of these inequalities separately. First, we solve

$$4 - \tfrac{1}{2}x \geq 7$$
$$-\tfrac{1}{2}x \geq 3$$
$$(-2)\left(-\tfrac{1}{2}x\right) \leq (-2)3 \quad \leftarrow \begin{cases} \text{multiplication by } -2 \text{ reverses} \\ \text{the direction of the inequality} \end{cases}$$
$$x \leq -6.$$

In interval notation the solution set of this inequality is $(-\infty, -6]$. Next, we solve

$$4 - \tfrac{1}{2}x \leq -7$$
$$-\tfrac{1}{2}x \leq -11$$
$$(-2)\left(-\tfrac{1}{2}\right)x \geq (-2)(-11) \quad \leftarrow \begin{cases} \text{multiplication by } -2 \text{ reverses} \\ \text{the direction of the inequality} \end{cases}$$
$$x \geq 22.$$

In interval notation the solution set is $[22, \infty)$.

Since the two intervals are disjoint, the solution set is the union of intervals: $(-\infty, -6] \cup [22, \infty)$. The graph of this solution set is shown in FIGURE 1.2.7. ∎

FIGURE 1.2.7 Solution set in Example 6

Note in Figure 1.2.4(a) that the number 0 is the midpoint of the solution interval for $|x| < a$ and in Figure 1.2.6 that the number b is the midpoint of the solution interval for the inequality $|x - b| < a$. With this in mind, work through the next example.

EXAMPLE 7   **Constructing an Inequality**

Find an inequality of the form $|x - b| < a$ for which the open interval $(4, 8)$ is its solution set.

Solution The midpoint of the interval $(4, 8)$ is $m = \dfrac{4 + 8}{2} = 6$. The distance between the midpoint m and one of the endpoints of the interval is $d(m, 8) = |8 - 6| = 2$. Therefore the required inequality is $|x - 6| < 2$. ◼

Exercises 1.2   Answers to selected odd-numbered problems begin on page ANS–1.

In Problems 1–6, write the given quantity without the absolute value symbols.

1. $|\pi - 4|$ **2.** $|\sqrt{5} - 3|$
3. $|8 - \sqrt{63}|$ **4.** $|\sqrt{5} - 2.3|$
5. $|-6| - |-2|$ **6.** $||-3| - |10||$

In Problems 7–12, write the given expression without the absolute value symbols.

7. $|h|$, if h is negative   **8.** $|-h|$, if h is negative
9. $|x - 6|$, if $x < 6$ **10.** $|2x - 1|$, if $x \geq \frac{1}{2}$

11. $|x - y| - |y - x|$ **12.** $\dfrac{|x - y|}{|y - x|}, x \neq y$

In Problems 13–16, write the expression $|x - 2| + |x - 5|$ without the absolute value symbols if x is in the given interval.

13. $(-\infty, 1)$ **14.** $(7, \infty)$ **15.** $(3, 4]$ **16.** $[2, 5]$

In Problems 17–20, write the expression $|x + 1| - |x - 3|$ without the absolute value symbols if x is in the given interval.

17. $[-1, 3)$ **18.** $(0, 1)$ **19.** (π, ∞) **20.** $(-\infty, -5)$

In Problems 21–24, find the distance between the given numbers and find the midpoint of the line segment between them.

21. $3, 7$ **22.** $-100, 255$ **23.** $-\frac{3}{2}, \frac{3}{2}$ **24.** $-\frac{1}{4}, \frac{7}{4}$

In Problems 25–28, m is the midpoint of the line segment joining a (the left endpoint) and b (the right endpoint). Use the given conditions to find the indicated quantities.

25. $m = 5, d(a, m) = 3$; a and b **26.** $m = -1, d(m, b) = 2$; a and b
27. $a = 4, d(a, m) = \pi$; m and b **28.** $a = 10, d(m, b) = 5$; m and b

In Problems 29–34, solve the given equation.

29. $|4x - 1| = 2$ **30.** $|5v - 4| = 7$
31. $|\frac{1}{4} - \frac{3}{2}y| = 1$ **32.** $|2 - 16t| = 0$

33. $\left|\dfrac{x}{x - 1}\right| = 2$ **34.** $\left|\dfrac{x + 1}{x - 2}\right| = 4$

In Problems 35–46, solve the given inequality. Write the solution set using interval notation. Graph the solution set.

35. $|-5x| < 4$

36. $|3x| > 18$

37. $|3 + x| > 7$

38. $|x - 4| \leq 9$

39. $|2x - 7| \leq 1$

40. $|5 - \frac{1}{3}x| < \frac{1}{2}$

41. $|x + \sqrt{2}| \geq 1$

42. $|6x + 4| > 4$

43. $\left|\dfrac{3x - 1}{-4}\right| < 2$

44. $\left|\dfrac{2 - 5x}{3}\right| \geq 5$

45. $|x - 5| < 0.01$

46. $|x - (-2)| < 0.001$

In Problems 47–50, proceed as in Example 7 and find an inequality $|x - b| < a$ or $|x - b| > a$ for which the given interval or union of intervals is its solution set.

47. $(-3, 11)$

48. $(1, 2)$

49. $(-\infty, 1) \cup (9, \infty)$

50. $(-\infty, -3) \cup (13, \infty)$

In Problems 51 and 52, find an inequality whose solution is the set of real numbers x satisfying the given condition. Express each set using interval notation.

51. Greater than or equal to 2 units from -3

52. Less than $\frac{1}{2}$ unit from 3.5

Applications

53. Comparing Ages Bill and Mary's ages, A_B and A_M, differ by at most 3 years. Write this fact as an inequality using absolute value symbols.

54. Survival Your score on the first exam is 72%. The midterm grade is the average of the first exam score with the midterm exam score. If the B range is from 80% to 89%, what score should you obtain on the midterm exam so that your mid-semester grade is B?

55. Weight of Coffee The weight w of the coffee in cans filled by a food processing company satisfies

$$\left|\frac{w - 12}{0.05}\right| \leq 1,$$

where w is measured in ounces. Determine the interval in which w lies.

56. Weight of Cans A grocery scale is designed to be accurate to within 0.25 oz. If two identical cans of soup placed on the scale have a combined weight of 33.15 oz, what are the largest and smallest possible weights of one of the cans?

For Discussion

57. Discuss how you might solve the given inequality and equation.

(a) $\left|\dfrac{x + 5}{x - 2}\right| \leq 3$

(b) $|5 - x| = |1 - 3x|$

Carry out your ideas.

58. The distance between the number x and 5 is $|x - 5|$.

(a) In words, describe the graphical interpretation of the inequalities
$0 < |x - 5|$ and $0 < |x - 5| < 3$.

(b) Solve each inequality in part (a) and write each solution set using interval notation.

59. (a) Interpret $|x - 3|$ as distance between the numbers x and 3. Sketch on the number line the set of real numbers that satisfy $2 < |x - 3| < 5$.

(b) Now solve the simultaneous inequality $2 < |x - 3| < 5$ by first solving $|x - 3| < 5$ and then $2 < |x - 3|$. Take the intersection of the two solution sets and compare with your sketch in part (a).

60. Here is a statement you may encounter in the beginning of a course in calculus. Express the following statement as best you can in words:

For every $\epsilon > 0$ there exists a $\delta > 0$ such that $|y - L| < \epsilon$ whenever $0 < |x - a| < \delta$.

Do not use the symbols $>$, $<$, or $|\ |$. The symbols ϵ and δ are the Greek letters *epsilon* and *delta* and represent real numbers.

1.3 The Rectangular Coordinate System

INTRODUCTION In Section 1.1 we saw that each real number can be associated with exactly one point on the number, or coordinate, line. We now examine a correspondence between points in a plane and ordered pairs of real numbers.

René Descartes

☐ **Coordinate Plane** A **rectangular coordinate system** is formed by two perpendicular number lines that intersect at the point corresponding to the number 0 on each line. This point of intersection is called the **origin** and is denoted by the symbol O. The horizontal and vertical number lines are called the ***x*-axis** and the ***y*-axis**, respectively. These axes divide the plane into four regions, called **quadrants**, which are numbered as shown in FIGURE 1.3.1(a). As we can see in Figure 1.3.1(b), the scales on the *x*- and *y*-axes need not be the same. Throughout this text, if tick marks are *not* labeled on the coordinates axes, as in Figure 1.3.1(a), then you may assume that one tick corresponds to one unit. A plane containing a rectangular coordinate system is called an ***xy*-plane**, a **coordinate plane**, or simply **2-space**.

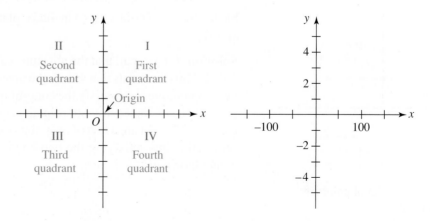

(a) Four quadrants   (b) Different scales on *x*- and *y*-axes

FIGURE 1.3.1 Coordinate plane

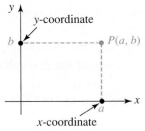

FIGURE 1.3.2 Point with coordinates (a, b)

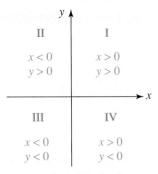

FIGURE 1.3.3 Algebraic signs of coordinates in the four quadrants

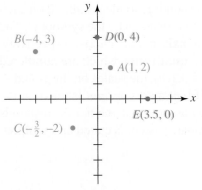

FIGURE 1.3.4 Plots of five points in Example 1

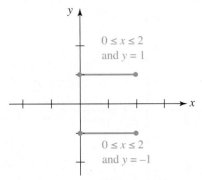

FIGURE 1.3.5 Set of points in Example 2

The rectangular coordinate system and the xy-plane are also called the **Cartesian coordinate system** and the **Cartesian plane** after the famous French mathematician and philosopher **René Descartes** (1596–1650).

▢ **Coordinates of a Point** Let P represent a point in the coordinate plane. We associate an ordered pair of real numbers with P by drawing a vertical line from P to the x-axis and a horizontal line from P to the y-axis. If the vertical line intersects the x-axis at the number a and the horizontal line intersects the y-axis at the number b, we associate the **ordered pair** of real numbers (a, b) with the point. Conversely, to each ordered pair (a, b) of real numbers there corresponds a point P in the plane. This point lies at the intersection of the vertical line through a on the x-axis and the horizontal line passing through b on the y-axis. Hereafter we will refer to an ordered pair as a **point** and denote it by either $P(a, b)$ or (a, b).* The number a is the **x-coordinate** of the point and the number b is the **y-coordinate** of the point and we say that P has **coordinates** (a, b). For example, the coordinates of the origin are $(0, 0)$. See FIGURE 1.3.2.

The algebraic signs of the x-coordinate and the y-coordinate of any point (x, y) in each of the four quadrants are indicated in FIGURE 1.3.3. Points on either of the two axes are not considered to be in any quadrant. Because a point on the x-axis has the form $(x, 0)$, an equation that describes the x-axis is $y = 0$. Similarly, a point on the y-axis has the form $(0, y)$ and so an equation of the y-axis is $x = 0$. When we locate a point in the coordinate plane corresponding to an ordered pair of numbers and represent it using a solid dot, we say that we **plot** or **graph** the point.

EXAMPLE 1   Plotting Points

Plot the points $A(1, 2)$, $B(-4, 3)$, $C\left(-\frac{3}{2}, -2\right)$, $D(0, 4)$, and $E(3.5, 0)$. Specify the quadrant in which each point lies.

Solution The five points are plotted in the coordinate plane in FIGURE 1.3.4. Point A lies in the first quadrant (quadrant I), B in the second quadrant (quadrant II), and C is in the third quadrant (quadrant III). Points D and E, which lie on the y- and x-axes, respectively, are not in any quadrant. ▪

EXAMPLE 2   Plotting Points

Sketch the set of points (x, y) in the xy-plane whose coordinates satisfy both $0 \le x \le 2$ and $|y| = 1$.

Solution First, recall that the absolute-value equation $|y| = 1$ implies that $y = -1$ or $y = 1$. Thus the points that satisfy the given conditions are the points whose coordinates (x, y) *simultaneously* satisfy the conditions: each x-coordinate is a number in the closed interval $[0, 2]$ and each y-coordinate is either $y = -1$ or $y = 1$. For example, $(1, 1)$, $\left(\frac{1}{2}, -1\right)$, $(2, -1)$ are a few of the points that satisfy the two conditions. Graphically, the set of all points satisfying the two conditions are points on the two parallel line segments shown in FIGURE 1.3.5. ▪

---

*This is the same notation used to denote an open interval. It should be clear from the context of the discussion whether we are considering a point (a, b) or an open interval (a, b).

| EXAMPLE 3 | **Regions Defined by Inequalities** |

Sketch the set of points (x, y) in the xy-plane whose coordinates satisfy each of the following conditions. **(a)** $xy < 0$ **(b)** $|y| \geq 2$

Solution **(a)** From (*ii*) of the sign properties of products in Theorem 1.1.2, we know that a product of two real numbers x and y is negative when one of the numbers is positive and the other is negative. Thus, $xy < 0$ when $x > 0$ and $y < 0$ *or* when $x < 0$ and $y > 0$. We see from Figure 1.3.3 that $xy < 0$ for all points (x, y) in the second and fourth quadrants. Hence we can represent the set of points for which $xy < 0$ by the shaded regions in FIGURE 1.3.6. The coordinate axes are shown as dashed lines to indicate that the points on these axes are not included in the solution set.

(b) In Section 1.2 we saw that $|y| \geq 2$ means that either $y \geq 2$ or $y \leq -2$. Since x is not restricted in any way it can be any real number, and so the points (x, y) for which

$$y \geq 2 \quad \text{and} \quad -\infty < x < \infty \quad \text{or} \quad y \leq -2 \quad \text{and} \quad -\infty < x < \infty$$

can be represented by the two shaded regions in FIGURE 1.3.7. We use solid lines to represent the boundaries $y = -2$ and $y = 2$ of the region to indicate that the points on these boundaries are included in the solution set.

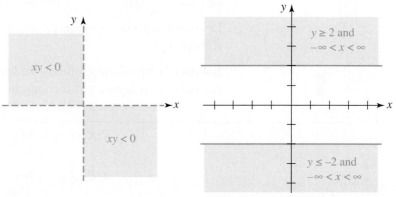

FIGURE 1.3.6 Region in the xy-plane satisfying the condition in (a) of Example 3

FIGURE 1.3.7 Region in the xy-plane satisfying the condition in (b) of Example 3

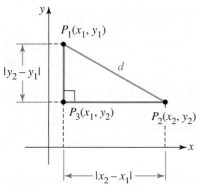

FIGURE 1.3.8 Distance between points P_1 and P_2

☐ **Distance Formula** Suppose $P_1(x_1, y_1)$ and $P_2(x_2, y_2)$ are two distinct points in the xy-plane that are not on a vertical line or on a horizontal line. As a consequence, P_1, P_2, and $P_3(x_1, y_2)$ are vertices of a right triangle, as shown in FIGURE 1.3.8. The length of the side P_3P_2 is $|x_2 - x_1|$ and the length of the side P_1P_3 is $|y_2 - y_1|$. If we denote the length of P_1P_2 by d, then

$$d^2 = |x_2 - x_1|^2 + |y_2 - y_1|^2 \tag{1}$$

by the Pythagorean theorem. Since the square of any real number is equal to the square of its absolute value, we can replace the absolute value signs in (1) with parentheses. The distance formula given next follows immediately from (1).

THEOREM 1.3.1 Distance Formula

The **distance** between any two points $P_1(x_1, y_1)$ and $P_2(x_2, y_2)$ in the xy-plane is given by

$$d(P_1, P_2) = \sqrt{(x_2 - x_1)^2 + (y_2 - y_1)^2} \qquad (2)$$

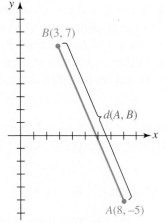

FIGURE 1.3.9 Distance between two points in Example 4

Although we derived this equation for two points not on a vertical or horizontal line, (2) holds in these cases as well. Also, because $(x_2 - x_1)^2 = (x_1 - x_2)^2$, it makes no difference which point is used first in the distance formula, that is, $d(P_1, P_2) = d(P_2, P_1)$.

EXAMPLE 4   **Distance Between Two Points**

Find the distance between the points $A(8, -5)$ and $B(3, 7)$.

Solution From (2), with A and B playing the parts of P_1 and P_2:

$$d(A, B) = \sqrt{(3 - 8)^2 + (7 - (-5))^2}$$
$$= \sqrt{(-5)^2 + (12)^2} = \sqrt{169} = 13.$$

The distance d is illustrated in FIGURE 1.3.9.

EXAMPLE 5   **Three Points Form a Triangle**

Determine whether the points $P_1(7, 1)$, $P_2(-4, -1)$, and $P_3(4, 5)$ are the vertices of a right triangle.

Solution From plane geometry we know that a triangle is a right triangle if and only if the sum of the squares of the lengths of two of its sides is equal to the square of the length of the remaining side. Now, from the distance formula (2), we have

$$d(P_1, P_2) = \sqrt{(-4 - 7)^2 + (-1 - 1)^2}$$
$$= \sqrt{121 + 4} = \sqrt{125},$$
$$d(P_2, P_3) = \sqrt{(4 - (-4))^2 + (5 - (-1))^2}$$
$$= \sqrt{64 + 36} = \sqrt{100} = 10,$$
$$d(P_3, P_1) = \sqrt{(7 - 4)^2 + (1 - 5)^2}$$
$$= \sqrt{9 + 16} = \sqrt{25} = 5.$$

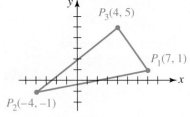

FIGURE 1.3.10 Triangle in Example 5

Since

$$[d(P_3, P_1)]^2 + [d(P_2, P_3)]^2 = 25 + 100 = 125 = [d(P_1, P_2)]^2,$$

we conclude that P_1, P_2, and P_3 are the vertices of a right triangle with the right angle at P_3. See FIGURE 1.3.10.

☐ **Midpoint Formula** In Section 1.2 we saw that the midpoint of a line segment between two numbers a and b on the number line is the average, $(a + b)/2$. In the xy-plane, each coordinate of the midpoint M of a line segment joining two points $P_1(x_1, y_1)$ and $P_2(x_2, y_2)$ shown in FIGURE 1.3.11 is the average of the corresponding coordinates of the endpoints of the intervals $[x_1, x_2]$ and $[y_1, y_2]$.

FIGURE 1.3.11 M is the midpoint of the line segment joining P_1 and P_2

To prove this, we note in Figure 1.3.11 that triangles P_1CM and MDP_2 are congruent because corresponding angles are equal and $d(P_1, M) = d(M, P_2)$. Hence, $d(P_1, C) = d(M, D)$, or $y - y_1 = y_2 - y$. Solving the last equation for y gives

$y = \dfrac{y_1 + y_2}{2}$. Similarly, $d(C, M) = d(D, P_2)$, so that $x - x_1 = x_2 - x$, and therefore $x = \dfrac{x_1 + x_2}{2}$. We have proved the following result.

THEOREM 1.3.2 Midpoint Formula

The coordinates of the **midpoint** M of the line segment joining the points $P_1(x_1, y_1)$ and $P_2(x_2, y_2)$ are given by

$$M = \left(\frac{x_1 + x_2}{2}, \frac{y_1 + y_2}{2} \right) \tag{3}$$

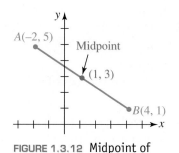

FIGURE 1.3.12 Midpoint of line segment in Example 6

EXAMPLE 6 Midpoint of a Line Segment

Find the coordinates of the midpoint of the line segment joining $A(-2, 5)$ and $B(4, 1)$.

Solution From the midpoint formula (3), the coordinates of the midpoint M are given by

$$\left(\frac{-2 + 4}{2}, \frac{5 + 1}{2} \right) \qquad \text{or} \qquad (1, 3).$$

This point is shown in red in FIGURE 1.3.12.

Exercises 1.3   Answers to selected odd-numbered problems begin on page ANS–2.

In Problems 1–4, plot the given points.

1. $(2, 3), (4, 5), (0, 2), (-1, -3)$ **2.** $(1, 4), (-3, 0), (-4, 2), (-1, -1)$
3. $\left(-\frac{1}{2}, -2\right), (0, 0), \left(-1, \frac{4}{3}\right), (3, 3)$ **4.** $(0, 0.8), (-2, 0), (1.2, -1.2), (-2, 2)$

In Problems 5–16, determine the quadrant in which the given point lies if (a, b) is in quadrant I.

5. $(-a, b)$ **6.** $(a, -b)$ **7.** $(-a, -b)$ **8.** (b, a)
9. $(-b, a)$ **10.** $(-b, -a)$ **11.** (a, a) **12.** $(b, -b)$
13. $(-a, -a)$ **14.** $(-a, a)$ **15.** $(b, -a)$ **16.** $(-b, b)$

17. Plot the points given in Problems 5–16 if (a, b) is the point shown in FIGURE 1.3.13.

18. Give the coordinates of the points shown in FIGURE 1.3.14.

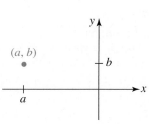

FIGURE 1.3.13 Point (a, b) in Problem 17

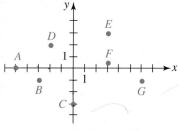

FIGURE 1.3.14 Points A–G in Problem 18

19. The points $(-2, 0)$, $(-2, 6)$, and $(3, 0)$ are vertices of a rectangle. Find the fourth vertex.

20. Describe the set of all points (x, x) in the coordinate plane. The set of all points $(x, -x)$.

In Problems 21–26, sketch the set of points (x, y) in the xy-plane whose coordinates satisfy the given conditions.

21. $xy = 0$ **22.** $xy > 0$

23. $|x| \leq 1$ and $|y| \leq 2$ **24.** $x \leq 2$ and $y \geq -1$

25. $|x| > 4$ **26.** $|y| \leq 1$

In Problems 27–32, find the distance between the given points.

27. $A(1, 2)$, $B(-3, 4)$ **28.** $A(-1, 3)$, $B(5, 0)$

29. $A(2, 4)$, $B(-4, -4)$ **30.** $A(-12, -3)$, $B(-5, -7)$

31. $A\left(-\frac{3}{2}, 1\right)$, $B\left(\frac{5}{2}, -2\right)$ **32.** $A\left(-\frac{5}{3}, 4\right)$, $B\left(-\frac{2}{3}, -1\right)$

In Problems 33–38, determine whether the points A, B, and C are vertices of a right triangle, an isosceles triangle, or both.

33. $A(8, 1)$, $B(-3, -1)$, $C(10, 5)$ **34.** $A(-2, -1)$, $B(8, 2)$, $C(1, -11)$

35. $A(2, 8)$, $B(0, -3)$, $C(6, 5)$ **36.** $A(4, 0)$, $B(1, 1)$, $C(2, 3)$

37. $A(-2, 1)$, $B(0, 9)$, $C(3, 4)$ **38.** $A(1, 1)$, $B(4, 5)$, $C(8, 8)$

39. Suppose the points $A(0, 0)$ and $B(0, 6)$ are vertices of a triangle. Find a third vertex C so that the triangle is equilateral.

40. Find all points on the y-axis that are 5 units from the point $(4, 4)$.

41. Consider the line segment joining the points $A(-1, 2)$ and $B(3, 4)$.

   (a) Find an equation that expresses the fact that a point $P(x, y)$ is equidistant from A and from B.

   (b) Describe geometrically the set of points described by the equation in part (a).

42. Use the distance formula to determine whether the points $A(-1, -5)$, $B(2, 4)$, and $C(4, 10)$ lie on a straight line.

43. Find all points each with x-coordinate 6 such that the distance from each point to $(-1, 2)$ is $\sqrt{85}$.

44. Which point, $\left(1/\sqrt{2}, 1/\sqrt{2}\right)$ or $(0.25, 0.97)$, is closer to the origin?

In Problems 45 and 46, find all points $P(x, x)$ that are the indicated distance from the given point.

45. $(-2, 0)$; $\sqrt{10}$ **46.** $(3, -5)$; $\sqrt{34}$

In Problems 47–52, find the midpoint M of the line segment joining the points A and B.

47. $A(4, 1)$, $B(-2, 4)$ **48.** $A\left(\frac{2}{3}, 1\right)$, $B\left(\frac{7}{3}, -3\right)$

49. $A(-1, 0)$, $B(-8, 5)$ **50.** $A\left(\frac{1}{2}, -\frac{3}{2}\right)$, $B\left(-\frac{5}{2}, 1\right)$

51. $A(2a, 3b)$, $B(4a, -6b)$ **52.** $A(x, x)$, $B(-x, x + 2)$

In Problems 53–56, find the point B if M is the midpoint of the line segment joining points A and B.

53. $A(-2, 1)$, $M\left(\frac{3}{2}, 0\right)$ **54.** $A\left(4, \frac{1}{2}\right)$, $M\left(7, -\frac{5}{2}\right)$

55. $A(5, 8)$, $M(-1, -1)$ **56.** $A(-10, 2)$, $M(5, 1)$

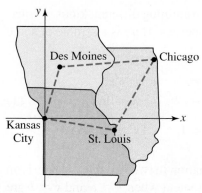

FIGURE 1.3.15 Map for Problem 62

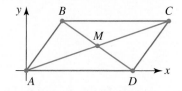

FIGURE 1.3.16 Parallelogram in Problem 63

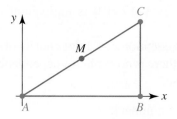

FIGURE 1.3.17 Right triangle in Problem 64

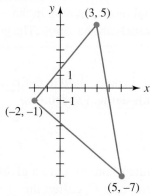

FIGURE 1.3.18 Triangle in Problem 66

57. Find the distance from the midpoint M_1 of the line segment joining $A(-1, 3)$ and $B(3, 5)$ to the midpoint M_2 of the line segment joining $C(4, 6)$ and $D(-2, -10)$.

58. Find all points on the x-axis that are 3 units from the midpoint of the line segment joining $(5, 2)$ and $(-5, -6)$.

59. The x-axis is the perpendicular bisector of the line segment through $A(2, 5)$ and $B(x, y)$. Find x and y.

60. Consider the line segment joining the points $A(0, 0)$ and $B(6, 0)$. Find a point $C(x, y)$ in the first quadrant such that A, B, and C are vertices of an equilateral triangle.

61. Find the points $P_1(x_1, y_1)$, $P_2(x_2, y_2)$, and $P_3(x_3, y_3)$ on the line segment joining $A(3, 6)$ and $B(5, 8)$ that divide the line segment into four equal parts.

Applications

62. **Going to Chicago** Kansas City and Chicago are not directly connected by an interstate highway, but each city is connected to St. Louis and Des Moines. See FIGURE 1.3.15. Des Moines is approximately 40 mi east and 180 mi north of Kansas City, St. Louis is approximately 230 mi east and 40 mi south of Kansas City, and Chicago is approximately 360 mi east and 200 mi north of Kansas City. Assume that this part of the Midwest is a flat plane and that the connecting highways are straight lines. Which route from Kansas City to Chicago, through St. Louis or through Des Moines, is shorter?

For Discussion

63. Consider the parallelogram in FIGURE 1.3.16. Assume that the coordinates of points B and D are (a, b) and $(c, 0)$, respectively. Discuss: How can it be shown that the diagonals AC and BD of the parallelogram (the blue line segments in the figure) bisect each other at a point M? Carry out your ideas.

64. The points $A(0, 0)$, $B(a, 0)$, and $C(a, b)$ are vertices of the right triangle shown in FIGURE 1.3.17. Discuss: How can it be shown that the midpoint M of the hypotenuse is equidistant from the vertices? Carry out your ideas.

65. Describe the set of points (x, y) in the xy-plane whose coordinates satisfy
    (a) $y = x$ (b) $y > x$ (c) $y < x$.

66. The triangle in FIGURE 1.3.18 is rigidly translated two units to the right and then four units upward. Find the coordinates of the three vertices of the translated triangle.

1.4   Circles and Graphs

INTRODUCTION An **equation in two variables**, say x and y, is simpl~
cal statement that asserts two quantities involving these variable~
of the physical sciences, engineering, and business, eq~
nication. For example, if a physicist wants to tell someo.
a great height travels in a certain time t, he/she will write s
look at $s = 16t^2$ and immediately classify it as a certain ty
cation of an equation carries with it information about prope.

of that kind. The remainder of this text is devoted to examining different kinds of equations involving two variables and studying their properties. Here is a sample of some of the equations you will see:

$$x = 1, \quad x^2 + y^2 = 1, \quad y = x^2, \quad y = \sqrt{x},$$
$$y = 5x - 1, \quad y = x^3 - 3x, \quad y = 2^x, \quad y = \ln x, \qquad (1)$$
$$y = \sin x, \quad y^2 = x - 1, \quad \frac{x^2}{4} + \frac{y^2}{9} = 1, \quad \tfrac{1}{2}x^2 - y^2 = 1.$$

☐ **Solutions of Equations** A **solution** of an equation in two variables x and y is an ordered pair of numbers (a, b) that yields a true statement when $x = a$ and $y = b$ are substituted into the equation. For example, $(-2, 4)$ is a solution of the equation $y = x^2$ because

$$\begin{array}{cc} y = 4 & x = -2 \\ \downarrow & \downarrow \end{array}$$
$$4 = (-2)^2$$

is a true statement. We also say that the coordinates $(-2, 4)$ **satisfy** the equation. The set of all solutions of an equation is called its **solution set**. Two equations are said to be **equivalent** if they have the same solution set. For example, we will see in Example 4 of this section that the equation $x^2 + y^2 + 10x - 2y + 17 = 0$ is equivalent to $(x + 5)^2 + (y - 1)^2 = 3^2$.

In the list given in (1), you might object that the first equation $x = 1$ does not involve two variables. It is a matter of interpretation! Because there is no explicit y dependence in the equation, $x = 1$ can be interpreted to mean the set

$$\{(x, y) \,|\, x = 1, \text{ where } y \text{ is any real number}\}.$$

The solutions of $x = 1$ are then ordered pairs $(1, y)$, where you are free to choose y arbitrarily so long as it is a real number. For example, $(1, 0)$ and $(1, 3)$ are solutions of the equation $x = 1$. The **graph** of an equation is the visual representation in the coordinate plane of the set of points whose coordinates (a, b) satisfy the equation. The graph of $x = 1$ is the vertical line shown in **FIGURE 1.4.1**.

FIGURE 1.4.1 Graph of equation $x = 1$

☐ **Circles** The distance formula discussed in Section 1.3 can be used to define a set of points in the coordinate plane. One such important set is defined as follows.

DEFINITION 1.4.1 Circle

A **circle** is the set of all points $P(x, y)$ in the coordinate plane that are a given fixed distance r, called the **radius**, from a given fixed point C, called the **center**.

If the center has coordinates $C(h, k)$, then from the preceding definition a point $P(x, y)$ lies on a circle of radius r if and only if

$$d(P, C) = r \quad \text{or} \quad \sqrt{(x - h)^2 + (y - k)^2} = r.$$

Since $(x - h)^2 + (y - k)^2$ is always nonnegative, we obtain an equivalent equation when both sides are squared. We conclude that a circle of radius r and center $C(h, k)$ has the equation

$$(x - h)^2 + (y - k)^2 = r^2. \qquad (2)$$

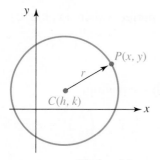

FIGURE 1.4.2 Circle with radius r and center (h, k)

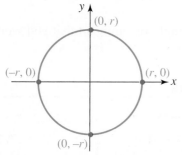

FIGURE 1.4.3 Circle with radius r and center (0, 0)

In FIGURE 1.4.2 we have sketched a typical graph of an equation of the form given in (2). Equation (2) is called the **standard form** of the equation of a circle. We note that the symbols h and k in (2) represent real numbers and as such can be positive, zero, or negative. When $h = 0$ and $k = 0$, we see that the standard form of the equation of a circle with center at the origin is

$$x^2 + y^2 = r^2. \tag{3}$$

See FIGURE 1.4.3. When $r = 1$ we say that (2) is an equation of a **unit circle**. For example, $x^2 + y^2 = 1$ is an equation of a unit circle centered at the origin.

EXAMPLE 1   Center and Radius

Find the center and radius of the circle whose equation is

$$(x - 8)^2 + (y + 2)^2 = 49. \tag{4}$$

Solution To obtain the standard form of the equation, we rewrite (4) as

$$(x - 8)^2 + (y - (-2))^2 = 7^2.$$

From this last form we identify $h = 8$, $k = -2$, and $r = 7$. Thus the circle is centered at $(8, -2)$ and has radius 7. ∎

EXAMPLE 2   Equation of a Circle

Find an equation of the circle with center $C(-5, 4)$ with radius $\sqrt{2}$.

Solution Substituting $h = -5$, $k = 4$, and $r = \sqrt{2}$ in (2), we obtain

$$(x - (-5))^2 + (y - 4)^2 = (\sqrt{2})^2 \quad \text{or} \quad (x + 5)^2 + (y - 4)^2 = 2. \quad ∎$$

EXAMPLE 3   Equation of a Circle

Find an equation of the circle with center $C(4, 3)$ and passing through $P(1, 4)$.

Solution With $h = 4$ and $k = 3$, we have from (2)

$$(x - 4)^2 + (y - 3)^2 = r^2. \tag{5}$$

Since the point $P(1, 4)$ lies on the circle as shown in FIGURE 1.4.4, its coordinates must satisfy equation (5). That is,

$$(1 - 4)^2 + (4 - 3)^2 = r^2 \quad \text{or} \quad 10 = r^2.$$

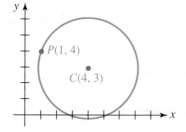

FIGURE 1.4.4 Circle in Example 3

Thus the required equation in standard form is $(x - 4)^2 + (y - 3)^2 = 10$. ∎

☐ **Completing the Square** If the terms $(x - h)^2$ and $(y - k)^2$ are expanded and the like terms grouped together, an equation of a circle in standard form can be written as

$$x^2 + y^2 + ax + by + c = 0. \tag{6}$$

Of course in this last form the center and radius are not apparent. To reverse the process—in other words, to go from (6) to the standard form (2)—we must **complete the square** in both x and y. Recall from algebra that adding $(a/2)^2$ to an expression such as $x^2 + ax$ yields $x^2 + ax + (a/2)^2$, which is the perfect square $(x + a/2)^2$. By rearranging the terms in (6),

$$(x^2 + ax \quad) + (y^2 + by \quad) = -c,$$

and then adding $(a/2)^2$ and $(b/2)^2$ to *both* sides of the last equation,

The terms in color added inside the parentheses on the left-hand side are also added to the right-hand side of the equality. This new equation is equivalent to (6).

$$\left(x^2 + ax + \left(\frac{a}{2}\right)^2\right) + \left(y^2 + by + \left(\frac{b}{2}\right)^2\right) = \left(\frac{a}{2}\right)^2 + \left(\frac{b}{2}\right)^2 - c,$$

we obtain the standard form of the equation of a circle:

$$\left(x + \frac{a}{2}\right)^2 + \left(y + \frac{b}{2}\right)^2 = \frac{1}{4}(a^2 + b^2 - 4c).$$

You should *not* memorize the last equation; we strongly recommend that you work through the process of completing the square each time.

▮ EXAMPLE 4   Completing the Square

Find the center and radius of the circle whose equation is

$$x^2 + y^2 + 10x - 2y + 17 = 0. \tag{7}$$

Solution To find the center and radius we rewrite equation (7) in the standard form (2). First, we rearrange the terms,

$$(x^2 + 10x \quad) + (y^2 - 2y \quad) = -17.$$

Then, we complete the square in x and y by adding, in turn, $(10/2)^2$ in the first set of parentheses and $(-2/2)^2$ in the second set of parentheses. Proceed carefully here because we must add these numbers to both sides of the equation:

$$\left(x^2 + 10x + \left(\tfrac{10}{2}\right)^2\right) + \left(y^2 - 2y + \left(\tfrac{-2}{2}\right)^2\right) = -17 + \left(\tfrac{10}{2}\right)^2 + \left(\tfrac{-2}{2}\right)^2$$
$$(x^2 + 10x + 25) + (y^2 - 2y + 1) = 9$$
$$(x + 5)^2 + (y - 1)^2 = 3^2.$$

From the last equation we see that the circle is centered at $(-5, 1)$ and has radius 3. See FIGURE 1.4.5. ▮

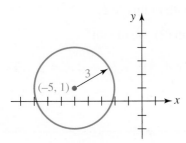

FIGURE 1.4.5 Circle in Example 4

It is possible that an expression for which we must complete the square has a leading coefficient other than 1. For example,

Note:

$$3x^2 + 3y^2 - 18x + 6y + 2 = 0$$

is an equation of circle. As in Example 4, we start by rearranging the equation:

$$(3x^2 - 18x \quad) + (3y^2 + 6y \quad) = -2.$$

Now, however, we must do one extra step before attempting completion of the square, that is, we must divide both sides of the equation by 3 so that the coefficients of x^2 and y^2 are each 1:

$$(x^2 - 6x \quad) + (y^2 + 2y \quad) = -\tfrac{2}{3}.$$

At this point we can now add the appropriate numbers within each set of parentheses *and* to the right-hand side of the equality. You should verify that the resulting standard form is $(x - 3)^2 + (y + 1)^2 = \frac{28}{3}$.

CHAPTER 1 INEQUALITIES, EQUATIONS, AND GRAPHS

Semicircles If we solve (3) for y we get $y^2 = r^2 - x^2$ or $y = \pm\sqrt{r^2 - x^2}$. This last expression is equivalent to two equations, $y = \sqrt{r^2 - x^2}$ and $y = -\sqrt{r^2 - x^2}$. In like manner if we solve (3) for x we obtain $x = \sqrt{r^2 - y^2}$ and $x = -\sqrt{r^2 - y^2}$.

By convention, the symbol $\sqrt{}$ denotes a nonnegative quantity, thus the y-values defined by an equation such as $y = \sqrt{r^2 - x^2}$ are nonnegative. The graphs of the four equations are, in turn, the upper half, lower half, right half, and left half of the circle given in Figure 1.4.3. Each graph in FIGURE 1.4.6 is called a **semicircle**.

(a) Upper half

(b) Lower half

(c) Right half

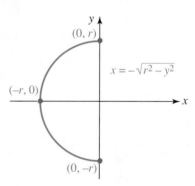

(d) Left half

FIGURE 1.4.6 Semicircles

Inequalities One last point about circles: On occasion we encounter problems where we must sketch the set of points in the xy-plane whose coordinates satisfy inequalities such as $x^2 + y^2 < r^2$ or $x^2 + y^2 \geq r^2$. The equation $x^2 + y^2 = r^2$ describes the set of points (x, y) whose distance to the origin $(0, 0)$ is exactly r. Therefore the inequality $x^2 + y^2 < r^2$ describes the set of points (x, y) whose distance to the origin is less than r. In other words, the points (x, y) whose coordinates satisfy the inequality $x^2 + y^2 < r^2$ are in the *interior* of the circle. Similarly, the points (x, y) whose coordinates satisfy $x^2 + y^2 \geq r^2$ lie either *on* the circle or are *exterior* to it. See FIGURE 1.4.7 and FIGURE 1.4.8.

Graphs It is difficult to read a newspaper, read a science or business text, surf the Internet, or even watch the news on TV without seeing graphical representations of data. It may even be impossible to get past the first page in a mathematics text without seeing some kind of graph. So many diverse quantities are connected by means of equations, and so many questions about the behavior of the quantities linked by the equation can be answered by means of a graph, that the ability to graph equations quickly and accurately—like the ability to do algebra quickly and accurately— is high on the list of skills essential to your success in a course in calculus. For the rest of this section we are going to talk about graphs in general, and more specifically about two important aspects of graphs of equations.

FIGURE 1.4.7 Set of points satisfying $x^2 + y^2 < r^2$

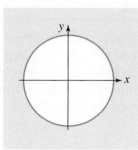

FIGURE 1.4.8 Set of points satisfying $x^2 + y^2 \geq r^2$

□ **Intercepts** Locating the points at which the graph of an equation crosses the coordinate axes can be helpful when sketching a graph by hand. The **x-intercepts** of a graph of an equation are the points at which the graph crosses the *x*-axis. Since every point on the *x*-axis has *y*-coordinate 0, the *x*-coordinates of these points (if there are any) can be found from the given equation by setting $y = 0$ and solving for *x*. In turn, the **y-intercepts** of the graph of an equation are the points at which its graph crosses the *y*-axis. The *y*-coordinates of these points can found by setting $x = 0$ in the equation and solving for *y*. See FIGURE 1.4.9.

(a) Five intercepts (b) Two *y*-intercepts (c) Graph has no intercepts

FIGURE 1.4.9 Intercepts of a graph

EXAMPLE 5 **Intercepts**

Find the intercepts of the graphs of the equations
(a) $x^2 - y^2 = 9$ (b) $y = 2x^2 + 5x - 12$.

Solution (a) To find the *x*-intercepts we set $y = 0$ and solve the resulting equation $x^2 = 9$ for *x*:

$$x^2 - 9 = 0 \quad \text{or} \quad (x + 3)(x - 3) = 0$$

gives $x = -3$ and $x = 3$. The *x*-intercepts of the graph are the points $(-3, 0)$ and $(3, 0)$. To find the *y*-intercepts we set $x = 0$ and solve $-y^2 = 9$ or $y^2 = -9$ for *y*. Because there are no real numbers whose square is negative we conclude the graph of the equation does not cross the *y*-axis.

(b) Setting $y = 0$ yields $2x^2 + 5x - 12 = 0$. This is a quadratic equation and can be solved either by factoring or by the quadratic formula. Factoring gives

$$(x + 4)(2x - 3) = 0,$$

and so $x = -4$ and $x = \frac{3}{2}$. The *x*-intercepts of the graph are the points $(-4, 0)$ and $\left(\frac{3}{2}, 0\right)$. Now, setting $x = 0$ in the equation $y = 2x^2 + 5x - 12$ immediately gives $y = -12$. The *y*-intercept of the graph is the point $(0, -12)$. ■

EXAMPLE 6 **Example 4 Revisited**

Let's return to the circle in Example 4 and determine its intercepts from the equation in (7). Setting $y = 0$ in $x^2 + y^2 + 10x - 2y + 17 = 0$ and using the quadratic formula to solve $x^2 + 10x + 17 = 0$ shows the *x*-intercepts of this circle are $(-5 - 2\sqrt{2}, 0)$ and $(-5 + 2\sqrt{2}, 0)$. If we let $x = 0$, then the quadratic formula shows that the roots of the equation $y^2 - 2y + 17 = 0$ are complex numbers. As seen in Figure 1.4.5, the circle does not cross the *y*-axis. ■

□ **Symmetry** A graph can also possess symmetry. You may already know that the graph of the equation $y = x^2$ is called a *parabola*. FIGURE 1.4.10 shows that the graph of $y = x^2$ is symmetric with respect to the *y*-axis since the portion of the graph that lies in the

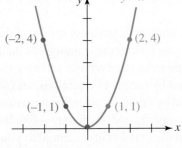

FIGURE 1.4.10 Graph with *y*-axis symmetry

CHAPTER 1 INEQUALITIES, EQUATIONS, AND GRAPHS

second quadrant is the *mirror image* or *reflection* of that portion of the graph in the first quadrant. In general, a graph is **symmetric with respect to the y-axis** if whenever (x, y) is a point on the graph, $(-x, y)$ is also a point on the graph. Note in Figure 1.4.10 that the points $(1, 1)$ and $(2, 4)$ are on the graph. Because the graph possesses y-axis symmetry, the points $(-1, 1)$ and $(-2, 4)$ must also be on the graph. A graph is said to be **symmetric with respect to the x-axis** if whenever (x, y) is a point on the graph, $(x, -y)$ is also a point on the graph. Finally, a graph is **symmetric with respect to the origin** if whenever (x, y) is on the graph, $(-x, -y)$ is also a point on the graph. FIGURE 1.4.11 illustrates these three types of symmetries.

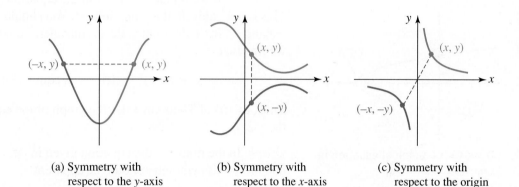

(a) Symmetry with respect to the y-axis

(b) Symmetry with respect to the x-axis

(c) Symmetry with respect to the origin

FIGURE 1.4.11 Symmetries of a graph

Observe that the graph of the circle given in Figure 1.4.3 possesses all three of these symmetries.

As a practical matter we would like to know whether a graph possesses any symmetry in advance of plotting it. This can be done by applying the following tests to the equation that defines the graph.

THEOREM 1.4.1   Tests for Symmetry

The graph of an equation is symmetric with respect to:

(*i*)   the **y-axis** if replacing x by $-x$ results in an equivalent equation
(*ii*) the **x-axis** if replacing y by $-y$ results in an equivalent equation
(*iii*) the **origin** if replacing x and y by $-x$ and $-y$ results in an equivalent equation

---

The advantage of using symmetry in graphing should be apparent: If, say, the graph of an equation is symmetric with respect to the x-axis, then we need only produce the graph for $y \geq 0$ since points on the graph for $y < 0$ are obtained by taking the mirror images, through the x-axis, of the points in the first and second quadrants.

EXAMPLE 7   Test for Symmetry

By replacing x by $-x$ in the equation $y = x^2$ and using $(-x)^2 = x^2$, we see that

$$y = (-x)^2 \quad \text{is equivalent to} \quad y = x^2.$$

By (*i*) of Theorem 1.4.1 this proves what is apparent in Figure 1.4.10; the graph of $y = x^2$ is symmetric with respect to the y-axis.

EXAMPLE 8   **Intercepts and Symmetry**

Determine the intercepts and any symmetry for the graph of

$$x + y^2 = 10. \tag{8}$$

Solution *Intercepts*: Setting $y = 0$ in equation (8) immediately gives $x = 10$. The graph of the equation has a single x-intercept, $(10, 0)$. When $x = 0$, we get $y^2 = 10$, which implies that $y = -\sqrt{10}$ or $y = \sqrt{10}$. Thus there are two y-intercepts, $(0, -\sqrt{10})$ and $(0, \sqrt{10})$.

Symmetry: If we replace x by $-x$ in the equation $x + y^2 = 10$ we get $-x + y^2 = 10$. This is not equivalent to equation (8). You should also verify that replacing x and y by $-x$ and $-y$ in (8) does not yield an equivalent equation. However, if we replace y by $-y$, we find that

$$x + (-y)^2 = 10 \quad \text{is equivalent to} \quad x + y^2 = 10.$$

Thus, by (*ii*) of Theorem 1.4.1 the graph of the equation is symmetric with respect to the x-axis.

Graph: In the graph of the equation given in FIGURE 1.4.12, the intercepts are indicated and the x-axis symmetry should be apparent.   ▬

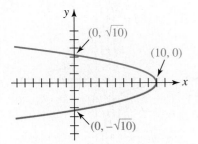

FIGURE 1.4.12 Graph of equation in Example 8

Exercises 1.4   Answers to selected odd-numbered problems begin on page ANS-2.

In Problems 1–6, find the center and the radius of the given circle. Sketch its graph.

1. $x^2 + y^2 = 5$
2. $x^2 + y^2 = 9$
3. $x^2 + (y - 3)^2 = 49$
4. $(x + 2)^2 + y^2 = 36$
5. $\left(x - \frac{1}{2}\right)^2 + \left(y - \frac{3}{2}\right)^2 = 1$
6. $(x + 3)^2 + (y - 5)^2 = 25$

In Problems 7–14, complete the square in x and y to find the center and the radius of the given circle.

7. $x^2 + y^2 + 8y = 0$
8. $x^2 + y^2 - 6x = 0$
9. $x^2 + y^2 + 2x - 4y - 4 = 0$
10. $x^2 + y^2 - 18x - 6y - 10 = 0$
11. $x^2 + y^2 - 20x + 16y + 128 = 0$
12. $x^2 + y^2 + 3x - 16y + 63 = 0$
13. $2x^2 + 2y^2 + 4x + 16y + 1 = 0$
14. $\frac{1}{2}x^2 + \frac{1}{2}y^2 + \frac{5}{2}x + 10y + 5 = 0$

In Problems 15–24, find an equation of the circle that satisfies the given conditions.

15. center $(0, 0)$, radius 1
16. center $(1, -3)$, radius 5
17. center $(0, 2)$, radius $\sqrt{2}$
18. center $(-9, -4)$, radius $\frac{3}{2}$
19. endpoints of a diameter at $(-1, 4)$ and $(3, 8)$
20. endpoints of a diameter at $(4, 2)$ and $(-3, 5)$
21. center $(0, 0)$, graph passes through $(-1, -2)$
22. center $(4, -5)$, graph passes through $(7, -3)$
23. center $(5, 6)$, graph tangent to the x-axis
24. center $(-4, 3)$, graph tangent to the y-axis

In Problems 25–28, sketch the semicircle defined by the given equation.

25. $y = \sqrt{4 - x^2}$

26. $x = 1 - \sqrt{1 - y^2}$

27. $x = \sqrt{1 - (y - 1)^2}$

28. $y = -\sqrt{9 - (x - 3)^2}$

29. Find an equation for the upper half of the circle $x^2 + (y - 3)^2 = 4$. Repeat for the right half of the circle.

30. Find an equation for the lower half of the circle $(x - 5)^2 + (y - 1)^2 = 9$. Repeat for the left half of the circle.

In Problems 31–34, sketch the set of points in the xy-plane whose coordinates satisfy the given inequality.

31. $x^2 + y^2 \geq 9$

32. $(x - 1)^2 + (y + 5)^2 \leq 25$

33. $1 \leq x^2 + y^2 \leq 4$

34. $x^2 + y^2 > 2y$

In Problems 35 and 36, give an inequality that describes the set of points (x, y) given in the figure.

35.

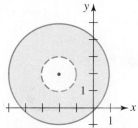

FIGURE 1.4.13 Set of points in Problem 35

36.

FIGURE 1.4.14 Set of points in Problem 36

In Problems 37–40, find the x- and y-intercepts of the given circle.

37. the circle with center $(3, -6)$ and radius 7

38. the circle $x^2 + y^2 + 5x - 6y = 0$

39. the circle in Problem 9

40. the circle in Problem 10

In Problems 41–66, find any intercepts of the graph of the given equation. Determine whether the graph of the equation possesses symmetry with respect to the x-axis, y-axis, or origin. Do not graph.

41. $y = -3x$

42. $y - 2x = 0$

43. $-x + 2y = 1$

44. $2x + 3y = 6$

45. $x = y^2$

46. $y = x^3$

47. $y = x^2 - 4$

48. $x = 2y^2 - 4$

49. $y = x^2 - 2x - 2$

50. $y^2 = 16(x + 4)$

51. $y = x(x^2 - 3)$

52. $y = (x - 2)^2(x + 2)^2$

53. $x = -\sqrt{y^2 - 16}$

54. $y^3 - 4x^2 + 8 = 0$

55. $4y^2 - x^2 = 36$

56. $\dfrac{x^2}{25} + \dfrac{y^2}{9} = 1$

57. $y = \dfrac{x^2 - 7}{x^3}$

58. $y = \dfrac{x^2 - 10}{x^2 + 10}$

59. $y = \dfrac{x^2 - x - 20}{x + 6}$

60. $y = \dfrac{(x + 2)(x - 8)}{x + 1}$

61. $y = \sqrt{x} - 3$ **62.** $y = 2 - \sqrt{x + 5}$

63. $y = |x - 9|$ **64.** $x = |y| - 4$

65. $|x| + |y| = 4$ **66.** $x + 3 = |y - 5|$

In Problems 67–70, state all the symmetries of the given graph.

67.

FIGURE 1.4.15 Graph for Problem 67

68.

FIGURE 1.4.16 Graph for Problem 68

69.

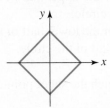

FIGURE 1.4.17 Graph for Problem 69

70.

FIGURE 1.4.18 Graph for Problem 70

In Problems 71–76, use symmetry to complete the given graph.

71. The graph is symmetric with respect to the *y*-axis.

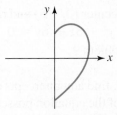

FIGURE 1.4.19 Graph for Problem 71

72. The graph is symmetric with respect to the *x*-axis.

FIGURE 1.4.20 Graph for Problem 72

73. The graph is symmetric with respect to the origin.

FIGURE 1.4.21 Graph for Problem 73

74. The graph is symmetric with respect to the *y*-axis.

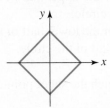

FIGURE 1.4.22 Graph for Problem 74

75. The graph is symmetric with respect to the *x*- and *y*-axes.

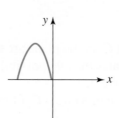

FIGURE 1.4.23 Graph for Problem 75

76. The graph is symmetric with respect to the origin.

FIGURE 1.4.24 Graph for Problem 76

77. The circle in FIGURE 1.4.25 has radius *r*. What is its equation in standard form?

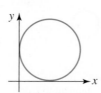

FIGURE 1.4.25 Graph for Problem 77

78. The circle in FIGURE 1.4.26 has center (h, k). What is its equation in standard form?

FIGURE 1.4.26 Graph for Problem 78

In Problems 79 and 80, find the areas of the shaded regions.

79.

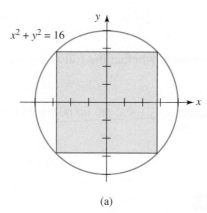

$x^2 + y^2 = 16$ (a)

$x^2 + y^2 = 16$ (b)

FIGURE 1.4.27 Graphs for Problem 79

80.

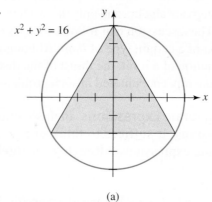

$x^2 + y^2 = 16$ (a)

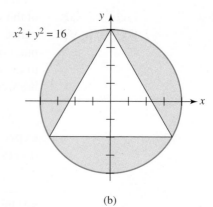

$x^2 + y^2 = 16$ (b)

FIGURE 1.4.28 Graphs for Problem 80

81. Show that the triangle in part (a) of Problem 80 is an equilateral triangle.

82. The equation $(x - 4)^2 + (y + 10)^2 = 0$ does not describe a circle. What is the graph of this equation?

For Discussion

83. Two circles C_1 and C_2 in the plane are **tangent** if they intersect at a single point. FIGURE 1.4.29 illustrates the two possibilities. Suppose circle C_1 has radius 4 and is centered at the origin.
  (a) Find an equation of a circle C_2 of radius 2 and center $(h, 3)$ that is externally tangent to C_1.
  (b) Find the point of tangency of the circles in part (a).

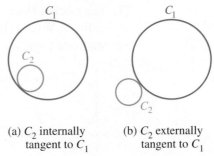

(a) C_2 internally tangent to C_1

(b) C_2 externally tangent to C_1

FIGURE 1.4.29 Tangent circles in Problem 83

84. Determine whether the following statement is true or false. Defend your answer.

If a graph has two of the three symmetries defined on page 29, then the graph necessarily possesses the third symmetry.

85. Determine whether the following statement is true or false. Defend your answer.

Every equation of the form $x^2 + y^2 + ax + by + c = 0$ is a circle.

86. Explain why there are no points $P(x, x)$ that are a distance $\sqrt{10}$ from the point $(3, -5)$.

1.5 Algebra and Limits

INTRODUCTION A calculus problem often consists of a sequence of steps, where most of the steps are algebra and only the last few—sometimes just the last step—involve calculus. The discussion that follows focuses on one kind of calculus problem: the computation of a certain type of *limit*. Although we give a brief and intuitive introduction to the notion of a limit, the thrust of the discussion is an overview of the type of algebra frequently encountered in such problems.

☐ **Algebraic Expressions** In this section we are concerned only with **fractional expressions**. It suffices to think of a fractional expression as a quotient of two *algebraic* expressions.* Roughly, an **algebraic expression** is one that is the result

---

*At this point we are excluding trigonometric, logarithmic, and exponential expressions. See Chapters 4 and 5.

of performing a finite number of additions, subtractions, multiplications, divisions, or roots on a collection of variables and real numbers. For example, some algebraic expressions in a single variable x are

$$5x^3 - 3x + 1, \qquad \frac{2x^2 - 18}{x + 3}, \qquad \text{and} \qquad x + \sqrt{x - 5}.$$

An area of algebra that causes difficulties in working calculus problems is the manipulation of fractional expressions.

☐ **Factoring** When the distributive law

$$a(b + c) = ab + ac$$

is read right to left,

$$ab + ac = a(b + c),$$

we say that the expression $ab + ac$ has been **factored**. We will see in Chapter 3 that factoring plays an important role in solving equations, as well as in graphing. But in the present context we are concerned only with using factoring to simplify fractional expressions.

The following three factorization formulas are important and are used as a matter of course throughout various fields of mathematics.

THEOREM 1.5.1   Factorizations Worth Knowing

Difference of two squares: $a^2 - b^2 = (a - b)(a + b)$	(1)
Difference of two cubes: $a^3 - b^3 = (a - b)(a^2 + ab + b^2)$	(2)
Sum of two cubes: $a^3 + b^3 = (a + b)(a^2 - ab + b^2)$	(3)

The symbols a and b in (1)–(3) are placeholders. For example, the expression $x^4 - 16$ is of the form given in (1). With the identifications $a = x^2$ and $b = 4$, we have

$$x^4 - 16 = (x^2)^2 - 4^2 = (x^2 - 4)(x^2 + 4). \tag{4}$$

Since the factor $x^2 - 4$ is also the difference of two squares, (4) continues as

$$x^4 - 16 = (x^2 - 4)(x^2 + 4) = (x - 2)(x + 2)(x^2 + 4). \tag{5}$$

The factorization in (5) is as far as we can go using real numbers and integer exponents; the sum of two squares $x^2 + 4$ does not factor using real numbers. As another example, consider the expression $2x^2 - 3$. Since any positive real number can be written as the square of its square root we have $2 = (\sqrt{2})^2$ and $3 = (\sqrt{3})^2$, and so from (1) the expression $2x^2 - 3$ factors in the following manner:

$$2x^2 - 3 = (\overset{\overset{a}{\downarrow}}{\sqrt{2}x})^2 - (\overset{\overset{b}{\downarrow}}{\sqrt{3}})^2 = (\sqrt{2}x - \sqrt{3})(\sqrt{2}x + \sqrt{3}).$$

▶ Cancellation Property: If a, b, and c are real numbers, then
$$\frac{ac}{bc} = \frac{a}{b}, \quad c \neq 0.$$

We use factoring and the cancellation property to simplify a fractional expression.

EXAMPLE 1   **Factoring and Canceling**

Simplify **(a)** $\dfrac{x^2 - 1}{x - 1}$ **(b)** $\dfrac{x + 3}{x^2 - 4x - 21}$.

Solution **(a)** From (1) we see

$$\frac{x^2 - 1}{x - 1} = \frac{(x - 1)(x + 1)}{x - 1} = x + 1.$$

The cancellation of $x - 1$ in the foregoing expression is only valid for $x \neq 1$. For $x = 1$ we would be dividing by 0.

(b) We look for factors $x - a$ and $x - b$ such that

$$x^2 - 4x - 21 = (x - a)(x - b).$$

This implies $ab = -21$, so a and b must be factors of -21 whose sum is $-(a + b) = -4$. The usual trial and error procedure leads to $a = 7$ and $b = -3$. Therefore,

$$\frac{x + 3}{x^2 - 4x - 21} = \frac{x + 3}{(x + 3)(x - 7)} = \frac{1}{x - 7}, \qquad x \neq -3. \qquad ■$$

☐ **Binomial Expansion** A two-term algebraic expression $a + b$ is called a **binomial**. You undoubtedly have worked problems where you had to expand powers of binomials such as $(a + b)^2$ and $(a + b)^3$. This occurs so often in mathematics courses that we recommend that you memorize the expansions given in (6) and (7) below.

THEOREM 1.5.2 Binomial Expansions Worth Knowing

Expansions of $(a + b)^n$ for $n = 2$ and $n = 3$ are, respectively,

$$(a + b)^2 = a^2 + 2ab + b^2 \qquad (6)$$
$$(a + b)^3 = a^3 + 3a^2b + 3ab^2 + b^3 \qquad (7)$$

Of course, formulas (6) and (7) work just as well for a binomial in the form of a difference $a - b$. Simply treat $a - b$ as the sum $a + (-b)$ and replace the symbol b in (6) and (7) with $-b$:

$$(a - b)^2 = (a + (-b))^2 = a^2 + 2a(-b) + (-b)^2 = a^2 - 2ab + b^2,$$

and
$$(a - b)^3 = (a + (-b))^3 = a^3 + 3a^2(-b) + 3a(-b)^2 + (-b)^3$$
$$= a^3 - 3a^2b + 3ab^2 - b^3.$$

There are ways of remembering how to obtain the coefficients in the expansion of higher powers such as $(a + b)^4$. **Pascal's triangle** is one such way and is reviewed in the *Student Resource Manual* that accompanies this text.

▆ EXAMPLE 2   Binomial Expansion

Simplify $\dfrac{(7 + h)^2 - 49}{h}$.

Solution We use the expansion of $(a + b)^2$ given in (6) with $a = 7$ and $b = h$:

$$\frac{(7 + h)^2 - 49}{h} = \frac{(7^2 + 2(7)h + h^2) - 49}{h}$$
$$= \frac{49 + 14h + h^2 - 49}{h} \qquad \leftarrow 49 - 49 = 0$$

$$= \frac{h(14 + h)}{h} \quad \leftarrow \text{cancel the } h\text{'s}$$

$$= 14 + h, \quad h \neq 0.$$ ▪

☐ **Addition of Fractional Expressions** Combining two or more fractional expressions, or simplification of a complex fraction where the numerator or denominator is itself a fraction, can be particularly troublesome for some students.

EXAMPLE 3 **Addition of Fractions**

Write as one fraction $\dfrac{10x}{2x^2 + 3x - 2} - \dfrac{4}{x + 2} + \dfrac{8}{2x - 1}$.

Solution Because $2x^2 + 3x - 2 = (2x - 1)(x + 2)$, the least common denominator of the three terms is $(2x - 1)(x + 2)$. Therefore, we multiply the second term by $(2x - 1)/(2x - 1)$ and the third term by $(x + 2)/(x + 2)$:

$$\frac{10x}{(2x - 1)(x + 2)} - \frac{4}{x + 2}\frac{2x - 1}{2x - 1} + \frac{8}{2x - 1}\frac{x + 2}{x + 2}.$$

Adding numerators and simplifying gives

$$\frac{10x - 4(2x - 1) + 8(x + 2)}{(2x - 1)(x + 2)} = \frac{10x - 8x + 4 + 8x + 16}{(2x - 1)(x + 2)}$$

$$= \frac{10x + 20}{(2x - 1)(x + 2)}$$

$$= \frac{10(x + 2)}{(2x - 1)(x + 2)}$$

$$= \frac{10}{2x - 1}.$$

The cancellation of $x + 2$ is permissible provided $x \neq -2$. ▪

We will illustrate the simplification of a complex fraction in Example 9.

☐ **Rationalization** You may have learned **rationalization of a denominator** in a previous mathematics course. Recall that rationalization of a denominator consists of multiplying an expression by a factor equal to 1 with the intent of clearing a radical from a denominator. For example, to rationalize the denominator in $1/\sqrt{2}$ we multiply the fraction by $\sqrt{2}/\sqrt{2}$:

$$\overset{\text{fraction is equal to 1}}{\underset{\downarrow}{}}$$

$$\frac{1}{\sqrt{2}} = \frac{1}{\sqrt{2}}\frac{\sqrt{2}}{\sqrt{2}} = \frac{\sqrt{2}}{(\sqrt{2})^2} = \frac{\sqrt{2}}{2}.$$

There is no rule in mathematics that says only denominators must be rationalized. There are times in calculus when we are interested in rationalization not only of denominators but numerators as well. The next example uses the factorization of the difference of two squares in a slightly different manner. For $a > 0$ and $b > 0$, we can write $(\sqrt{a})^2 = a$, $(\sqrt{b})^2 = b$, and so we can write $a - b = (\sqrt{a})^2 - (\sqrt{b})^2$. It then follows from (1) that

$$a - b = (\sqrt{a} - \sqrt{b})(\sqrt{a} + \sqrt{b}). \tag{8}$$

A variation of (8) is $a^2 - b = (a - \sqrt{b})(a + \sqrt{b})$. Thus if a numerator or denominator of a fractional expression contains a binomial term that includes at least one radical, such as

$$a - \sqrt{b},\, a + \sqrt{b},\, \sqrt{a} - b,\, \sqrt{a} + b,\, \sqrt{a} - \sqrt{b},\, \text{or } \sqrt{a} + \sqrt{b},$$

we multiply the numerator and the denominator of the fraction by the corresponding **conjugate factor**

$$a + \sqrt{b},\, a - \sqrt{b},\, \sqrt{a} + b,\, \sqrt{a} - b,\, \sqrt{a} + \sqrt{b},\, \text{or } \sqrt{a} - \sqrt{b}.$$

For example, to rationalize the denominator of $3/(\sqrt{2} - \sqrt{5})$ we use (8) to write

$$\frac{3}{\sqrt{2} - \sqrt{5}} = \frac{3}{\sqrt{2} - \sqrt{5}} \underset{\underset{\text{conjugate factor of denominator}}{\uparrow}}{\overset{\overset{\text{fraction is equal to 1}}{\downarrow}}{\frac{\sqrt{2} + \sqrt{5}}{\sqrt{2} + \sqrt{5}}}} = \frac{3(\sqrt{2} + \sqrt{5})}{(\sqrt{2})^2 - (\sqrt{5})^2}$$

$$= \frac{3(\sqrt{2} + \sqrt{5})}{-3} = -(\sqrt{2} + \sqrt{5}) = -\sqrt{2} - \sqrt{5}.$$

▮ EXAMPLE 4   Rationalization of a Numerator

Rationalize the numerator in $\dfrac{\sqrt{4 + x} - 2}{x}$.

Solution Think of the numerator as $\sqrt{a} - b$ where $a = 4 + x$ and $b = 2$. Because the conjugate factor of $\sqrt{a} - b$ is $\sqrt{a} + b$, we are able to clear the radical in the numerator by multiplying the numerator and denominator of the given fractional expression by $\sqrt{4 + x} + 2$:

$$\frac{\sqrt{4 + x} - 2}{x} = \frac{\sqrt{4 + x} - 2}{x} \frac{\sqrt{4 + x} + 2}{\sqrt{4 + x} + 2} = \frac{(\sqrt{4 + x})^2 - 2^2}{x(\sqrt{4 + x} + 2)}$$

$$= \frac{4 + x - 4}{x(\sqrt{4 + x} + 2)} = \frac{x}{x(\sqrt{4 + x} + 2)}.$$

After canceling the x's in the numerator and the denominator in the last term the rationalization is complete:

$$\frac{\sqrt{4 + x} - 2}{x} = \frac{1}{\sqrt{4 + x} + 2}, \qquad x \neq 0. \qquad\blacksquare$$

☐ **Limits—The Calculus Connection**   Consider the fractional algebraic expression $\dfrac{x^2 - 1}{x - 1}$. Observe that this fraction cannot be evaluated at $x = 1$ because substituting 1 into the expression results in the undefined quantity 0/0. However, the fractional expression can be evaluated at any other real number; in particular, it can be evaluated at numbers that are very *close* to 1. The numerical values of the fractional expression given in the following two tables are easily obtained using the simplification in part (a) of Example 1:

$$\frac{x^2 - 1}{x - 1} = x + 1, \qquad \text{for } x \neq 1.$$

x	0.9	0.99	0.999
$\dfrac{x^2 - 1}{x - 1}$	1.9	1.99	1.999

x	1.1	1.01	1.001
$\dfrac{x^2 - 1}{x - 1}$	2.1	2.01	2.001

(9)

☐ **Arrow Notation** The discussion of the limit concept is facilitated by using a special notation. If we let the **arrow symbol** $\rightarrow$ represent the word *approach*, then the symbolism

$$x \rightarrow a^- \text{ indicates that } x \text{ approaches a number } a \text{ from the left,}$$

that is, through numbers that are less than a, and

$$x \rightarrow a^+ \text{ indicates that } x \text{ approaches a number } a \text{ from the right,}$$

that is, through numbers that are greater than a. Finally, the notation

$$x \rightarrow a \text{ signifies that } x \text{ approaches a number } a \text{ from both sides,}$$

in other words, from the left and the right sides of a on the number line. In the left-hand table in (9) we are letting $x \rightarrow 1^-$, and in the right-hand table $x \rightarrow 1^+$. Each table in (9) shows that the fractional expression $\dfrac{x^2 - 1}{x - 1}$ is close to the number 2 when x is close to 1, that is,

$$\frac{x^2 - 1}{x - 1} \rightarrow 2 \text{ as } x \rightarrow 1^- \qquad \text{and} \qquad \frac{x^2 - 1}{x - 1} \rightarrow 2 \text{ as } x \rightarrow 1^+. \qquad (10)$$

We say that 2 is the **limit** of $\dfrac{x^2 - 1}{x - 1}$ as x approaches 1 and write

$$\lim_{x \rightarrow 1} \frac{x^2 - 1}{x - 1} = 2. \qquad (11)$$

Before proceeding any further, we should make it clear that *a limit of an expression need not exist*. In the next two tables, consider $1/x$ as x approaches zero:

$x \rightarrow 0^-$	-0.1	-0.01	-0.001
$1/x$	-10	-100	-1000

$x \rightarrow 0^+$	0.1	0.01	0.001
$1/x$	10	100	1000

As can be seen in the tables, as x gets closer and closer to 0, the values of $1/x$ are becoming larger and larger in absolute value. In other words, $1/x$ is becoming unbounded. In this case we write

$$\frac{1}{x} \rightarrow -\infty \text{ as } x \rightarrow 0^- \qquad \text{and} \qquad \frac{1}{x} \rightarrow \infty \text{ as } x \rightarrow 0^+,$$

where ∞ is the infinity symbol. We say that $\lim\limits_{x \rightarrow 0} (1/x)$ does not exist.

☐ **Existence of a Limit** Suppose the symbol $f(x)$ denotes an expression involving a single variable x and that the symbols a and L represent real numbers. If, as illustrated in (10),

$$f(x) \rightarrow L \text{ as } x \rightarrow a^- \qquad \text{and} \qquad f(x) \rightarrow L \text{ as } x \rightarrow a^+, \qquad (12)$$

then we say that $\lim\limits_{x \rightarrow a} f(x)$ **exists** and write

$$\lim_{x \rightarrow a} f(x) = L. \qquad (13)$$

In calculus, you will not be asked to find a limit by constructing tables of numerical values, although you surely will be asked to construct such tables because they are useful in convincing yourself of either the existence or the nonexistence of a limit.

(See Problems 47 and 48 in Exercises 1.5.) Limits are either found or are proved to exist using analytical methods, in many cases using proven laws or properties of limits. Because it is not our goal to delve into theoretical or geometrical interpretations of a limit, and because we want to make the point that the calculus part of *some* problems is often the least significant part of the solution, we will accept three results from calculus without proof: If a and c are real numbers, then

$$\lim_{x \to a} c = c, \qquad \lim_{x \to a} x = a, \qquad \text{and} \qquad \lim_{x \to a} x^n = a^n, \tag{14}$$

where n is a positive integer. For example, (14) allows us to write*

$$\lim_{x \to 3} (5x + 4) = 5(3) + 4 = 19$$

and

$$\lim_{x \to 3} (2x^2 + x + 1) = 2(3)^2 + 3 + 1 = 22.$$

In the preceding line we used $\lim_{x \to 3} x = 3$, $\lim_{x \to 3} x^2 = 9$, $\lim_{x \to 3} 4 = 4$, and $\lim_{x \to 3} 1 = 1$.

☐ **Indeterminate Form** The limit concept is the foundation of calculus, and one kind of limit is of particular significance in calculus: the limit of a fractional expression where *both* the numerator and the denominator are approaching 0. Such a limit is said to have the **indeterminate form 0/0**. For example, in view of the results in (14), $\lim_{x \to 1} (x - 1) = 0$ and $\lim_{x \to 1} (x^2 - 1) = 0$. Therefore $\lim_{x \to 1} \dfrac{x^2 - 1}{x - 1}$ has the indeterminate form 0/0. Of course, not all limit problems have this indeterminate form, but because of their importance (see Section 2.10), the limits in the remaining five examples, as well as *all* the limits in Exercises 1.5, have the form 0/0. Moreover, for simplicity we will only consider limits that actually exist.

We now show you how to find $\lim_{x \to 1} \dfrac{x^2 - 1}{x - 1}$ without the help of numerical tables:

algebra from Example 1(a)
↓

$$\lim_{x \to 1} \frac{x^2 - 1}{x - 1} = \lim_{x \to 1} (x + 1) = 1 + 1 = 2.$$

Done! The intermediate steps were all algebra performed to rewrite the expression in a more tractable form, a form where the actual limit can be computed with minimal effort.

EXAMPLE 5   Example 1 Revisited

Find $\lim_{x \to -3} \dfrac{x + 3}{x^2 - 4x - 21}$.

Solution This is the fractional expression in part (b) of Example 1. Observe that as $x \to -3$, the given limit has the indeterminate form 0/0. Now, using the algebraic simplification of this expression done in Example 1, we find that

algebra from Example 1(b)
↓

$$\lim_{x \to -3} \frac{x + 3}{x^2 - 4x - 21} = \lim_{x \to -3} \frac{1}{x - 7} = \frac{1}{-10} = -\frac{1}{10}. \qquad ▬$$

EXAMPLE 6   Example 2 Revisited

Find $\lim_{h \to 0} \dfrac{(7 + h)^2 - 49}{h}$.

*We are actually using several other properties of limits here. However, we do not feel this is the place to discuss all the properties of the limit concept.

Solution Using the algebra from Example 2,

$$\lim_{h \to 0} \frac{(7 + h)^2 - 49}{h} = \lim_{h \to 0} (14 + h) = 14.$$ ▬

EXAMPLE 7   Example 3 Revisited

Find $\displaystyle\lim_{x \to -2} \left[\frac{10x}{2x^2 + 3x - 2} - \frac{4}{x + 2} + \frac{8}{2x - 1} \right]$.

Solution Were this a calculus course, you should observe that the first and second terms are of the form $1/0$ as $x \to -2$. You may think that this is the situation $\infty - \infty$, and so gives 0. No. Remember we never treat ∞ as we would a number. The observation that the given algebraic expression contains these undefined quantities should trigger the idea that combining the fractions into *one* fractional expression would be a way to proceed. After carrying out the algebra, as done in Example 3, you would then finish the problem as follows:

algebra from Example 3
↓

$$\lim_{x \to -2} \left[\frac{10x}{2x^2 + 3x - 2} - \frac{4}{x + 2} + \frac{8}{2x - 1} \right] = \lim_{x \to -2} \frac{10}{2x - 1} = \frac{10}{-5} = -2.$$ ▬

EXAMPLE 8   Example 4 Revisited

Find $\displaystyle\lim_{x \to 0} \frac{\sqrt{4 + x} - 2}{x}$.

Solution Using the algebra from Example 4, we find

algebra from Example 4
↓

$$\lim_{x \to 0} \frac{\sqrt{4 + x} - 2}{x} = \lim_{x \to 0} \frac{1}{\sqrt{4 + x} + 2} = \frac{1}{\sqrt{4} + 2} = \frac{1}{2 + 2} = \frac{1}{4}.$$ ▬

When finding the value of a limit, the algebra can be done as a side problem (as we have done in Examples 1–4), and then making use of your work, completing the problem as we have illustrated in Examples 5–8. In our last example, we combine the algebra with computing the limit. We recommend that you work through this example rather than just read it.

EXAMPLE 9   Limit of a Complex Fraction

Find $\displaystyle\lim_{x \to 0} \frac{\dfrac{1}{(2 + x)^3} - \dfrac{1}{8}}{x}$.

Solution The given expression is an example of a complex fraction, that is, a quotient where either the numerator or denominator is a fractional expression. We begin by finding a common denominator in the numerator:

$$\lim_{x \to 0} \frac{\dfrac{1}{(2 + x)^3} - \dfrac{1}{8}}{x} = \lim_{x \to 0} \frac{\dfrac{1}{(2 + x)^3}\dfrac{8}{8} - \dfrac{1}{8}\dfrac{(2 + x)^3}{(2 + x)^3}}{x}$$

$$= \lim_{x \to 0} \frac{\dfrac{8 - (2 + x)^3}{8(2 + x)^3}}{x}$$

To continue we use the expansion of $(a + b)^3$ given in (7) with $a = 2$ and $b = x$:

$$\lim_{x \to 0} \frac{\dfrac{1}{(2 + x)^3} - \dfrac{1}{8}}{x} = \lim_{x \to 0} \frac{\dfrac{8 - (2^3 + 3(2)^2 x + 3(2)x^2 + x^3)}{8(2 + x)^3}}{x}$$

$$= \lim_{x \to 0} \frac{\dfrac{8 - 8 - 12x - 6x^2 - x^3}{8(2 + x)^3}}{x} \quad \leftarrow 8 - 8 = 0$$

$$= \lim_{x \to 0} \frac{\dfrac{-12x - 6x^2 - x^3}{8(2 + x)^3}}{x}.$$

Because the x in the denominator of the last complex fraction is equivalent to the fraction $x/1$ we invert and multiply:

Recall that division of fractions is converted into multiplication of fractions:

$$\frac{\dfrac{a}{b}}{\dfrac{c}{d}} = \frac{a}{b} \times \frac{d}{c} = \frac{ad}{bc}, \quad bc \neq 0$$

$$\lim_{x \to 0} \frac{\dfrac{1}{(2 + x)^3} - \dfrac{1}{8}}{x} = \lim_{x \to 0} \frac{\dfrac{-12x - 6x^2 - x^3}{8(2 + x)^3}}{\dfrac{x}{1}}$$

$$= \lim_{x \to 0} \frac{-12x - 6x^2 - x^3}{8(2 + x)^3} \frac{1}{x}$$

$$= \lim_{x \to 0} \frac{x(-12 - 6x - x^2)}{8(2 + x)^3} \frac{1}{x} \quad \leftarrow \begin{cases} \text{factor } x \text{ from numerator} \\ \text{and cancel } x\text{'s} \end{cases}$$

$$= \lim_{x \to 0} \frac{-12 - 6x - x^2}{8(2 + x)^3}.$$

Finally, we have

$$\lim_{x \to 0} \frac{\dfrac{1}{(2 + x)^3} - \dfrac{1}{8}}{x} = \lim_{x \to 0} \frac{-12 - 6x - x^2}{8(2 + x)^3} = \frac{-12}{8 \cdot 2^3} = -\frac{3}{16},$$

since $\lim\limits_{x \to 0} x = 0$ and $\lim\limits_{x \to 0} x^2 = 0$ by (14).

NOTES FROM THE CLASSROOM

(*i*) On tests we see students carrying out the expansion of $(a + b)^3$ by brute force, multiplying out $(a + b)(a + b)(a + b)$. This procedure is not recommended; it is slow and you are prone to errors. Instead, you should memorize (6) and (7).

(*ii*) In *any* mathematics course—not just calculus—do not erase or leave out important steps of your work. Most mathematics instructors want to see all work. Presenting that work in a neat and orderly fashion is also to your advantage. Finally, in the case of a limit problem such as Example 9, be sure to write down the symbol $\lim\limits_{x \to a}$ at each step. For example, we frequently see *incorrect* statements like this:

$$\lim_{x \to 1} \frac{x - 1}{x^2 - 1} = \frac{1}{x + 1} = \frac{1}{2}$$

on students' papers. The *correct* version of the preceding line is

$$\lim_{x \to 1} \frac{x - 1}{x^2 - 1} = \lim_{x \to 1} \frac{1}{x + 1} = \frac{1}{2}.$$

(*iii*) The symbolic statements in (12) can be written as **one-sided limits**:

$$\lim_{x \to a^-} f(x) = L \quad \text{and} \quad \lim_{x \to a^+} f(x) = L, \tag{15}$$

Then $\lim_{x \to a} f(x)$ *exists* when both $\lim_{x \to a^-} f(x)$ and $\lim_{x \to a^+} f(x)$ exist and are equal. On the other hand, we can say that $\lim_{x \to a} f(x)$ *does not exist* when both $\lim_{x \to a^-} f(x)$ and $\lim_{x \to a^+} f(x)$ exist and are not equal. In addition, $\lim_{x \to a} f(x)$ does not exist if either $\lim_{x \to a^-} f(x)$ or $\lim_{x \to a^+} f(x)$ fails to exist.

Exercises 1.5   Answers to selected odd-numbered problems begin on page ANS–3.

In Problems 1–12, use factorization to simplify the given expression in part (a). Then, if instructed, find the indicated limit in part (b).

1. (a) $\dfrac{x^2 - 25}{x - 5}$ 　　　　　　　(b) $\lim\limits_{x \to 5} \dfrac{x^2 - 25}{x - 5}$

2. (a) $\dfrac{y - 3}{y^2 - 9}$ 　　　　　　　(b) $\lim\limits_{y \to 3} \dfrac{y - 3}{y^2 - 9}$

3. (a) $\dfrac{x^2 - 7x + 6}{x - 1}$ 　　　　　(b) $\lim\limits_{x \to 1} \dfrac{x^2 - 7x + 6}{x - 1}$

4. (a) $\dfrac{2x + 10}{x^2 + 7x + 10}$ 　　　　(b) $\lim\limits_{x \to -5} \dfrac{2x + 10}{x^2 + 7x + 10}$

5. (a) $\dfrac{x^2 + x - 6}{x^2 - 5x + 6}$ 　　　　(b) $\lim\limits_{x \to 2} \dfrac{x^2 + x - 6}{x^2 - 5x + 6}$

6. (a) $\dfrac{x^2 - 8x}{x^2 - 6x - 16}$ 　　　　(b) $\lim\limits_{x \to 8} \dfrac{x^2 - 8x}{x^2 - 6x - 16}$

7. (a) $\dfrac{x^3 - 1}{x - 1}$ 　　　　　　　(b) $\lim\limits_{x \to 1} \dfrac{x^3 - 1}{x - 1}$

8. (a) $\dfrac{x^2 - 4}{x^3 + 8}$ 　　　　　　　(b) $\lim\limits_{x \to -2} \dfrac{x^2 - 4}{x^3 + 8}$

9. (a) $\dfrac{x^3 - 1}{x^2 + 3x - 4}$ 　　　　　(b) $\lim\limits_{x \to 1} \dfrac{x^3 - 1}{x^2 + 3x - 4}$

10. (a) $\dfrac{x^5 + 2x^4 + x^3}{x^4 - 2x^2 + 1}$ 　　　(b) $\lim\limits_{x \to -1} \dfrac{x^5 + 2x^4 + x^3}{x^4 - 2x^2 + 1}$

11. (a) $\dfrac{x^3 + 3x^2 + 3x + 1}{x^4 + x^3 + x + 1}$ 　(b) $\lim\limits_{x \to -1} \dfrac{x^3 + 3x^2 + 3x + 1}{x^4 + x^3 + x + 1}$

12. (a) $\dfrac{x^4 - 5x^3 + 4x - 20}{x^4 - 5x^3 + x - 5}$ 　(b) $\lim\limits_{x \to 5} \dfrac{x^4 - 5x^3 + 4x - 20}{x^4 - 5x^3 + x - 5}$

In Problems 13–20, use binomial expansion to simplify the given expression in part (a). Then, if instructed, find the indicated limit in part (b).

13. (a) $\dfrac{(2 + h)^2 - 4}{h}$ (b) $\lim\limits_{h \to 0} \dfrac{(2 + h)^2 - 4}{h}$

14. (a) $\dfrac{5 - 5(h + 1)^2}{h}$ (b) $\lim\limits_{h \to 0} \dfrac{5 - 5(h + 1)^2}{h}$

15. (a) $\dfrac{(2x + 1)^2 - 9}{x - 1}$ (b) $\lim\limits_{x \to 1} \dfrac{(2x + 1)^2 - 9}{x - 1}$

16. (a) $\dfrac{2(x - 1)^2 - 4(x - 1) - 6}{x}$ (b) $\lim\limits_{x \to 0} \dfrac{2(x - 1)^2 - 4(x - 1) - 6}{x}$

17. (a) $\dfrac{(1 + x)^3 - 1}{x}$ (b) $\lim\limits_{x \to 0} \dfrac{(1 + x)^3 - 1}{x}$

18. (a) $\dfrac{(x + 1)^3 + (x - 1)^3}{x}$ (b) $\lim\limits_{x \to 0} \dfrac{(x + 1)^3 + (x - 1)^3}{x}$

19. (a) $\dfrac{2(h + 1)^3 - 5(h + 1)^2 + 3}{h}$ (b) $\lim\limits_{h \to 0} \dfrac{2(h + 1)^3 - 5(h + 1)^2 + 3}{h}$

20. (a) $\dfrac{(x + 2)^4 - 16}{x}$ (b) $\lim\limits_{x \to 0} \dfrac{(x + 2)^4 - 16}{x}$

In Problems 21–26, use addition of algebraic fractions to simplify the given expression in part (a). Then, if instructed, find the indicated limit in part (b).

21. (a) $\dfrac{1}{x - 2} - \dfrac{6}{x^2 + 2x - 8}$ (b) $\lim\limits_{x \to 2} \left[\dfrac{1}{x - 2} - \dfrac{6}{x^2 + 2x - 8} \right]$

22. (a) $\dfrac{x^2 + 3x - 1}{x} + \dfrac{1}{x}$ (b) $\lim\limits_{x \to 0} \left[\dfrac{x^2 + 3x - 1}{x} + \dfrac{1}{x} \right]$

23. (a) $\dfrac{1}{x - 10} - \dfrac{20}{x^2 - 100}$ (b) $\lim\limits_{x \to 10} \left[\dfrac{1}{x - 10} - \dfrac{20}{x^2 - 100} \right]$

24. (a) $\dfrac{1}{x} \left[\dfrac{1}{9} - \dfrac{1}{x + 9} \right]$ (b) $\lim\limits_{x \to 0} \dfrac{1}{x} \left[\dfrac{1}{9} - \dfrac{1}{x + 9} \right]$

25. (a) $\dfrac{\dfrac{1}{(2 + h)^2} - \dfrac{1}{4}}{h}$ (b) $\lim\limits_{h \to 0} \dfrac{\dfrac{1}{(2 + h)^2} - \dfrac{1}{4}}{h}$

26. (a) $\dfrac{1}{t - 1} \left[\dfrac{1}{(t + 3)^2} - \dfrac{1}{16} \right]$ (b) $\lim\limits_{t \to 1} \dfrac{1}{t - 1} \left[\dfrac{1}{(t + 3)^2} - \dfrac{1}{16} \right]$

In Problems 27–34, use rationalization to simplify the given expression in part (a). Then, if instructed, find the indicated limit in part (b).

27. (a) $\dfrac{\sqrt{x} - 3}{x - 9}$ (b) $\lim\limits_{x \to 9} \dfrac{\sqrt{x} - 3}{x - 9}$

28. (a) $\dfrac{\dfrac{1}{\sqrt{x}} - \dfrac{1}{\sqrt{2}}}{x - 2}$ (b) $\lim\limits_{x \to 2} \dfrac{\dfrac{1}{\sqrt{x}} - \dfrac{1}{\sqrt{2}}}{x - 2}$

29. (a) $\dfrac{x}{\sqrt{7 + x} - \sqrt{7}}$ (b) $\lim\limits_{x \to 0} \dfrac{x}{\sqrt{7 + x} - \sqrt{7}}$

30. (a) $\dfrac{\sqrt{u + 4} - 3}{u - 5}$ (b) $\lim\limits_{u \to 5} \dfrac{\sqrt{u + 4} - 3}{u - 5}$

CHAPTER 1 INEQUALITIES, EQUATIONS, AND GRAPHS

31. (a) $\dfrac{25 - t}{5 - \sqrt{t}}$
 (b) $\displaystyle\lim_{t \to 25} \dfrac{25 - t}{5 - \sqrt{t}}$

32. (a) $\dfrac{1}{h}\left[1 - \dfrac{1}{\sqrt{1 + h}}\right]$
 (b) $\displaystyle\lim_{h \to 0} \dfrac{1}{h}\left[1 - \dfrac{1}{\sqrt{1 + h}}\right]$

33. (a) $\dfrac{4y^2}{\sqrt{y^2 + y + 1} - \sqrt{y + 1}}$
 (b) $\displaystyle\lim_{y \to 0} \dfrac{4y^2}{\sqrt{y^2 + y + 1} - \sqrt{y + 1}}$

34. (a) $\dfrac{9t^2}{t + 2 - 2\sqrt{t + 1}}$
 (b) $\displaystyle\lim_{t \to 0} \dfrac{9t^2}{t + 2 - 2\sqrt{t + 1}}$

Calculus-Related Problems

In Problems 35–40, the given algebraic expression is an unsimplified answer to a calculus problem. Simplify the expression.

35. $\dfrac{x + \dfrac{1}{x} - a - \dfrac{1}{a}}{x - a}$

36. $\dfrac{\dfrac{3}{(x + 1)^2} - \dfrac{3}{(a + 1)^2}}{x - a}$

37. $(3x^2 + 4x - 1)(4)(2x - 3)^3(2) + (2x - 3)^4(6x + 4)$

38. $(12x - 1)^{1/3}(2)(x^2 - 1)(2x) + (x^2 - 1)^2(\tfrac{1}{3})(12x - 1)^{-2/3}(12)$

39. $\dfrac{2x(-4x + 6)^{1/2} - x^2(\tfrac{1}{2})(-4x + 6)^{-1/2}(-4)}{[(-4x + 6)^{1/2}]^2}$

40. $\dfrac{1}{2}\left(\dfrac{2x - 1}{4x + 1}\right)^{-1/2} \cdot \dfrac{(4x + 1)2 - (2x - 1)4}{(4x + 1)^2}$

In Problems 41–46, the given equation is a partial answer to a calculus problem. Solve the equation for the symbol y'.

41. $3y^2 y' - y - xy' = x$

42. $y' = 2(x - y)(1 - y')$

43. $2yy' + 2x = y'$

44. $2xy^2 + x^2(2y)y' - 2 = -3y'$

45. $\dfrac{(x - y)(1 + y') - (x + y)(1 - y')}{(x - y)^2} = 1$

46. $\dfrac{1}{1 + x^2 y^2}(xy' + y) = 2xyy' + y^2$

Calculator/Computer Problems

In Problems 47 and 48, use a calculator or computer to estimate the given limit by completing each table. Round the entries in each table to eight decimal places.

47. $\displaystyle\lim_{x \to 1} \dfrac{x^3 - 1}{\sqrt[3]{x} - 1}$;

$x \to 0^+$	1.1	1.01	1.001	1.0001	1.00001
$\dfrac{x^3 - 1}{\sqrt[3]{x} - 1}$					

$x \to 0^-$	0.9	0.99	0.999	0.9999	0.99999
$\dfrac{x^3 - 1}{\sqrt[3]{x} - 1}$					

48. $\displaystyle\lim_{x \to 0} (1 + x)^{1/x}$;

$x \to 0^+$	0.1	0.01	0.001	0.0001	0.00001
$(1 + x)^{1/x}$					

$x \to 0^-$	-0.1	-0.01	-0.001	-0.0001	-0.00001
$(1 + x)^{1/x}$					

For Discussion

In Problems 49 and 50, discuss what algebra is necessary to evaluate the given limit. Carry out your ideas.

49. $\displaystyle\lim_{x \to 1} \frac{x - 1}{x^8 - 1}$

50. $\displaystyle\lim_{x \to 0} \frac{\sqrt[3]{x + 27} - 3}{x}$

Chapter 1 Review Exercises   Answers to selected odd-numbered problems begin on page ANS-3.

A. Fill in the Blanks

In Problems 1–22, fill in the blanks.

1. An inequality with $(-\infty, 9]$ as its solution set is _____.
2. The solution set of the inequality $-3 < x \le 8$ as an interval is _____.
3. If the point (a, b) lies in quadrant IV, then (b, a) lies in quadrant _____.
4. The point $(x, -3x)$ in the second quadrant that is 5 units from $(2, -1)$ is _____.
5. If the graph of an equation contains the point $(2, 3)$ and is symmetric with respect to the x-axis, then the graph also contains the point _____.
6. If the graph of an equation contains the point $(-1, 6)$ and is symmetric with respect to the origin, then the graph also contains the point _____.
7. An equation of a circle with center $(-2, -5)$ and radius 6 is _____.
8. If $|2 - x| = 15$, then $x =$ _____.
9. The distance from the midpoint of the line segment joining $(4, -6)$ and $(-2, 0)$ to the origin is _____.
10. The graph of $y = 2|x| - 5$ is symmetric with respect to _____.
11. The intercepts of the graph of $y = 2|x| - 5$ are _____.
12. The circle $x^2 - 16x + y^2 = 0$ is symmetric with respect to _____.
13. The center and radius of the circle $x^2 - 16x + y^2 = 0$ are _____.
14. The intercepts of the circle $(x - 1)^2 + (y - 2)^2 = 10$ are _____.
15. Two points on the circle $x^2 + y^2 = 25$ with the same x-coordinate -3 are _____.
16. The graph of $y = -\sqrt{100 - x^2}$ is a _____.
17. The inequality _____ describes the set of points in the xy-plane outside the circle $x^2 + y^2 = 36$.
18. The distance from the center of the circle $x^2 + 6x + y^2 - 9x = 0$ to the origin is _____.
19. An equation of a circle centered at the origin passing through the point $(-\sqrt{2}, 5)$ is _____.
20. If $(a, a + \sqrt{3})$ lies on the graph of $y = 2x$, then $a =$ _____.

21. The set of real numbers x whose distance between x and $\sqrt{2}$ is greater than 3 is defined by the absolute-value inequality _____.

22. A point (x, y) in the xy-plane whose coordinates satisfy $xy < 0$ lies in quadrant(s) _____ or _____.

B. True/False

In Problems 1–22, answer true or false.

1. The word *nonnegative* means the same as the word *positive*. _____

2. The number 0 is neither positive nor negative. _____

3. -3 is not greater than -1. _____

4. If $a < b$, then $b - a$ is a positive number. _____

5. If $a < b$, then $a^2 < b^2$. _____

6. For any real number a, $-a \leq a$. _____

7. If $a < 0$, then $\dfrac{a}{-a} < 0$. _____

8. If $a^2 < a$, then $a < 1$. _____

9. If x is a negative number, then $-x$ is a positive number. _____

10. The solution set of $|4x - 6| \geq -1$ is $(-\infty, \infty)$. _____

11. $|-3t + 6| = 3|t - 2|$ _____

12. $|-x| = x$ _____

13. There are exactly two points (x, y) on a circle centered at the origin at which $y = x$. _____

14. The point $(5, 0)$ is in quadrant I. _____

15. The point $(-3, 7)$ is in quadrant III. _____

16. The distance between the points $(0, 0)$ and $(3, 6)$ is 9. _____

17. To find y-intercepts of the graph of an equation we let $x = 0$ and solve for y. _____

18. There is no point on the circle $x^2 + y^2 - 10x + 22 = 0$ with x-coordinate 2. _____

19. A circle whose equation can be put into the form $x^2 + y^2 + ax + by = 0$ must pass through the origin. _____

20. The points $(0, 0)$, $(a, 0)$, $a > 0$, and $(0, b)$, $b < 0$, are vertices of a right triangle. _____

21. The graph of the equation $x^2y + 4y = x$ is symmetric with respect to the origin. _____

22. The inequality $\dfrac{100}{x^2 + 64} \leq 0$ has no solution. _____

C. Review Exercises

In Problems 1–4, assume that $0 < a < b$. Compare the given expressions using inequality symbols.

1. a^2 and ab **2.** $-a$ and $-b$ **3.** a and $a + b$ **4.** $\dfrac{1}{a}$ and $\dfrac{1}{a + b}$

In Problems 5–10, fill in the blank with either an appropriate inequality symbol or a number.

5. If $x - 10 > 5$, then $x +$ _____ > 25.

6. If $x - 2 \leq 7$, then x _____ 9.

7. If $-\frac{1}{3}x \geq 4$, then x _____ -12.

8. If $3x - 6 \leq 4x - 4$, then x _____ 2.

9. If $-2 \leq 1 - x \leq 5$, then _____ $\leq x \leq$ _____.

10. If $-3 < x < 9$, then _____ $< -2x <$ _____.

11. On the number line, $m = 5$ is the midpoint of the line segment joining the number a (left endpoint) and the number b (right endpoint). Use the fact that $d(a, b) = 2$ to find a and b.

12. In the xy-plane, find an equation that describes the set of points (x, y) that are equidistant from $(0, 5)$ and $(x, -5)$.

In Problems 13–16, describe the given interval on the real number line using **(a)** an inequality, **(b)** interval notation.

13.

FIGURE 1.R.1 Graph for Problem 13

14.

FIGURE 1.R.2 Graph for Problem 14

15.

FIGURE 1.R.3 Graph for Problem 15

16.

FIGURE 1.R.4 Graph for Problem 16

In Problems 17–30, solve the given inequality. Write the solution set using interval notation.

17. $2x - 5 \geq 6x + 7$

18. $\frac{1}{4}x - 3 < \frac{1}{2}x + 1$

19. $-4 < x - 8 < 4$

20. $7 \leq 3 - 2x < 11$

21. $|x| > 10$

22. $|-6x| \leq 42$

23. $|3x - 4| < 5$

24. $|5 - 2x| \geq 7$

25. $3x \geq 2x^2 - 5$

26. $x^2 > 6x - 9$

27. $x^3 > x$

28. $(x^2 - x)(x^2 + x) \leq 0$

29. $\dfrac{1}{x} + x > 2$

30. $\dfrac{2x - 6}{x - 1} \geq 1$

In Problems 31–34, find equations of two different circles so that each circle satisfies the given conditions.

31. center in the first quadrant and graph is tangent to both the x- and y-axes

32. x-intercepts of the graph are $(-6, 0)$ and $(-2, 0)$

33. passes through the origin and center is on the negative y-axis

34. center on the x-axis and graph is tangent to the horizontal line through the point $(0, 3)$

35. Lens Equation The lens equation

$$\frac{1}{f} = \frac{1}{d_o} + \frac{1}{d_i},$$

discovered by Carl Friedrich Gauss in 1841, relates the distance d_o from an object to a thin convex lens (in meters) to the distance d_i from the lens to its image (in meters), where f is the focal length of the lens and $d_o > f$. See FIGURE 1.R.5. If $f = 0.30$ m, then what distances d_o correspond to $d_i > 0.5$ m? Write the solution as a simultaneous inequality.

36. Solve the inequality $|x - 3| + |x + 1| < 10$. Write the solution set using interval notation.

FIGURE 1.R.5 Convex lens in Problem 35

CHAPTER 1 INEQUALITIES, EQUATIONS, AND GRAPHS

In Problems 37–40, simplify the given expression in part (a). Then, if instructed, find the indicated limit in part (b).

37. (a) $\dfrac{2x - 1}{4x^2 - 1}$ **(b)** $\displaystyle\lim_{x \to \frac{1}{2}} \dfrac{2x - 1}{4x^2 - 1}$

38. (a) $\dfrac{x^2 - 6x + 5}{x - 5}$ **(b)** $\displaystyle\lim_{x \to 5} \dfrac{x^2 - 6x + 5}{x - 5}$

39. (a) $\dfrac{x^2 - 16}{\sqrt{x} - 2}$ **(b)** $\displaystyle\lim_{x \to 4} \dfrac{x^2 - 16}{\sqrt{x} - 2}$

40. (a) $\dfrac{1}{h}\left(\dfrac{1}{3 + h} - \dfrac{1}{3}\right)$ **(b)** $\displaystyle\lim_{h \to 0} \dfrac{1}{h}\left(\dfrac{1}{3 + h} - \dfrac{1}{3}\right)$

© Joshua Haviv/ShutterStock, Inc.

2 Functions

Chapter Contents

2.1 Functions and Graphs

Student/desk correspondence

INTRODUCTION Using the objects and the persons around us, it is easy to make up a rule of correspondence that associates, or pairs, the members, or elements, of one set with the members of another set. For example, to each social security number there is a person, to each car registered in the state of California there is a license plate number, to each book there corresponds at least one author, to each state there is a governor, and so on. A natural correspondence occurs between a set of 20 students and a set of, say, 25 desks in a classroom when each student selects and sits in a different desk. In mathematics we are interested in a special type of correspondence, a *single-valued correspondence*, called a function.

DEFINITION 2.1.1 Function

A **function** from a set X to a set Y is a rule of correspondence that assigns to each element x in X exactly one element y in Y.

In the student/desk correspondence above suppose the set of 20 students is the set X and the set of 25 desks is the set Y. This correspondence is a function from the set X to the set Y provided no student sits in two desks at the same time.

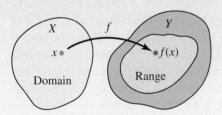

FIGURE 2.1.1 Domain and range of a function f

☐ **Terminology** A function is usually denoted by a letter such as f, g, or h. We can then represent a function f from a set X to a set Y by the notation $f: X \to Y$. The set X is called the **domain** of f. The set of corresponding elements y in the set Y is called the **range** of the function. For our student/desk function, the set of students is the domain and the set of 20 desks actually occupied by the students constitutes the range. Notice that the range of f need not be the entire set Y. The unique element y in the range that corresponds to a selected element x in the domain X is called the **value** of the function at x, or the **image** of x, and is written $f(x)$. The latter symbol is read "f of x" or "f at x," and we write $y = f(x)$.* See FIGURE 2.1.1. Since the value of y depends on the choice of x, y is called the **dependent variable**; x is called the **independent variable**. Unless otherwise stated, we will assume hereafter that the sets X and Y consist of real numbers.

EXAMPLE 1   The Squaring Function

The rule for squaring a real number is given by the equation $y = x^2$ or $f(x) = x^2$. The values of f at $x = -5$ and $x = \sqrt{7}$ are obtained by replacing x, in turn, by the numbers -5 and $\sqrt{7}$:

$$f(-5) = (-5)^2 = 25 \quad \text{and} \quad f(\sqrt{7}) = (\sqrt{7})^2 = 7. \qquad \blacksquare$$

Occasionally for emphasis we will write a function using parentheses in place of the symbol x. For example, we can write the squaring function $f(x) = x^2$ as

$$f(\ \ ) = (\ \ )^2. \tag{1}$$

---

*Many instructors like to call x the *input* of the function and $f(x)$ the *output*.

This illustrates the fact that x is a *placeholder* for any number in the domain of the function $y = f(x)$. Thus, if we wish to evaluate (1) at, say, $3 + h$, where h represents a real number, we put $3 + h$ into the parentheses and carry out the appropriate algebra:

See (6) of Section 1.5. ▶

$$f(3 + h) = (3 + h)^2 = 9 + 6h + h^2.$$

If a function f is defined by means of a formula or an equation, then typically the domain of $y = f(x)$ is not expressly stated. We will see that we can usually deduce the domain of $y = f(x)$ either from the structure of the equation or from the context of the problem.

![EXAMPLE 2] **Domain and Range**

In Example 1, since any real number x can be squared and the result x^2 is another real number, $f(x) = x^2$ is a function from R to R, that is, $f: R \to R$. In other words, the domain of f is the set R of real numbers. Using interval notation, we also write the domain as $(-\infty, \infty)$. The range of f is the set of nonnegative real numbers or $[0, \infty)$; this follows from the fact that $x^2 \geq 0$ for every real number x. ■

☐ **Domain of a Function** As mentioned earlier, the domain of a function $y = f(x)$ that is defined by a formula is usually not specified. Unless stated or implied to the contrary, it is understood that:

> *The domain of a function f is the largest subset of the set of real numbers for which $f(x)$ is a real number.*

The term **natural domain** is also used. ▶

This set is sometimes referred to as the **implicit domain** of the function. For example, we cannot compute $f(0)$ for the reciprocal function $f(x) = 1/x$ since $1/0$ is not a real number. In this case we say that f is **undefined** at $x = 0$. Since every nonzero real number has a reciprocal, the domain of $f(x) = 1/x$ is the set of real numbers except 0. By the same reasoning, the function $g(x) = 1/(x^2 - 4)$ is not defined at either $x = -2$ or $x = 2$, and so its domain is the set of real numbers with -2 and 2 excluded. The square root function $h(x) = \sqrt{x}$ is not defined at $x = -1$ because $\sqrt{-1}$ is not a real number. In order for $h(x) = \sqrt{x}$ to be defined in the real number system we must require the **radicand**, in this case simply x, to be nonnegative. From the inequality $x \geq 0$ we see that the domain of the function h is the interval $[0, \infty)$.

![EXAMPLE 3] **Domain and Range**

Determine the domain and range of $f(x) = 4 + \sqrt{x - 3}$.

Solution The radicand $x - 3$ must be nonnegative. By solving the inequality $x - 3 \geq 0$ we get $x \geq 3$, and so the domain of f is $[3, \infty)$. Now, since the symbol $\sqrt{}$ denotes the nonnegative square root of a number, $\sqrt{x - 3} \geq 0$ for $x \geq 3$ and consequently $4 + \sqrt{x - 3} \geq 4$. The least value of $f(x)$ occurs at $x = 3$ and is $f(3) = 4 + \sqrt{0} = 4$. Moreover, because $x - 3$ and $\sqrt{x - 3}$ increase as x takes on increasing larger values, we conclude that $y \geq 4$. Consequently the range of f is the interval $[4, \infty)$. ■

![EXAMPLE 4] **Domain of f**

Determine the domain of $f(x) = \sqrt{x^2 + 2x - 15}$.

Solution As in Example 3, the expression under the radical symbol—the radicand—must be nonnegative, that is, the domain of f is the set of real numbers x for which $x^2 + 2x - 15 \geq 0$ or $(x + 5)(x - 3) \geq 0$. We have already solved the last inequality by means of a sign chart in Example 3 of Section 1.1. The solution set $(-\infty, -5] \cup [3, \infty)$ of the inequality is also the domain of f. ■

EXAMPLE 5   **Domains of Two Functions**

Determine the domain of the given function.

(a) $g(x) = \dfrac{1}{\sqrt{x^2 + 2x - 15}}$ **(b)** $h(x) = \dfrac{5x}{x^2 - 3x - 4}$

Solution A function that is given by a fractional expression is not defined at the *x*-values for which its denominator is equal to 0.

(a) The expression under the radical is the same as in Example 4. Since $x^2 + 2x - 15$ is in the denominator we must have $x^2 + 2x - 15 \neq 0$. This excludes $x = -5$ and $x = 3$. In addition, since $x^2 + 2x - 15$ appears under a radical, we must have $x^2 + 2x - 15 > 0$ for all other values of x. Thus the domain of the function g is the union of two open intervals $(-\infty, -5) \cup (3, \infty)$.

(b) Since the denominator of $h(x)$ factors,

$$x^2 - 3x - 4 = (x + 1)(x - 4)$$

we see that $(x + 1)(x - 4) = 0$ for $x = -1$ and $x = 4$. In contrast to the function in part (a), these are the *only* numbers for which h is not defined. Hence, the domain of the function h is the set of real numbers with $x = -1$ and $x = 4$ excluded. ∎

Using interval notation, the domain of the function h in part (b) of Example 5 can be written as

$$(-\infty, -1) \cup (-1, 4) \cup (4, \infty).$$

As an alternative to this ungainly union of disjoint intervals, this domain can also be written using set-builder notation as $\{x \mid x \neq -1 \text{ and } x \neq 4\}$.

☐ **Graphs** A function is often used to describe phenomena in fields such as science, engineering, and business. In order to interpret and utilize data, it is useful to display this data in the form of a graph. The graph of a function f is the graph of the set of ordered pairs $(x, f(x))$, where x is in the domain of f. In the *xy*-plane an ordered pair $(x, f(x))$ is a point, so that the graph of a function is a set of points. If a function is defined by an equation $y = f(x)$, then the graph of f is the graph of the equation. To obtain points on the graph of an equation $y = f(x)$, we judiciously choose numbers $x_1, x_2, x_3, \ldots$ in its domain, compute $f(x_1), f(x_2), f(x_3), \ldots$, plot the corresponding points $(x_1, f(x_1))$, $(x_2, f(x_2)), (x_3, f(x_3)), \ldots$, and then connect these points with a curve. See FIGURE 2.1.2. Keep in mind that:

- a value of x is a directed distance from the *y*-axis, and
- a function value $f(x)$ is a directed distance from the *x*-axis.

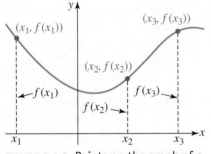

FIGURE 2.1.2 Points on the graph of a function $y = f(x)$

☐ **End Behavior** A word about the figures in this text is in order. With a few exceptions, it is usually impossible to display the complete graph of a function, and so we often display only the more important features of the graph. In FIGURE 2.1.3(a), notice that the graph goes down on its left and right sides. Unless indicated to the contrary, we may assume that there are no major surprises beyond what we have shown and the graph simply continues in the manner indicated. The graph in Figure 2.1.3(a) indicates the so-called **end behavior** or **global behavior** of the function: For a point (x, y) on the graph, the values of the *y*-coordinate become unbounded in magnitude in the downward or negative direction as the *x*-coordinate becomes unbounded in magnitude in both the negative and positive directions on the number line. It is convenient to describe this end behavior using the arrow symbols

$$y \to -\infty \text{ as } x \to -\infty \quad \text{and} \quad y \to -\infty \text{ as } x \to \infty.$$

See Section 1.5. ▶
The symbol $\to$ is read "approaches." Thus, for example, $y \to -\infty$ as $x \to \infty$ is read "y approaches negative infinity as x approaches infinity." More will be said about the concept of global behavior of a function in Chapter 3. If a graph terminates at either its right or left end, we will indicate this by a dot when clarity demands it. See FIGURE 2.1.4. We will use a solid dot to represent the fact that the endpoint is included on the graph and an open dot to signify that the endpoint is not included on the graph.

☐ **Vertical Line Test** From the definition of a function we know that for each x in the domain of f there corresponds only one value $f(x)$ in the range. This means a vertical line that intersects the graph of a function $y = f(x)$ (this is equivalent to choosing an x) can do so in at most one point. Conversely, if *every* vertical line that intersects a graph of an equation does so in at most one point, then the graph is the graph of a function. The last statement is called the **vertical line test** for a function. See Figure 2.1.3(a). On the other hand, if *some* vertical line intersects a graph of an equation more than once, then the graph is not that of a function. See Figures 2.1.3(b) and 2.1.3(c). When a vertical line intersects a graph in several points, the same number x corresponds to different values of y in contradiction to the definition of a function.

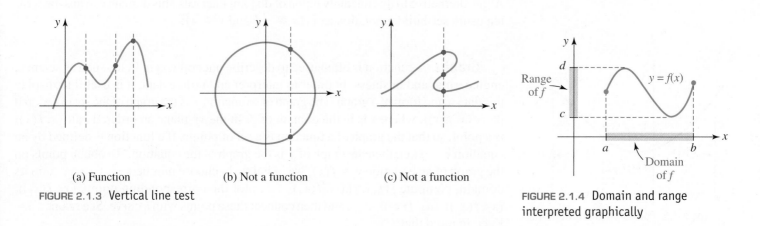

| (a) Function | (b) Not a function | (c) Not a function |

FIGURE 2.1.3 Vertical line test

FIGURE 2.1.4 Domain and range interpreted graphically

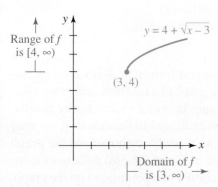

FIGURE 2.1.5 Graph of function in Example 6

If you have an accurate graph of a function $y = f(x)$ it is often possible to *see* the domain and range of f. In Figure 2.1.4 assume that the blue curve is the entire, or complete, graph of some function f. The domain of f then is the interval $[a, b]$ on the x-axis and the range is the interval $[c, d]$ on the y-axis.

▌EXAMPLE 6   **Example 3 Revisited**

From the graph of $f(x) = 4 + \sqrt{x - 3}$ given in FIGURE 2.1.5, we can see that the domain and range of f are, respectively, the interval $[3, \infty)$ on the x-axis and the interval $[4, \infty)$ on the y-axis. This agrees with the results in Example 3. ▬

CHAPTER 2 FUNCTIONS

☐ **Intercepts** To graph a function defined by an equation $y = f(x)$, it is usually a good idea to first determine whether the graph of f has any intercepts. Recall that all points on the y-axis are of the form $(0, y)$. Thus, if 0 is the domain of a function f, the **y-intercept** is the point on the y-axis whose y-coordinate is $f(0)$, in other words, $(0, f(0))$. See FIGURE 2.1.6(a). Similarly, all points on the x-axis have the form $(x, 0)$. This means that to find the **x-intercepts** of the graph of $y = f(x)$, we determine the values of x that make $y = 0$. That is, we must solve the equation $f(x) = 0$ for x. A number c for which

$$f(c) = 0$$

is referred to as either a **zero** of the function f or a **solution** (or **root**) of the equation $f(x) = 0$.

- The *real* zeros of a function f are the x-coordinates of the x-intercepts of the graph of f.

In Figure 2.1.6(b), we have illustrated a function that has three zeros x_1, x_2, and x_3 because $f(x_1) = 0, f(x_2) = 0$, and $f(x_3) = 0$. The corresponding three x-intercepts are the points $(x_1, 0)$, $(x_2, 0)$, and $(x_3, 0)$. Of course, the graph of the function may have no intercepts. This case is illustrated in Figure 2.1.5.

(a) y-intercept (b) Three x-intercepts (c) y-intercept, two x-intercepts

FIGURE 2.1.6 Intercepts of the graph of a function

More will be said about this in Chapter 3. ▶

A graph does not necessarily have to *cross* a coordinate axis at an intercept; a graph could simply be tangent to, or *touch*, an axis. In Figure 2.1.6(c) the graph of $y = f(x)$ is tangent to the x-axis at $(x_1, 0)$. Also, the graph of a function f can have at most one y-intercept since, if 0 is the domain of f, there can correspond only one y-value, namely, $y = f(0)$.

EXAMPLE 7 Intercepts

Find, if possible, the x- and y-intercepts of the given function.

(a) $f(x) = x^2 + 2x - 2$ **(b)** $f(x) = \dfrac{x^2 - 2x - 3}{x}$

Solution (a) Since 0 is in the domain of f, $f(0) = -2$ is the y-coordinate of the y-intercept of the graph of f. The y-intercept is the point $(0, -2)$. To obtain the x-intercepts we must determine whether f has any real zeros, that is, real solutions of the equation $f(x) = 0$. Since the left-hand side of the equation $x^2 + 2x - 2 = 0$ has no obvious factors, we use the quadratic formula to obtain $x = \frac{1}{2}(-2 \pm \sqrt{12})$. Since $\sqrt{12} = \sqrt{4 \cdot 3} = 2\sqrt{3}$ the zeros of f are the irrational numbers $-1 - \sqrt{3}$ and $-1 + \sqrt{3}$. The x-intercepts are the points $(-1 - \sqrt{3}, 0)$ and $(-1 + \sqrt{3}, 0)$.

(b) Because 0 is not in the domain of f ($f(0) = -3/0$ is not defined), the graph of f possesses no y-intercept. Now since f is a fractional expression, the only way we can have

$f(x) = 0$ is to have the numerator equal zero. Factoring the left-hand side of $x^2 - 2x - 3 = 0$ gives $(x + 1)(x - 3) = 0$. Therefore the numbers -1 and 3 are the zeros of f. The x-intercepts are the points $(-1, 0)$ and $(3, 0)$. ▬

☐ **Approximating Zeros** Even when it is obvious that the graph of a function $y = f(x)$ possesses x-intercepts it is not always a straightforward matter to solve the equation $f(x) = 0$. In fact, it is *impossible* to solve some equations exactly; sometimes the best we can do is to **approximate** the zeros of the function. One way of doing this is to obtain a very accurate graph of f.

FIGURE 2.1.7 Graph of function in Example 8

EXAMPLE 8　　Approximate Intercepts

With the aid of a graphing utility the graph of the function $f(x) = x^3 - x + 4$ is given in FIGURE 2.1.7. From $f(0) = 4$ we see that the y-intercept is $(0, 4)$. As we see in the figure, there appears to be only one x-intercept with x-coordinate close to -1.7 or -1.8. But there is no convenient way of finding the roots of the equation $x^3 - x + 4 = 0$. We can however approximate the real root of this equation with the aid of the *find root* feature of either a graphing calculator or computer algebra system. We find that $x \approx -1.796$ and so the approximate x-intercept is $(-1.796, 0)$. As a check, note that the function value

$$f(-1.796) = (-1.796)^3 - (-1.796) + 4 \approx 0.0028$$

is nearly 0. ▬

NOTES FROM THE CLASSROOM

(*i*) When sketching the graph of a function, you should never resort to plotting a lot of points by hand. That is something a **graphing utility**, that is, a calculator or a computer algebra system (CAS), does so well. On the other hand, you should not become dependent on a calculator to obtain a graph. Believe it or not, there are precalculus and calculus instructors who do not allow the use of graphing calculators on quizzes or tests. Usually there is no objection to your using calculators or computers as an aid in checking homework problems, but in the classroom instructors want to see the product of your own mind, namely, the ability to analyze. So you are strongly encouraged to develop your graphing skills to the point where you are able to quickly sketch by hand the graph of a function from a basic familiarity of types of functions and by plotting a minimum of well-chosen points such as intercepts.

(*ii*) A function can involve several independent variables. As a simple example, the perimeter P and area A of a rectangle are functions of its length x and width y, that is, $P = 2x + 2y$ and $A = xy$. In a general discussion, a function of, say, two independent variables x and y is written $f(x, y)$. A major part of the third term of a typical calculus course is devoted to the study of functions of several variables.

In Problems 1–6, find the indicated function values.

1. If $f(x) = x^2 - 1$; $f(-5), f(-\sqrt{3}), f(3)$, and $f(6)$

2. If $f(x) = -2x^2 + x$; $f(-5), f(-\frac{1}{2}), f(2)$, and $f(7)$

3. If $f(x) = \sqrt{x + 1}$; $f(-1), f(0), f(3)$, and $f(5)$

4. If $f(x) = \sqrt{2x + 4}$; $f(-\frac{1}{2}), f(\frac{1}{2}), f(\frac{5}{2})$, and $f(4)$

5. If $f(x) = \dfrac{3x}{x^2 + 1}$; $f(-1), f(0), f(1)$, and $f(\sqrt{2})$

6. If $f(x) = \dfrac{x^2}{x^3 - 2}$; $f(-\sqrt{2}), f(-1), f(0)$, and $f(\frac{1}{2})$

In Problems 7 and 8, find

$$f(x), f(2a), f(a^2), f(-5x), f(2a + 1), f(x + h)$$

for the given function f and simplify as much as possible.

7. $f(\ ) = -2(\ )^2 + 3(\ )$ \hspace{2em} **8.** $f(\ ) = (\ )^3 - 2(\ )^2 + 20$

9. For what values of x is $f(x) = 6x^2 - 1$ equal to 23?

10. For what values of x is $f(x) = \sqrt{x} - 4$ equal to 4?

In Problems 11–20, find the domain of the given function f.

11. $f(x) = \sqrt{4x - 2}$ \hspace{4em} **12.** $f(x) = \sqrt{15 - 5x}$

13. $f(x) = \dfrac{10}{\sqrt{1 - x}}$ \hspace{4em} **14.** $f(x) = \dfrac{2x}{\sqrt{3x - 1}}$

15. $f(x) = \dfrac{2x - 5}{x(x - 3)}$ \hspace{4em} **16.** $f(x) = \dfrac{x}{x^2 - 1}$

17. $f(x) = \dfrac{1}{x^2 - 10x + 25}$ \hspace{2em} **18.** $f(x) = \dfrac{x + 1}{x^2 - 4x - 12}$

19. $f(x) = \dfrac{x}{x^2 - x + 1}$ \hspace{4em} **20.** $f(x) = \dfrac{x^2 - 9}{x^2 - 2x - 1}$

In Problems 21–26, use the sign-chart method to find the domain of the given function f.

21. $f(x) = \sqrt{25 - x^2}$ \hspace{4em} **22.** $f(x) = \sqrt{x(4 - x)}$

23. $f(x) = \sqrt{x^2 - 5x}$ \hspace{4em} **24.** $f(x) = \sqrt{x^2 - 3x - 10}$

25. $f(x) = \sqrt{\dfrac{3 - x}{x + 2}}$ \hspace{4em} **26.** $f(x) = \sqrt{\dfrac{5 - x}{x}}$

In Problems 27–30, determine whether the graph in the figure is the graph of a function.

27.

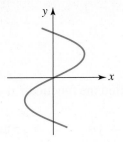

FIGURE 2.1.8 Graph for Problem 27

28.

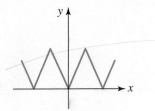

FIGURE 2.1.9 Graph for Problem 28

29.

FIGURE 2.1.10 Graph for
Problem 29

30.

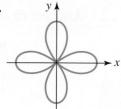

FIGURE 2.1.11 Graph for
Problem 30

In Problems 31–34, use the graph of the function f given in the figure to find its
domain and range.

31.

FIGURE 2.1.12 Graph for
Problem 31

32.

FIGURE 2.1.13 Graph for
Problem 32

33.

FIGURE 2.1.14 Graph for
Problem 33

34.

FIGURE 2.1.15 Graph for
Problem 34

In Problems 35–42, find the real zeros of the given function f.

35. $f(x) = 5x + 6$

36. $f(x) = -2x + 9$

37. $f(x) = x^2 - 5x + 6$

38. $f(x) = x^2 - 2x - 1$

39. $f(x) = x(3x - 1)(x + 9)$

40. $f(x) = x^3 - x^2 - 2x$

41. $f(x) = x^4 - 1$

42. $f(x) = 2 - \sqrt{4 - x^2}$

In Problems 43–50, find the x- and y-intercepts, if any, of the graph of the given
function f. Do not graph.

43. $f(x) = \frac{1}{2}x - 4$

44. $f(x) = x^2 - 6x + 5$

45. $f(x) = 4(x - 2)^2 - 1$

46. $f(x) = (2x - 3)(x^2 + 8x + 16)$

47. $f(x) = \dfrac{x^2 + 4}{x^2 - 16}$

48. $f(x) = \dfrac{x(x + 1)(x - 6)}{x + 8}$

49. $f(x) = \frac{3}{2}\sqrt{4 - x^2}$

50. $f(x) = \frac{1}{2}\sqrt{x^2 - 2x - 3}$

In Problems 51–56, find the function f if $f(2) = -3$ in each case.

51. $f(x) = 4x + k$

52. $f(x) = -2x^2 + kx$

53. $f(x) = kx^3 - x + 1$

54. $f(x) = x^4 + x^3 + kx^2 - x - 1$

55. $f(x) = \dfrac{2x - k}{x}$

56. $f(x) = \dfrac{x + k}{x - k}$

In Problems 57 and 58, use the graph of the function f given in the figure to estimate the values of $f(-3), f(-2), f(-1), f(1), f(2),$ and $f(3)$. Estimate the y-intercept.

57.

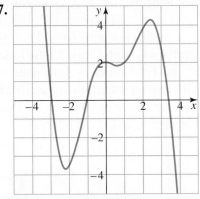

FIGURE 2.1.16 Graph for Problem 57

58.

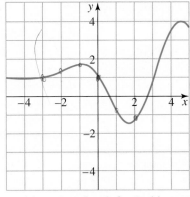

FIGURE 2.1.17 Graph for Problem 58

In Problems 59 and 60, use the graph of the function f given in the figure to estimate the values of $f(-2), f(-1.5), f(0.5), f(1), f(2),$ and $f(3.2)$. Estimate the x-intercepts.

59.

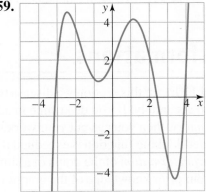

FIGURE 2.1.18 Graph for Problem 59

60.

FIGURE 2.1.19 Graph for Problem 60

Calculus-Related Problems

61. In calculus some of the functions that you will encounter have as their domain the set of positive integers n. The **factorial function** $f(n) = n!$ is defined as the product of the first n positive integers, that is,

$$f(n) = n! = 1 \cdot 2 \cdot 3 \cdots (n-1) \cdot n.$$

(a) Evaluate $f(2), f(3), f(5),$ and $f(7)$.
(b) Show that $f(n + 1) = f(n) \cdot (n + 1)$.
(c) Simplify $f(n + 2)/f(n)$.

62. Another function of a positive integer n gives the sum of the first n squared positive integers:

$$S(n) = \tfrac{1}{6}n(n + 1)(2n + 1) = 1^2 + 2^2 + \cdots + n^2.$$

(a) Find the value of the sum $1^2 + 2^2 + \cdots + 99^2 + 100^2$.
(b) Find n such that $300 < S(n) < 400$. [*Hint*: Use a calculator.]

For Discussion

63. Determine an equation of a function $y = f(x)$ whose domain is **(a)** $[3, \infty)$, **(b)** $(3, \infty)$.

64. Determine an equation of a function $y = f(x)$ whose range is **(a)** $[3, \infty)$, **(b)** $(3, \infty)$.

65. Find a function $S(n)$ analogous to that given in Problem 62 for the sum of the first n positive integers. [*Hint*: Add the corresponding terms of
$$S(n) = 1 + 2 + 3 + \cdots + (n - 1) + n$$
and $\qquad S(n) = n + (n - 1) + (n - 2) + \cdots + 2 + 1.]$

66. Area Under a Graph Consider the function $y = f(x)$ whose graph is shown in FIGURE 2.1.20. Find the areas $A(1)$, $A(2)$, and $A(3)$ of the blue isosceles triangles bounded between the graph of f and the intervals $\left[\frac{1}{2}, \frac{3}{2}\right]$, $\left[\frac{7}{4}, \frac{9}{4}\right]$ and $\left[\frac{23}{8}, \frac{25}{8}\right]$ on the x-axis. Discern the pattern of the intervals and find the area function $A(n)$ of the triangular region on the nth interval.

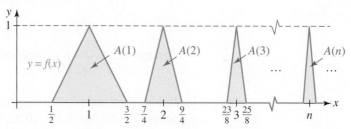

FIGURE 2.1.20 Graph for Problem 66

2.2 Symmetry and Transformations

INTRODUCTION In this section we discuss two aids in sketching graphs of functions quickly and accurately. If you determine in advance that the graph of a function possesses *symmetry*, then you can cut your work in half. In addition, sketching a graph of a complicated-looking function is expedited if you recognize that the required graph is actually a *transformation* of the graph of a simpler function. This latter graphing aid is based on your prior knowledge of the graphs of some basic functions.

☐ **Power Functions** A function of the form

$$f(x) = x^n, \qquad (1)$$

where n represents a real number, is called a **power function**. The domain of a power function (1) depends on the power n. For example, we have already seen in Section 2.1 for $n = 2$, $n = \frac{1}{2}$, and $n = -1$, respectively, that:

- the domain of $f(x) = x^2$ is the set R of real numbers or $(-\infty, \infty)$,
- the domain of $f(x) = x^{1/2} = \sqrt{x}$ is $[0, \infty)$,
- the domain of $f(x) = x^{-1} = \dfrac{1}{x}$ is the set R of real numbers except $x = 0$.

Simple power functions, or modified versions of these functions, occur so often in problems in calculus that you do not want to spend valuable time plotting their graphs. We suggest that you know (memorize) the short catalogue of graphs of power functions given in FIGURE 2.2.1 on the next page. You might already know that the graph in part (a) of that figure is a **line** and the graph in part (b) is called a **parabola**.

☐ **Symmetry** In Section 1.4 we discussed symmetry of a graph with respect to the y-axis, the x-axis, and the origin. Of those three types of symmetries, the graph of a function f can be symmetric with respect to the y-axis or with respect to the origin, but

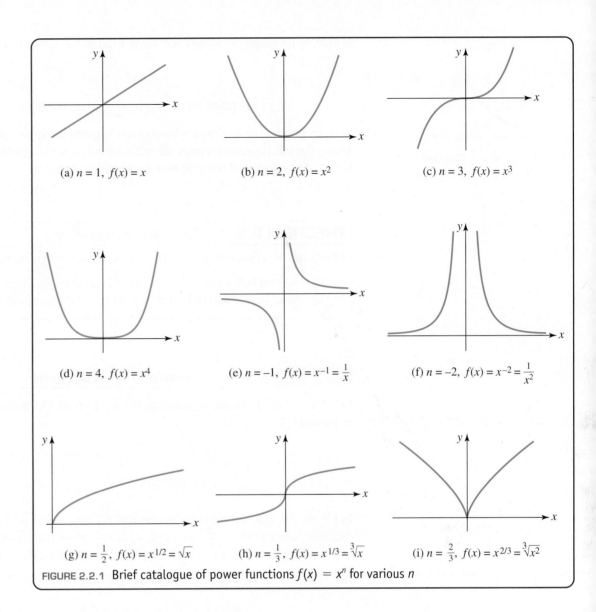

FIGURE 2.2.1 Brief catalogue of power functions $f(x) = x^n$ for various n

Can you explain why the graph of a function cannot have symmetry with respect to the *x*-axis? See Problem 47 in Exercises 2.2.

the graph of a nonzero function *f cannot* be symmetric with respect to the *x*-axis. Before proceeding with the discussion of symmetry of graphs of functions we need the following definition.

FIGURE 2.2.2 Even function

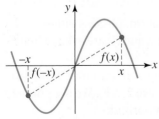

FIGURE 2.2.3 Odd function

DEFINITION 2.2.1 Even and Odd Functions

(*i*) A function f with domain X is said to be an **even function** if
$f(-x) = f(x)$ for every x in X
(*ii*) A function f with domain X is said to be an **odd function** if
$f(-x) = -f(x)$ for every x in X

☐ **Graphical Interpretation** The graphical interpretation of Definition 2.2.1 is illustrated in FIGURES 2.2.2 and 2.2.3. In Figure 2.2.2, observe that if f is an even function and

$$f(x)$$
$$\downarrow$$
(x, y) is a point on its graph, then necessarily $(-x, y)$

$$f(-x) = f(x)$$
$$\downarrow$$

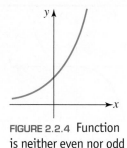

FIGURE 2.2.4 Function is neither even nor odd

is also on its graph. Similarly we see in Figure 2.2.3, that if f is an odd function and

$$\underset{\downarrow}{f(x)} \qquad\qquad \underset{\downarrow}{f(-x) = -f(x)}$$

(x, y) is a point on its graph, then necessarily $(-x, -y)$

is on its graph. The function whose graph is given in FIGURE 2.2.4 is neither even or odd. Using the information on page 29 of Section 1.4, we summarize these observations in terms of symmetry of the graph in the next theorem.

THEOREM 2.2.1 Tests for Symmetry

The graph of a function $y = f(x)$ with domain X is symmetric with respect to:

(*i*) the **y-axis** if and only if $y = f(x)$ is an even function, or
(*ii*) the **origin** if and only if $y = f(x)$ is an odd function

---

EXAMPLE 1 **Even and Odd Functions**

(a) $f(x) = x^{2/3}$ is an even function since by (*i*) of Definition 2.2.1 and the laws of exponents

$$\overset{\text{cube root of } -1 \text{ is } -1}{\underset{\downarrow}{}}$$

$$f(-x) = (-x)^{2/3} = (-1)^{2/3}x^{2/3} = \left(\sqrt[3]{-1}\right)^2 x^{2/3} = (-1)^2 x^{2/3} = x^{2/3} = f(x).$$

In Figure 2.2.1(i), we see that the graph of f is symmetric with respect to the y-axis. For example, since $f(8) = 8^{2/3} = 4$, $(8, 4)$ is a point on the graph of $y = x^{2/3}$. Because f is an even function, $f(-8) = f(8)$ implies $(-8, 4)$ is on the same graph.

(b) $f(x) = x^3$ is an odd function since by (*ii*) of Definition 2.2.1,

$$f(-x) = (-x)^3 = (-1)^3 x^3 = -x^3 = -f(x).$$

Inspection of Figure 2.2.1(c) shows that the graph of f is symmetric with respect to the origin. For example, since $f(1) = 1$, $(1, 1)$ is a point on the graph of $y = x^3$. Because f is an odd function, $f(-1) = -f(1)$ implies $(-1, -1)$ is on the same graph.

(c) $f(x) = x^3 + 1$ is neither even nor odd. From

$$f(-x) = (-x)^3 + 1 = -x^3 + 1$$

we see that $f(-x) \neq f(x)$, and $f(-x) \neq -f(x)$. The graph of f has neither y-axis nor origin symmetry. ▬

The graphs in Figure 2.2.1, with part (g) the only exception, possess either y-axis or origin symmetry. The functions in Figures 2.2.1(b), (d), (f), and (i) are even, whereas the functions in Figures 2.2.1(a), (c), (e), and (h) are odd.

Often we can sketch the graph of a function by applying a certain transformation to the graph of a simpler function (such as those given in Figure 2.2.1). We are going to consider two kinds of graphical transformations, rigid and nonrigid.

☐ **Rigid Transformations-Shifts** A **rigid transformation** of a graph of function f is one that changes the *position* of the graph or one that changes the *orientation* of the graph in the xy-plane but does change its basic shape. For example, the circle $(x - 2)^2 + (y - 3)^2 = 1$ with center $(2, 3)$ and radius $r = 1$, has *exactly* the same shape as the circle $x^2 + y^2 = 1$ with center at the origin. We can think of the graph of $(x - 2)^2 + (y - 3)^2 = 1$ as the graph of $x^2 + y^2 = 1$ shifted horizontally 2 units to the right followed by an upward vertical shift of 3 units. For the graph of a function $y = f(x)$ we examine four kinds of **shifts** or **translations**.

THEOREM 2.2.2   Vertical and Horizontal Shifts

Suppose $y = f(x)$ is a function and c is a positive constant. Then the graph of

    (*i*) $y = f(x) + c$ is the graph of f shifted vertically **up** c units
    (*ii*) $y = f(x) - c$ is the graph of f shifted vertically **down** c units
    (*iii*) $y = f(x + c)$ is the graph of f shifted horizontally to the **left** c units
    (*iv*) $y = f(x - c)$ is the graph of f shifted horizontally to the **right** c units

---

FIGURE 2.2.5 Graph of $y = f(x)$

Consider the graph of a function $y = f(x)$ given in FIGURE 2.2.5. The shifts of this graph described in (*i*)–(*iv*) of Theorem 2.2.2 are the graphs in red in parts (a)–(d) of FIGURE 2.2.6. If (x, y) is a point on the graph of $y = f(x)$ and the graph of f is shifted, say, upward by $c > 0$ units, then $(x, y + c)$ is a point on the new graph. In general, the x-coordinates do not change as a result of a vertical shift. See Figures 2.2.6(a) and 2.2.6(b). Similarly, in a horizontal shift the y-coordinates of points on the shifted graph are the same as on the original graph. See Figures 2.2.6(c) and 2.2.6(d).

(a) Vertical shift up

(b) Vertical shift down

(c) Horizontal shift left

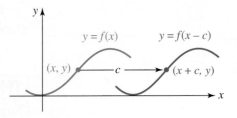

(d) Horizontal shift right

FIGURE 2.2.6 Vertical and horizontal shifts of the graph of $y = f(x)$ by an amount $c > 0$

EXAMPLE 2　　　　　　　**Vertical and Horizontal Shifts**

The graphs of $y = x^2 + 1$, $y = x^2 - 1$, $y = (x + 1)^2$, and $y = (x - 1)^2$ are obtained from the graph of $f(x) = x^2$ in **FIGURE 2.2.7(a)** by shifting this graph, in turn, 1 unit up (Figure 2.2.7(b)), 1 unit down (Figure 2.2.7(c)), 1 unit to the left (Figure 2.2.7(d)), and 1 unit to the right (Figure 2.2.7(e)).

    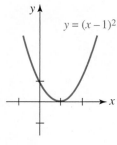

(a) Starting point　　(b) Shift up　　(c) Shift down　　(d) Shift left　　(e) Shift right

FIGURE 2.2.7 Shifted graphs in Example 2

☐ **Combining Shifts** In general, the graph of a function

$$y = f(x \pm c_1) \pm c_2, \qquad (2)$$

where c_1 and c_2 are positive constants, combines a horizontal shift (left or right) with a vertical shift (up or down). For example, the graph of $y = f(x - c_1) + c_2$ is the graph of $y = f(x)$ shifted c_1 units to the right and then c_2 units up.

► The order in which the shifts are done is irrelevant. We could do the upward shift first followed by the shift to the right.

EXAMPLE 3　　　　　　**Graph Shifted Vertically and Horizontally**

Graph $y = (x + 1)^2 - 1$.

Solution From the preceding paragraph we identify in (2) the form $y = f(x + c_1) - c_2$ with $c_1 = 1$ and $c_2 = 1$. Thus, the graph of $y = (x + 1)^2 - 1$ is the graph of $f(x) = x^2$ shifted 1 unit to the left followed by a downward shift of 1 unit. The graph is given in **FIGURE 2.2.8**.

$y = (x + 1)^2 - 1$

FIGURE 2.2.8 Shifted graph in Example 3

From the graph in Figure 2.2.8 we see immediately that the range of the function $y = (x + 1)^2 - 1 = x^2 + 2x$ is the interval $[-1, \infty)$ on the y-axis. Note also that the graph has x-intercepts $(0, 0)$ and $(-2, 0)$; you should verify this by solving $x^2 + 2x = 0$. Also, if you reexamine Figure 2.1.5 in Section 2.1 you will see that the graph of $y = 4 + \sqrt{x - 3}$ is the graph of the square root function $f(x) = \sqrt{x}$ (Figure 2.2.1(g)) shifted 3 units to the right and then 4 units up.

☐ **Rigid Transformations-Reflections** Another way of rigidly transforming the graph of a function is by a **reflection** in a coordinate axis.

THEOREM 2.2.3　Reflections

Suppose $y = f(x)$ is a function. Then the graph of

　　(*i*) $y = -f(x)$ is the graph of f reflected in the **x-axis**
　　(*ii*) $y = f(-x)$ is the graph of f reflected in the **y-axis**

Reflection or mirror image
in a horizontal axis

Reflection or mirror image
in a vertical axis

In part (a) of **FIGURE 2.2.9** we have reproduced the graph of a function $y = f(x)$ given in Figure 2.2.5. The reflections of this graph described in (*i*)–(*ii*) of Theorem 2.2.3 are illustrated in Figures 2.2.9(b) and 2.2.9(c). If (x, y) denotes a point on the graph of $y = f(x)$, then the point $(x, -y)$ is on the graph of $y = -f(x)$, and $(-x, y)$ is on the graph of $y = f(-x)$. Each of these reflections is a mirror image of the graph of $y = f(x)$ in the respective coordinate axis.

(a) Starting point   (b) Reflection in x-axis   (c) Reflection in y-axis

FIGURE 2.2.9 Reflections in the coordinate axes

EXAMPLE 4   Reflections

Graph **(a)** $y = -\sqrt{x}$ **(b)** $y = \sqrt{-x}$

Solution The starting point is the graph of $f(x) = \sqrt{x}$ given in **FIGURE 2.2.10(a)**.
(a) The graph of $y = -\sqrt{x}$ is the reflection of the graph of $f(x) = \sqrt{x}$ in the x-axis. Observe in Figure 2.2.10(b) that since $(1, 1)$ is on the graph of f, the point $(1, -1)$ is on the graph of $y = -\sqrt{x}$.

(b) The graph of $y = \sqrt{-x}$ is the reflection of the graph of $f(x) = \sqrt{x}$ in the y-axis. Observe in Figure 2.2.10(c) that since $(1, 1)$ is on the graph of f, the point $(-1, 1)$ is on the graph of $y = \sqrt{-x}$. The function $y = \sqrt{-x}$ looks a little strange, but bear in mind that its domain is determined by the requirement that $-x \geq 0$, or equivalently $x \leq 0$, and so the reflected graph is defined on the interval $(-\infty, 0]$.

(a) Starting point   (b) Reflection in x-axis   (c) Reflection in y-axis

FIGURE 2.2.10 Graphs in Example 4

If a function f is even, then $f(-x) = f(x)$ shows that a reflection in the y-axis would give precisely the same graph. If a function is odd, then from $f(-x) = -f(x)$ we see that a reflection of the graph of f in the y-axis is identical to the graph of f reflected in the x-axis. In **FIGURE 2.2.11** the blue curve is the graph of the odd function $f(x) = x^3$; the red curve is the graph of $y = f(-x) = (-x)^3 = -x^3$. Notice that if the blue curve is reflected in either the y-axis or the x-axis, we get the red curve.

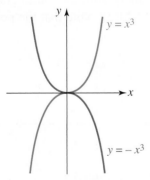

FIGURE 2.2.11 Reflection (red) of an odd function (blue) in y-axis

☐ **Nonrigid Transformations** A **nonrigid transformation** of a graph of function f is one that that distorts the shape of its graph. Stretching or compressing a graph are examples of nonrigid transformations. If a function f is multiplied by a

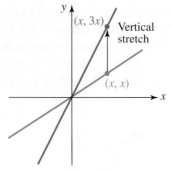

FIGURE 2.2.12 Vertical stretch (red) of the graph of $f(x) = x$ (blue)

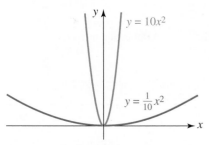

FIGURE 2.2.13 Vertical stretch (blue) and vertical compression (red) of the graph of $f(x) = x^2$

constant $c > 0$ the shape of the graph is changed but retains, *roughly*, its original shape. The graph of $y = cf(x)$ is the graph of $y = f(x)$ distorted vertically; the graph of f is either stretched (or elongated) vertically or is compressed (or flattened) vertically depending on the value of c.

THEOREM 2.2.4 Vertical Stretches and Compressions

Suppose $y = f(x)$ is a function and c a positive constant. Then the graph of $y = cf(x)$ is the graph of f

    (i) vertically stretched by a factor of c units if $c > 1$

    (ii) vertically compressed by a factor of c units if $0 < c < 1$

If (x, y) represents a point on the graph of f, then the point (x, cy) is on the graph of cf. The graphs of $y = x$ and $y = 3x$ are compared in FIGURE 2.2.12; the y-coordinate of a point on the graph of $y = 3x$ is 3 times as large as the y-coordinate of the point with the same x-coordinate on the graph of $y = x$. The comparison of the graphs of $y = 10x^2$ (blue graph) and $y = \frac{1}{10}x^2$ (red graph) in FIGURE 2.1.13 is a little more dramatic; the graph of $y = \frac{1}{10}x^2$ exhibits considerable vertical flattening, especially in a neighborhood of the origin. Note that c is positive in this discussion. To sketch the graph of $y = -10x^2$ we think of it as $y = -(10x^2)$, which means we first stretch the graph of $y = x^2$ vertically by a factor of 10 units and then reflect that graph in the x-axis.

The next example illustrates shifting, reflecting, and stretching of a graph.

■ EXAMPLE 5   **Combining Transformations**

Graph $y = 2 - 2\sqrt{x - 3}$.

Solution You should recognize that the given function consists of four transformations of the basic function $f(x) = \sqrt{x}$:

$$
\begin{array}{cc}
\text{vertical shift up} & \text{horizontal shift to right} \\
\downarrow & \downarrow
\end{array}
$$
$$
y = 2 - 2\sqrt{x - 3}
$$
$$
\begin{array}{cc}
\uparrow\ \uparrow & \\
\text{reflection in } x\text{-axis}\quad \text{vertical stretch}
\end{array}
$$

We start with the graph of $f(x) = \sqrt{x}$ in FIGURE 2.2.14(a). Then stretch this graph vertically by a factor of 2 to obtain $y = 2\sqrt{x}$ in Figure 2.2.14(b). Reflect this second graph in the x-axis to obtain $y = -2\sqrt{x}$ in Figure 2.2.14(c). Shift this third graph 3 units to the right to obtain $y = -2\sqrt{x - 3}$ in Figure 2.2.14(d). Finally, shift the fourth graph upward 2 units to obtain $y = 2 - 2\sqrt{x - 3}$ in Figure 2.2.14(e). Note that the point

 (a) Starting point   (b) Vertical stretch   (c) Reflection in x-axis   (d) Shift right   (e) Shift up

FIGURE 2.2.14 Graph of function in Example 5

(0, 0) on the graph of $f(x) = \sqrt{x}$ remains fixed in the vertical stretch and the reflection in the x-axis, but under the first (horizontal) shift (0, 0) moves to (3, 0) and under the second (vertical) shift (3, 0) moves to (3, 2).

NOTES FROM THE CLASSROOM
··

In this section we have seen how more complicated functions can be built up from functions $y = f(x)$ of the type given in Figure 2.2.1 by transformations. In some texts the beginning, or simpler, function is called the **parent function**. In Example 5 the square root function $f(x) = \sqrt{x}$ is the parent function for $y = 2 - 2\sqrt{x - 3}$.

Exercises 2.2 Answers to selected odd-numbered problems begin on page ANS–3.

In Problems 1–10, determine whether the given function $y = f(x)$ is even, odd, or neither even nor odd. Do not graph.

1. $f(x) = 4 - x^2$

2. $f(x) = x^2 + 2x$

3. $f(x) = x^3 - x + 4$

4. $f(x) = x^5 + x^3 + x$

5. $f(x) = 3x - \dfrac{1}{x}$

6. $f(x) = \dfrac{x}{x^2 + 1}$

7. $f(x) = 1 - \sqrt{1 - x^2}$

8. $f(x) = \sqrt[3]{x^3 + x}$

9. $f(x) = |x^3|$

10. $f(x) = x|x|$

In Problems 11–14, classify the function $y = f(x)$ whose graph is given as even, odd, or neither even nor odd.

11.

FIGURE 2.2.15 Graph for Problem 11

12.

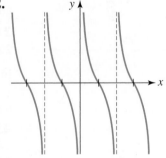

FIGURE 2.2.16 Graph for Problem 12

13.

FIGURE 2.2.17 Graph for Problem 13

14.

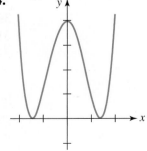

FIGURE 2.2.18 Graph for Problem 14

In Problems 15–18, complete the graph of the given function $y = f(x)$ if **(a)** f is an even function and **(b)** f is an odd function.

15.

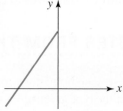

FIGURE 2.2.19 Graph for Problem 15

16.

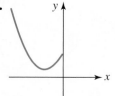

FIGURE 2.2.20 Graph for Problem 16

17.

FIGURE 2.2.21 Graph for Problem 17

18.

FIGURE 2.2.22 Graph for Problem 18

In Problems 19 and 20, suppose that $f(-2) = 4$ and $f(3) = 7$. Determine $f(2)$ and $f(-3)$.

19. If f is an even function.
20. If f is an odd function.

In Problems 21 and 22, suppose that $g(-1) = -5$ and $g(4) = 8$. Determine $g(1)$ and $g(-4)$.

21. If g is an odd function.
22. If g is an even function.

In Problems 23–32, the points $(-2, 1)$ and $(3, -4)$ are on the graph of the function $y = f(x)$. Find the corresponding points on the graph obtained by the given transformations.

23. the graph of f shifted up 2 units
24. the graph of f shifted down 5 units
25. the graph of f shifted to the left 6 units
26. the graph of f shifted to the right 1 unit
27. the graph of f shifted up 1 unit and to the left 4 units
28. the graph of f shifted down 3 units and to the right 5 units
29. the graph of f reflected in the y-axis
30. the graph of f reflected in the x-axis
31. the graph of f stretched vertically by a factor of 15 units
32. the graph of f compressed vertically by a factor of $\frac{1}{4}$ unit, then reflected in the x-axis

In Problems 33–36, use the graph of the function $y = f(x)$ given in the figure to graph the following functions.

(a) $y = f(x) + 2$
(b) $y = f(x) - 2$
(c) $y = f(x + 2)$
(d) $y = f(x - 5)$
(e) $y = -f(x)$
(f) $y = f(-x)$

33.

FIGURE 2.2.23 Graph for Problem 33

34.

FIGURE 2.2.24 Graph for Problem 34

35.

FIGURE 2.2.25 Graph for Problem 35

36.

FIGURE 2.2.26 Graph for Problem 36

In Problems 37 and 38, use the graph of the function $y = f(x)$ given in the figure to graph the following functions.

(a) $y = f(x) + 1$ **(b)** $y = f(x) - 1$
(c) $y = f(x + \pi)$ **(d)** $y = f(x - \pi/2)$
(e) $y = -f(x)$ **(f)** $y = f(-x)$
(g) $y = 3f(x)$ **(h)** $y = -\frac{1}{2}f(x)$

37.

FIGURE 2.2.27 Graph for Problem 37

38.

FIGURE 2.2.28 Graph for Problem 38

In Problems 39–42, find an equation of the final graph after the given transformations are applied to the graph of $y = f(x)$.

39. the graph of $f(x) = x^3$ shifted up 5 units and right 1 unit
40. the graph of $f(x) = x^{2/3}$ stretched vertically by a factor of 3 units, then shifted right 2 units
41. the graph of $f(x) = x^4$ reflected in the x-axis, then shifted left 7 units
42. the graph of $f(x) = 1/x$ reflected in the y-axis, then shifted left 5 units and down 10 units

In Problems 43–46, describe in words how the graph of the first function is obtained from the graph of the second function using rigid and nonrigid transformations. Carefully graph the first function.

43. $y = -1 + 2\sqrt{-x + 2}$; $y = \sqrt{x}$ **44.** $y = 2 + \frac{1}{2}(-x)^3$; $y = x^3$

45. $y = 2 - \dfrac{2}{x - 1}$; $y = \dfrac{1}{x}$ **46.** $y = -1 - \dfrac{1}{(x - 2)^2}$; $y = \dfrac{1}{x^2}$

2.2 Symmetry and Transformations

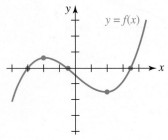

FIGURE 2.2.29 Graph for Problem 49

47. Explain why the graph of a nonzero function cannot be symmetric with respect to the x-axis.

48. What points, if any, on the graph of $y = f(x)$ remain fixed, that is, the same on the resulting graph after a vertical stretch or compression? After a reflection in the x-axis? After a reflection in the y-axis?

49. Copy the graph of $y = f(x)$ in FIGURE 2.2.29 on a piece of paper. By paying close attention to the five blue dots on the graph of $y = f(x)$ draw a representative graph of a vertical stretch and a vertical compression defined by **(a)** $y = cf(x), c > 1$ and **(b)** $y = cf(x), 0 < c < 1$. [*Hint*: See the first question in Problem 48.]

50. Discuss the relationship between the graphs of $y = f(x)$ and $y = f(|x|)$.

51. Discuss the relationship between the graphs of $y = f(x)$ and $y = f(cx)$, where $c > 0$ is a constant. Consider two cases: $0 < c < 1$ and $c > 1$.

52. Review the graphs of $y = x$ and $y = 1/x$ in Figure 2.2.1. Then discuss how to obtain the graph of the reciprocal $y = 1/f(x)$ from the graph of $y = f(x)$. Sketch the graph of $y = 1/f(x)$ for the function f whose graph is given in Figure 2.2.26.

2.3 Linear Functions

INTRODUCTION The notion of a line plays an important role in the study of differential calculus. There are three types of lines in the xy- or Cartesian plane: horizontal lines, vertical lines, and slant or oblique lines. We will see in this section that an equation of each of these lines stems from a **linear equation in two variables**

$$Ax + By + C = 0, \tag{1}$$

where A, B, and C are real constants. The characteristic that gives (1) its name *linear* is that the variables x and y appear only to the first power. We will refer back to (1) when we review lines and their equations, but let's note the cases of special interest:

$$A = 0, B \neq 0, \text{ gives } y = -\frac{C}{B}, \tag{2}$$

$$A \neq 0, B = 0, \text{ gives } x = -\frac{C}{A}, \tag{3}$$

$$A \neq 0, B \neq 0, \text{ gives } y = -\frac{A}{B}x - \frac{C}{B}. \tag{4}$$

The first and the third of these three equations define functions. By relabeling $-C/B$ in (2) as b we get a constant function.

DEFINITION 2.3.1 Constant Function

A **constant function** $y = f(x)$ is a function of the form

$$y = b \tag{5}$$

where b is a constant.

The **domain** of a constant function is the set of real numbers $(-\infty, \infty)$. In the definition of a function we are pairing each real number x with the same value of y, that is, (x, b). In our student/desk example of a function in Section 2.1 this is equivalent to having all the students in a classroom sit in one desk. On the other hand, the equation in (3) does not define a function. We cannot have one student (the fixed value of x) sit in all the desks in a classroom.

By relabeling $-A/B$ and $-C/B$ in (4) as a and b, respectively, we get the form of a linear function.

DEFINITION 2.3.2   Linear Function

A **linear function** $y = f(x)$ is a function of the form

$$f(x) = ax + b \tag{6}$$

where $a \neq 0$ and b are constants.

The **domain** of a linear function is the set of real numbers $(-\infty, \infty)$.

☐ **Graphs** Since the graphs of constant and linear functions are straight lines, it is appropriate that we will review equations of all lines. We begin with the recollection from plane geometry that through any two distinct points (x_1, y_1) and (x_2, y_2) in the plane there passes only one line L. If $x_1 \neq x_2$, then the number

$$m = \frac{y_2 - y_1}{x_2 - x_1} \tag{7}$$

is called the **slope** of the line determined by these two points. It is customary to call $y_2 - y_1$ the **change in y** or the **rise** of the line; $x_2 - x_1$ is the **change in x** or the **run** of the line. Therefore (7) is

$$m = \frac{\text{rise}}{\text{run}}.$$

See FIGURE 2.3.1(a). Any pair of distinct points on a line will determine the same slope. To see why this is so, consider the two similar right triangles in Figure 2.3.1(b). Since we know that the ratios of corresponding sides in similar triangles are equal we have

$$\frac{y_2 - y_1}{x_2 - x_1} = \frac{y_4 - y_3}{x_4 - x_3}.$$

Hence the slope of a line is independent of the choice of points on the line.

In FIGURE 2.3.2 we compare the graphs of lines with positive, negative, zero, and undefined slopes. In Figure 2.3.2(a) we see, reading the graph from left to right, that a line with positive slope ($m > 0$) rises as x increases. Figure 2.3.2(b) shows that a line with negative slope ($m < 0$) falls as x increases. If (x_1, y_1) and (x_2, y_2) are points on a horizontal line, then $y_1 = y_2$ and so its rise is $y_2 - y_1 = 0$. Hence from (7) the slope is zero ($m = 0$). See Figure 2.3.2(c). If (x_1, y_1) and (x_2, y_2) are points on a vertical line, then $x_1 = x_2$ and so its run is $x_2 - x_1 = 0$. In this case we say that the slope of the line is **undefined** or that the line has no slope. See Figure 2.3.2(d).

(a) Rise and run

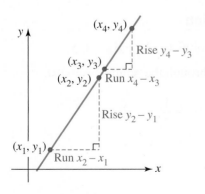

(b) Similar triangles

FIGURE 2.3.1 Slope of a line

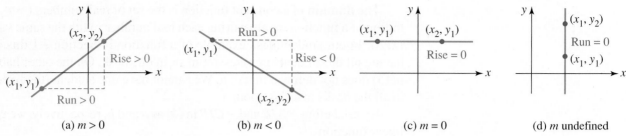

(a) $m > 0$ (b) $m < 0$ (c) $m = 0$ (d) m undefined

FIGURE 2.3.2 Lines with slope (a)–(c); line with no slope (d)

So when you see the phrase *line with slope* in a discussion you know that vertical lines are excluded.

☐ **Point-Slope Equation** We are now in a position to find an equation of a line L. To begin, suppose L has slope m and that (x_1, y_1) is on the line. If (x, y) represents any other point on L, then (7) gives

$$m = \frac{y - y_1}{x - x_1}.$$

Multiplying both sides of the last equality by $x - x_1$ gives an important equation.

THEOREM 2.3.1   Point-Slope Equation of a Line

The **point-slope equation** of the line through (x_1, y_1) with slope m is

$$y - y_1 = m(x - x_1) \tag{8}$$

EXAMPLE 1   **Point-Slope Equation**

Find an equation of the line with slope 6 and passing through $\left(-\frac{1}{2}, 2\right)$.

Solution Letting $m = 6$, $x_1 = -\frac{1}{2}$, and $y_1 = 2$ we obtain from (8)

$$y - 2 = 6\left[x - \left(-\frac{1}{2}\right)\right].$$

Simplifying gives $y - 2 = 6\left(x + \frac{1}{2}\right)$ or $y = 6x + 5$.   ▬

EXAMPLE 2   **Point-Slope Equation**

Find an equation of the line passing through the points $(4, 3)$ and $(-2, 5)$.

Solution First we compute the slope of the line through the points. From (7),

$$m = \frac{5 - 3}{-2 - 4} = \frac{2}{-6} = -\frac{1}{3}.$$

The point-slope equation (8) then gives

the distributive law
↓   ↓

$$y - 3 = -\frac{1}{3}(x - 4) \qquad \text{or} \qquad y = -\frac{1}{3}x + \frac{13}{3}. \qquad ▬$$

The distributive law $a(b + c) = ab + ac$ is the source of many errors on students' papers. A common error goes something like this:

$$-(2x - 3) = -2x - 3.$$

The correct result is:

$$-(2x - 3) = (-1)(2x - 3)$$
$$= (-1)2x - (-1)3$$
$$= -2x + 3.$$

CHAPTER 2 FUNCTIONS

☐ **Slope-Intercept Equation** Any line with slope (that is, any line that is not vertical) must cross the y-axis. If this y-intercept is $(0, b)$, then with $x_1 = 0, y_1 = b$, the point-slope form (8) gives $y - b = m(x - 0)$. The last equation simplifies to the next result.

THEOREM 2.3.2 Slope-Intercept Equation of a Line

The **slope-intercept equation** of the line with slope m and y-intercept $(0, b)$ is

$$y = mx + b \tag{9}$$

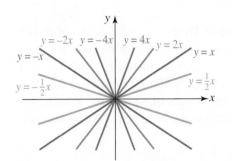

FIGURE 2.3.3 Lines with slope through the origin are $y = mx$

☐ **Family of Lines** For $m \neq 0$, (8) and (9) give us the form of the linear function in (6). The coefficient a in (6) is, of course, the slope m of the line. When $b = 0$ in (9), the equation $y = mx$ represents a **family of lines** that pass through the origin $(0, 0)$. In FIGURE 2.3.3 we have drawn a few of the members of that family.

EXAMPLE 3 **Example 2 Revisited**

We can also use the slope-intercept from (9) to obtain the equation of the line through two points in Example 2. As in that example, we start by finding the slope $m = -\frac{1}{3}$. The equation of the line is then $y = -\frac{1}{3}x + b$. Substituting the coordinates of either point $(4, 3)$ or $(-2, 5)$ into the last equation enables us to determine b. If we use $x = 4$ and $y = 3$, then $3 = -\frac{1}{3} \cdot 4 + b$ and so $b = 3 + \frac{4}{3} = \frac{13}{3}$. The equation of the line is $y = -\frac{1}{3}x + \frac{13}{3}$. ▬

☐ **Horizontal and Vertical Lines** We saw in Figure 2.3.2(c) that a horizontal line has slope $m = 0$. An equation of a horizontal line passing through a point (a, b) can be obtained from (8), that is, $y - b = 0(x - a)$. The **equation of a horizontal line** is then

$$y = b. \tag{10}$$

We have already seen this in (5) and in (2) where $-C/B$ played the part of the symbol b. A vertical line through (a, b) has undefined slope and all points on the line have the same x-coordinate. The **equation of a vertical line** is then

$$x = a. \tag{11}$$

Equation (11) is (3) with $-C/A$ replaced by the symbol a.

EXAMPLE 4 **Vertical and Horizontal Lines**

FIGURE 2.3.4 Horizontal and vertical lines in Example 4

Find equations for the vertical and horizontal lines through $(3, -1)$. Graph these lines.

Solution Any point on the vertical line through $(3, -1)$ has x-coordinate 3. The equation of this line is then $x = 3$. Similarly, any point on the horizontal line through $(3, -1)$ has y-coordinate -1. The equation of this line is $y = -1$. Both lines are graphed in FIGURE 2.3.4. Don't forget, only $y = -1$ is a function. ▬

Parallel lines

☐ **Parallel and Perpendicular Lines** Suppose L_1 and L_2 are two distinct lines with slope. This assumption means that both L_1 and L_2 are nonvertical lines. Then necessarily L_1 and L_2 are either parallel or they intersect. If the lines intersect at a right angle they are said to be perpendicular. We can determine whether two lines are parallel or are perpendicular by examining their slopes.

THEOREM 2.3.3   Slopes of Parallel and Perpendicular Lines

If L_1 and L_2 are lines with slopes m_1 and m_2, respectively, then

- L_1 is **parallel** to L_2 if and only if $m_1 = m_2$, and   (12)
- L_1 is **perpendicular** to L_2 if and only if $m_1 m_2 = -1$ (13)

FIGURE 2.3.5 Parallel lines

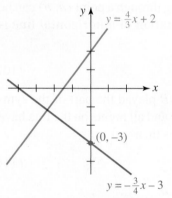

FIGURE 2.3.6 Parallel lines in Example 5

FIGURE 2.3.7 Perpendicular lines in Example 6

FIGURE 2.3.8 Graph of equation in Example 7

There are several ways of proving the two parts of Theorem 2.3.3. The proof of (12) can be obtained using similar right triangles, as in FIGURE 2.3.5, and the fact that the ratios of corresponding sides in such triangles are equal. We leave the justification of (13) as an exercise. See Problem 64 in Exercises 2.3. Note that the condition $m_1 m_2 = -1$ implies that $m_2 = -1/m_1$, that is, the slopes are negative reciprocals of each other. A horizontal line $y = b$ and a vertical line $x = a$ are perpendicular, but the latter is a line with no slope.

EXAMPLE 5   Parallel Lines

The linear equations $3x + y = 2$ and $6x + 2y = 15$ can be rewritten in the slope-intercept forms

$$y = -3x + 2 \qquad \text{and} \qquad y = -3x + \tfrac{15}{2},$$

respectively. As noted in red in the preceding line the slope of each line is -3. Therefore the lines are parallel. The graphs of these equations are shown in FIGURE 2.3.6. ∎

EXAMPLE 6   Perpendicular Lines

Find an equation of the line through $(0, -3)$ that is perpendicular to the graph of $4x - 3y + 6 = 0$.

Solution We express the given linear equation in slope-intercept form:

$$4x - 3y + 6 = 0 \qquad \text{implies} \qquad 3y = 4x + 6.$$

Dividing by 3 gives $y = \tfrac{4}{3}x + 2$. This line, whose graph is given in blue in FIGURE 2.3.7, has slope $\tfrac{4}{3}$. The slope of any line perpendicular to it is the negative reciprocal of $\tfrac{4}{3}$, namely, $-\tfrac{3}{4}$. Since $(0, -3)$ is the y-intercept of the required line, it follows from (9) that its equation is $y = -\tfrac{3}{4}x - 3$. The graph of the last equation is the red line in FIGURE 2.3.7. ∎

☐ **Graphs** As mentioned in the earlier sections of this chapter, when graphing an equation it is always a good habit to try to find x- and y-intercepts of its graph. Except in the cases of horizontal and vertical lines, and lines through the origin, a line will have distinct x- and y-intercepts. Of course, that is all we need to draw a line: two points.

EXAMPLE 7   Graph of a Linear Equation

Graph the linear equation $3x - 2y + 8 = 0$.

Solution There is no need to rewrite the linear equation in the form $y = mx + b$. We simply find the intercepts of the graph.

y-intercept: Setting $x = 0$ gives $-2y + 8 = 0$ or $y = 4$. The y-intercept is $(0, 4)$.
x-intercept: Setting $y = 0$ gives $3x + 8 = 0$ or $x = -\tfrac{8}{3}$. The x-intercept is $\left(-\tfrac{8}{3}, 0\right)$.

As shown in FIGURE 2.3.8, the line is drawn through the two intercepts $(0, 4)$ and $\left(-\tfrac{8}{3}, 0\right)$. ∎

(a) $f(x_1) < f(x_2)$

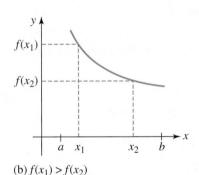

(b) $f(x_1) > f(x_2)$

FIGURE 2.3.9 Increasing function in (a); decreasing function in (b)

Increasing-Decreasing Functions We have just seen in Figures 2.3.2(a) and 2.3.2(b) that if $a > 0$ (which, as we have just seen plays the part of m) the values of a linear function $f(x) = ax + b$ increase as x increases, whereas for $a < 0$, the values $f(x)$ decrease as x increases. The notions of increasing and decreasing can be extended to *any* function. The ability to determine intervals over which a function f is either increasing or decreasing plays an important role in applications of calculus.

DEFINITION 2.3.3 Increasing/Decreasing

Suppose $y = f(x)$ is a function defined on an interval $[a, b]$, and x_1 and x_2 are any two numbers in the interval such that $x_1 < x_2$. Then the function f is

- **increasing** on the interval if $f(x_1) < f(x_2)$ (14)
- **decreasing** on the interval if $f(x_1) > f(x_2)$ (15)

In FIGURE 2.3.9(a) the function f is increasing on the interval $[a, b]$, whereas f is decreasing on $[a, b]$ in Figure 2.3.9(b). A linear function $f(x) = ax + b$ increases on the interval $(-\infty, \infty)$ for $a > 0$ and decreases on the interval $(-\infty, \infty)$ for $a < 0$.

Points of Intersection We are often interested in finding the points where the graphs of two functions intersect. The x-intercepts of the graph of a function f can be interpreted as the points where the graph of f intersects the graph of the constant function $y = 0$. In general, at a point P of intersection of the graphs of two functions f and g, the coordinates (x, y) of P must satisfy both equations $y = f(x)$ and $y = g(x)$, and so $f(x) = g(x)$. Two lines can intersect at a single point.

EXAMPLE 8   Intersecting Lines

Find the point where the two lines in Figure 2.3.7 intersect.

Solution We equate $y = \frac{4}{3}x + 2$ and $y = -\frac{3}{4}x - 3$ and solve for x:

$$\frac{4}{3}x + 2 = -\frac{3}{4}x - 3$$
$$\left(\frac{4}{3} + \frac{3}{4}\right)x = -5$$
$$\frac{25}{12}x = -5$$
$$x = -\frac{12}{5}.$$

By substituting $x = -\frac{12}{5}$ into either equation we find that $y = -\frac{6}{5}$. The point of intersection of the lines is then $\left(-\frac{12}{5}, -\frac{6}{5}\right)$.

Exercises 2.3   Answers to selected odd-numbered problems begin on page ANS–4.

In Problems 1–6, find the slope of the line through the given points. Graph the line through the points.

     1. $(3, -7), (1, 0)$ **2.** $(-4, -1), (1, -1)$
     3. $(5, 2), (4, -3)$ **4.** $(1, 4), (6, -2)$
     5. $(-1, 2), (3, -2)$ **6.** $\left(8, -\frac{1}{2}\right), \left(2, \frac{5}{2}\right)$

In Problems 7 and 8, use the graph of the given line to estimate its slope.

7.

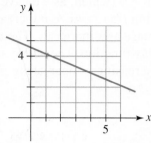

FIGURE 2.3.10 Graph for
Problem 7

8.

FIGURE 2.3.11 Graph for
Problem 8

In Problems 9–16, find the slope and the x- and y-intercepts of the given line. Graph the line.

9. $3x - 4y + 12 = 0$

10. $\frac{1}{2}x - 3y = 3$

11. $2x - 3y = 9$

12. $-4x - 2y + 6 = 0$

13. $2x + 5y - 8 = 0$

14. $\dfrac{y}{2} - \dfrac{x}{10} - 1 = 0$

15. $y + \frac{2}{3}x = 1$

16. $y = 2x + 6$

In Problems 17–22, find an equation of the line through $(1, 2)$ with the indicated slope.

17. $\frac{2}{3}$

18. $\frac{1}{10}$

19. 0

20. -2

21. -1

22. undefined

In Problems 23–36, find an equation of the line that satisfies the given conditions.

23. through $(2, 3)$ and $(6, -5)$

24. through $(5, -6)$ and $(4, 0)$

25. through $(8, 1)$ and $(-3, 1)$

26. through $(2, 2)$ and $(-2, -2)$

27. through $(-2, 0)$ and $(-2, 6)$

28. through $(0, 0)$ and (a, b)

29. through $(-2, 4)$ parallel to $3x + y - 5 = 0$

30. through $(1, -3)$ parallel to $2x - 5y + 4 = 0$

31. through $(5, -7)$ parallel to the y-axis

32. through the origin parallel to the line through $(1, 0)$ and $(-2, 6)$

33. through $(2, 3)$ perpendicular to $x - 4y + 1 = 0$

34. through $(0, -2)$ perpendicular to $3x + 4y + 5 = 0$

35. through $(-5, -4)$ perpendicular to the line through $(1, 1)$ and $(3, 11)$

36. through the origin perpendicular to every line with slope 2

37. Find the coordinates of the point P shown in FIGURE 2.3.12.

38. A line through $(2, 4)$ has slope 8. Without finding an equation of the line, determine whether the point $(1, -5)$ is on the line.

FIGURE 2.3.12 Lines in
Problem 37

In Problems 39–42, determine which of the given lines are parallel to each other and which are perpendicular to each other.

39. (a) $3x - 5y + 9 = 0$

(b) $5x = -3y$

(c) $-3x + 5y = 2$

(d) $3x + 5y + 4 = 0$

(e) $-5x - 3y + 8 = 0$

(f) $5x - 3y - 2 = 0$

40. (a) $2x + 4y + 3 = 0$ **(b)** $2x - y = 2$
(c) $x + 9 = 0$ **(d)** $x = 4$
(e) $y - 6 = 0$ **(f)** $-x - 2y + 6 = 0$
41. (a) $3x - y - 1 = 0$ **(b)** $x - 3y + 9 = 0$
(c) $3x + y = 0$ **(d)** $x + 3y = 1$
(e) $6x - 3y + 10 = 0$ **(f)** $x + 2y = -8$
42. (a) $y + 5 = 0$ **(b)** $x = 7$
(c) $4x + 6y = 3$ **(d)** $12x - 9y + 7 = 0$
(e) $2x - 3y - 2 = 0$ **(f)** $3x + 4y - 11 = 0$

In Problems 43 and 44, find a linear function (6) that satisfies both of the given conditions.

43. $f(-1) = 5, f(1) = 6$ **44.** $f(-1) = 1 + f(2), f(3) = 4f(1)$

In Problems 45–48, find the point of intersection of the graphs of the given linear functions. Sketch both lines.

45. $f(x) = -2x + 1, g(x) = 4x + 6$ **46.** $f(x) = 2x + 5, g(x) = \frac{3}{2}x + 5$
47. $f(x) = 4x + 7, g(x) = \frac{1}{3}x + \frac{10}{3}$ **48.** $f(x) = 2x - 10, g(x) = -3x$

In Problems 49 and 50, for the given linear function compute the quotient

$$\frac{f(x + h) - f(x)}{h},$$

where h is a constant.

49. $f(x) = -9x + 12$ **50.** $f(x) = \frac{4}{3}x - 5$

51. Find an equation of the red line L shown in FIGURE 2.3.13 if an equation of the blue curve is $y = x^2 + 1$.

52. A tangent line L to a circle at a point P on the circle is perpendicular to the line through P and the center of the circle. Find an equation of the red line L shown in FIGURE 2.3.14.

FIGURE 2.3.13 Graphs for Problem 51

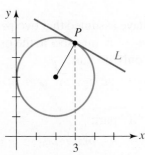

FIGURE 2.3.14 Circle and tangent line in Problem 52

FIGURE 2.3.15 Thermometers in Problems 53 and 54

Applications

53. Thermometers The functional relationship between degrees Celsius T_C and degrees Fahrenheit T_F is linear.
   (a) Express T_F as a function of T_C if $(0°C, 32°F)$ and $(60°C, 140°F)$ are on the graph of T_F.
   (b) Show that $100°C$ is equivalent to the Fahrenheit boiling point $212°F$. See FIGURE 2.3.15.

54. Thermometers—Continued The functional relationship between degrees Celsius T_C and temperature measured in Kelvin units T_K is linear.
   (a) Express T_K as a function of T_C if $(0°C, 273K)$ and $(27°C, 300K)$ are on the graph of T_K.
   (b) Express the boiling point of water $100°C$ in Kelvin units. See Figure 2.3.15.
   (c) Absolute zero is defined as $0K$. What is $0K$ in degrees Celsius?
   (d) Express T_K as a linear function of T_F.
   (e) What is $0K$ in degrees Fahrenheit?

55. Simple Interest In simple interest, the amount A accrued over time is the linear function $A = P + Prt$, where P is the principal, t is measured in years, and r is the annual interest rate (expressed as a decimal). Compute A after 20 years if the principal is $P = 1000$ and the annual interest rate is 3.4%. At what time is $A = 2200$?

56. Linear Depreciation Straight line, or linear, depreciation consists of an item losing all its initial worth of A dollars over a period of n years by an amount A/n

each year. If an item costing \$20,000 when new is depreciated linearly over 25 years, determine a linear function giving its value V after x years, where $0 \leq x \leq 25$. What is the value of the item after 10 years?

For Discussion

57. Consider the linear function $f(x) = \frac{5}{2}x - 4$. If x is changed by 1 unit, how many units will y change? If x is changed by 2 units? If x is changed by n (n a positive integer) units?

58. Consider the interval $[x_1, x_2]$ and the linear function $f(x) = ax + b$, $a \neq 0$. Show that

$$f\left(\frac{x_1 + x_2}{2}\right) = \frac{f(x_1) + f(x_2)}{2}$$

and interpret this result geometrically for $a > 0$.

59. How would you find an equation of the line that is the perpendicular bisector of the line segment through $\left(\frac{1}{2}, 10\right)$ and $\left(\frac{3}{2}, 4\right)$?

60. Using only the concepts of this section, how would you prove or disprove that the triangle with vertices $(2, 3)$, $(-1, -3)$, and $(4, 2)$ is a right triangle?

61. Using only the concepts of this section, how would you prove or disprove that the quadrilateral with vertices $(0, 4)$, $(-1, 3)$, $(-2, 8)$, and $(-3, 7)$ is a parallelogram?

62. If C is an arbitrary real constant, the equation $2x - 3y = C$ defines a family of lines. Choose four different values of C and plot the corresponding lines on the same coordinate axes. What is true about the lines that are members of this family?

63. Find the equations of the lines through $(0, 4)$ that are tangent to the circle $x^2 + y^2 = 4$.

64. To prove (13) you have to prove two things, the "only if" and the "if" parts of the theorem.

   (a) In FIGURE 2.3.16, without loss of generality, we have assumed that two perpendicular lines, $y = m_1 x$, $m_1 > 0$ and $y = m_2 x$, $m_2 < 0$, intersect at the origin. Use the information in the figure to prove the "only if" part:

      If L_1 and L_2 are perpendicular lines with slopes m_1 and m_2, then $m_1 m_2 = -1$.

   (b) Reverse your argument in part (a) to prove the "if" part:

      If L_1 and L_2 are lines with slopes m_1 and m_2 such that $m_1 m_2 = -1$, then L_1 and L_2 are perpendicular.

FIGURE 2.3.16 Lines through origin in Problem 64

2.4 Quadratic Functions

INTRODUCTION The squaring function $y = x^2$ that played an important role in Section 2.2 is a member of a family of functions called **quadratic functions**.

DEFINITION 2.4.1 Quadratic Function

A **quadratic function** $y = f(x)$ is a function of the form

$$f(x) = ax^2 + bx + c \tag{1}$$

where $a \neq 0$, b, and c are constants.

The **domain** of a quadratic function f is the set of real numbers $(-\infty, \infty)$.

FIGURE 2.4.1 Graph of simplest parabola

☐ **Graphs and Transformations** The graph of any quadratic function is called a **parabola**. The graph of a quadratic function has the same basic shape of the squaring function $y = x^2$ shown in FIGURE 2.4.1. In the examples that follow we will see that the graphs of quadratic functions (1) are simply transformations of the graph of $y = x^2$:

* The graph of $f(x) = ax^2, a > 0$, is the graph of $y = x^2$ **stretched** vertically when $a > 1$, and **compressed** vertically when $0 < a < 1$.
* The graph of $f(x) = ax^2, a < 0$, is the graph of $y = ax^2, a > 0$, **reflected** in the x-axis.
* The graph of $f(x) = ax^2 + bx + c, b \neq 0$, is the graph of $y = ax^2$ **shifted** horizontally or vertically.

From the first two items in the bulleted list, we conclude that the graph of a quadratic function opens upward (as in Figure 2.4.1) if $a > 0$ and opens downward if $a < 0$.

EXAMPLE 1   **Stretch, Compression, and Reflection**

(a) The graphs of $y = 4x^2$ and $y = \frac{1}{10}x^2$ are, respectively, a vertical stretch and a vertical compression of the graph of $y = x^2$. The graphs of these functions are shown in FIGURE 2.4.2(a); the graph of $y = 4x^2$ is shown in red, the graph of $y = \frac{1}{10}x^2$ is green, and the graph of $y = x^2$ is blue.

(b) The graphs of $y = -4x^2$, $y = -\frac{1}{10}x^2$, $y = -x^2$ are obtained from the graphs of the functions in part (a) by reflecting their graphs in the x-axis. See Figure 2.4.2(b).

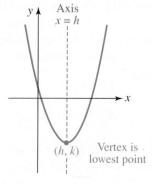

(a) Red graph is a vertical stretch of blue graph; green graph is a vertical compression of blue graph

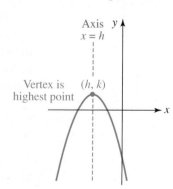

(b) Reflections in x-axis

FIGURE 2.4.2 Graphs of quadratic functions in Example 1

☐ **Vertex and Axis** If the graph of a quadratic function opens upward $a > 0$ (or downward $a < 0$), the lowest (highest) point (h, k) on the parabola is called its **vertex**. All parabolas are symmetric with respect to a vertical line through the vertex (h, k). The line $x = h$ is called the **axis of symmetry** or simply the **axis** of the parabola. See FIGURE 2.4.3.

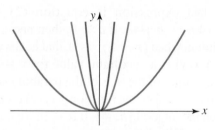

(a) $y = ax^2 + bx + c, a > 0$ (b) $y = ax^2 + bx + c, a < 0$

FIGURE 2.4.3 Vertex and axis of a parabola

☐ **Standard Form** The vertex (h, k) of a parabola can be determined by recasting the equation $f(x) = ax^2 + bx + c$ into the **standard form**

$$f(x) = a(x - h)^2 + k. \tag{2}$$

See Section 1.4. ▶

The form (2) is obtained from the equation (1) by completing the square in x. Completing the square in (1) starts with factoring the number a from all terms involving the variable x:

$$f(x) = ax^2 + bx + c$$
$$= a\left(x^2 + \frac{b}{a}x\right) + c.$$

Within the parentheses we add and subtract the square of one-half the coefficient of x:

$$\overset{\text{square of } \frac{b}{2a}}{\downarrow}$$

$$f(x) = a\left(x^2 + \frac{b}{a}x + \frac{b^2}{4a^2} - \frac{b^2}{4a^2}\right) + c \quad \leftarrow \text{ terms in color add to 0}$$
$$= a\left(x^2 + \frac{b}{a}x + \frac{b^2}{4a^2}\right) - \frac{b^2}{4a} + c \quad \leftarrow \text{ note that } a \cdot \left(-\frac{b^2}{4a^2}\right) = -\frac{b^2}{4a} \tag{3}$$
$$= a\left(x + \frac{b}{2a}\right)^2 + \frac{4ac - b^2}{4a}$$

The last expression is equation (2) with the identifications $h = -b/2a$ and $k = (4ac - b^2)/4a$. If $a > 0$, then necessarily $a(x - h)^2 \geq 0$. Hence $f(x)$ in (2) is a minimum when $(x - h)^2 = 0$, that is, for $x = h$. A similar argument shows that if $a < 0$ in (2), $f(x)$ is a maximum value for $x = h$. Thus (h, k) is the vertex of the parabola. The equation of the axis of the parabola is $x = h$ or $x = -b/2a$.

If $a > 0$, then the function f in (2) is decreasing on the interval $(-\infty, h]$ and increasing on the interval $[h, \infty)$. If $a < 0$, we have just the opposite, that is, f is increasing on $(-\infty, h]$ followed by decreasing on $[h, \infty)$.

☐ **Coordinates of Vertex** We strongly suggest that you *do not memorize* the result in the last line of (3), but practice completing the square each time. However, if memorization is permitted by your instructor to save time, then the vertex can be found by computing the coordinates of the point

$$\left(-\frac{b}{2a}, f\left(-\frac{b}{2a}\right)\right). \tag{4}$$

☐ **Intercepts** The graph of (1) always has a **y-intercept** since $f(0) = c$, and so the y-intercept is $(0, c)$. To determine whether the graph has **x-intercepts** we must solve the equation $f(x) = 0$. The last equation can be solved either by factoring or by using the quadratic formula. Recall that a quadratic equation $ax^2 + bx + c = 0$, $a \neq 0$, has the solutions

$$x_1 = \frac{-b - \sqrt{b^2 - 4ac}}{2a}, \qquad x_2 = \frac{-b + \sqrt{b^2 - 4ac}}{2a}.$$

We distinguish three cases according to the algebraic sign of the discriminant $b^2 - 4ac$.

- If $b^2 - 4ac > 0$, then there are two distinct real solutions x_1 and x_2. The parabola crosses the x-axis at the points $(x_1, 0)$ and $(x_2, 0)$.

- If $b^2 - 4ac = 0$, then there is a single real solution x_1. The vertex of the parabola is located on the x-axis at $(x_1, 0)$. The parabola is tangent to, or touches, the x-axis at this point.
- If $b^2 - 4ac < 0$, then there are no real solutions. The parabola does not cross the x-axis.

As the next example shows, a reasonable sketch of a parabola can be obtained by plotting the intercepts and the vertex.

▮ EXAMPLE 2   Graph Using Intercepts and Vertex

Graph $f(x) = x^2 - 2x - 3$.

Solution Since $a = 1 > 0$ we know that the parabola will open upward. From $f(0) = -3$ we get the y-intercept $(0, -3)$. To see whether there are any x-intercepts we solve $x^2 - 2x - 3 = 0$. By factoring

$$(x + 1)(x - 3) = 0,$$

we find the solutions $x = -1$ and $x = 3$. The x-intercepts are $(-1, 0)$ and $(3, 0)$. To locate the vertex we complete the square:

$$f(x) = (x^2 - 2x + 1) - 1 - 3 = (x^2 - 2x + 1) - 4.$$

Thus the standard form is $f(x) = (x - 1)^2 - 4$. With the identifications $h = 1$ and $k = -4$, we conclude that the vertex is $(1, -4)$. Using this information we draw a parabola through these four points as shown in FIGURE 2.4.4.

One last observation. By finding the vertex we automatically determine the range of a quadratic function. In our current example, $y = -4$ is the smallest number in the range of f and so the range of f is the interval $[-4, \infty)$ on the y-axis. ▮

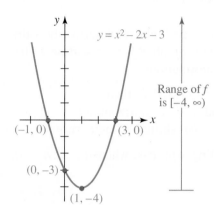

FIGURE 2.4.4 Parabola in Example 2

▮ EXAMPLE 3   Vertex Is the x-intercept

Graph $f(x) = -4x^2 + 12x - 9$.

Solution The graph of this quadratic function is a parabola that opens downward because $a = -4 < 0$. To complete the square we start by factoring -4 from the two x-terms:

$$
\begin{aligned}
f(x) &= -4x^2 + 12x - 9 \\
&= -4(x^2 - 3x) - 9 \\
&= -4\left(x^2 - 3x + \frac{9}{4} - \frac{9}{4}\right) - 9 \\
&= -4\left(x^2 - 3x + \frac{9}{4}\right) + 9 - 9 \\
&= -4\left(x^2 - 3x + \frac{9}{4}\right).
\end{aligned}
$$

FIGURE 2.4.5 Parabola in Example 3

Thus the standard form is $f(x) = -4\left(x - \frac{3}{2}\right)^2$. With $h = \frac{3}{2}$ and $k = 0$ we see that the vertex is $\left(\frac{3}{2}, 0\right)$. The y-intercept is $(0, f(0)) = (0, -9)$. Solving $-4x^2 + 12x - 9 = 0$, we find that there is only one x-intercept, namely, $\left(\frac{3}{2}, 0\right)$. Of course, this was to be expected because the vertex $\left(\frac{3}{2}, 0\right)$ is on the x-axis. As shown in FIGURE 2.4.5 a rough sketch can be obtained from these two points alone. The parabola is tangent to the x-axis at $\left(\frac{3}{2}, 0\right)$. ▮

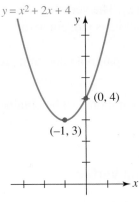

$y = x^2 + 2x + 4$

$(0, 4)$

$(-1, 3)$

FIGURE 2.4.6 Parabola in Example 4

EXAMPLE 4   **Using (4) to Find the Vertex**

Graph $f(x) = x^2 + 2x + 4$.

Solution The graph is a parabola that opens upward because $a = 1 > 0$. For the sake of illustration we will use (4) this time to find the vertex. With $b = 2$, $-b/2a = -2/2 = -1$ and

$$f(-1) = (-1)^2 + 2(-1) + 4 = 3,$$

the vertex is $(-1, f(-1)) = (-1, 3)$. Now the y-intercept is $(0, f(0)) = (0, 4)$ but the quadratic formula shows that the equation $f(x) = 0$ or $x^2 + 2x + 4 = 0$ has no real solutions. Therefore the graph has no x-intercepts. Since the vertex is above the x-axis and the parabola opens upward, the graph must lie entirely above the x-axis. See FIGURE 2.4.6. ■

☐ **Graphs and Transformations** The standard form (2) clearly describes how the graph of any quadratic function is constructed from the graph of $y = x^2$ starting with a nonrigid transformation followed by two rigid transformations:

- $y = ax^2$ is the graph of $y = x^2$ stretched or compressed vertically.
- $y = a(x - h)^2$ is the graph of $y = ax^2$ shifted $|h|$ units horizontally.
- $y = a(x - h)^2 + k$ is the graph of $y = a(x - h)^2$ shifted $|k|$ units vertically.

FIGURE 2.4.7 illustrates the horizontal and vertical shifting in the case where $a > 0, h > 0$, and $k > 0$.

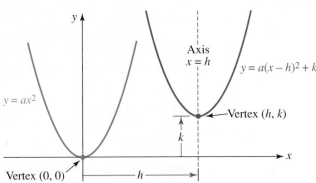

FIGURE 2.4.7 The red graph is obtained by shifting the blue graph h units to the right and k units upward.

EXAMPLE 5   **Horizontally Shifted Graphs**

Compare the graphs of **(a)** $y = (x - 2)^2$ and **(b)** $y = (x + 3)^2$.

Solution The blue dashed graph in FIGURE 2.4.8 is the graph of $y = x^2$. Matching the given functions with (2) shows in each case that $a = 1$ and $k = 0$. This means that neither graph undergoes a vertical stretch or compression, and neither graph is shifted vertically.

(a) With the identification $h = 2$, the graph of $y = (x - 2)^2$ is the graph of $y = x^2$ shifted horizontally 2 units to the right. The vertex $(0, 0)$ for $y = x^2$ becomes the vertex $(2, 0)$ for $y = (x - 2)^2$. See the red graph in Figure 2.4.8.

(b) With the identification $h = -3$, the graph of $y = (x + 3)^2$ is the graph of $y = x^2$ shifted horizontally $|-3| = 3$ units to the left. The vertex $(0, 0)$ for $y = x^2$ becomes the vertex $(-3, 0)$ for $y = (x + 3)^2$. See the green graph in Figure 2.4.8.

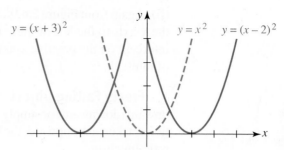

FIGURE 2.4.8 Shifted graphs (red and green) in Example 5

EXAMPLE 6   **Shifted Graph**

Graph $y = 2(x - 1)^2 - 6$.

Solution The graph is the graph of $y = x^2$ stretched vertically upward, followed by a horizontal shift to the right of 1 unit, followed by a vertical shift downward of 6 units. In FIGURE 2.4.9, you should note how the vertex $(0, 0)$ on the graph of $y = x^2$ is moved to $(1, -6)$ on the graph of $y = 2(x - 1)^2 - 6$ as a result of these transformations. You should also follow how the point $(1, 1)$ shown in Figure 2.4.9(a) ends up as $(2, -4)$ in Figure 2.4.9(d).

(a) Basic parabola

(b) Vertical stretch

(c) Horizontal shift

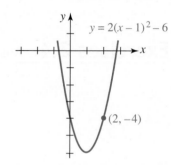

(d) Vertical shift

FIGURE 2.4.9 Graphs in Example 6

☐ **Graphs and Inequalities** A graph of a function f can be of help in solving certain inequalities such as $f(x) > 0$ or $f(x) \le 0$. In the case of the inequality $f(x) > 0$, the solution set is the set of all numbers x in the domain of f for which the y-coordinates of points on the graph are positive, in other words, where the graph of f is *above* the x-axis $(y = 0)$. For a quadratic function $f(x) = ax^2 + bx + c$, the ability to see the solution set is especially useful when the sign-chart method is not convenient because the factors of $ax^2 + bx + c$ are not obvious.

The sign-chart method for solving ▶ nonlinear inequalities was discussed in Section 1.1.

EXAMPLE 7   **Example 6 Revisited**

Solve **(a)** $2x^2 - 4x - 4 < 0$ **(b)** $2x^2 - 4x - 4 \ge 0$.

Solution You should verify that $y = 2x^2 - 4x - 4$ is a equivalent form of the function $y = 2(x - 1)^2 - 6$ in Example 6. To solve the given inequalities we only need the graph and its x-intercepts (if there are any). In this case, the x-coordinates of these intercepts, obtained by solving the equation $y = 0$ or $2x^2 - 4x - 4 = 0$ by the quadratic formula, are $1 - \sqrt{3}$ and $1 + \sqrt{3}$.

(a) From the graph in Figure 2.4.9(d) we see that solution set of the inequality $y < 0$ is $(1 - \sqrt{3}, 1 + \sqrt{3})$ because the graph of the function is below the x-axis on this interval.

(b) Again from Figure 2.4.9(d), we see in that the graph of the function is above ($y > 0$) the x-axis to the left of the x-intercept on the negative x-axis and to the right of the x-intercept on the positive x-axis. Thus the solution of the inequality $y \geq 0$ is the union of intervals $(-\infty, 1 - \sqrt{3}] \cup [1 + \sqrt{3}, \infty)$. ▪

☐ **Freely Falling Object** Suppose an object, such as a ball, is either thrown straight upward (downward) or simply dropped from an initial height s_0. Then if the positive direction is taken to be upward, the height $s(t)$ of the object above ground is given by the quadratic function

$$s(t) = -\tfrac{1}{2}gt^2 + v_0 t + s_0, \tag{5}$$

where g is the acceleration due to gravity (32 ft/s^2 or 9.8 m/s^2), v_0 is the initial velocity imparted to the object, and t is time measured in seconds. See **FIGURE 2.4.10**. If the object is dropped, then $v_0 = 0$. An assumption in the derivation of (5), a straightforward exercise in integral calculus, is that the motion takes place close to the surface of Earth and so the retarding effects of air resistance is ignored. Also, the velocity of the object while it is in the air is given by the linear function

$$v(t) = -gt + v_0. \tag{6}$$

See Problems 55–58 in Exercises 2.4.

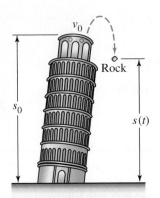

FIGURE 2.4.10 Rock thrown upward from an initial height s_0

Exercises 2.4 Answers to selected odd-numbered problems begin on page ANS-5.

In Problems 1–6, sketch the graph of the given function f.

1. $f(x) = 2x^2$

2. $f(x) = -2x^2$

3. $f(x) = 2x^2 - 2$

4. $f(x) = 2x^2 + 5$

5. $f(x) = -2x^2 + 1$

6. $f(x) = -2x^2 - 3$

In Problems 7–18, consider the quadratic function f.
(a) Find all intercepts of the graph of f.
(b) Express the function f in standard form.
(c) Find the vertex and axis of symmetry.
(d) Sketch the graph of f.

7. $f(x) = x(x + 5)$

8. $f(x) = -x^2 + 4x$

9. $f(x) = (3 - x)(x + 1)$

10. $f(x) = (x - 2)(x - 6)$

11. $f(x) = x^2 - 3x + 2$

12. $f(x) = -x^2 + 6x - 5$

13. $f(x) = 4x^2 - 4x + 3$

14. $f(x) = -x^2 + 6x - 10$

15. $f(x) = -\tfrac{1}{2}x^2 + x + 1$

16. $f(x) = x^2 - 2x - 7$

17. $f(x) = x^2 - 10x + 25$

18. $f(x) = -x^2 + 6x - 9$

In Problems 19 and 20, find the maximum or the minimum value of the function f. Give the range of the function f.

19. $f(x) = 3x^2 - 8x + 1$

20. $f(x) = -2x^2 - 6x + 3$

In Problems 21–24, find the largest interval on which the function f is increasing and the largest interval on which f is decreasing.

21. $f(x) = \tfrac{1}{3}x^2 - 25$

22. $f(x) = -(x + 10)^2$

23. $f(x) = -2x^2 - 12x$

24. $f(x) = x^2 + 8x - 1$

CHAPTER 2 FUNCTIONS

In Problems 25–30, describe in words how the graph of the given function can be obtained from the graph of $y = x^2$ by rigid or nonrigid transformations.

25. $f(x) = (x - 10)^2$ **26.** $f(x) = (x + 6)^2$

27. $f(x) = -\frac{1}{3}(x + 4)^2 + 9$ **28.** $f(x) = 10(x - 2)^2 - 1$

29. $f(x) = (-x - 6)^2 - 4$ **30.** $f(x) = -(1 - x)^2 + 1$

In Problems 31–36, the given graph is the graph of $y = x^2$ shifted/reflected in the xy-plane. Give an equation of the graph.

31.

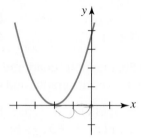

FIGURE 2.4.11 Graph for Problem 31

32.

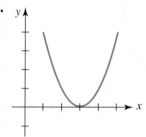

FIGURE 2.4.12 Graph for Problem 32

33.

FIGURE 2.4.13 Graph for Problem 33

34.

FIGURE 2.4.14 Graph for Problem 34

35.

FIGURE 2.4.15 Graph for Problem 35

36.

FIGURE 2.4.16 Graph for Problem 36

In Problems 37 and 38, find a quadratic function $f(x) = ax^2 + bx + c$ that satisfies the given conditions.

37. f has the values $f(0) = 5, f(1) = 10, f(-1) = 4$
38. graph passes through $(2, -1)$, zeros of f are 1 and 3

In Problems 39 and 40, find a quadratic function in standard form $f(x) = a(x - h)^2 + k$ that satisfies the given conditions.

39. the vertex of the graph of f is $(1, 2)$, graph passes through $(2, 6)$
40. the maximum value of f is 10, axis of symmetry is $x = -1$, and y-intercept is $(0, 8)$

In Problems 41–44, sketch the region in the *xy*-plane that is bounded between the graphs of the given functions. Find the points of intersection of the graphs.

41. $y = -x + 4, \quad y = x^2 + 2x$
42. $y = 2x - 2, \quad y = 1 - x^2$
43. $y = x^2 + 2x + 2, \quad y = -x^2 - 2x + 2$
44. $y = x^2 - 6x + 1, \quad y = -x^2 + 2x + 1$

In Problems 45–48, proceed as in Example 7 and use a graph as an aid in solving the given inequality.

45. $-x^2 + 6x < 7$
46. $x^2 + x \leq 1$
47. $x^2 + 2\sqrt{3}x + 3 \leq 0$
48. $x^2 + 2x + 3 \geq 0$

In Problems 49 and 50, graph the quadratic function *f* and the linear function *g*. Find the points of intersection of the graphs and solve the given inequality.

49. $f(x) = x^2 - x - 3, g(x) = x + 5; \quad f(x) \geq g(x)$
50. $f(x) = -x^2 + 2x, g(x) = -3x - 6; \quad g(x) < f(x)$

51. Find the maximum value of $f(x) = -x + 6\sqrt{x} + 10$. [*Hint*: See Problems 19 and 20. Let $t = \sqrt{x}$.]

52. Consider the graphs shown in FIGURE 2.4.17. Find the points on both graphs for $1 \leq x \leq 6$ such that the vertical distance *d* between the graphs, indicated by the black dashed line, is a maximum. What is the maximum vertical distance?

53. **(a)** Express the square of the distance *d* from the point (x, y) on the graph of $y = 2x$ to the point $(5, 0)$ shown in FIGURE 2.4.18 as a function of *x*.
(b) Use the function in part (a) to find the point (x, y) that is closest to $(5, 0)$.

54. As shown in FIGURE 2.4.19, an arrow that is shot at a 45° angle with the horizontal travels along a parabolic arc defined by the equation $y = ax^2 + x + c$. Use the fact that the arrow is launched at a vertical height of 6 ft and travels a horizontal distance of 200 ft to find the coefficients *a* and *c*. What is the maximum height attained by the arrow?

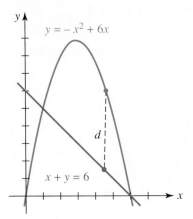

FIGURE 2.4.17 Graphs for Problem 52

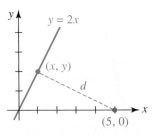

FIGURE 2.4.18 Distance in Problem 53

FIGURE 2.4.19 Arrow in Problem 54

55. An arrow is shot vertically upward with an initial velocity of 64 ft/s from a point 6 ft above the ground. See FIGURE 2.4.20.
(a) Find the height $s(t)$ and the velocity $v(t)$ of the arrow at time $t \geq 0$.
(b) What is the maximum height attained by the arrow? What is the velocity of the arrow at the time the arrow attains its maximum height?
(c) At what time does the arrow fall back to the 6-ft level? What is its velocity at this time?

FIGURE 2.4.20 Arrow in Problem 55

CHAPTER 2 FUNCTIONS

56. The height above ground of a toy rocket launched upward from the top of a building is given by $s(t) = -16t^2 + 96t + 256$.
  (a) What is the height of the building?
  (b) What is the maximum height attained by the rocket?
  (c) Find the time when the rocket strikes the ground.

57. A ball is dropped from the roof of a building that is 122.5 meters above ground level.
  (a) What is the height and velocity of the ball at $t = 1$ s?
  (b) At what time does the ball hit the ground?
  (c) What is the impact velocity of the ball when it hits the ground?

58. A few years ago a newspaper in the Midwest reported that an escape artist was planning to jump off a bridge into the Mississippi River wearing 70 lb of chains and manacles. The newspaper article stated that the height of the bridge was 48 ft and predicted that the escape artist's impact velocity on hitting the water would be 85 mi/h. Assuming that he simply dropped from the bridge, then his height (in feet) and velocity (in feet/second) t seconds after jumping off the bridge are given by the functions $s(t) = -16t^2 + 48$ and $v(t) = -32t$, respectively. Determine whether the newspaper's estimate of his impact velocity was accurate.

Applications

59. Spread of a Disease One model for the spread of a flu virus assumes that within a population of P persons the rate at which a disease spreads is jointly proportional to the number D of persons already carrying the disease and the number $P - D$ of persons not yet infected. Mathematically, the model is given by the quadratic function

$$R(D) = kD(P - D),$$

Spreading a virus

where $R(D)$ is the rate of spread of the flu virus (in cases per day) and $k > 0$ is a constant of proportionality.
  (a) Show that if the population P is a constant, then the disease spreads most rapidly when exactly one-half the population is carrying the flu.
  (b) Suppose that in a town of 10,000 persons, 125 are sick on Sunday, and 37 new cases occur on Monday. Estimate the constant k.
  (c) Use the result of part (b) to estimate the number of new cases on Tuesday. [*Hint*: The number of persons carrying the flu on Monday is $162 = 125 + 37$.]
  (d) Estimate the number of new cases on Wednesday, Thursday, Friday, and Saturday.

For Discussion

60. In Problems 56 and 58, what is the domain of the function $s(t)$? [*Hint*: It is *not* $(-\infty, \infty)$.]

61. On the Moon the acceleration due to gravity is one-sixth the acceleration due to gravity on Earth. If a ball is tossed vertically upward from the surface of the Moon, would it attain a maximum height six times that on Earth when the same initial velocity is used? Defend your answer.

62. Suppose the quadratic function $f(x) = ax^2 + bx + c$ has two distinct real zeros. How would you prove that the x-coordinate of the vertex is the midpoint of the line segment between the x-coordinates of the intercepts? Carry out your ideas.

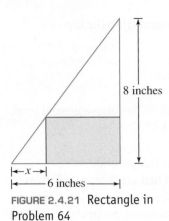

FIGURE 2.4.21 Rectangle in Problem 64

63. Carefully graph the quadratic function $f(x) = x^2 - 4x + 3$. Use a calculator or CAS if necessary.

(a) Simplify $m(x) = \dfrac{f(x) - f(1)}{x - 1}$.

(b) Use $m(x)$ in part (a) to find $m(-5)$ and $m(4)$.

(c) The numbers $m(-5)$ and $m(4)$ in part (b) represent the slopes of two lines. Find equations of these lines and then graph the lines superimposed on the graph of $f(x)$. What are these lines called?

64. In FIGURE 2.4.21 a rectangle is inscribed in a right triangle whose base has length 6 inches and height has length 8 inches. Express the area A of the rectangle as a quadratic function of the variable x shown in the figure and give its domain. Find the value of x for which the area A is a maximum. Discuss: Why does the maximum area of the rectangle occur at the same value of x for all heights $h > 0$ of the right triangle?

2.5 Piecewise-Defined Functions

INTRODUCTION A function f may involve two or more equations, or formulas, with each equation defined on different parts of the domain of f. A function defined in this manner is called a **piecewise-defined function**, or simply, a **piecewise function**. For example,

$$f(x) = \begin{cases} x^2, & x < 0 \\ x + 1, & x \geq 0 \end{cases} \tag{1}$$

is not two functions, but a single function in which the rule of correspondence is given in two pieces. In this case, one piece is used for the negative real numbers ($x < 0$) and the other part on the nonnegative numbers ($x \geq 0$); the domain of f is the union of the intervals $(-\infty, 0) \cup [0, \infty) = (-\infty, \infty)$. For example, since $-4 < 0$, the rule given in (1) indicates that we square the number:

$$f(-4) = (-4)^2 = 16;$$

on the other hand, since $6 \geq 0$ we add 1 to the number:

$$f(6) = 6 + 1 = 7.$$

EXAMPLE 1   Graph of a Piecewise-Defined Function

Graph the piecewise-defined function

$$f(x) = \begin{cases} -1, & x < 0 \\ 0, & x = 0 \\ x + 1, & x > 0. \end{cases} \tag{2}$$

Solution Although the domain of f consists of all real numbers $(-\infty, \infty)$, each piece of the function is defined on a different part of this domain. We draw

- the horizontal line $y = -1$ for $x < 0$,
- the point $(0, 0)$ for $x = 0$, and
- the line $y = x + 1$ for $x > 0$.

The graph is given in FIGURE 2.5.1.

FIGURE 2.5.1 Graph of piecewise-defined function in Example 1

FIGURE 2.5.2 Graph of function g defined in (3)

The solid dot at the origin in Figure 2.5.1 indicates that the function in (2) is defined at $x = 0$ only by $f(0) = 0$; the open dots indicate that the formulas corresponding to $x < 0$ and to $x > 0$ do not define f at $x = 0$. Since we are making up a function, consider the definition:

$$g(x) = \begin{cases} -1, & x \le 0 \\ x + 1, & x > 0. \end{cases} \tag{3}$$

The graph of g shown in FIGURE 2.5.2 is very similar to the graph of (2), but (2) and (3) are not the same function since $f(0) = 0$ but $g(0) = -1$.

☐ **Greatest Integer Function** We consider next a special kind of piecewise-defined function known as a *step function*, that is, a function which is constant on an interval and then jumps to another constant value on the next abutting interval. This new function, which has many notations, will be denoted here by $f(x) = [\![x]\!]$, and is defined by the rule

$$[\![x]\!] = n, \quad \text{where } n \text{ is an integer satisfying } n \le x < n + 1. \tag{4}$$

The function f is called the **greatest integer function** because (4), translated into words, means that:

$f(x)$ is the greatest integer n that is less than or equal to x.

For example,

$f(6) = 6$ since $6 \le x = 6$, $f(-1.5) = -2$ since $-2 \le x = -1.5$,
$f(0.4) = 0$ since $0 \le x = 0.4$, $f(7.6) = 7$ since $7 \le x = 7.6$,
$f(\pi) = 3$ since $3 \le x = \pi$, $f(-\sqrt{2}) = -2$ since $-2 \le x = -\sqrt{2}$,

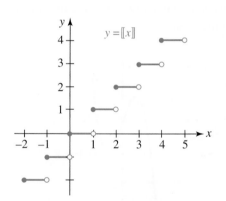

FIGURE 2.5.3 Greatest integer function

and so on. The domain of f is the set of real numbers and consists of the union of an infinite number of disjoint intervals; in other words, $f(x) = [\![x]\!]$ is a piecewise-defined function given by

$$f(x) = [\![x]\!] = \begin{cases} \vdots & \\ -2, & -2 \le x < -1 \\ -1, & -1 \le x < 0 \\ 0, & 0 \le x < 1 \\ 1, & 1 \le x < 2 \\ 2, & 2 \le x < 3 \\ \vdots & \end{cases} \tag{5}$$

The range of f is the set of integers. A portion of the graph of f is given on the closed interval $[-2, 5]$ in FIGURE 2.5.3.

There are several alternative names and notations for the greatest integer function. In some texts $f(x) = [\![x]\!]$ is written $f(x) = \text{int}(x)$. In computer science the greatest integer function is known as the **floor function** and is denoted by $f(x) = \lfloor x \rfloor$. See Problems 49, 50, and 55 in Exercises 2.5.

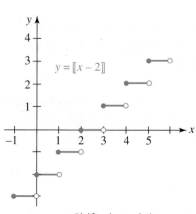

FIGURE 2.5.4 Shifted graph in Example 2

EXAMPLE 2   Shifted Graph

Graph $y = [\![x - 2]\!]$.

Solution The function is $y = f(x - 2)$, where $f(x) = [\![x]\!]$. Thus the graph in Figure 2.5.3 is shifted horizontally 2 units to the right. Note in Figure 2.5.3 that if n is an integer, then $f(n) = [\![n]\!] = n$. But in FIGURE 2.5.4, for $x = n, y = n - 2$. ■

☐ **Continuous Functions** A **continuous function** is one whose graph has no holes, finite gaps, or infinite breaks. In calculus, the formal definition of continuity of a function involves the limit concept first introduced in Section 1.5. If you are curious about how this is done, then read the *Notes from the Classroom* at the end of this section. In this course it suffices to think in informal terms. A continuous function is often characterized by saying that its graph can be drawn "without lifting pencil from paper." When a function is not continuous at a number a we say that it is **discontinuous** at a.

EXAMPLE 3   Discontinuous/Continuous

(a) Each of the functions

$$f(x) = \frac{x^2 - 4}{x - 2} = x + 2, \quad x \neq 2,$$

$$g(x) = \frac{|x - 2|}{x - 2} = \begin{cases} -1, & x < 2 \\ 1, & x > 2, \end{cases}$$

$$h(x) = \frac{1}{x - 2},$$

is discontinuous at the number 2. In FIGURE 2.5.5(a) we see that there is a hole in the graph f at 2, that is, there is no point $(2, f(2))$. In Figure 2.5.5(b) there is a finite gap or jump in the graph of g at 2. The graph of the function h in Figure 2.5.5(c) has an infinite break at 2.

(b) The function $f(x) = x^3 - 3x + 2$ is continuous at every real number; its graph given in Figure 2.5.5(d) has no holes, gaps, or infinite breaks.

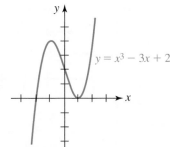

(a) Hole in graph   (b) Finite gap in graph   (c) Infinite break in graph   (d) No holes, gaps, or breaks

FIGURE 2.5.5 Graphs of functions in Example 3

You should be aware that constant functions, linear functions, and quadratic functions are continuous. Piecewise-defined functions can be continuous or discontinuous. The functions given in (2), (3), and (4) are discontinuous.

☐ **Absolute-Value Function** The function $y = |x|$, called the **absolute-value function**, appears frequently in the study of calculus. To obtain the graph, we graph its two pieces consisting of perpendicular half lines:

$$y = |x| = \begin{cases} -x, & \text{if } x < 0 \\ x, & \text{if } x \geq 0. \end{cases} \tag{6}$$

See FIGURE 2.5.6(a). Since $y \geq 0$ for all x, another way of graphing (6) is simply to sketch the line $y = x$ and then reflect in the x-axis that portion of the line that is below the

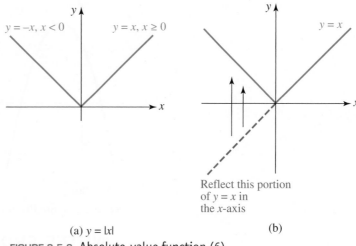

(a) $y = |x|$ (b)

FIGURE 2.5.6 Absolute-value function (6)

x-axis. See Figure 2.5.6(b). The domain of (6) is the set of real numbers $(-\infty, \infty)$, and as is seen in Figure 2.5.6(a), the absolute-value function is an even function, decreasing on the interval $(-\infty, 0)$, increasing on the interval $(0, \infty)$, and is continuous.

In some applications we are interested in the graph of the absolute value of an arbitrary function $y = f(x)$; in other words, $y = |f(x)|$. Since $|f(x)|$ is nonnegative for all numbers x in the domain of f, the graph of $y = |f(x)|$ does not extend below the x-axis. Moreover, the definition of the absolute value of $f(x)$,

$$|f(x)| = \begin{cases} -f(x), & \text{if } f(x) < 0 \\ f(x), & \text{if } f(x) \geq 0, \end{cases} \tag{7}$$

shows that we must negate $f(x)$ whenever $f(x)$ is negative. There is no need to worry about solving the inequalities in (7); to obtain the graph of $y = |f(x)|$, we can proceed just as we did in Figure 2.5.6(b): Carefully draw the graph of $y = f(x)$ and then reflect in the x-axis all portions of the graph that are below the x-axis.

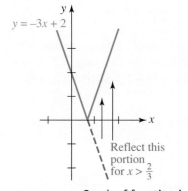

FIGURE 2.5.7 Graph of function in Example 4

EXAMPLE 4   Absolute Value of a Function

Graph $y = |-3x + 2|$.

Solution We first draw the graph of the linear function $f(x) = -3x + 2$. Note that since the slope is negative, f is decreasing and its graph crosses the x-axis at $\left(\frac{2}{3}, 0\right)$. We dash the graph for $x > \frac{2}{3}$ since that portion is below the x-axis. Finally, we reflect that portion upward in the x-axis to obtain the solid blue v-shaped graph in FIGURE 2.5.7. Since $f(x) = x$ is a simple linear function, it is not surprising that the graph of the absolute value of any linear function $f(x) = ax + b$, $a \neq 0$, will result in a graph similar to that of the absolute-value function shown in Figure 2.5.6(a). ∎

EXAMPLE 5   Absolute Value of a Function

Graph $y = |-x^2 + 2x + 3|$.

Solution As in Example 4, we begin by drawing the graph of the function $f(x) = -x^2 + 2x + 3$ by finding its intercepts $(-1, 0)$, $(3, 0)$, $(0, 3)$ and, since f is a quadratic function, its vertex $(1, 4)$. Observe in FIGURE 2.5.8(a) that $y < 0$ for $x < -1$ and for $x > 3$. These portions of the graph of f are reflected in the x-axis to obtain the graph of $y = |-x^2 + 2x + 3|$ shown in red in Figure 2.5.8(b).

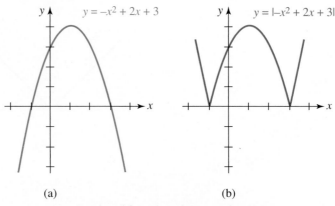

$y = -x^2 + 2x + 3$

$y = |-x^2 + 2x + 3|$

(a)   (b)

FIGURE 2.5.8 Graphs of functions in Example 5

NOTES FROM THE CLASSROOM

The precise definition of continuity of a function is given in terms of the limit concept introduced in Section 1.5. A function f is said to be **continuous** at a number a if

$$(i)\ f(a) \text{ is defined}, \quad (ii)\ \lim_{x \to a} f(x) \text{ exists}, \quad \text{and} \quad (iii)\ \lim_{x \to a} f(x) = f(a). \qquad (8)$$

It is sufficient to say that f is discontinuous, or is not continuous, at a number a if any *one* of the three conditions in (8) fails to be satisfied. For example, in Figure 2.5.5(a), the function is discontinuous at 2 because $f(2)$ is not defined. In other words, 2 is not in the domain of f. But observe that the limit as $x \to 2$ exists, that is, $\lim_{x \to 2} f(x) = 4$. In Figure 2.5.5(b), the function is discontinuous at 2 because $\lim_{x \to 2} f(x)$ does not exist. Note that the one-sided limits exist but are not equal:

$$\lim_{x \to 2^-} f(x) = -1 \quad \text{and} \quad \lim_{x \to 2^+} f(x) = 1.$$

In this case it also happens that $f(2)$ is not defined. In Figure 2.5.5(c), we can conclude that the function is not continuous at 2 either from the fact that $f(2)$ is not defined or from the fact that $\lim_{x \to 2} f(x)$ does not exist. In Figure 2.5.5(d), the function *is* continuous at every real number a because

$$(i)\, f(a) = a^3 - 3a + 2, \quad (ii)\, \lim_{x \to a} f(x) = a^3 - 3a + 2, \text{ and } (iii)\, \lim_{x \to a} f(x) = f(a).$$

Exercises 2.5   Answers to selected odd-numbered problems begin on page ANS–6.

In Problems 1–4, find the indicated values of the given piecewise-defined function f.

1. $f(x) = \begin{cases} \dfrac{x^2 - 4}{x - 2}, & x \neq 2 \\ 4, & x = 2; \end{cases}$ $f(0), f(2), f(-7)$

2. $f(x) = \begin{cases} \dfrac{x^4 - 1}{x^2 - 1}, & x \neq \pm 1 \\ 3, & x = -1 \\ 5, & x = 1; \end{cases}$ $f(-1), f(1), f(3)$

3. $f(x) = \begin{cases} x^2 + 2x, & x \geq 1 \\ -x^3, & x < 1; \end{cases}$ $\quad f(1), f(0), f(-2), f(\sqrt{2})$

4. $f(x) = \begin{cases} 0, & x < 0 \\ x, & 0 < x < 1 \\ x + 1, & x \geq 1; \end{cases}$ $\quad f(-\frac{1}{2}), f(\frac{1}{3}), f(4), f(6.2)$

5. The piecewise-defined function

$$f(x) = \begin{cases} 1, & x \text{ a rational number} \\ 0, & x \text{ an irrational number} \end{cases}$$

is called the **Dirichlet function** after the German mathematician **Johann Peter Gustav Lejeune Dirichlet** (1805–1859). Dirichlet is responsible for the definition of a function as we know it today. Find each of the following function values.

(a) $f(\frac{1}{3})$ **(b)** $f(-1)$ **(c)** $f(\sqrt{2})$
(d) $f(1.\overline{12})$ **(e)** $f(5.72)$ **(f)** $f(\pi)$

6. Suppose f is the Dirichlet function f in Problem 5.
  (a) If x_1 and x_2 are rational numbers, then what is $f(x_1 + x_2)$?
  (b) If x_1 and x_2 are irrational numbers, then what is $f(x_1 + x_2)$?
  (c) If x_1 and x_2 are rational and irrational numbers, respectively, then what is
      $f(x_1 + x_2)$?
  (d) If r is a positive rational number, show that f is r-periodic, that is,
      $f(x + r) = f(x)$ for any real number x.
  (e) Describe what the graph of f looks like. What is the y-intercept of the graph?

7. Determine the values of x for which the piecewise-defined function

$$f(x) = \begin{cases} x^3 + 1, & x < 0 \\ x^2 - 2, & x \geq 0, \end{cases}$$

is equal to the given number.
  (a) 7   **(b)** 0   **(c)** -1
  (d) -2 **(e)** 1   **(f)** -7

8. Determine the values of x for which the piecewise-defined function

$$f(x) = \begin{cases} x + 1, & x < 0 \\ 2, & x = 0 \\ x^2, & x > 0, \end{cases}$$

is equal to the given number.
  (a) 1   **(b)** 0   **(c)** 4
  (d) $\frac{1}{2}$ **(e)** 2   **(f)** -4

In Problems 9–34, sketch the graph of the given piecewise-defined function. Find any x- and y-intercepts of the graph. Give any numbers at which the function is discontinuous.

9. $y = \begin{cases} -x, & x \leq 1 \\ -1, & x > 1 \end{cases}$ **10.** $y = \begin{cases} x - 1, & x < 0 \\ x + 1, & x \geq 0 \end{cases}$

11. $y = \begin{cases} -3, & x < -3 \\ x, & -3 \leq x \leq 3 \\ 3, & x > 3 \end{cases}$ **12.** $y = \begin{cases} -x^2 - 1, & x < 0 \\ 0, & x = 0 \\ x^2 + 1, & x > 0 \end{cases}$

13. $y = [\![x + 2]\!]$ **14.** $y = 2 + [\![x]\!]$
15. $y = -[\![x]\!]$ **16.** $y = [\![-x]\!]$
17. $y = |x + 3|$ **18.** $y = -|x - 4|$

19. $y = 2 - |x|$ **20.** $y = -1 - |x|$

21. $y = -2 + |x + 1|$ **22.** $y = 1 - \frac{1}{2}|x - 2|$

23. $y = -|5 - 3x|$ **24.** $y = |2x - 5|$

25. $y = |x^2 - 1|$ **26.** $y = |4 - x^2|$

27. $y = |x^2 - 2x|$ **28.** $y = |-x^2 - 4x + 5|$

29. $y = ||x| - 2|$ **30.** $y = |\sqrt{x} - 2|$

31. $y = |x^3 - 1|$ **32.** $y = |[\![x]\!]|$

33. $y = \begin{cases} 1, & x < 0 \\ |x - 1|, & 0 \le x \le 2 \\ 1, & x > 2 \end{cases}$ **34.** $y = \begin{cases} -x, & x < 0 \\ 1 - |x - 1|, & 0 \le x \le 2 \\ x - 2, & x > 2 \end{cases}$

In Problems 35 and 36, sketch the graph of the given piecewise-defined function f for $c = -2, -1, 1, 2$.

35. $f(x) = \begin{cases} x^2 + c, & x < 0 \\ -x^2 + c, & x \ge 0 \end{cases}$ **36.** $f(x) = \begin{cases} -(x - c)^2, & x \le c \\ (x - c)^2, & x > c \end{cases}$

37. Without graphing, give the range of the function $f(x) = (-1)^{[\![x]\!]}$.

38. Compare the graphs of $y = 2[\![x]\!]$ and $y = [\![2x]\!]$.

In Problems 39–42, find a piecewise-defined function f whose graph is given. Assume that the domain of f is $(-\infty, \infty)$.

39.

FIGURE 2.5.9 Graph for Problem 39

40.

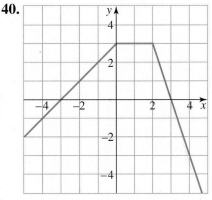

FIGURE 2.5.10 Graph for Problem 40

41.

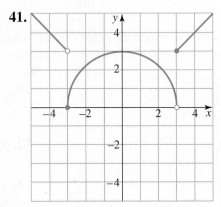

FIGURE 2.5.11 Graph for Problem 41

42.

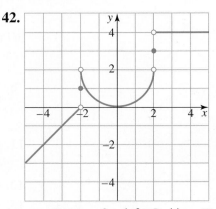

FIGURE 2.5.12 Graph for Problem 42

In Problems 43 and 44, sketch the graph of $y = |f(x)|$.

43. f is the function whose graph is given in Figure 2.5.9.

44. f is the function whose graph is given in Figure 2.5.10.

In Problems 45 and 46, use the definition of absolute value and express the given function f as a piecewise-defined function.

45. $f(x) = \dfrac{|x|}{x}$

46. $f(x) = \dfrac{x - 3}{|x - 3|}$

In Problems 47 and 48, find the value of the constant k such that the given piecewise-defined function f is continuous at $x = 2$. That is, the graph of f has no holes, gaps, or breaks in its graph at $x = 2$.

47. $f(x) = \begin{cases} \frac{1}{2}x + 1, & x \le 2 \\ kx, & x > 2 \end{cases}$

48. $f(x) = \begin{cases} kx + 2, & x < 2 \\ x^2 + 1, & x \ge 2 \end{cases}$

49. The **ceiling function** $g(x) = \lceil x \rceil$ is defined to be the least integer n that is greater than or equal to x. Fill in the blanks.

$$g(x) = \lceil x \rceil = \begin{cases} \vdots & \\ \underline{\hspace{1cm}}, & -3 < x \le -2 \\ \underline{\hspace{1cm}}, & -2 < x \le -1 \\ \underline{\hspace{1cm}}, & -1 < x \le 0 \\ \underline{\hspace{1cm}}, & 0 < x \le 1 \\ \underline{\hspace{1cm}}, & 1 < x \le 2 \\ \underline{\hspace{1cm}}, & 2 < x \le 3 \\ \vdots & \end{cases}$$

50. Graph the ceiling function $g(x) = \lceil x \rceil$ defined in Problem 49.

For Discussion

In Problems 51–54, describe in words how the graphs of the given functions differ. [*Hint*: Factor and cancel.]

51. $f(x) = \dfrac{x^2 - 9}{x - 3}$, $\qquad g(x) = \begin{cases} \dfrac{x^2 - 9}{x - 3}, & x \ne 3 \\ 4, & x = 3, \end{cases}$

$h(x) = \begin{cases} \dfrac{x^2 - 9}{x - 3}, & x \ne 3 \\ 6, & x = 3 \end{cases}$

52. $f(x) = -\dfrac{x^2 - 7x + 6}{x - 1}$, $\qquad g(x) = \begin{cases} -\dfrac{x^2 - 7x + 6}{x - 1}, & x \ne 1 \\ 8, & x = 1, \end{cases}$

$h(x) = \begin{cases} -\dfrac{x^2 - 7x + 6}{x - 1}, & x \ne 1 \\ 5, & x = 1 \end{cases}$

53. $f(x) = \dfrac{x^4 - 1}{x^2 - 1}, \qquad g(x) = \begin{cases} \dfrac{x^4 - 1}{x^2 - 1}, & x \neq 1 \\ 0, & x = 1, \end{cases}$

$h(x) = \begin{cases} \dfrac{x^4 - 1}{x^2 - 1}, & x \neq 1 \\ 2, & x = 1 \end{cases}$

54. $f(x) = \dfrac{x^3 - 8}{x - 2}, \qquad g(x) = \begin{cases} \dfrac{x^3 - 8}{x - 2}, & x \neq 2 \\ 5, & x = 2, \end{cases}$

$h(x) = \begin{cases} \dfrac{x^3 - 8}{x - 2}, & x \neq 2 \\ 12, & x = 2 \end{cases}$

55. Using the notion of a reflection of a graph in an axis, express the ceiling function $g(x) = \lceil x \rceil$ in terms of the floor function $f(x) = \lfloor x \rfloor$ (see page 89).

56. Discuss how to graph the function $y = |x| + |x - 3|$. Carry out your ideas.

57. In the computer algebra system *Mathematica*, frac(x) is the piecewise-defined function

$$\text{frac}(x) = \begin{cases} x - \lfloor x \rfloor, & x \geq 0 \\ x - \lceil x \rceil, & x < 0, \end{cases}$$

where $\lfloor x \rfloor$ is the floor function (see page 89) and $\lceil x \rceil$ is the ceiling function in defined in Problem 49. Find each of the following function values.

(a) $\text{frac}\left(\tfrac{1}{2}\right)$ (b) $\text{frac}(1)$ (c) $\text{frac}\left(\tfrac{5}{4}\right)$

(d) $\text{frac}\left(-\tfrac{1}{3}\right)$ (e) $\text{frac}(-2)$ (f) $\text{frac}\left(-\tfrac{14}{5}\right)$

(g) $\text{frac}(\pi)$ (h) $\text{frac}(\pi^2)$ (i) $\text{frac}(-4\pi)$

58. Graph the function $y = \text{frac}(x)$, defined in Problem 57, on the interval $[-5, 5]$.

2.6 Combining Functions

INTRODUCTION Two functions f and g can be combined in several ways to create new functions. In this section we will examine two such ways in which functions can be combined: through arithmetic operations, and through the operation of function composition.

▢ **Arithmetic Combinations** Two functions can be combined through the familiar four arithmetic operations of addition, subtraction, multiplication, and division.

DEFINITION 2.6.1   Arithmetic Combinations

If f and g are two functions, then the **sum** $f + g$, the **difference** $f - g$, the **product** fg, and the **quotient** f/g are defined as follows:

$$(f + g)(x) = f(x) + g(x) \qquad (1)$$
$$(f - g)(x) = f(x) - g(x) \qquad (2)$$
$$(fg)(x) = f(x)g(x) \qquad (3)$$
$$\left(\frac{f}{g}\right)(x) = \frac{f(x)}{g(x)}, \quad \text{provided } g(x) \neq 0 \qquad (4)$$

EXAMPLE 1   **Sum, Difference, Product, and Quotient**

Consider the functions $f(x) = x^2 + 4x$ and $g(x) = x^2 - 9$. From (1)–(4) we can produce four new functions:

$$(f + g)(x) = f(x) + g(x) = (x^2 + 4x) + (x^2 - 9) = 2x^2 + 4x - 9,$$
$$(f - g)(x) = f(x) - g(x) = (x^2 + 4x) - (x^2 - 9) = 4x + 9,$$
$$(fg)(x) = f(x)g(x) = (x^2 + 4x)(x^2 - 9) = x^4 + 4x^3 - 9x^2 - 36x,$$

and
$$\left(\frac{f}{g}\right)(x) = \frac{f(x)}{g(x)} = \frac{x^2 + 4x}{x^2 - 9}.$$

☐ **Domain of an Arithmetic Combination** When combining two functions arithmetically it is necessary that both f and g be defined at a same number x. Hence the **domain** of the functions $f + g$, $f - g$, and fg is the set of real numbers that are *common* to both domains, that is, the domain is the *intersection* of the domain of f with the domain of g. In the case of the quotient f/g, the domain is also the intersection of the two domains, *but* we must also exclude any values of x for which the denominator $g(x)$ is zero. In Example 1 the domain of f and the domain of g is the set of real numbers $(-\infty, \infty)$, and so the domain of $f + g$, $f - g$, and fg is also $(-\infty, \infty)$. However, since $g(-3) = 0$ and $g(3) = 0$, the domain of the quotient $(f/g)(x)$ is $(-\infty, \infty)$ with $x = 3$ and $x = -3$ excluded, in other words, $(-\infty, -3) \cup (-3, 3) \cup (3, \infty)$. In summary, if the domain of f is the set X_1 and the domain of g is the set X_2, then:

- the domain of $f + g$, $f - g$, and fg is $X_1 \cap X_2$, and
- the domain of f/g is the set $\{x | x \in X_1 \cap X_2, g(x) \neq 0\}$.

EXAMPLE 2   **Domain of $f + g$**

By solving the inequality $1 - x \geq 0$, it is seen that the domain of $f(x) = \sqrt{1 - x}$ is the interval $(-\infty, 1]$. Similarly, the domain of the function $g(x) = \sqrt{x + 2}$ is the interval $[-2, \infty)$. Hence, the domain of the sum

$$(f + g)(x) = f(x) + g(x) = \sqrt{1 - x} + \sqrt{x + 2}$$

is the intersection $(-\infty, 1] \cap [-2, \infty)$. You should verify this result by sketching these intervals on the number line that this intersection, or the set of numbers common to both domains, is the closed interval $[-2, 1]$.

☐ **Composition of Functions** Another method of combining functions f and g is called **function composition**. To illustrate the idea, let's suppose that for a given x in the domain of g the function value $g(x)$ is a number in the domain of the function f. This means we are able to evaluate f at $g(x)$, in other words, $f(g(x))$. For example, suppose $f(x) = x^2$ and $g(x) = x + 2$. Then for $x = 1$, $g(1) = 3$, and since 3 is the domain of f, we can write $f(g(1)) = f(3) = 3^2 = 9$. Indeed, for these two particular functions it turns out that we can evaluate f at any function value $g(x)$, that is,

$$f(g(x)) = f(x + 2) = (x + 2)^2.$$

The resulting function, called the composition of f and g, is defined next.

DEFINITION 2.6.2 Function Composition

If f and g are two functions, then the **composition** of f and g, denoted by $f \circ g$, is the function defined by

$$(f \circ g)(x) = f(g(x)) \tag{5}$$

The **composition** of g and f, denoted by $g \circ f$, is the function defined by

$$(g \circ f)(x) = g(f(x)) \tag{6}$$

When computing a composition such as $(f \circ g)(x) = f(g(x))$, be sure to substitute $g(x)$ for every x that appears in $f(x)$. See part (a) of the next example.

■ EXAMPLE 3 Two Compositions

If $f(x) = x^2 + 3x - 1$ and $g(x) = 2x^2 + 1$, find **(a)** $(f \circ g)(x)$ and **(b)** $(g \circ f)(x)$.

Solution (a) For emphasis we replace x by the set of parentheses ( ) and write f in the form

$$f(\ ) = (\ )^2 + 3(\ ) - 1.$$

Thus to evaluate $(f \circ g)(x)$ we fill each set of parentheses with $g(x)$. We find

$$
\begin{aligned}
(f \circ g)(x) = f(g(x)) &= f(2x^2 + 1) \\
&= (2x^2 + 1)^2 + 3(2x^2 + 1) - 1 \qquad \leftarrow \begin{cases} \text{use } (a+b)^2 = a^2 + 2ab + b^2 \\ \text{and the distributive law} \end{cases} \\
&= 4x^4 + 4x^2 + 1 + 3 \cdot 2x^2 + 3 \cdot 1 - 1 \\
&= 4x^4 + 10x^2 + 3.
\end{aligned}
$$

(b) In this case write g in the form

$$g(\ ) = 2(\ )^2 + 1.$$

Then

$$
\begin{aligned}
(g \circ f)(x) = g(f(x)) &= g(x^2 + 3x - 1) \\
&= 2(x^2 + 3x - 1)^2 + 1 \quad \leftarrow \text{use } (a+b+c)^2 = ((a+b)+c)^2 \\
&= 2(x^4 + 6x^3 + 7x^2 - 6x + 1) + 1 \qquad = (a+b)^2 + 2(a+b)c + c^2 \text{ etc.} \\
&= 2 \cdot x^4 + 2 \cdot 6x^3 + 2 \cdot 7x^2 - 2 \cdot 6x + 2 \cdot 1 + 1 \\
&= 2x^4 + 12x^3 + 14x^2 - 12x + 3. \qquad \blacksquare
\end{aligned}
$$

Parts (a) and (b) of Example 3 illustrate that function composition is not commutative. That is, in general

$$f \circ g \neq g \circ f.$$

The next example shows that a function can be composed with itself.

■ EXAMPLE 4 f Composed with f

If $f(x) = 5x - 1$, then the composition $f \circ f$ is given by

$$(f \circ f)(x) = f(f(x)) = f(5x - 1) = 5(5x - 1) - 1 = 25x - 6. \qquad \blacksquare$$

■ EXAMPLE 5 Writing a Function as a Composition

Express $F(x) = \sqrt{6x^3 + 8}$ as the composition of two functions f and g.

Solution If we define f and g as $f(x) = \sqrt{x}$ and $g(x) = 6x^3 + 8$, then

$$F(x) = (f \circ g)(x) = f(g(x)) = f(6x^3 + 8) = \sqrt{6x^3 + 8}. \qquad \blacksquare$$

There are other solutions to Example 5. For instance, if the functions f and g are defined by $f(x) = \sqrt{6x + 8}$ and $g(x) = x^3$, then observe $(f \circ g)(x) = f(x^3) = \sqrt{6x^3 + 8}$.

☐ **Domain of a Composition** As stated in the introductory example to this discussion, to evaluate the composition $(f \circ g)(x) = f(g(x))$ the number $g(x)$ must be in the domain of f. For example, the domain $f(x) = \sqrt{x}$ is $x \geq 0$, and the domain of $g(x) = x - 2$ is the set of real numbers $(-\infty, \infty)$. Observe that we cannot evaluate $f(g(1))$ because $g(1) = -1$ and -1 is not in the domain of f. In order to substitute $g(x)$ into $f(x)$, $g(x)$ must satisfy the inequality that defines the domain of f, namely, $g(x) \geq 0$. This last inequality is the same as $x - 2 \geq 0$ or $x \geq 2$. The domain of the composition $f(g(x)) = \sqrt{g(x)} = \sqrt{x - 2}$ is $[2, \infty)$, which is only a portion of the original domain $(-\infty, \infty)$ of g. In general:

Read this sentence several times. ▶

- The domain of the composition $f \circ g$ consists of the numbers x in the domain of g such that $g(x)$ is in the domain of f.

EXAMPLE 6　　Domain of a Composition

Consider the function $f(x) = \sqrt{x - 3}$. From the requirement that $x - 3 \geq 0$ we see that whatever number x is substituted into f must satisfy $x \geq 3$. Now suppose $g(x) = x^2 + 2$ and we want to evaluate $f(g(x))$. Although the domain of g is the set of all real numbers, in order to substitute $g(x)$ into $f(x)$ we require that x be a number in that domain so that $g(x) \geq 3$. From FIGURE 2.6.1 we see that the last inequality is satisfied whenever $x \leq -1$ or $x \geq 1$. In other words, the domain of the composition

$$f(g(x)) = f(x^2 + 2) = \sqrt{(x^2 + 2) - 3} = \sqrt{x^2 - 1}$$

is the union of intervals $(-\infty, -1] \cup [1, \infty)$.　　　　　■

FIGURE 2.6.1 Domain of $(f \circ g)(x)$ in Example 6

In certain applications a quantity y is given as a function of a variable x which in turn is a function of another variable t. By means of function composition we can express y as a function of t. The next example illustrates the idea; the symbol V plays the part of y and r plays the part of x.

EXAMPLE 7　　Inflating a Balloon

A weather balloon is being inflated with a gas. If the radius of the balloon is increasing at a rate of 5 cm/s, express the volume of the balloon as a function of time t in seconds.

Solution Let's assume that as the balloon is inflated, its shape is that of a sphere. If r denotes the radius of the balloon, then $r(t) = 5t$. Since the volume of a sphere is $V = \frac{4}{3}\pi r^3$, the composition is $(V \circ r)(t) = V(r(t)) = V(5t)$ or

$$V = \frac{4}{3}\pi(5t)^3 = \frac{500}{3}\pi t^3.$$　　　　　■

Courtesy of NASA.

Weather balloon

☐ **Transformations** The rigid and nonrigid transformations that were studied in Section 2.2 are examples of the operations on functions just discussed. For $c > 0$, a constant, the rigid transformations defined by $y = f(x) + c$ and $y = f(x) - c$ are the *sum* and *difference* of the function $f(x)$ and the constant function $g(x) = c$. The nonrigid transformation $y = cf(x)$ is the *product* of $f(x)$ and the constant function $g(x) = c$. The rigid transformations defined by $y = f(x + c)$ and $y = f(x - c)$ are *compositions* of $f(x)$ with the linear functions $g(x) = x + c$ and $g(x) = x - c$, respectively.

In Problems 1–10, find the functions $f + g, f - g, fg,$ and $f/g,$ and give their domains.

1. $f(x) = x^2 + 1, \quad g(x) = 2x^2 - x$ **2.** $f(x) = x^2 - 4, \quad g(x) = x + 3$

3. $f(x) = x, \quad g(x) = \sqrt{x - 1}$ **4.** $f(x) = x - 2, \quad g(x) = \dfrac{1}{x + 8}$

5. $f(x) = 3x^3 - 4x^2 + 5x, \quad g(x) = (1 - x)^2$

6. $f(x) = \dfrac{4}{x - 6}, \quad g(x) = \dfrac{x}{x - 3}$

7. $f(x) = \sqrt{x + 2}, \quad g(x) = \sqrt{5 - 5x}$

8. $f(x) = \dfrac{1}{x^2 - 9}, \quad g(x) = \dfrac{\sqrt{x + 4}}{x}$

9. $f(x) = x^2 - x - 6, \quad g(x) = -x^2 + 9$

10. $f(x) = \dfrac{10}{x}, \quad g(x) = 2 + \dfrac{10}{x}$

11. Fill in the table.

x	0	1	2	3	4
$f(x)$	-1	2	10	8	0
$g(x)$	2	3	0	1	4
$(f \circ g)(x)$					

12. Fill in the table where g is an odd function.

x	0	1	2	3	4
$f(x)$	-2	-3	0	-1	-4
$g(x)$	9	7	-6	-5	13
$(g \circ f)(x)$					

In Problems 13–16, find the functions $f \circ g$ and $g \circ f$ and give their domains.

13. $f(x) = x^2 + 1, \quad g(x) = \sqrt{x - 1}$
14. $f(x) = x^2 - x + 5, \quad g(x) = -x + 4$
15. $f(x) = \dfrac{1}{2x - 1}, \quad g(x) = x^2 + 1$
16. $f(x) = \dfrac{x + 1}{x}, \quad g(x) = \dfrac{1}{x}$

In Problems 17–22, find the functions $f \circ g$ and $g \circ f.$

17. $f(x) = 2x - 3, \quad g(x) = \frac{1}{2}(x + 3)$ **18.** $f(x) = x - 1, \quad g(x) = x^3$

19. $f(x) = x + \dfrac{1}{x^2}, \quad g(x) = \dfrac{1}{x}$ **20.** $f(x) = \sqrt{x - 4}, \quad g(x) = x^2$

21. $f(x) = x + 1, \quad g(x) = x + \sqrt{x - 1}$ **22.** $f(x) = x^3 - 4, \quad g(x) = \sqrt[3]{x + 3}$

In Problems 23–26, find $f \circ f$ and $f \circ (1/f)$.

23. $f(x) = 2x + 6$

24. $f(x) = x^2 + 1$

25. $f(x) = \dfrac{1}{x^2}$

26. $f(x) = \dfrac{x + 4}{x}$

In Problems 27 and 28, find $(f \circ g \circ h)(x) = f(g(h(x)))$.

27. $f(x) = \sqrt{x}, \quad g(x) = x^2, \quad h(x) = x - 1$

28. $f(x) = x^2, \quad g(x) = x^2 + 3x, \quad h(x) = 2x$

29. For the functions $f(x) = 2x + 7$, $g(x) = 3x^2$, find $(f \circ g \circ g)(x)$.

30. For the functions $f(x) = -x + 5$, $g(x) = -4x^2 + x$, find $(f \circ g \circ f)(x)$.

In Problems 31 and 32, find $(f \circ f \circ f)(x) = f(f(f(x)))$.

31. $f(x) = 2x - 5$

32. $f(x) = x^2 - 1$

In Problems 33–36, find functions f and g such that $F(x) = f \circ g$.

33. $F(x) = (x^2 - 4x)^5$

34. $F(x) = \sqrt{9x^2 + 16}$

35. $F(x) = (x - 3)^2 + 4\sqrt{x - 3}$

36. $F(x) = 1 + |2x + 9|$

In Problems 37 and 38, sketch the graphs of the compositions $f \circ g$ and $g \circ f$.

37. $f(x) = |x| - 2, \quad g(x) = |x - 2|$ **38.** $f(x) = [\![x - 1]\!], \quad g(x) = |x|$

39. Consider the function $y = f(x) + g(x)$, where $f(x) = x$ and $g(x) = -[\![x]\!]$. Fill in the blanks and then sketch the graph of the sum $f + g$ on the indicated intervals.

$$y = \begin{cases} \;\;\vdots \\ \underline{\hspace{1.5em}}, & -3 \le x < -2 \\ \underline{\hspace{1.5em}}, & -2 \le x < -1 \\ \underline{\hspace{1.5em}}, & -1 \le x < 0 \\ \underline{\hspace{1.5em}}, & 0 \le x < 1 \\ \underline{\hspace{1.5em}}, & 1 \le x < 2 \\ \underline{\hspace{1.5em}}, & 2 \le x < 3 \\ \;\;\vdots \end{cases}$$

40. Consider the function $y = f(x) + g(x)$, where $f(x) = |x|$ and $g(x) = [\![x]\!]$. Proceed as in Problem 39 and then sketch the graph of the sum $f + g$.

In Problems 41 and 42, sketch the graph of the sum $f + g$.

41. $f(x) = |x - 1|, \quad g(x) = |x|$ **42.** $f(x) = x, \quad g(x) = |x|$

In Problems 43 and 44, sketch the graph of the product fg.

43. $f(x) = x, \quad g(x) = |x|$ **44.** $f(x) = x, \quad g(x) = [\![x]\!]$

In Problems 45 and 46, sketch the graph of the reciprocal $1/f$.

45. $f(x) = |x|$

46. $f(x) = x - 3$

2.6 Combining Functions

Calculus-Related Problems

In Problems 47 and 48,

(a) find the points of intersection of the graphs of the given functions,

(b) find the vertical distance d between the graphs on the interval I determined by the x-coordinates of their points of intersection,

(c) use the concept of a vertex of a parabola to find the maximum value of d on the interval I.

47.

FIGURE 2.6.2 Graph for Problem 47

48.

FIGURE 2.6.3 Graph for Problem 48

Applications

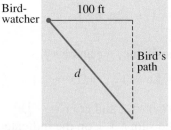

FIGURE 2.6.4 Birdwatcher in Problem 49

49. For the Birds A birdwatcher sights a bird 100 ft due east of her position. If the bird is flying due south at a rate of 500 ft/min, express the distance d from the birdwatcher to the bird as a function of time t. Find the distance 5 minutes after the sighting. See FIGURE 2.6.4.

50. Bacteria Growth A certain bacteria when cultured grows in a circular shape. The radius of the circle, measured in centimeters, is given by

$$r(t) = 4 - \frac{4}{t^2 + 1},$$

where time t is measured in hours.

(a) Express the area covered by the bacteria as a function of time t.

(b) Express the circumference of the area covered as a function of time t.

For Discussion

51. Suppose $f(x) = x^2 + 1$ and $g(x) = \sqrt{x}$. Discuss: Why is the domain of

$$(f \circ g)(x) = f(g(x)) = (\sqrt{x})^2 + 1 = x + 1$$

not $(-\infty, \infty)$?

52. Suppose $f(x) = \dfrac{2}{x - 1}$ and $g(x) = \dfrac{5}{x + 3}$. Discuss: Why is the domain of

$$(f \circ g)(x) = f(g(x)) = \frac{2}{g(x) - 1} = \frac{2}{\dfrac{5}{x + 3} - 1} = \frac{2x + 6}{2 - x}$$

not the set $\{x \mid x \neq 2\}$?

53. Find the error in the following reasoning: If $f(x) = 1/(x - 2)$ and $g(x) = 1/\sqrt{x + 1}$, then

$$\left(\frac{f}{g}\right)(x) = \frac{1/(x - 2)}{1/\sqrt{x + 1}} = \frac{\sqrt{x + 1}}{x - 2} \qquad \text{and so} \qquad \left(\frac{f}{g}\right)(-1) = \frac{\sqrt{0}}{-3} = 0.$$

54. Suppose $f_1(x) = \sqrt{x + 2}, f_2(x) = \dfrac{x}{\sqrt{x(x - 10)}}$, and $f_3(x) = \dfrac{x + 1}{x}$. What is the domain of the function $y = f_1(x) + f_2(x) + f_3(x)$?

55. Suppose $f(x) = x^3 + 4x$, $g(x) = x - 2$, and $h(x) = -x$. Discuss: Without actually graphing, how are the graphs of $f \circ g$, $g \circ f$, $f \circ h$, and $h \circ f$ related to the graph of f?

56. The domain of each piecewise-defined function,

$$f(x) = \begin{cases} x, & x < 0 \\ x + 1, & x \geq 0, \end{cases}$$
$$g(x) = \begin{cases} x^2, & x \leq -1 \\ x - 2, & x > -1, \end{cases}$$

is $(-\infty, \infty)$. Discuss how to find $f + g$, $f - g$, and fg. Carry out your ideas.

57. Discuss how the graph of $y = \frac{1}{2}\{f(x) + |f(x)|\}$ is related to the graph of $y = f(x)$. Illustrate your ideas using $f(x) = x^2 - 6x + 5$.

58. Discuss:
   (a) Is the sum of two even functions f and g even?
   (b) Is the sum of two odd functions f and g odd?
   (c) Is the product of an even function f with an odd function g even, odd, or neither?
   (d) Is the product of an odd function f with an odd function g even, odd, or neither?

59. The product fg of two linear functions with real coefficients, $f(x) = ax + b$ and $g(x) = cx + d$, is a quadratic function. Discuss: Why must the graph of this quadratic function have at least one x-intercept?

60. Make up two different functions f and g so that the domain of $F(x) = f \circ g$ is $[-2, 0) \cup (0, 2]$.

61. Suppose $y = f(x)$ is a function with domain the set X. Discuss: Is the domain of the composition $f \circ f$ also X?

62. The **Heaviside function**

$$U(x - a) = \begin{cases} 0, & x < a \\ 1, & x \geq a, \end{cases}$$

is frequently combined with other functions by either addition or multiplication. Given that $f(x) = x^2$, compare the graphs of $y = f(x - 3)$ and $y = f(x - 3)\,U(x - 3)$.

63. If U is the Heaviside function defined in Problem 62, then sketch the following functions.
   (a) $y = 2U(x - 1) + U(x - 2)$ **(b)** $y = U\left(x + \frac{1}{2}\right) - U\left(x - \frac{1}{2}\right)$

64. Find an equation for the function f illustrated in FIGURE 2.6.5 in terms of the Heaviside function $U(x - a)$. [*Hint*: Think addition.]

65. Use the identity

$$f(x) = \frac{1}{2}[f(x) + f(-x)] + \frac{1}{2}[f(x) - f(-x)]$$

to show that every function f can be written as the sum of an even function g and an odd function h.

66. A square with a side of length x is inscribed in a circle of radius r. See FIGURE 2.6.6. If $A(x)$ is the area of the square, find $x(r)$ and $(A \circ x)(r)$.

FIGURE 2.6.5 Graph for Problem 64

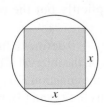

FIGURE 2.6.6 Inscribed square in Problem 66

2.7 Functions Defined Implicitly

INTRODUCTION The graphs of many equations that we study in mathematics are not the graphs of functions. For example, the equation

$$x^2 + y^2 = 9 \qquad (1)$$

describes a circle of radius 3 centered at the origin. Equation (1) is not a function because for any choice of x satisfying $-3 < x < 3$ there corresponds two values of y. See FIGURE 2.7.1(a). However, equation (1) defines *at least* two functions f and g on the interval $[-3, 3]$. To obtain these functions we solve the equation $x^2 + y^2 = 9$ for y in terms of x. From $y = \pm\sqrt{9 - x^2}$ we obtain

$$y = f(x) = \sqrt{9 - x^2} \qquad (2)$$

and
$$y = g(x) = -\sqrt{9 - x^2}. \qquad (3)$$

The graphs of equations (2) and (3) are, in turn, the *upper semicircle* and the *lower semicircle* shown in Figures 2.7.1(b) and 2.7.1(c).

(a) Not a function (b) Function (c) Function

FIGURE 2.7.1 Not a function (a); functions (b) and (c)

Functions Defined Explicitly and Implicitly An equation of the form $y = f(x)$ is said to define a function **explicitly**, or is an **explicit function**, because the variable y is expressed directly in terms of the variable x by means of a formula $f(x)$ which gives a single value of y for each appropriate value of x. All the functions that we have studied in the earlier sections of this chapter were defined explicitly. In contrast, an equation of the form

$$F(x, y) = 0 \qquad (4)$$

may define one or more functions **implicitly**. It may be possible to determine explicit functions by solving $F(x, y) = 0$ for y in terms of x. For example, the explicit functions given in (2) and (3) are said to be defined implicitly by the equation

$$\overbrace{x^2 + y^2 - 9}^{F(x,\, y)} = 0.$$

As another example: $y = \frac{1}{2}x^3 - 1$ defines a function explicitly but the equation $2y - x^3 + 2 = 0$ defines the same function implicitly.

In general, if an equation involving two variables $F(x, y) = 0$ defines a function f implicitly on some interval, then $F(x, f(x)) = 0$ is an identity on the interval. The graph of f is either a portion (that is, one or more arcs) of the graph of the equation $F(x, y) = 0$, or all of the graph of this equation. In the case of the functions in (2) and (3), note that both equations

$$x^2 + [f(x)]^2 - 9 = 0 \quad \text{and} \quad x^2 + [g(x)]^2 - 9 = 0$$

$y = \sqrt{9 - x^2}, \, [-3, 0]$

FIGURE 2.7.2 Another function defined implicitly by (1)

See Example 5 in Section 2.2. ▶

are identities on the interval $[-3, 3]$. The equation $x^2 + y^2 = 9$ actually defines many more functions implicitly. For example, the graph of the function $y = h(x) = \sqrt{9 - x^2}$ with domain $[-3, 0]$ is one quarter of the circle in Figure 2.7.1(a) and $x^2 + [h(x)]^2 - 9 = 0$ on the interval $[-3, 0]$. See **FIGURE 2.7.2** and Problem 36 in Exercises 2.7.

■ EXAMPLE 1   Functions Defined Implicitly

Find functions defined implicitly by the equation $1 - x - (y - 1)^2 = 0$.

Solution By isolating the term containing the variable y on one side of the equation, we see that

$$(y - 1)^2 = 1 - x \text{ implies } y = 1 \pm \sqrt{1 - x}.$$

From the last equation we obtain two explicit functions:

$$y = f(x) = 1 + \sqrt{1 - x} \tag{5}$$

and
$$y = g(x) = 1 - \sqrt{1 - x}. \tag{6}$$

The domain of each function is $(-\infty, 1]$. The graphs of both f and g can be obtained using vertical and horizontal shifts and a reflection of the graph $y = \sqrt{x}$. The graphs of f and g are shown, respectively, in parts (a) and (b) of **FIGURE 2.7.3**. If the graphs of f and g are plotted on the same rectangular coordinate system, we then obtain the complete graph of the original equation $1 - x - (y - 1)^2 = 0$. See Figure 2.7.3(c).

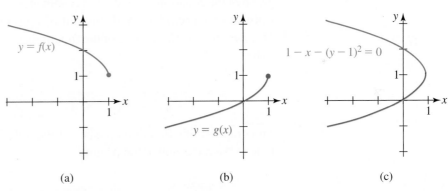

(a)   (b)   (c)

FIGURE 2.7.3 Two functions in Example 1

■ EXAMPLE 2   Functions Defined Implicitly

Find functions defined implicitly by the equation $x^2 - 2xy + 2y^2 - 2 = 0$.

Solution The algebra in this case is not as straightforward as in Example 1. But if we write the given equation as

$$y^2 + (-x)y + (\tfrac{1}{2}x^2 - 1) = 0,$$

we can then identify it as quadratic equation of the form $ay^2 + by + c = 0$ with $a = 1$, $b = -x$, and $c = \tfrac{1}{2}x^2 - 1$. The quadratic formula then gives $y = \tfrac{1}{2}(x \pm \sqrt{4 - x^2})$. Thus we get two explicit functions

$$y = f(x) = \tfrac{1}{2}(x + \sqrt{4 - x^2}) \tag{7}$$

and
$$y = g(x) = \tfrac{1}{2}(x - \sqrt{4 - x^2}). \tag{8}$$

The domain of each function is the interval $[-2, 2]$. With the aid of a graphing utility we get the graphs shown in parts (a) and (b) of FIGURE 2.7.4. As in Example 1, by plotting these two graphs on the same coordinate system we obtain the complete graph of $x^2 - 2xy + 2y^2 = 2$ given in Figure 2.7.4(c).

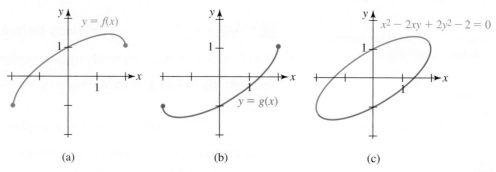

(a)   (b)   (c)

FIGURE 2.7.4 Two functions in Example 2

☐ **Algebra** Do not jump to the conclusion from the preceding examples that an equation $F(x, y) = 0$ can readily be solved for y. For example, it is known that the equation

$$x^3 + y^3 = 3xy \qquad (9)$$

defines several functions implicitly, but very few instructors of mathematics (including the authors) can solve equation (9) for y without looking up a specialized formula. The graph of the equation $x^3 + y^3 = 3xy$ shown in FIGURE 2.7.5(a) is a famous curve called the **Folium of Descartes**. With the aid of a computer algebra system (CAS) such as *Mathematica* or *Maple,* one of the functions $y = f(x)$ defined implicitly by (9) is found to be

$$y = \frac{2x}{\sqrt[3]{-4x^3 + 4\sqrt{x^6 - 4x^3}}} + \frac{1}{2}\sqrt[3]{-4x^3 + 4\sqrt{x^6 - 4x^3}}, \quad x \neq 0. \qquad (10)$$

The graph of this function consists of the red arcs shown in Figure 2.7.5(b). Note the hole in the red graph at the origin.

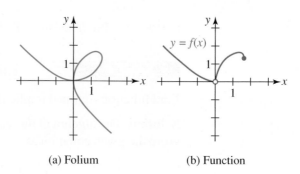

(a) Folium   (b) Function

FIGURE 2.7.5 Graph of (9) in (a); Graph of (10) in (b)

One last point: solving an equation $F(x, y) = 0$ for y in terms of x is often more than just an exercise in challenging algebra or a lesson in the use of the correct syntax of a CAS. Sometimes it is *impossible*! For example, the equation

$$x^4 + x^2y^3 - y^5 - 2x - y = 0 \qquad (11)$$

cannot be solved for y.

(i) If we can solve an equation $F(x, y) = 0$ for y in terms of x and determine a function $y = f(x)$, then, as discussed above, the graph of f is a portion, or perhaps, all of the graph of $F(x, y) = 0$. Stated in a different manner, this means that the graph of f cannot contain points that are *not* on the graph of $F(x, y) = 0$. Bear this in mind when you work Problems 17 and 18 in Exercises 2.7.

(ii) We will see in the next section, as well as in Sections 4.7 and 5.2, the notion of an implicitly defined function plays an important role in defining the inverse of a certain kind of function.

Exercises 2.7   Answers to selected odd-numbered problems begin on page ANS–8.

In Problems 1–10, find two functions defined implicitly by the given equation. Graph each function.

1. $4x^2 - y^2 = 0$
2. $2y^2 - xy = 0$
3. $x + y^2 = 5$
4. $(y - 1)^2 = 4(x + 2)$
5. $(x + 1)^2 + y^2 = 1$
6. $x^2 + (y - 3)^2 = 9$
7. $(y - 1)(y - x^3) = 0$
8. $x + y^2 = 0$
9. $y^2 + 4y + x = 0$
10. $(y - x)^2 - y + x = 0$

In Problems 11–18, find a single function defined implicitly by the given equation. Graph the function and give its domain. Use a graphing utility if necessary.

11. $y + x = xy$
12. $xy + 2y + 3 = x$
13. $3x^2 + 4xy = 16$
14. $x + 2y - x^2 + xy = 0$
15. $x + 1 - y^3 = 0$
16. $\sqrt[3]{y} - x + 3 = 0$
17. $\sqrt{x} + \sqrt{y} = 1$
18. $-x + \sqrt{y} + 2 = 0$

In Problems 19 and 20, find at least two functions defined implicitly by the given equation. Graph each function and give its domain.

19. $|y| = x$
20. $|x| + |y| = 4$

In Problems 21–30, find at least two functions defined implicitly by the given equation. Use a graphing utility to obtain the graph of each function and give its domain.

21. $x^2 - y^2 = 4$
22. $4y^2 - 9x^2 - 36 = 0$
23. $x^2 + xy + y^2 = 3$
24. $x^2 - 2xy + y^2 = x + 2$
25. $x^2(x^2 + y^2) = y^2$
26. $y^2 + x^3 = 2x^2$
27. $(y^2 - x^2)^2 = x$
28. $x^2 + 4xy - 2y^2 = 6$
29. $x^4 + y^4 = 16$
30. $x^{2/3} + y^{2/3} = 1$

In Problems 31 and 32, find two functions defined implicitly by the given equation. Use the sign-chart method in Section 1.1 to find the domain of each function. Use a graphing utility to obtain the graph of each function.

31. $y^2 = x(x^2 - 3)$
32. $(x + 2)y^2 = x - 1$

Calculator/Computer Problems

33. If $y = f(x)$ is the function given in (10), use a CAS to carry out the algebra to verify the identity:

$$x^3 - 3xf(x) + [f(x)]^3 = 0$$

for every x in the domain of f.

34. The graph of the equation

$$(x^2 + y^2 - 2x)^2 = x^2 + y^2$$

is given in FIGURE 2.7.6.

(a) Use a CAS, or straightforward algebra, to find four functions defined implicitly by this equation. By either method, it might help to start with the substitution $Y = y^2$.

(b) Use a graphing utility to obtain the graphs of these four functions.

(c) Give the domain of each of the functions found in part (a).

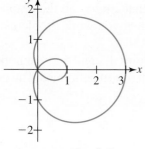

FIGURE 2.7.6 Graph for Problem 34

For Discussion

35. Give two piecewise-defined functions f and g, with the same domain, defined implicitly by the equation in Problem 31.

36. Give three piecewise-defined functions f, g, and h, with domain $[-3, 3]$, defined implicitly by the equation $x^2 + y^2 = 9$.

2.8 Inverse Functions

INTRODUCTION Recall that a function f is a rule of correspondence that assigns to each value x in its domain X, a single or unique value y in its range. This rule does not preclude having the same number y associated with several *different* values of x. For example, for $f(x) = x^2 + 1$, the value $y = 5$ occurs at either $x = -2$ or $x = 2$. On the other hand, for the function $g(x) = x^3$, the value $y = 64$ occurs only at $x = 4$. Indeed, for every value y in the range of $g(x) = x^3$, there corresponds only one value of x in the domain. Functions of this last kind are given a special name.

DEFINITION 2.8.1 One-to-One Function

A function f is said to be **one-to-one** if each number in the range of f is associated with exactly one number in its domain X.

☐ **Horizontal Line Test** Interpreted geometrically, this means that a horizontal line ($y = $ constant) can intersect the graph of a one-to-one function in at most one point. Furthermore, if *every* horizontal line that intersects the graph of a function does so in at most one point, then the function is necessarily one-to-one. A function is *not* one-to-one if *some* horizontal line intersects its graph more than once.

    Horizontal Line Test

(a) The graph of the function $f(x) = x^2 + 1$ and a horizontal line $y = c$, $c > 1$, that intersects its graph are given in FIGURE 2.8.1. The figure indicates that there are two numbers x_1 and x_2 in the domain of f for which $f(x_1) = f(x_2) = c$. Hence the function f is not one-to-one.

(b) Inspection of FIGURE 2.8.2 shows that every horizontal line $y = c$ intersects the graph of the function $f(x) = x^3$. But for any given value of c there is only one number x_1 in the domain of f such that $f(x_1) = c$. Hence the function f is one-to-one.

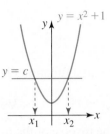

FIGURE 2.8.1 Function f is not one-to-one

FIGURE 2.8.2 Function f is one-to-one

A one-to-one function can be defined in several different ways. Based on the preceding discussion, the following statement should make sense.

> *A function f is one-to-one if and only if $f(x_1) = f(x_2)$ implies $x_1 = x_2$ for all x_1 and x_2 in the domain of f.* (1)

Stated in a negative way, (1) indicates that a function f is *not* one-to-one if different numbers x_1 and x_2 (that is, $x_1 \neq x_2$) can be found in the domain of f such that $f(x_1) = f(x_2)$. You will see this formulation of the one-to-one concept when we solve certain kinds of equations in Chapter 5.

Consider (1) as a way of determining whether a function f is one-to-one without the benefit of a graph.

■ EXAMPLE 2   Checking for One-to-One

(a) Consider the function $f(x) = x^4 - 8x + 6$. Now $0 \neq 2$ but observe that $f(0) = f(2) = 6$. Therefore f is not one-to-one.

(b) Consider the function $f(x) = \dfrac{1}{2x - 3}$, and let x_1 and x_2 be numbers in the domain of f. If we assume $f(x_1) = f(x_2)$, that is, $\dfrac{1}{2x_1 - 3} = \dfrac{1}{2x_2 - 3}$, then by taking the reciprocal of both sides we see

$$2x_1 - 3 = 2x_2 - 3 \quad \text{implies} \quad 2x_1 = 2x_2 \quad \text{or} \quad x_1 = x_2.$$

We conclude from (1) that f is one-to-one. ■

□ **Inverse of a One-to-One Function** Suppose f is a one-to-one function with domain X and range Y. Since every number y in Y corresponds to precisely one number x in X, the function f must actually determine a "reverse" function f^{-1} whose domain is Y and range is X. As shown in FIGURE 2.8.3, f and f^{-1} must satisfy

$$f(x) = y \quad \text{and} \quad f^{-1}(y) = x. \tag{2}$$

The equations in (2) are actually the compositions of the functions f and f^{-1}:

$$f(f^{-1}(y)) = y \quad \text{and} \quad f^{-1}(f(x)) = x. \tag{3}$$

The function f^{-1} is called the **inverse** of f or the **inverse function** for f. Following the convention that each domain element be denoted by the symbol x, the first equation in (3) is rewritten as $f(f^{-1}(x)) = x$. We summarize the results in (3).

Note of Caution: The symbol f^{-1} does *not* ▶ mean the reciprocal $1/f$. The number -1 is **not** an exponent.

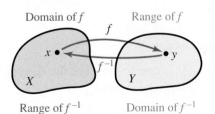

FIGURE 2.8.3 Functions f and f^{-1}

DEFINITION 2.8.2 Inverse Function

Let f be a one-to-one function with domain X and range Y. The **inverse** of f is the function f^{-1} with domain Y and range X for which

$$f(f^{-1}(x)) = x \text{ for every } x \text{ in } Y \qquad (4)$$

and
$$f^{-1}(f(x)) = x \text{ for every } x \text{ in } X \qquad (5)$$

Of course, if a function f is not one-to-one, then it has no inverse function.

EXAMPLE 3   Verifying an Inverse

Verify that the inverse of the one-to-one function $f(x) = \frac{1}{2}x + 7$ is $g(x) = 2x - 14$.

Solution First observe that the domain and range of both functions is the entire set of real numbers $(-\infty, \infty)$. Now we can use (4) and (5).

First from (4) we see that

$$f(g(x)) = f(2x - 14) = \tfrac{1}{2}(2x - 14) + 7 = x - 7 + 7 = x$$

for every real number x. Similarly, from (5)

$$g(f(x)) = g(\tfrac{1}{2}x + 7) = 2(\tfrac{1}{2}x + 7) - 14 = x + 14 - 14 = x$$

for every real number x. This shows that $g = f^{-1}$.   ■

☐ **Properties** Before we actually examine a method for finding the inverse of a one-to-one function f, let's list some important properties about f and its inverse f^{-1}.

THEOREM 2.8.1 Properties of Inverse Functions

   (*i*) The domain of f^{-1} = range of f
  (*ii*) The range of f^{-1} = domain of f
 (*iii*) $y = f(x)$ is equivalent to $x = f^{-1}(y)$
 (*iv*) An inverse function f^{-1} is one-to-one
  (*v*) The inverse of f^{-1} is f, that is, $(f^{-1})^{-1} = f$
 (*vi*) The inverse of f is unique

☐ **Algebraic Method for Finding f^{-1}** If a function f is one-to-one then it must have an inverse. If $y = f^{-1}(x)$ denotes the inverse of f, then we know from (4) that $x = f(f^{-1}(x))$ or $x = f(y)$. In the terminology of Section 2.7, f^{-1} is a function *defined implicitly* by the equation $x = f(y)$. Thus we need to do the following two things to find an explicit function f^{-1}:

- Interchange the variables x and y in $y = f(x)$. This gives $x = f(y)$.   (6)
- Solve $x = f(y)$ for the symbol y in terms of x (if possible). This gives $y = f^{-1}(x)$.

EXAMPLE 4   Inverse of a Function

(a) Find the inverse of $f(x) = \dfrac{1}{2x - 3}$.

(b) Find the domain and range of f^{-1}. Find the range of f.

Solution (a) We proved in part (b) of Example 2 that f is one-to-one. For the algebra that follows it is convenient to replace the symbol $f(x)$ with the symbol y:

$$y = \frac{1}{2x - 3}.$$

We then interchange the symbols x and y:

$$x = \frac{1}{2y - 3} \qquad \leftarrow \text{solve this equation for } y$$

By taking the reciprocal of both sides of the last equation we get,

$$2y - 3 = \frac{1}{x}$$

$$2y = 3 + \frac{1}{x} = \frac{3x + 1}{x}. \qquad \leftarrow \text{common denominator}$$

Dividing both sides of this equation by 2 yields the explicit function

$$y = \frac{3x + 1}{2x}.$$

The last expression is the inverse of f and it is customary to write the inverse function using the symbol $f^{-1}(x)$:

$$f^{-1}(x) = \frac{3x + 1}{2x}.$$

(b) Inspection of the given function f reveals that its domain is the set of real numbers except $\frac{3}{2}$, that is, $\{x \mid x \neq \frac{3}{2}\}$. Moreover, from the inverse just found we see that the domain of f^{-1} is $\{x \mid x \neq 0\}$. Because range of $f^{-1} = $ domain of f we then know that the range of f^{-1} is $\{y \mid y \neq \frac{3}{2}\}$. From domain of $f^{-1} = $ range of f we have also discovered that the range of f is $\{y \mid y \neq 0\}$. ∎

■ EXAMPLE 5   Inverse of a Function

Find the inverse of $f(x) = x^3$.

Solution We saw that the function f was one-to-one in part (b) of Example 1. As in Example 4, we begin by replacing the symbol $f(x)$ with y, that is, write the given function as $y = x^3$. Next we interchange x and y to obtain $x = y^3$. Solving the last equation for y then gives $y = x^{1/3}$. Thus $f^{-1}(x) = x^{1/3}$ or equivalently $f^{-1}(x) = \sqrt[3]{x}$. ∎

Finding the inverse of a one-to-one function $y = f(x)$ is sometimes difficult and at times impossible. For example, the function $f(x) = x^3 + x + 3$ is one-to-one and so we know f has an inverse f^{-1} defined implicitly by the equation $x = f(y)$ or $x = y^3 + y + 3$. But solving this equation for y is difficult for everyone (including your instructor). As seen in FIGURE 2.8.4, the domain and range of f are $(-\infty, \infty)$ and consequently the domain and range of f^{-1} are $(-\infty, \infty)$. Even though we don't know f^{-1} explicitly it makes complete sense to talk about the values such as $f^{-1}(3)$ and $f^{-1}(5)$. In the case of $f^{-1}(3)$ we simply observe that $f(0) = 3$. This means that $f^{-1}(3) = 0$. Can you figure out the value of $f^{-1}(5)$? See Problems 43–48 in Exercises 2.8.

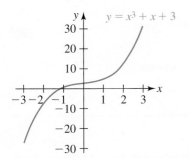

FIGURE 2.8.4 Function is one-to-one

☐ **Graphs of f and f^{-1}** Suppose that (a, b) represents any point on the graph of a one-to-one function f. Then $f(a) = b$ and

$$f^{-1}(b) = f^{-1}(f(a)) = a$$

implies that (b, a) is a point on the graph of f^{-1}. As shown in FIGURE 2.8.5(a), the points (a, b) and (b, a) are reflections of each other in the line $y = x$. This means that the line $y = x$ is the perpendicular bisector of the line segment from (a, b) to (b, a). Because

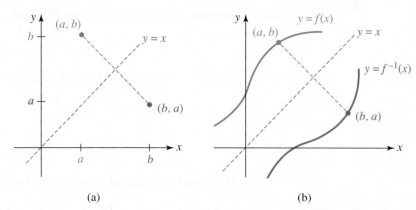

FIGURE 2.8.5 Graphs of f and f^{-1} are reflections in the line $y = x$.

each point on one graph is the reflection of a corresponding point on the other graph, we see in Figure 2.8.5(b) that the graphs of f^{-1} and f are **reflections** of each other in the line $y = x$. We also say that the graphs of f^{-1} and f are symmetric with respect to the line $y = x$.

EXAMPLE 6   Graphs of f and f^{-1}

In Example 5 we saw that the inverse of $y = x^3$ is $y = x^{1/3}$. In FIGURES 2.8.6(a) and 2.8.6(b) we show the graphs of these functions; in Figure 2.8.6(c) the graphs are superimposed on the same coordinate system to illustrate that the graphs are reflections of each other in the line $y = x$.

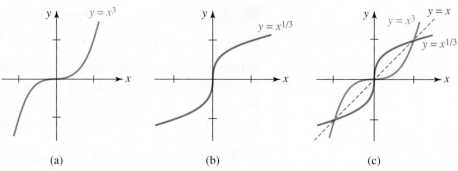

FIGURE 2.8.6 Graphs of f and f^{-1} in Example 6

Every linear function $f(x) = ax + b$, $a \neq 0$, is one-to-one.

EXAMPLE 7   Inverse of a Function

Find the inverse of the linear function $f(x) = 5x - 7$.

Solution Because the graph of $y = 5x - 7$ is a nonhorizontal line, it follows from the horizontal line test that f is a one-to-one function. To find f^{-1} we interchange x and y and solve $x = 5y - 7$ for y:

$$5y = x + 7 \qquad \text{implies} \qquad y = \frac{1}{5}x + \frac{7}{5}.$$

Therefore $f^{-1}(x) = \frac{1}{5}x + \frac{7}{5}$. The graphs of f and f^{-1} are compared in FIGURE 2.8.7.

FIGURE 2.8.7 Graphs of f and f^{-1} in Example 7

Every quadratic function $f(x) = ax^2 + bx + c$, $a \neq 0$, is *not* one-to-one.

☐ **Restricted Domains** For a function f that is not one-to-one, it may be possible to restrict its domain in such a manner so that the new function consisting of f defined on this restricted domain is one-to-one and so has an inverse. In most cases we want to restrict the domain so that the new function retains its original range. The next example illustrates this concept.

■ EXAMPLE 8   Restricted Domain

In Example 1 we showed graphically that the quadratic function $f(x) = x^2 + 1$ is not one-to-one. The domain of f is $(-\infty, \infty)$, and as seen in FIGURE 2.8.8(a), the range of f is $[1, \infty)$. Now by defining $f(x) = x^2 + 1$ only on the interval $[0, \infty)$, we see two things in Figure 2.8.8(b): the range of f is preserved and $f(x) = x^2 + 1$ confined to the domain $[0, \infty)$ passes the horizontal line test, in other words, is one-to-one. The inverse of this new one-to-one function is obtained in the usual manner. Solving $x = y^2 + 1$ implies

$$y^2 = x - 1 \quad \text{and so} \quad y = \pm\sqrt{x - 1}.$$

The appropriate algebraic sign in the last equation is determined from the fact that the domain and range of f^{-1} are $[1, \infty)$ and $[0, \infty)$, respectively. This forces us to choose $f^{-1}(x) = \sqrt{x - 1}$ as the inverse of f. See Figure 2.8.8(c).

$y = x^2 + 1$
on $(-\infty, \infty)$

$y = x^2 + 1$
on $[0, \infty)$

$y = \sqrt{x - 1}$
on $[1, \infty)$

(a) Not a one-to-one function   (b) One-to-one function   (c) Inverse of function in part (b)

FIGURE 2.8.8 Inverse function in Example 8

Exercises 2.8   Answers to selected odd-numbered problems begin on page ANS–9.

In Problems 1–6, the graph of a function f is given. Use the horizontal line test to determine whether f is one-to-one.

1.

FIGURE 2.8.9 Graph for Problem 1

2.

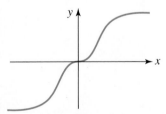

FIGURE 2.8.10 Graph for Problem 2

3.

FIGURE 2.8.11 Graph for Problem 3

4.

FIGURE 2.8.12 Graph for Problem 4

5.

FIGURE 2.8.13 Graph for Problem 5

6.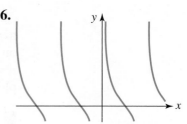

FIGURE 2.8.14 Graph for Problem 6

In Problems 7–10, sketch the graph of the given piecewise-defined function f to determine whether it is one-to-one.

7. $f(x) = \begin{cases} x - 2, & x < 0 \\ \sqrt{x}, & x \geq 0 \end{cases}$

8. $f(x) = \begin{cases} -\sqrt{-x}, & x < 0 \\ \sqrt{x}, & x \geq 0 \end{cases}$

9. $f(x) = \begin{cases} -x - 1, & x < 0 \\ x^2, & x \geq 0 \end{cases}$

10. $f(x) = \begin{cases} x^2 + x, & x < 0 \\ x^2 - x, & x \geq 0 \end{cases}$

In Problems 11–14, proceed as in Example 2(a) to show that the given function f is not one-to-one.

11. $f(x) = x^2 - 6x$

12. $f(x) = (x - 2)(x + 1)$

13. $f(x) = \dfrac{x^2}{4x^2 + 1}$

14. $f(x) = |x + 10|$

In Problems 15–18, proceed as in Example 2(b) to show that the given function f is one-to-one.

15. $f(x) = \dfrac{2}{5x + 8}$

16. $f(x) = \dfrac{2x - 5}{x - 1}$

17. $f(x) = \sqrt{4 - x}$

18. $f(x) = \dfrac{1}{x^3 + 1}$

In Problems 19–24, proceed as in Example 3 and verify that the inverse of the one-to-one function f is the function g by showing $f(g(x)) = x$ and $g(f(x)) = x$.

19. $f(x) = x + 5$; $g(x) = x - 5$

20. $f(x) = 5x - 10$; $g(x) = \frac{1}{5}x + 2$

21. $f(x) = \dfrac{1}{x^3}$; $g(x) = \dfrac{1}{\sqrt[3]{x}}$

22. $f(x) = \sqrt[3]{\frac{1}{3}x + 9}$; $g(x) = 3x^3 - 27$

23. $f(x) = \dfrac{1}{x - 4}$; $g(x) = \dfrac{1}{x} + 4$

24. $f(x) = \dfrac{x - 3}{x + 1}$; $g(x) = \dfrac{x + 3}{1 - x}$

In Problems 25 and 26, the given function f is one-to-one. Without finding f^{-1} find its domain and range.

25. $f(x) = 4 + \sqrt{x}$

26. $f(x) = 5 - \sqrt{x + 8}$

In Problems 27 and 28, the given function f is one-to-one. The domain and range of f is given. Find f^{-1} and give its domain and range.

27. $f(x) = \dfrac{2}{\sqrt{x}}$, $x > 0, y > 0$

28. $f(x) = 2 + \dfrac{3}{\sqrt{x}}$, $x > 0, y > 2$

In Problems 29–36, the given function f is one-to-one. Find f^{-1}. Sketch the graphs of f and f^{-1} on the same rectangular coordinate system.

29. $f(x) = -2x + 6$

30. $f(x) = -2x + 1$

31. $f(x) = x^3 + 2$

32. $f(x) = 1 - x^3$

CHAPTER 2 FUNCTIONS

33. $f(x) = 2 - \sqrt{x}$

34. $f(x) = \sqrt{x - 7}$

35. $f(x) = \sqrt[3]{7x}$

36. $f(x) = \sqrt[3]{4x - 5}$

In Problems 37–42, the given function f is one-to-one. Find f^{-1}. Proceed as in part (b) of Example 4 and find the domain and range of f^{-1}. Then find the range of f.

37. $f(x) = \dfrac{3}{x} - 1$

38. $f(x) = 10 - \dfrac{1}{5x}$

39. $f(x) = \dfrac{1}{2x - 1}$

40. $f(x) = \dfrac{2}{5x + 8}$

41. $f(x) = \dfrac{7x}{2x - 3}$

42. $f(x) = \dfrac{1 - x}{x - 2}$

In Problems 43 and 44, the function $f(x) = 8x^5 + 2x^3 + 5x - 4$ is one-to-one. Without finding f^{-1}, verify the given function value.

43. $f^{-1}(-19) = -1$

44. $f^{-1}(-1) = \frac{1}{2}$

In Problems 45–48, the given function f is one-to-one. Without finding f^{-1}, determine the indicated function value.

45. $f(x) = x^3 + 3x + 4;\quad f^{-1}(8)$

46. $f(x) = x^3 + x - 1;\quad f^{-1}(1)$

47. $f(x) = 2x^5 - 20;\quad f^{-1}(-20)$

48. $f(x) = \dfrac{2}{x^3 + 4};\quad f^{-1}(-\frac{1}{2})$

In Problems 49–52, the given function f is one-to-one. Without finding f^{-1}, find the point on the graph of f^{-1} corresponding to the indicated value of x in the domain of f.

49. $f(x) = 2x^3 + 2x;\quad x = 2$

50. $f(x) = 8x - 3;\quad x = 5$

51. $f(x) = x + \sqrt{x};\quad x = 9$

52. $f(x) = \dfrac{4x}{x + 1};\quad x = \frac{1}{2}$

In Problems 53 and 54, sketch the graph of f^{-1} from the graph of f.

53.

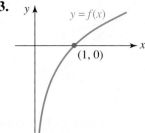

FIGURE 2.8.15 Graph for Problem 53

54.

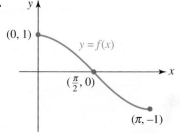

FIGURE 2.8.16 Graph for Problem 54

In Problems 55 and 56, sketch the graph of f from the graph of f^{-1}.

55.

FIGURE 2.8.17 Graph for Problem 55

56.

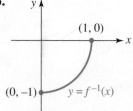

FIGURE 2.8.18 Graph for Problem 56

In Problems 57–60, the function f is not one-to-one on the given domain but is one-to-one on the restricted domain (the second interval). Find the inverse of the one-to-one function and give its domain. Sketch the graph of f on the restricted domain and the graph of f^{-1} on the same coordinate axes.

57. $f(x) = 4x^2 + 2, (-\infty, \infty); [0, \infty)$ **58.** $f(x) = (3 - 2x)^2, (-\infty, \infty); [\frac{3}{2}, \infty)$

59. $f(x) = \frac{1}{2}\sqrt{4 - x^2}, [-2, 2]; [0, 2]$ **60.** $f(x) = \sqrt{1 - x^2}, [-1, 1]; [0, 1]$

61. If the functions f and g have inverses, then it can be proved that
$$(f \circ g)^{-1} = g^{-1} \circ f^{-1}.$$
Verify this property for the one-to-one functions $f(x) = x^3$ and $g(x) = 4x + 5$.

62. It can be shown that the equation $y = \sqrt[3]{x} - \sqrt[3]{y}$ defines implicitly a one-to-one function $y = f(x)$. Without finding f, find f^{-1}.

For Discussion

63. Suppose f is a continuous function that is increasing (or decreasing) for all x in its domain. Explain why f is necessarily one-to-one.

64. Explain why the graph of a one-to-one function f can have at most one x-intercept.

65. The function $f(x) = |2x - 4|$ is not one-to-one. How should the domain of f be restricted so that the new function has an inverse? Find f^{-1} and give its domain and range. Sketch the graph of f on the restricted domain and the graph of f^{-1} on the same coordinate axes.

66. What property do the one-to-one functions $y = f(x)$ shown in FIGURES 2.8.19(a) and 2.8.19(b) have in common? Find two more explicit functions with this same property. Be very explicit about what this property has to do with f^{-1}.

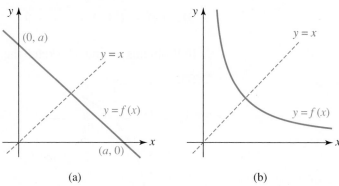

(a)   (b)

FIGURE 2.8.19 Graphs for Problem 66

67. The piecewise-defined function
$$f(x) = \begin{cases} x + 1, & -4 \le x < 0 \\ \frac{1}{4}(x + 4), & 0 \le x \le 4, \end{cases}$$
shown in FIGURE 2.8.20 is one-to-one. Sketch the graph of f^{-1}.

68. Find f^{-1} for the piecewise-defined function f in Problem 67. Give the domain and range of f^{-1}.

69. Suppose a function f is one-to-one. For c a constant, we know that the graph of the function $g(x) = f(x) + c$ is the graph of f shifted vertically and so g is also a one-to-one function. What is the inverse of the function g? Describe the graph of g^{-1} in terms of the graph of f^{-1}.

70. Suppose that f is one-to-one and that the domain of f and f^{-1} is $(-\infty, \infty)$. If $f^{-1}(5) = 10$, find x such that $3 + f(2x - 4) = 8$.

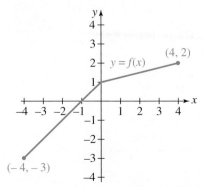

FIGURE 2.8.20 Graph for Problem 67

2.9 Building a Function from Words

INTRODUCTION In calculus there will be several instances when you will be expected to translate the words that describe a problem into mathematical symbols and then set up or construct either an *equation* or a *function*.

In this section we focus on problems that involve functions. We begin with a verbal description about the product of two numbers.

▌EXAMPLE 1   Product of Two Numbers

The sum of two nonnegative numbers is 15. Express the product of one and the square of the other as a function of one of the numbers.

Solution We first represent the two numbers by the symbols x and y and recall that "nonnegative" means that $x \geq 0$ and $y \geq 0$. The first sentence then says that $x + y = 15$; this is *not* the function we are seeking. The second sentence describes the function we want; it is called "the product." Let's denote "the product" by the symbol P. Now P is the product of one of the numbers, say, x and the square of the other, that is, y^2:

$$P = xy^2. \tag{1}$$

No, we are not finished because P is supposed to be a "function of *one* of the numbers." We now use the fact that the numbers x and y are related by $x + y = 15$. From this last equation we substitute $y = 15 - x$ into (1) to obtain the desired result:

$$P(x) = x(15 - x)^2. \tag{2} \quad �ી$$

Here is a symbolic summary of the analysis of the problem given in Example 1:

$$
\underbrace{\text{The sum of two nonnegative numbers is 15}}_{\substack{\text{let the numbers be } x \geq 0 \text{ and } y \geq 0 \\ x + y = 15}}. \text{ Express the } \underbrace{\text{product}}_{P} \text{ of } \underbrace{\text{one}}_{x} \text{ and the } \underbrace{\text{square of the other}}_{y^2} \text{ as a function of } \underbrace{\text{one of the numbers}}_{\text{use } x}. \tag{3}
$$

Notice that the second sentence is vague about which number is squared. This means that it really doesn't matter; (1) could also be written as $P = yx^2$. Also, we could have used $x = 15 - y$ in (1) to arrive at $P(y) = (15 - y)y^2$. In a calculus setting it would not have mattered whether we worked with $P(x)$ or with $P(y)$ because by finding *one* of the numbers we automatically find the other from the equation $x + y = 15$. This last equation is commonly called a **constraint**. A constraint not only defines the relationship between the variables x and y but often puts a limitation on how x and y can vary. As we see in the next example, the constraint helps in determining the domain of the function that you have just constructed.

▌EXAMPLE 2   Example 1 Continued

What is the domain of the function $P(x)$ in (2)?

Solution Taken out of the context of the statement of the problem in Example 1, one would have to conclude from the discussion on page 52 of Section 2.1 that the domain of the function

$$P(x) = x(15 - x)^2 = 225x - 30x^2 + x^3$$

is the set of real numbers $(-\infty, \infty)$. *But* in the context of the original problem, the numbers were to be nonnegative. From the requirement that $x \geq 0$ *and* $y = 15 - x \geq 0$ we get $x \geq 0$ and $x \leq 15$, which means that x must satisfy the simultaneous inequality $0 \leq x \leq 15$. Using interval notation, the domain of the product function P in (2) is the closed interval $[0, 15]$. ∎

Another way of looking at the conclusion of Example 2 is this: The constraint $x + y = 15$ dictates that $y = 15 - x$. Thus *if* x were allowed to be larger than 15 (say, $x = 17.5$), then $y = 15 - x$ would be a negative number, which contradicts the initial assumption that $y \geq 0$.

☐ **Optimization Problems** For the remainder of this section we are going to examine "word problems" taken directly from a calculus text. These problems, variously called "optimization problems" or "applied maximum and minimum problems," consist of two parts, the "precalculus part" where you set up the function to be optimized and the "calculus part" where you perform calculus-specific operations on the function that you have just found to find its maximum or minimum value. The calculus part is usually identifiable by words such as "maximum (or minimum)," "least," "greatest," "large as possible," "find the dimensions," and so on. For example, the actual statement of Example 1 as it appears in a calculus text is:

> *Find two nonnegative numbers whose sum is 15 such that the product of one and the square of the other is a maximum.*

The big hurdle for many students of calculus is separating out the words that define the function to be optimized from all the words in the statement of the problem.

Before proceeding with the examples, you are encouraged to read the *Notes from the Classroom* at the end of this section.

The next example describes a geometric problem that asks for a "largest rectangle." Remember, you are not expected to work the entire problem by trying to actually find the "largest rectangle," which you would do in a calculus course. Right now your only job is to pick out the words, as we illustrated in (3), that tell you what the function is and then construct it using the variables introduced. In calculus, the function to be optimized is called the **objective function**. See Problems 29–50 in Exercises 2.9.

Please note. ▶

EXAMPLE 3   Largest Rectangle

Find the objective function in the following calculus problem:

> *A rectangle has two vertices on the x-axis and two vertices on the semicircle whose equation is* $y = \sqrt{25 - x^2}$. *See* FIGURE 2.9.1(a). *Find the dimensions of the largest rectangle.*

Solution In calculus the words "largest rectangle" mean that we are seeking *the* rectangle, of the many that can be drawn in the semicircle, that has the greatest or maximum *area*. Hence, the function we must construct is the area A of the rectangle. If (x, y), $x > 0$, $y > 0$, denotes the vertex of the rectangle on the circle in the first quadrant, then as shown in Figure 2.9.1(b) the area A is length × width, or

$$A = (2x) \times y = 2xy. \tag{4}$$

The constraint in this problem is the equation $y = \sqrt{25 - x^2}$ of the semicircle. We use the constraint equation to eliminate y in (4) and obtain the area of the rectangle or the objective function,

$$A(x) = 2x\sqrt{25 - x^2}. \tag{5}$$

This ends the "precalculus part" of the problem.

(a)

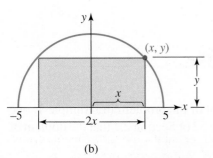

(b)

FIGURE 2.9.1 Rectangle in Example 3

The next step would be calculus procedures to determine the value of x for which the objective function $A(x)$ takes on its largest value. ■

Were we again to consider the function $A(x)$ out of the context of the problem in Example 3, its domain would be $[-5, 5]$. Because we assumed that $x > 0$ the domain of $A(x)$ in (4) is actually the open interval $(0, 5)$. But in calculus we would use the closed interval $[0, 5]$ even though when $x = 0$ and $x = 5$ the area would be $A(0) = 0$ and $A(5) = 0$, respectively. Do not worry about this last technicality.

■ EXAMPLE 4   Least Amount of Fencing

Find the objective function and its domain in the following calculus problem:

A rancher intends to mark off a rectangular plot of land that will have an area of 1000 m². The plot will be fenced and divided into two equal portions by an additional fence parallel to two sides. Find the dimensions of the land that require the least amount of fence.

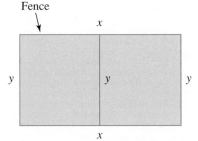

Fence

FIGURE 2.9.2 Rectangular plot of land in Example 4

Solution Your drawing should be a rectangle with a line drawn down its middle, similar to that given in FIGURE 2.9.2. As shown in the figure, let $x > 0$ be the length of the rectangular plot of land and let $y > 0$ denote its width. The function we seek is the "amount of fence." If the symbol F represents this amount, then the sum of the lengths of the *five* portions—two horizontal and three vertical—of the fence is

$$F = 2x + 3y. \tag{6}$$

But the fenced-in land is to have an area of 1000 m², and so x and y must be related by the constraint $xy = 1000$. From the last equation we get $y = 1000/x$ which can be used to eliminate y in (6). Thus, the amount of fence F as a function of x is $F(x) = 2x + 3(1000/x)$ or

$$F(x) = 2x + \frac{3000}{x}. \tag{7}$$

Since x represents a physical dimension that satisfies $xy = 1000$, we conclude that it is positive. But other than that, there is no restriction on x. Thus, unlike the previous example, the objective function (7) is not defined on a closed interval. The domain of $F(x)$ is $(0, \infty)$. ■

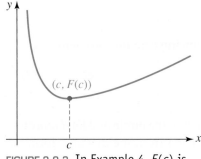

FIGURE 2.9.3 In Example 4, $F(c)$ is the smallest value of F for $x > 0$

As can be seen from the graph of (7) given in FIGURE 2.9.3, F has a minimum at some value of x, say $x = c$. With a graphing calculator or computer we can approximate c and $F(c)$, but with calculus we can often find their exact values.

If a problem involves triangles, you should study the problem carefully and determine whether the Pythagorean theorem, similar triangles, or trigonometry is applicable.

■ EXAMPLE 5   Shortest Ladder

Find the objective function and its domain in the following calculus problem:

A 10-ft wall stands 5 ft from a building. Find the length of the shortest ladder, supported by the wall, that reaches from the ground to the building.

FIGURE 2.9.4 Ladder in Example 5

Solution The words "shortest ladder" indicate that we want a function that describes the length of the ladder. Let L denote this length. With x and y defined in FIGURE 2.9.4, we see that there are two right triangles, the larger triangle has three sides with lengths L, y, and $x + 5$, and the smaller triangle has two sides of lengths x and 10. Now the ladder is the hypotenuse of the larger right triangle, so by the Pythagorean theorem,

$$L^2 = (x + 5)^2 + y^2. \tag{8}$$

The right triangles in Figure 2.9.4 are similar because they both contain a right angle and share the common acute angle the ladder makes with the ground. We then use the fact that the ratios of corresponding sides of similar triangles are equal. This enables us to write

$$\frac{y}{x + 5} = \frac{10}{x} \qquad \text{so that} \qquad y = \frac{10(x + 5)}{x}.$$

Using the last result, (8) becomes

$$L^2 = (x + 5)^2 + \left(\frac{10(x + 5)}{x}\right)^2$$

$$= (x + 5)^2 \left(1 + \frac{100}{x^2}\right) \qquad \leftarrow \text{factoring } (x + 5)^2$$

$$= (x + 5)^2 \left(\frac{x^2 + 100}{x^2}\right) \qquad \leftarrow \text{common denominator}$$

Taking the square root gives us L as a function of x,

$$L(x) = \frac{x + 5}{x}\sqrt{x^2 + 100}. \qquad \leftarrow \left\{\begin{array}{l}\text{square root of a product} \\ \text{is the product of the square roots}\end{array}\right.$$

Because $L(0)$ is not defined, the domain of the objective function $L(x)$ is the set of positive numbers, that is, $(0, \infty)$. ▮

▮ EXAMPLE 6   Closest Point

Find the objective function and its domain in the following calculus problem:

Find the point in the first quadrant on the circle $x^2 + y^2 = 1$ that is closest to the point $(2, 4)$.

Solution Let (x, y) denote a point in the first quadrant on the circle and let d represent the distance from (x, y) to $(2, 4)$. See FIGURE 2.9.5. Then from the distance formula, (2) of Section 1.3,

$$d = \sqrt{(x - 2)^2 + (y - 4)^2} = \sqrt{x^2 + y^2 - 4x - 8y + 20}. \tag{9}$$

The constraint in this problem is the equation of the circle $x^2 + y^2 = 1$. From this we can immediately replace $x^2 + y^2$ in (9) by the number 1. Moreover, using the constraint to write $y = \sqrt{1 - x^2}$ allows us to eliminate y in (9). Thus the distance d as a function of x is:

$$d(x) = \sqrt{21 - 4x - 8\sqrt{1 - x^2}}. \tag{10}$$

FIGURE 2.9.5 Distance d in Example 6

Since (x, y) is a point on the circle in the first quadrant it suffices to let the variable x range from 0 to 1, that is, the domain of the objective function in (10) is the closed interval $[0, 1]$. ▮

NOTES FROM THE CLASSROOM

When we get to the sections in a calculus text devoted to word problems students often react with groans, ambivalence, and dismay. While not guaranteeing anything, the following suggestions might help you to get through the problems in Exercises 2.9.

- At least try to develop a positive attitude. Try to be neat and organized.
- Read the problem slowly. Then read the problem several more times.
- Pay attention to words such as "maximum," "least," "greatest," and "closest" because they may provide a clue about the nature of the function you are seeking. For example, if a problem asks for "closest," then the function you are seeking most probably involves *distance*; if a problem asks for "least material," then the function you want may be *surface area*. See Problems 37 and 44 in Exercises 2.9.
- Whenever possible, sketch a curve or a picture and identify given quantities in your sketch. Keep your sketch simple.
- Introduce variables and note any constraint or relationship between the variables (such as $x + y = 15$ in Example 1).
- Identify the domain of the function just constructed. Keep in mind that if the problem mentions "dimensions" then the variables representing those quantities must be nonnegative.

Exercises 2.9   Answers to selected odd-numbered problems begin on page ANS–10.

In Problems 1–28, proceed as in Example 1 and translate the words into an appropriate function. Give the domain of the function.

FIGURE 2.9.6 Rectangle in Problem 7

FIGURE 2.9.7 Line segment in Problem 8

1. The product of two positive numbers is 50. Express their sum as a function of one of the numbers.
2. Express the sum of a nonzero number and its reciprocal as a function of the number.
3. The sum of two nonnegative numbers is 1. Express the sum of the square of one and twice the square of the other as a function of one of the numbers.
4. Let m and n be positive integers. The sum of two nonnegative numbers is S. Express the product of the mth power of one and the nth power of the other as a function of one of the numbers.
5. A rectangle has a perimeter of 200 in. Express the area of the rectangle as a function of the length of one of its sides.
6. A rectangle has an area of 400 in^2. Express the perimeter of the rectangle as a function of the length of one of its sides.
7. Express the area of the shaded rectangle in FIGURE 2.9.6 as a function of x.
8. Express the length of the line segment containing the point $(2, 4)$ shown in FIGURE 2.9.7 as a function of x.
9. Express the distance from a point (x, y) on the graph of $x + y = 1$ to the point $(2, 3)$ as a function of x.
10. Express the distance from a point (x, y) on the graph of $y = 4 - x^2$ to the point $(0, 1)$ as a function of x.

11. Consider the segment of the parabola $y = x^2$ between the points $A(-2, 4)$ and $B(2, 4)$ shown in FIGURE 2.9.8. Express the sum S of the squares of the distances $d(A, P)$ and $d(P, B)$ as a function of x.

12. The red horizontal line shown in FIGURE 2.9.9 is a chord of a circle of radius r. Express the length d of the chord AB as a function of its distance y from the center C of the circle.

FIGURE 2.9.8 Parabola in Problem 11

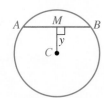

FIGURE 2.9.9 Circle and chord in Problem 12

13. Express the perimeter of a square as a function of its area A.
14. Express the area of a circle as a function of its diameter d.
15. Express the diameter of a circle as a function of its circumference C.
16. Express the volume of a cube as a function of the area A of its base.
17. Express the area of an equilateral triangle as a function of its height h.
18. Express the area of an equilateral triangle as a function of the length s of one of its sides.
19. A wire of length x is bent into the shape of a circle. Express the area of the circle as a function of x.
20. A wire of length L is cut x units from its left end. As shown in FIGURE 2.9.10, the piece of wire of length x (blue in the figure) is bent into the shape of a circle, whereas the remaining piece wire of length $L - x$ (red in the figure) is bent into the shape of a square. Express the sum of the areas as a function of x.
21. A tree is planted 30 ft from the base of a street lamp that is 25 ft tall. Express the length of the tree's shadow as a function of its height. What happens to the length of its shadow as the height of the tree approaches 25 ft?
22. The frame of a kite consists of six pieces of lightweight plastic. The outer frame of the kite consists of four precut pieces, two pieces of length 2 ft, and two pieces of length 3 ft. Express the area of the kite as a function of x, where $2x$ is the length of the horizontal crossbar piece shown in FIGURE 2.9.11.

FIGURE 2.9.10 Wire of length L in Problem 20

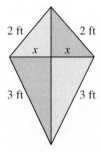

FIGURE 2.9.11 Kite in Problem 22

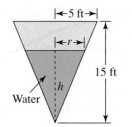

FIGURE 2.9.12 Conical tank in Problem 24

23. A company wants to construct an open rectangular box with a volume of 450 in³ so that the length of its base is 3 times its width. Express the surface area of the box as a function of the width.

24. A conical tank, with vertex down, has a radius of 5 ft and a height of 15 ft. Water is pumped into the tank. Express the volume of the water as a function of its depth. See FIGURE 2.9.12. [*Hint*: The volume of a cone is $V = \frac{1}{3}\pi r^2 h$. Although the tank is a three-dimensional object, examine it in cross section as a two-dimensional triangle.]

25. Car A passes point O heading east at a constant rate of 40 mi/h; car B passes the same point 1 hour later heading north at a constant rate of 60 mi/h. Express the distance between the cars as a function of time t, where t is measured starting when car B passes point O. See FIGURE 2.9.13.

26. At time $t = 0$ (measured in hours), two airliners with a vertical separation of 1 mile, pass each other going in opposite directions. See FIGURE 2.9.14. Suppose the planes are flying horizontally at rates of 500 mi/h and 550 mi/h.
(a) Express the horizontal distance between them as a function of time t. [*Hint*: distance = rate × time.]
(b) Express the diagonal distance between them as a function of time t.

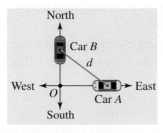

FIGURE 2.9.13 Cars in Problem 25

(a) $t = 0$ (b) $t > 0$

FIGURE 2.9.14 Planes in Problem 26

27. The swimming pool shown in FIGURE 2.9.15 is 3 ft deep at the shallow end, 8 ft deep at the deepest end, 40 ft long, 30 ft wide, and the bottom is an inclined plane. Water is pumped into the pool. Express the volume of the water in the pool as a function of height h of the water above the deep end. [*Hint*: The volume will be a piecewise-defined function.]

28. USPS regulations for parcel post stipulate that the length plus girth (the perimeter of one end) of a package must not exceed 108 inches. Express the volume of the package as a function of the width x shown in FIGURE 2.9.16.

FIGURE 2.9.15 Swimming pool in Problem 27

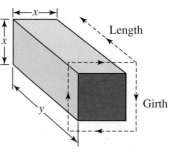

FIGURE 2.9.16 Package in Problem 28

2.9 Building a Function from Words

In Problems 29–50, proceed as in Examples 3–6 and find the objective function for the given calculus problem. Give the domain of the objective function but **do not actually attempt to solve the problem**. It would be good idea to reread the three paragraphs under the heading *Optimization Problems* on page 118.

29. Find a number that exceeds its square by the greatest amount.

30. Of all rectangles with perimeter 20 inches, find the one with the shortest diagonal.

31. A rectangular plot of land will be fenced into three equal portions by two dividing fences parallel to two sides. If the area to be enclosed is 4000 m², find the dimensions of the land that require the least amount of fence.

32. A rectangular plot of land will be fenced into three equal portions by two dividing fences parallel to two sides. If the total fence to be used is 8000 m, find the dimensions of the land that has the greatest area.

33. A rancher wishes to build a rectangular corral with an area of 128,000 ft² with one side along a straight river. The fencing along the river costs $1.50 per foot, whereas along the other three sides the fencing costs $2.50 per foot. Find the dimensions of the corral so that the cost of construction is a minimum. [*Hint*: Along the river the cost of x ft of fence is $1.50x$.]

34. A rectangular yard is to be enclosed with a fence by attaching it to a house whose length is 40 feet. See FIGURE 2.9.17. The amount of fencing to be used is 160 feet. Find the dimensions of the yard so that the greatest area is enclosed.

35. Consider all rectangles that have the same perimeter p. (Here p represents a constant.) Of these rectangles, show that the one with the largest area is a square.

36. Find the vertices $(x, 0)$ and $(0, y)$ of the shaded triangular region in FIGURE 2.9.18 so that its area is a minimum.

37. (a) An open rectangular box is to be constructed with a square base and a volume of 32,000 cm³. Find the dimensions of the box that require the least amount of material. See FIGURE 2.9.19.

(b) If the rectangular box in part (a) is closed, find the dimensions that require the least amount of material.

FIGURE 2.9.17 House and yard in Problem 34

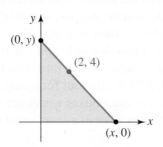

FIGURE 2.9.18 Line segment in Problem 36

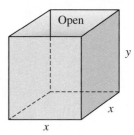

FIGURE 2.9.19 Box in Problem 37

38. A closed rectangular box is to be constructed with a square base. The material for the top costs $2 per square foot whereas the material for the remaining sides costs $1 per square foot. If the total cost to construct each box is $36, find the dimensions of the box of greatest volume that can be made.

39. A rain gutter with a rectangular cross section is made from a 1 ft × 20 ft piece of metal by bending up equal amounts from the 1-ft side. See FIGURE 2.9.20. How should the metal be bent up on each side in order to make the capacity of the gutter a maximum? [*Hint*: Capacity = volume.]

FIGURE 2.9.20 Rain gutter in Problem 39

FIGURE 2.9.21 Norman window in Problem 40

40. A Norman window consists of a rectangle surmounted by a semicircle as shown in FIGURE 2.9.21. If the total perimeter of the window is 10 m, find the dimensions of the window with the largest area.

41. A printed page will have 2-in. margins of white space on the sides and 1-in. margins of white space on the top and bottom. The area of the printed portion is 32 in². Determine the dimensions of the page so that the least amount of paper is used.

42. Find the dimensions of the right circular cylinder with greatest volume that can be inscribed in a right circular cone of radius 8 in. and height 12 in. See FIGURE 2.9.22.

43. Find the maximum length L of a thin board that can be carried horizontally around the right-angle corner shown in FIGURE 2.9.23. [*Hint*: Use similar triangles.]

FIGURE 2.9.22 Inscribed cylinder in Problem 42

FIGURE 2.9.23 Board in Problem 43

44. A juice can is to be made in the form of a right circular cylinder and have a volume of 32 in³. See FIGURE 2.9.24. Find the dimensions of the can so that the least amount of material is used in its construction. [*Hint*: Material = total surface area of can = area of top + area of bottom + area of lateral side. If the circular top and bottom covers are removed and the cylinder is cut straight up its side and flattened out, the result is the rectangle shown in Figure 2.9.24(c).]

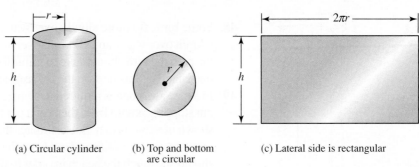

(a) Circular cylinder

(b) Top and bottom are circular

(c) Lateral side is rectangular

FIGURE 2.9.24 Juice can in Problem 44

45. The lateral side of a cylinder is to be made from a rectangle of flimsy sheet plastic. Because the plastic material cannot support itself, a thin, stiff wire is embedded in the material as shown in FIGURE 2.9.25(a). Find the dimensions of the cylinder of

(a) Rectangular sheet of plastic material

(b) Lateral side of cylinder

FIGURE 2.9.25 Cylinder in Problem 45

largest volume that can be constructed if the wire has a fixed length L. [*Hint*: There are two constraints in this problem. In Figure 2.9.25(b), the circumference of a circular end of the cylinder is y.]

46. Many medications are packaged in capsules as shown in the accompanying photo. Assume that a capsule is formed by adjoining two hemispheres to the ends of a right circular cylinder as shown in FIGURE 2.9.26. If the total volume of the capsule is to be 0.007 in.^{3}, find the dimensions of the capsule so that the least amount of material is used in its construction. [*Hint*: The volume of a sphere is $\frac{4}{3}\pi r^3$ and its surface area is $4\pi r^2$.]

47. A 20-ft long water trough has ends in the form of isosceles triangles with sides that are 4 ft long. See FIGURE 2.9.27. Determine the dimension across the top of the triangular end so that the volume of the trough is a maximum. [*Hint*: A *right cylinder* is not necessarily a *circular cylinder* where the top and bottom are circles. The top and bottom of a right cylinder are the same but could be a triangle, a pentagon, a trapezoid, and so on. The volume of a right cylinder is the area of the base $\times$ the height.]

FIGURE 2.9.26 Model of a capsule in Problem 46

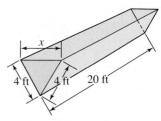
Capsule

FIGURE 2.9.27 Water trough in Problem 47

FIGURE 2.9.28 The bird in Problem 48

48. Some birds fly more slowly over water than over land. A bird flies at constant rates 6 km/h over water and 10 km/h over land. Use the information in FIGURE 2.9.28 to find the path the bird should take to minimize the total flying time between the shore of one island and its nest on the shore of another island. [*Hint*: distance = rate $\times$ time.]

49. In a race a woman is required to swim from a floating dock A to the beach and, without stopping, swim from the beach out to another floating dock C. The distances are shown in FIGURE 2.9.29. She estimates that she can swim from dock A to the beach at a constant rate of 3 mi/h and out from the beach to dock C at a rate of 2 mi/h. Where should she touch the beach in order to minimize the total swimming time from A to C?

50. Two flag poles are secured by wires that are attached at a single point between the poles. See FIGURE 2.9.30. Where should the point be located to minimize the total length of wire used?

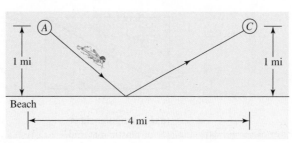

FIGURE 2.9.29 Swimmer in Problem 49

FIGURE 2.9.30 Flag poles in Problem 50

2.10 The Tangent Line Problem

INTRODUCTION In a calculus course you will study many different things, but roughly, the subject "calculus" is divided into two broad but related areas known as **differential calculus** and **integral calculus**. The discussion of each of these topics invariably begins with a motivating problem involving the graph of a function. Differential calculus is motivated by the problem

Find a tangent line to the graph of a function f,

whereas integral calculus is motivated by the problem

Find the area under the graph of a function f.

The first problem will be addressed in this section; the second problem will be discussed in Section 3.7.

FIGURE 2.10.1 Tangent line L touches a circle at point P

☐ **Tangent Line to a Graph** The word *tangent* stems from the Latin verb *tangere*, meaning "to touch." You might remember from the study of plane geometry that a tangent to a circle is a line L that intersects, or touches, the circle in exactly one point P. See FIGURE 2.10.1. It is not quite as easy to define a tangent line to the graph of a function f. The idea of *touching* carries over to the notion of a tangent line to the graph of a function, but the idea of *intersecting the graph in one point* does not carry over.*

FIGURE 2.10.2 Tangent line L to a graph at point P

☐ **Using Secant Lines** Suppose $y = f(x)$ is a continuous function. If, as shown in FIGURE 2.10.2, f possesses a line L tangent to its graph at a point P, then what is the equation of this line? To answer this question, we need the coordinates of P and the slope m_{tan} of L. The coordinates of P pose no difficulty, since a point on the graph of a function f is obtained by specifying a value of x in the domain of f. The coordinates of the point of tangency at $x = a$ are then $(a, f(a))$. As a means of approximating the slope m_{tan}, we can readily find the slopes m_{sec} of *secant lines* that pass through the point P and any other point Q on the graph. See FIGURE 2.10.3.

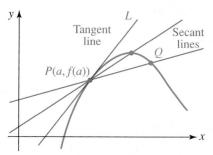

FIGURE 2.10.3 Slopes of secant lines approximate the slope m_{tan} of L

☐ **Definition of a Tangent Line** If P has coordinates $(a, f(a))$ and if Q has coordinates $(a + h, f(a + h))$, then as shown in FIGURE 2.10.4, the slope of the secant line through P and Q is

$$m_{sec} = \frac{\text{rise}}{\text{run}} = \frac{f(a + h) - f(a)}{(a + h) - a}$$

or

$$m_{sec} = \frac{f(a + h) - f(a)}{h}. \tag{1}$$

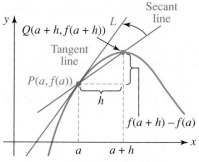

FIGURE 2.10.4 Secant lines swing into the tangent line L as $h \to 0$

The expression on the right-hand side of the equality in (1) is called a **difference quotient**. When we let h take on values that are closer and closer to zero, that is, as $h \to 0$, the sequence of points $Q(a + h, f(a + h))$ move along the curve closer and closer to the point $P(a, f(a))$. Intuitively, we expect the secant lines to approach the tangent line L, and that $m_{sec} \to m_{tan}$ as $h \to 0$. Using the idea of a limit introduced in Section 1.5 we write $m_{tan} = \lim_{h \to 0} m_{sec}$. We summarize this discussion using the difference quotient (1).

---

* We leave the discussion of the many subtleties and questions surrounding the tangent line problem to a course in calculus.

DEFINITION 2.10.1 Tangent Line with Slope

Let $y = f(x)$ be continuous at the number a. If the limit,

$$m_{\tan} = \lim_{h \to 0} \frac{f(a + h) - f(a)}{h} \tag{2}$$

exists, then the **tangent line** to the graph of f at $(a, f(a))$ is that line passing through the point $(a, f(a))$ with slope $m_{\tan}$.

Just like the problems discussed in Section 1.5, observe that the limit in (2) has the indeterminate form $0/0$ as $h \to 0$.

We are not going to delve into any theoretical details about when the limit (2) exists or does not exist—that discussion properly belongs in a calculus course. So to simplify the discussion, we will drop the phrase "provided the limit exists." For this course it suffices simply to be aware of the fact that the limit (2) may not exist for certain values of a See Problem 43 in Exercises 2.10.

It is very likely that early on in your calculus course you will be asked to compute the limit of a difference quotient such as (2). The computation of (2) is essentially a *four-step process*, and three of these steps involve only precalculus mathematics: algebra and trigonometry. Getting over the hurdles of algebraic or trigonometric manipulations in these first three steps is your primary goal. If done accurately, the fourth step, or the calculus step, may be the easiest part of the problem. In preparation for calculus we recommend that you be able to carry out the calculation of (2) for functions involving

review $(a + b)^n$ for $n = 2$ and 3 ▶

- positive integer powers of x such as x^n for $n = 1, 2$, and 3,

review adding symbolic fractions ▶

- division of functions such as $\dfrac{1}{x}$ and $\dfrac{x}{x + 1}$, and

review rationalization of numerators and ▶
denominators

- radicals such as $\sqrt{x}$.

See Problems 1–10 in Exercises 2.10.

EXAMPLE 1 **The Four-Step Process**

Find the slope of the tangent line to the graph of $y = x^2 + 2$ at $x = 1$.

Solution We first compute the difference quotient in (2) with the identification that $a = 1$.
(*i*) The initial step is the computation of $f(a + h)$. Because functions can be complicated, it might help in this step to think of x wherever it appears in the function $f(x)$ as a set of parentheses ( ). For the given function we write $f(\ \ ) = (\ \ )^2 + 2$. The idea is to substitute $1 + h$ into those parentheses and carry out the required algebra:

$$\begin{aligned} f(1 + h) &= (1 + h)^2 + 2 \\ &= (1 + 2h + h^2) + 2 \\ &= 3 + 2h + h^2. \end{aligned}$$

(*ii*) The computation of the difference $f(a + h) - f(a)$ is the most important step. It is imperative that you simplify this step as much as possible. Here is a tip: In many of the problems that you will be required to do in calculus you will be able to factor h from the difference $f(a + h) - f(a)$. To begin, compute $f(a)$, which in this case is $f(1) = 1^2 + 2 = 3$. Next, you can use the result from the preceding step:

$$
\begin{aligned}
f(1 + h) - f(1) &= 3 + 2h + h^2 - 3 \\
&= 2h + h^2 \\
&= h(2 + h). \quad \leftarrow \text{notice the factor of } h
\end{aligned}
$$

(*iii*) The computation of the difference quotient $\dfrac{f(a + h) - f(a)}{h}$ is now straightforward. Again, we use the results from the preceding step:

$$
\frac{f(1 + h) - f(1)}{h} = \frac{h(2 + h)}{h} = 2 + h. \quad \leftarrow \text{cancel the } h\text{'s}
$$

(*iv*) The calculus step is now easy. From (2) we have

$$
m_{\tan} = \lim_{h \to 0} \frac{f(1 + h) - f(1)}{h} = \lim_{h \to 0} (2 + h) = 2.
$$

from the result in step (*iii*)

The slope of the tangent line to the graph of $y = x^2 + 2$ at $(1, 3)$ is 2. ■

EXAMPLE 2   Equation of Tangent Line

Find an equation of the tangent line whose slope was found in Example 1.

Solution We know a point $(1, 3)$ and a slope $m_{\tan} = 2$, and so from the point-slope equation of a line we find

$$
y - 3 = 2(x - 1) \qquad \text{or} \qquad y = 2x + 1.
$$

Observe that the last equation is consistent with the *x*- and *y*-intercepts of the red line in FIGURE 2.10.5. ■

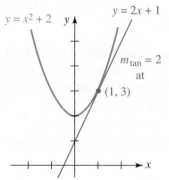

FIGURE 2.10.5 Tangent line in Example 2

☐ **The Derivative** As you inspect Figure 2.10.5, imagine tangent lines at various points on the graph of $f(x) = x^2 + 2$. This particular function is known to have a tangent line at every point on its graph. The tangent lines to the left of the origin have negative slope, the tangent line at $(0, 2)$ has zero slope, and the tangent lines to the right of the origin have positive slope (as seen in Example 1). In other words, for a function f the value of $m_{\tan}$ at a point $(a, f(a))$ depends on the choice of the number a. Roughly speaking, there is at most *one* value of $m_{\tan}$ for each number a in the domain of a function f. More specifically, $m_{\tan}$ is itself a *function* with a domain that is a subset of the domain of the function f. Furthermore, it is usually possible to obtain a formula for this *slope function*. This is accomplished by computing the limit of the difference quotient $\dfrac{f(x + h) - f(x)}{h}$ as $h \to 0$. We then substitute a value of x *after* the limit has been found. The slope function derived in this manner from f is said to be the **derivative of** f and (instead of $m_{\tan}$) is denoted by the symbol f'.

DEFINITION 2.10.2   The Derivative

The **derivative** of a function $y = f(x)$ is the function f' defined by

$$
f'(x) = \lim_{h \to 0} \frac{f(x + h) - f(x)}{h} \tag{3}
$$

EXAMPLE 3 **Example 1 Revisited**

Compute the derivative of $f(x) = x^2 + 2$.

Solution We proceed exactly as in Example 1 except that we find $f(x + h)$ instead of $f(1 + h)$. In the first three steps we calculate the difference quotient; in steps (*ii*) and (*iii*) we use the results in the preceding step. In step (*iv*) we compute the limit of the difference quotient.

(*i*) $f(x + h) = (x + h)^2 + 2 = x^2 + 2xh + h^2 + 2$

(*ii*) $\begin{aligned} f(x + h) - f(x) &= x^2 + 2xh + h^2 + 2 - (x^2 + 2) \\ &= x^2 + 2xh + h^2 + 2 - x^2 - 2 \\ &= 2xh + h^2 \\ &= h(2x + h) \end{aligned}$

(*iii*) $\dfrac{f(x + h) - f(x)}{h} = \dfrac{h(2x + h)}{h} = 2x + h$ ← cancel h's

(*iv*) From (3) the derivative of f is the limit as $h \to 0$ of the result in (*iii*). During the process of shrinking h smaller and smaller, x is held fixed. Hence

$$f'(x) = \lim_{h \to 0}(2x + h) = 2x.$$

So now we have two functions; from $f(x) = x^2 + 2$ we have obtained the derivative $f'(x) = 2x$. When evaluated at a number x, the function f gives the y-coordinate of a point on the graph and the derived function f' gives the slope of the tangent line at that point. We have already seen in Example 1 that $f(1) = 3$ and $f'(1) = 2$. ∎

With the aid of the derivative $f'(x) = 2x$ we can find slopes at other points on the graph of $f(x) = x^2 + 2$. For example,

at $x = 0$, $\begin{cases} f(0) = 2 \\ f'(0) = 0 \end{cases}$ ← point of tangency is $(0, 2)$
  ← slope of tangent line at $(0, 2)$ is $m = 0$

at $x = -3$, $\begin{cases} f(-3) = 11 \\ f'(-3) = -6 \end{cases}$ ← point of tangency is $(-3, 11)$
  ← slope of tangent line at $(-3, 11)$ is $m = -6$

The fact that $f'(0) = 0$ means that the tangent line to the graph of f is horizontal at the point $(0, 2)$.

EXAMPLE 4 **Derivative of a Function**

Compute the derivative of $f(x) = 2x^3 - 4x + 5$.

Solution

(*i*) The function is $f(\ ) = 2(\ )^3 - 4(\ ) + 5$ and so

$$f(x + h) = 2(x + h)^3 - 4(x + h) + 5.$$

The algebra here is a bit more complicated than in the previous example. We will use the binomial expansion for $(a + b)^3$ and the distributive law. Continuing,

See (7) in Section 1.5. ▶

$\begin{aligned} f(x + h) &= 2(x^3 + 3x^2h + 3xh^2 + h^3) - 4(x + h) + 5 \\ &= 2x^3 + 6x^2h + 6xh^2 + 2h^3 - 4x - 4h + 5 \end{aligned}$ ← $\begin{cases} \text{two applications} \\ \text{of the distributive law} \end{cases}$

(*ii*) As mentioned previously, in this step we are looking for a factor of h:

$$\begin{aligned}
f(x + h) - f(x) &= 2x^3 + 6x^2h + 6xh^2 + 2h^3 - 4x \\
&\quad - 4h + 5 - (2x^3 - 4x + 5) \\
&= 2x^3 + 6x^2h + 6xh^2 + 2h^3 - 4x \\
&\quad - 4h + 5 - 2x^3 + 4x - 5 \quad \leftarrow \text{terms in red add to 0} \\
&= 6x^2h + 6xh^2 + 2h^3 - 4h \\
&= h(6x^2 + 6xh + 2h^2 - 4) \quad \leftarrow \text{factor out } h
\end{aligned}$$

(*iii*) We use the last result:

$$\frac{f(x + h) - f(x)}{h} = \frac{h(6x^2 + 6xh + 2h^2 - 4)}{h} \quad \leftarrow \text{cancel } h\text{'s}$$
$$= 6x^2 + 6xh + 2h^2 - 4$$

(*iv*) From (3) and the preceding step the derivative of f is

$$f'(x) = \lim_{h \to 0}(6x^2 + 6xh + 2h^2 - 4) = 6x^2 - 4.$$

■

▌EXAMPLE 5   Equation of Tangent Line

Find an equation of the tangent line to the graph of $f(x) = 2/x$ at $x = 2$.

Solution We start by finding the derivative of f. In the second of the four steps we will have to combine two symbolic fractions by means of a common denominator.

(*i*) $f(x + h) = \dfrac{2}{x + h}$

(*ii*) $f(x + h) - f(x) = \dfrac{2}{x + h} - \dfrac{2}{x}$

$$= \frac{2}{x + h}\frac{x}{x} - \frac{2}{x}\frac{x + h}{x + h} \quad \leftarrow \text{a common denominator is } x(x + h)$$
$$= \frac{2x - 2x - 2h}{x(x + h)} \quad \leftarrow 2x - 2x = 0$$
$$= \frac{-2h}{x(x + h)} \quad \leftarrow \text{there is the factor of } h$$

(*iii*) The last result is to be divided by h, or more precisely $\dfrac{h}{1}$. We invert and multiply by $\dfrac{1}{h}$:

$$\frac{f(x + h) - f(x)}{h} = \frac{\dfrac{-2h}{x(x + h)}}{\dfrac{h}{1}} = \frac{-2h}{x(x + h)}\frac{1}{h} = \frac{-2}{x(x + h)} \quad \leftarrow \text{cancel } h\text{'s}$$

(*iv*) From (3) the derivative of f is

$$f'(x) = \lim_{h \to 0}\frac{-2}{x(x + h)} = \frac{-2}{x^2}.$$

We are now in a position to find an equation of the tangent line at the point corresponding to $x = 2$. From $f(2) = 2/2 = 1$, we get the point of tangency $(2, 1)$. Then

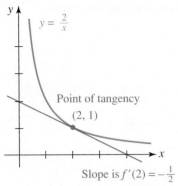

$y = \frac{2}{x}$

Point of tangency
(2, 1)

Slope is $f'(2) = -\frac{1}{2}$

FIGURE 2.10.6 Tangent line in Example 5

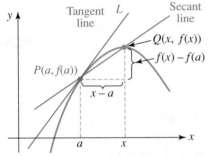

Tangent line *L*

Secant line

$Q(x, f(x))$

$f(x) - f(a)$

$P(a, f(a))$

$x - a$

a x

FIGURE 2.10.7 Secant line and tangent line at $(a, f(a))$

Review (1) and (2) of Section 1.5. ▶

from the derivative $f'(x) = -2/x^2$ we see that $f'(2) = -2/4$, and so the slope of the tangent line at (2, 1) is $-\frac{1}{2}$. From the point-slope equation of a line, the tangent line is

$$y - 1 = -\tfrac{1}{2}(x - 2) \qquad \text{or} \qquad y = -\tfrac{1}{2}x + 2.$$

The graph of $y = 2/x$ is the graph of $y = 1/x$ stretched vertically. (See Figure 2.2.1(e).) The tangent line at (2, 1) is shown in red in FIGURE 2.10.6. ∎

☐ **Alternative Definition** There is an alternative definition of the derivative. If we let $x = a + h$ in (2), then $h = x - a$. Consequently the slope of the secant line through $P(a, f(a))$ and $Q(x, f(x))$, as shown in FIGURE 2.10.7, is $\dfrac{f(x) - f(a)}{x - a}$. As $h \to 0$ we must have $x \to a$, and so the derivative (3) takes on the form

$$f'(a) = \lim_{x \to a} \frac{f(x) - f(a)}{x - a}. \tag{4}$$

EXAMPLE 6   Using (4)

Use (4) to compute the derivative of $f(x) = 4x^2 - 5x + 9$.

Solution We use the four-step process exactly as in Examples 3 and 4. The algebra is a slightly different; the analogue of the tip in (*ii*) of Example 1 is that we look for the factor $x - a$ in the difference $f(x) - f(a)$. Thus step (*ii*) will often require factoring the difference of two squares, the difference of two cubes, and so on.

(*i*) $f(a) = 4a^2 - 5a + 9$

(*ii*) $f(x) - f(a) = 4x^2 - 5x + 9 - (4a^2 - 5a + 9)$

$\qquad\qquad\qquad = 4x^2 - 5x + 9 - 4a^2 + 5a - 9$

$\qquad\qquad\qquad = 4x^2 - 5x - 4a^2 + 5a$ $\qquad \leftarrow \begin{cases} \text{regroup terms in} \\ \text{preparation for factoring} \end{cases}$

$\qquad\qquad\qquad = 4x^2 - 4a^2 - 5x + 5a$

$\qquad\qquad\qquad = 4(x^2 - a^2) - 5(x - a)$ $\qquad \leftarrow \begin{cases} \text{first term is the} \\ \text{difference of two squares} \end{cases}$

$\qquad\qquad\qquad = 4(x - a)(x + a) - 5(x - a)$ $\qquad \leftarrow$ notice the factor of $x - a$

$\qquad\qquad\qquad = (x - a)[4(x + a) - 5]$

$\qquad\qquad\qquad = (x - a)(4x + 4a - 5)$

(*iii*) $\dfrac{f(x) - f(a)}{x - a} = \dfrac{(x - a)(4x + 4a - 5)}{x - a}$ $\qquad \leftarrow$ cancel $x - a$

$\qquad\qquad\qquad\qquad = 4x + 4a - 5$

(*iv*) In the limit process indicated in (4), a is held fixed. Hence

$$f'(a) = \lim_{x \to a}(4x + 4a - 5) = 8a - 5. \quad \leftarrow \text{the limit of } 4x \text{ as } x \to a \text{ is } 4a$$ ∎

As you can see in (4) and the final line in Example 6, the derivative comes out a function of the symbol a rather than x, that is, $f'(a) = 8a - 5$. As a consequence, (4) is not used as often as (3) to compute a derivative. See Problems 33–40 in Exercises 2.10. Nevertheless, (4) is important because it is convenient to use in some theoretical aspects of differential calculus.

CHAPTER 2 FUNCTIONS

In Problems 1–10, proceed as in Example 1.

(a) Compute the difference quotient $\dfrac{f(a+h)-f(a)}{h}$ at the given value of a.

(b) Then, if instructed, compute $m_{\tan} = \lim\limits_{h\to 0}\dfrac{f(a+h)-f(a)}{h}$.

(c) Use the result of part (b) to find an equation of the tangent line at the point of tangency.

1. $f(x) = x^2 - 6, \quad a = 3$

2. $f(x) = -3x^2 + 10, \quad a = -1$

3. $f(x) = x^2 - 3x, \quad a = 1$

4. $f(x) = -x^2 + 5x - 3, \quad a = -2$

5. $f(x) = -2x^3 + x, \quad a = 2$

6. $f(x) = 8x^3 - 4, \quad a = \frac{1}{2}$

7. $f(x) = \dfrac{1}{2x}, \quad a = -1$

8. $f(x) = \dfrac{4}{x-1}, \quad a = 2$

9. $f(x) = \sqrt{x}, \quad a = 4$

10. $f(x) = \dfrac{1}{\sqrt{x}}, \quad a = 1$

In Problems 11–26, proceed as in Examples 3 and 4.

(a) Compute the difference quotient $\dfrac{f(x+h)-f(x)}{h}$ for the given function.

(b) Then, if instructed, compute the derivative $f'(x) = \lim\limits_{h\to 0}\dfrac{f(x+h)-f(x)}{h}$.

11. $f(x) = 10$

12. $f(x) = -3x + 8$

13. $f(x) = -4x^2$

14. $f(x) = x^2 - x$

15. $f(x) = 3x^2 - x + 7$

16. $f(x) = 2x^2 + x - 1$

17. $f(x) = x^3 + 5x - 4$

18. $f(x) = 2x^3 + x^2$

19. $f(x) = \dfrac{1}{4-x}$

20. $f(x) = \dfrac{3}{2x-4}$

21. $f(x) = \dfrac{x}{x-1}$

22. $f(x) = \dfrac{2x+3}{x+5}$

23. $f(x) = x + \dfrac{1}{x}$

24. $f(x) = \dfrac{1}{x^2}$

25. $f(x) = 2\sqrt{x}$

26. $f(x) = \sqrt{2x+1}$

In Problems 27–32, use the appropriate derivatives obtained in Problems 11–26. For the given function, find the point of tangency and slope of the tangent line at the indicated value of x. Find an equation of the tangent line at that point.

27. $f(x) = 3x^2 - x + 7, \quad x = 2$

28. $f(x) = x^2 - x, \quad x = 3$

29. $f(x) = x^3 + 5x - 4, \quad x = 1$

30. $f(x) = 2x^3 + x^2, \quad x = -\frac{1}{2}$

31. $f(x) = x + \dfrac{1}{x}, \quad x = \frac{1}{2}$

32. $f(x) = \dfrac{3}{2x-4}, \quad x = -1$

In Problems 33–40, proceed as in Example 6.

(a) Compute the difference quotient $\dfrac{f(x)-f(a)}{x-a}$ for the given function.

(b) Then, if instructed, compute the derivative $f'(a) = \lim\limits_{x\to a}\dfrac{f(x)-f(a)}{x-a}$.

33. $f(x) = 3x^2 + 1$

34. $f(x) = x^2 - 8x - 3$

35. $f(x) = 10x^3$

36. $f(x) = x^4$

37. $f(x) = \dfrac{1}{x}$

38. $f(x) = \dfrac{3x - 1}{x}$

39. $f(x) = \sqrt{7x}$

40. $f(x) = -\sqrt{x + 9}$

For Discussion

In Problems 41 and 42, use either (3) or (4) to compute the derivative of the given function. Find the points on the graph of f at which $f'(x) = 0$. Interpret your answers geometrically.

41. $f(x) = x^3 - 3x^2 - 9x$

42. $f(x) = x^4 - \frac{4}{3}x^3 + 2$

43. Use (2) to show that the graph of $f(x) = |x|$ possesses no tangent line at the point $(0, 0)$.

44. Use either (3) or (4) to compute the derivative of $f(x) = x^{1/3}$. [*Hint*: Recall from (3) of Section 1.5, $a^3 - b^3 = (a - b)(a^2 + ab + b^2)$.]

45. What is the tangent line to the graph of a linear function $f(x) = ax + b$?

46. If $f'(x) > 0$ for every x in an interval, then what can be said about f on the interval? If $f'(x) < 0$ for every x in an interval, then what can be said about f on the interval? [*Hint*: Draw a graph.]

47. If f is an even function and if (x, y) is on the graph of f, then $(-x, y)$ is also on the graph of f. How are the slopes of the tangent lines at (x, y) and $(-x, y)$ related?

48. If f is an odd function and if (x, y) is on the graph of f, then $(-x, -y)$ is also on the graph of f. How are the slopes of the tangent lines at (x, y) and $(-x, -y)$ related?

FIGURE 2.10.8 Graph for Problem 51

49. Consider the semicircle whose equation is $f(x) = \sqrt{1 - x^2}$. Discuss: How can the derivative $f'(x)$ be found using only the geometric fact that the radius of a circle is perpendicular to the tangent line at a point (x, y) on the circle?

50. Consider the semicircle whose equation is $f(x) = \sqrt{1 - x^2}$. Use (3) to find the derivative $f'(x)$ and compare your result with that in Problem 49.

51. Find an equation of the tangent line, shown in red in FIGURE 2.10.8, to the graph of $y = f(x)$ at point P. What are $f(-3)$ and $f'(-3)$?

52. Find an equation of the tangent line, shown in red in FIGURE 2.10.9, to the graph of $y = f(x)$ at point P. What is $f'(3)$? What is the y-intercept of the tangent line?

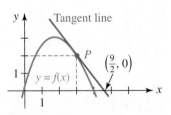

FIGURE 2.10.9 Graph for Problem 52

Chapter 2 Review Exercises Answers to selected odd-numbered problems begin on page ANS-10.

A. Fill in the Blanks

In Problems 1–34, fill in the blanks.

1. If $f(x) = \dfrac{2x^3 - 1}{x^2 + 2}$, then $\left(\frac{1}{2}, \underline{\hspace{1cm}}\right)$ is a point on the graph of f.

2. If $f(x) = \dfrac{Ax}{10x - 2}$ and $f(2) = 3$, then $A = \underline{\hspace{1cm}}$.

3. The domain of the function $f(x) = \dfrac{1}{\sqrt{5 - x}}$ is _____.

4. The range of the function $f(x) = |x| - 10$ is _____.

5. The zeros of the function $f(x) = \sqrt{x^2 - 2x}$ are _____.

6. If the graph of f is symmetric with respect to the y-axis, $f(-x) =$ _____.

7. The graph of $y = \dfrac{x - 1}{x}$ is the graph of $f(x) =$ _____ shifted 1 unit to the right.

8. The point (_____, 3) lies on the graph of $f(x) = \sqrt{x}$ reflected in the y-axis.

9. The lines $2x - 5y = 1$ and $kx + 3y + 3 = 0$ are parallel if $k =$ _____.

10. The x- and y-intercepts of the line $-4x + 3y - 48 = 0$ are _____.

11. The graph of a linear function for which $f(-2) = 0$ and $f(0) = -3$ has slope $m =$ _____.

12. An equation of a line through $(1, 2)$ that is perpendicular to $y = 3x - 5$ is _____.

13. The x- and y-intercepts of the parabola $f(x) = x^2 - 2x - 1$ are _____.

14. The range of the function $f(x) = -x^2 + 6x - 21$ is _____.

15. The quadratic function $f(x) = ax^2 + bx + c$ for which $f(0) = 7$ and whose only x-intercept is $(-2, 0)$ is $f(x) =$ _____.

16. If $f(x) = x + 2$ and $g(x) = x^2 - 2x$, then $(f \circ g)(-1) =$ _____.

17. The vertex of the graph of $f(x) = x^2$ is $(0, 0)$. Therefore, the vertex of the graph of $y = -5(x - 10)^2 + 2$ is _____.

18. Given that $f^{-1}(x) = \sqrt{x} - 4$ is the inverse of a one-to-one function f, and without finding f, the domain of f is _____ and range of f is _____.

19. The x-intercept of a one-to-one function f is $(5, 0)$, and so the y-intercept of f^{-1} is _____.

20. The inverse of $f(x) = \dfrac{x - 5}{2x + 1}$ is $f^{-1} =$ _____.

21. The point $(a, 16a)$ lies on the graph of

$$f(x) = \begin{cases} 4x - 3, & x < 0 \\ x^3, & 0 \le x \le 1 \\ x^2 + 64, & x > 1 \end{cases}$$

for $a =$ _____.

22. If $f(x) = [\![x + 2]\!] - 4$, then $f(-5.3) =$ _____.

23. If the entire graph of a one-to-one function f lies in the fourth quadrant, then graph of f^{-1} lies in the _____ quadrant.

24. The point $(3, 1)$ lies on the graph of a one-to-one function f. If $f^{-1}(2x) = 3$, then $x =$ _____.

In Problems 25–34, refer to FIGURE 2.R.1. Use approximation if necessary.

25. The domain of f is _____.

26. The range of f is _____.

27. x-intercepts of the graph of f are _____.

28. f is decreasing on the intervals _____.

29. f is increasing on the intervals _____.

30. $f(x) > 0$ on the intervals _____.

31. $f(x) < 0$ on the intervals _____.

32. $f(1) =$ _____.

33. The greatest function value on the interval $[2, 6]$ is _____.

34. If $f(x) = 0.5$, then $x =$ _____.

y, (0.5, 3)

$y = f(x)$

(4, 1.2)

(7, 0.5)

1

1

−1

(2, −0.6)

FIGURE 2.R.1 Graph for Problems 25–34

B. True/False

In Problems 1–24, answer true or false.

1. The points $(0, 3)$, $(2, 2)$, and $(6, 0)$ are collinear. _____
2. The graph of a function can have only one y-intercept. _____
3. If f is a function such that $f(a) = f(b)$, then $a = b$. _____
4. The graph of no nonzero function f can be symmetric with respect to the x-axis. _____
5. The domain of $f(x) = (x - 1)^{1/3}$ is $(-\infty, \infty)$. _____
6. If $f(x) = x$ and $g(x) = \sqrt{x + 2}$, then the domain of g/f is $[-2, \infty)$. _____
7. The graph of $y = (x + 2)^2 - 2x - 4$ is the graph of $f(x) = x^2 - 2x$ shifted 2 units to the left. _____
8. The graph of $y = |-x| - 1$ is a reflection of the graph of $f(x) = |x| - 1$ in the x-axis. _____
9. A function f is one-to-one if it never takes on the same value twice. _____
10. Two lines with positive slopes cannot be perpendicular. _____
11. The equation of a vertical line through $(2, -5)$ is $x = 2$. _____
12. A point of intersection of the graphs of f and f^{-1} must lie on the line $y = x$. _____
13. The one-to-one function $f(x) = 1/x$ has the property that $f = f^{-1}$. _____
14. The function $f(x) = 2x^2 + 16x - 2$ decreases on the interval $[-7, -2]$. _____
15. No even function defined on the interval $(-a, a)$, $a > 0$, can be one-to-one. _____
16. All odd functions are one-to-one. _____
17. If a function f is one-to-one, then $f^{-1}(x) = \dfrac{1}{f(x)}$. _____
18. If f is an increasing function on an interval containing $x_1 < x_2$, then $f(x_1) < f(x_2)$. _____
19. The function $f(x) = |x| - 1$ is decreasing on the interval $[0, \infty)$. _____
20. For function composition, $f \circ (g + h) = f \circ g + f \circ h$. _____
21. If the y-intercept for the graph of a function f is $(0, 1)$, then the y-intercept for the graph of $y = 4 - 3f(x)$ is $(0, 1)$. _____
22. For any function f, $f(x_1 + x_2) = f(x_1) + f(x_2)$. _____
23. The graph of $y = x^2 + 4x + 4$ is the graph of $f(x) = x^2$ shifted horizontally to the right. _____
24. The graph of $y = \sqrt{3 + x}$ is the graph of $f(x) = \sqrt{3 - x}$ reflected in the y-axis. _____

C. Review Exercises

In Problems 1 and 2, identify two functions f and g so that $h = f \circ g$.

1. $h(x) = \dfrac{(3x - 5)^2}{x^2}$

2. $h(x) = 4(x + 1) - \sqrt{x + 1}$

3. Write the equation of each new function if the graph of $f(x) = x^3 - 2$ is
   (a) shifted 3 units to the left,
   (b) shifted 5 units down,
   (c) shifted 1 unit to the right and 2 units up,
   (d) reflected in the x-axis,
   (e) reflected in the y-axis,
   (f) vertically stretched by a factor of 3.

FIGURE 2.R.2 Graph for Problem 4

4. FIGURE 2.R.2 shows the graph of a function f whose domain is $(-\infty, \infty)$. Sketch the graph of the following functions.

(a) $y = f(x) - \pi$ (b) $y = f(x - 2)$

(c) $y = f(x + 3) + \dfrac{\pi}{2}$ (d) $y = -f(x)$

(e) $y = f(-x)$ (f) $y = 2f(x)$

In Problems 5 and 6, use the graph of the one-to-one function f in Figure 2.R.2.

5. Give the domain and range of f^{-1}.
6. Sketch the graph of f^{-1}.

In Problems 7–10, the given graph is a rigidly transformed graph of a power function f. Identify the function and then write the equation of the graph.

7.

FIGURE 2.R.3 Graph for Problem 7

8.

FIGURE 2.R.4 Graph for Problem 8

9.

FIGURE 2.R.5 Graph for Problem 9

10.

FIGURE 2.R.6 Graph for Problem 10

11. Express $y = x - |x| + |x - 1|$ as a piecewise-defined function. Sketch the graph of the function.
12. Sketch the graph of the function $y = [\![x]\!] + [\![-x]\!]$. Give the numbers at which the function is discontinuous.

In Problems 13 and 14, by examining the graph of the function f give the domain of the function g.

13. $f(x) = x^2 - 6x + 10$, $g(x) = \sqrt{x^2 - 6x + 10}$

14. $f(x) = -x^2 + 7x - 6$, $g(x) = \dfrac{1}{\sqrt{-x^2 + 7x - 6}}$

In Problems 15 and 16, $f(x) = \dfrac{1}{x + 1}$, $g(x) = \dfrac{5}{x - 2}$. Give the domain of the indicated composition.

15. $f \circ g$ **16.** $g \circ f$

In Problems 17 and 18, the given function f is one-to-one. Find f^{-1}.

17. $f(x) = (x + 1)^3$ 　　　　　　　　　　　**18.** $f(x) = x + \sqrt{x}$

19. Express the area of the yellow region in FIGURE 2.R.7 as a function of h.

20. Determine a quadratic function that describes the parabolic arch shown in FIGURE 2.R.8.

FIGURE 2.R.7 Circle in
Problem 19

FIGURE 2.R.8 Arch in
Problem 20

21. The diameter d of a cube is the distance between opposite vertices as shown in FIGURE 2.R.9. Express the diameter d as a function of the length s of a side of the cube by first expressing the length y of the diagonal in Figure 2.R.9 as a function of s.

22. A circular cylinder of height h is inscribed in a sphere of radius 1 as shown in FIGURE 2.R.10. Express the volume of the cylinder as a function of h.

23. A baseball diamond is a square that is 90 ft on a side. See FIGURE 2.R.11. After a player hits a home run, he jogs around the bases at a rate of 6 ft/s.

　(a) As the player jogs between home base and first base, express his distance from home base as a function of time t, where $t = 0$ corresponds to the time he left home base—that is, $0 \le t \le 15$.

　(b) As the player jogs between home base and first base, express his distance from second base as a function of time t, where $0 \le t \le 15$.

24. Consider the four circles shown in FIGURE 2.R.12. Express the area of the blue region between them as a function of h.

FIGURE 2.R.9 Cube in
Problem 21

FIGURE 2.R.10 Cylinder in
Problem 22

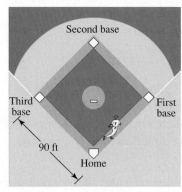

FIGURE 2.R.11 Baseball player in
Problem 23

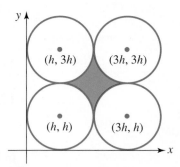

FIGURE 2.R.12 Circles in
Problem 24

In Problems 25–28, find the objective function for the given calculus problem. Do **not** actually attempt to solve the problem.

25. Find the minimum value of the sum of 20 times a positive number and 5 times the reciprocal of that number.

26. A rancher wants to use 100 m of fence to construct a diagonal fence connecting two existing walls that meet at a right angle. How should this be done so that the area enclosed by the walls and fence is a maximum?

27. The running track shown as the red curve in FIGURE 2.R.13 is to consist of two parallel straight parts and two semicircular parts. The length of the track is to be 2 km. Find the design of the track so that the rectangular plot of land enclosed by the track is a maximum.

28. A pipeline is to be constructed from a refinery across a swamp to storage tanks. See FIGURE 2.R.14. The cost of construction is $25,000 per mile over the swamp and $20,000 per mile over land. How should the pipeline be made so that the cost of construction is a minimum?

FIGURE 2.R.13 Running track in Problem 27

FIGURE 2.R.14 Pipeline in Problem 28

In Problems 29–32, compute $f'(x) = \lim\limits_{h \to 0} \dfrac{f(x + h) - f(x)}{h}$ for the given function.

Find an equation of the tangent line to the graph of f at the indicated value of x.

29. $f(x) = -3x^2 + 16x + 12, \quad x = 2$

30. $f(x) = x^3 - x^2, \quad x = -1$

31. $f(x) = \dfrac{-1}{2x^2}, \quad x = \frac{1}{2}$

32. $f(x) = x + 4\sqrt{x}, \quad x = 4$

In Problems 33 and 34, use of the derivative to find the points (if any) on the graph of the function f where the tangent line is horizontal.

33. f in Problem 30

34. f in Problem 32

3 Polynomial and Rational Functions

Chapter Contents

3.1 Polynomial Functions

INTRODUCTION In Chapter 2 we graphed functions such as $y = 3$, $y = 2x - 1$, $y = 5x^2 - 2x + 4$, and $y = x^3$. These functions, in which the variable x is raised to a *nonnegative integer power*, are examples of a more general type of function called a **polynomial function**. Our goal in this section is to examine some of the properties of polynomial functions and to present some general guidelines for graphing such functions. First we state the formal definition of a polynomial function.

DEFINITION 3.1.1 Polynomial Function

A **polynomial function** $y = f(x)$ is a function of the form

$$f(x) = a_n x^n + a_{n-1} x^{n-1} + \cdots + a_2 x^2 + a_1 x + a_0 \qquad (1)$$

where the coefficients $a_n, a_{n-1}, \ldots, a_2, a_1, a_0$ are real numbers and n is a nonnegative integer.

The **domain** of any polynomial function f is the set of all real numbers $(-\infty, \infty)$.

The following functions are *not* polynomial functions:

$$\underset{\text{not a nonnegative integer}}{\downarrow} \qquad \qquad \underset{\text{not a nonnegative integer}}{\downarrow}$$

$$y = 5x^2 - 3x^{-1} \qquad \text{and} \qquad y = 2x^{1/2} - 4.$$

The function

$$\underset{\text{nonnegative integer powers}}{\downarrow \quad \downarrow \quad \downarrow \quad \downarrow \quad \downarrow \quad \downarrow}$$

$$y = 8x^5 - \tfrac{1}{2}x^4 - 10x^3 + 7x^2 + 6x + 4$$

is a polynomial, where we interpret the number 4 as the coefficient of x^0. Since 0 is a nonnegative integer, a constant function such as $y = 3$ is a polynomial function because it is the same as $y = 3x^0$.

☐ **Degree** Polynomial functions are classified by their degree. The highest power of x in a polynomial is said to be its **degree**. So if $a_n \neq 0$, then we say that $f(x)$ in (1) has **degree** n. The number a_n in (1) is called the **leading coefficient** and a_0 is called the **constant term** of the polynomial. For example,

$$\overset{\text{degree}}{\downarrow}$$

$$f(x) = 3x^5 - 4x^3 - 3x + 8,$$

$$\underset{\text{leading coefficient}}{\uparrow} \qquad \qquad \underset{\text{constant term}}{\uparrow}$$

is a polynomial function of degree 5. We have already studied special polynomial functions in Sections 2.3 and 2.4. Polynomial functions of degrees $n = 0$, $n = 1$, and $n = 2$ are, respectively,

$$f(x) = a_0, \qquad\qquad \textbf{constant function}$$
$$f(x) = a_1 x + a_0, \qquad \textbf{linear function} \qquad \text{Section 2.3}$$
$$f(x) = a_2 x^2 + a_1 x + a_0, \quad \textbf{quadratic function} \} \quad \text{Section 2.4}$$

Polynomials of degrees $n = 3$, $n = 4$, and $n = 5$ are, in turn, commonly referred to as **cubic**, **quartic**, and **quintic functions**. The constant function $f(x) = 0$ is called the **zero polynomial**.

☐ **Graphs** Recall that the graph of a constant function $f(x) = a_0$ is a **horizontal line**, the graph of a linear function $f(x) = a_1x + a_0$ is a **line with slope** $m = a_1$, and the graph of a quadratic function $f(x) = a_2x^2 + a_1x + a_0$ is a **parabola**. See Sections 2.3 and 2.4. Such descriptive statements cannot be made about the graph of a higher-degree polynomial function. What is the shape of the graph of a fifth-degree polynomial function? It turns out that the graph of a polynomial function of degree $n \geq 3$ can have several possible shapes. In general, graphing a polynomial function f of degree $n \geq 3$ often demands the use of either calculus or a graphing utility. However, we will see in the discussion that follows that by determining

- shifting,
- end behavior,
- symmetry,
- intercepts, and
- local behavior

of the function we can, in some instances, quickly sketch a reasonable graph of a higher-degree polynomial function while keeping point-plotting to a minimum. Before elaborating on each of these concepts we return to the notion of a power function first introduced in Section 2.2.

☐ **Power Function** A special case of the power function is the **single-term polynomial function** or **monomial**,

$$f(x) = x^n, \quad n \text{ a positive integer.} \tag{2}$$

The graphs of f for degrees $n = 1, 2, 3, 4, 5,$ and 6 are given in FIGURE 3.1.1. The interesting fact about (2) is that all the graphs for n odd are basically the same. The notable characteristics are that the graphs are symmetric about the origin and become increasingly flatter near the origin as the degree n increases. See Figures 3.1.1(a)–3.1.1(c). A similar observation is true for the graphs of (2) for n even, except, of course, the graphs are symmetric with respect to the y-axis. See Figures 3.1.1(d)–3.1.1(f).

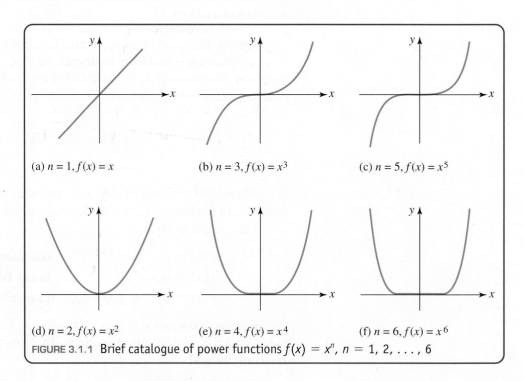

(a) $n = 1, f(x) = x$

(b) $n = 3, f(x) = x^3$

(c) $n = 5, f(x) = x^5$

(d) $n = 2, f(x) = x^2$

(e) $n = 4, f(x) = x^4$

(f) $n = 6, f(x) = x^6$

FIGURE 3.1.1 Brief catalogue of power functions $f(x) = x^n$, $n = 1, 2, \ldots, 6$

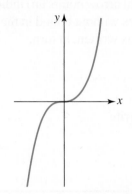

$y = -(x + 2)^3 - 1$

FIGURE 3.1.2 Reflected and shifted graph in Example 1

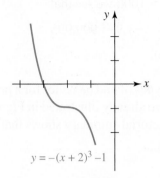

FIGURE 3.1.3 Graph of a polynomial function with $a_n > 0$ of odd degree on $[-1000, 1000]$

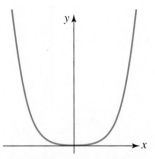

FIGURE 3.1.4 Graph of a polynomial function with $a_n > 0$ of even degree on $[-1000, 1000]$

☐ **Shifted Graphs** Recall from Section 2.2 that for $c > 0$, the graphs of polynomial functions of the form

$$y = ax^n + c, \qquad y = ax^n - c$$

and

$$y = a(x + c)^n, \qquad y = a(x - c)^n$$

can be obtained by vertical and horizontal shifts of the graph of $y = ax^n$. Also, if the leading coefficient a is positive, the graph of $y = ax^n$ is either a vertical stretch or a vertical compression of the graph of the basic single-term polynomial function $f(x) = x^n$. When a is negative we also carry out a reflection in the x-axis.

◼ EXAMPLE 1   **Graphing a Shifted Polynomial Function**

The graph of $y = -(x + 2)^3 - 1$ is the graph of $f(x) = x^3$ reflected in the x-axis, shifted 2 units to the left, and then shifted vertically downward 1 unit. First review Figure 3.1.1(b) and then see FIGURE 3.1.2.   ◼

☐ **End Behavior** The knowledge of the shape of a single-term polynomial function $f(x) = x^n$ is important for another reason. Before reading further, examine the computer-generated graphs given in FIGURE 3.1.3 and FIGURE 3.1.4. Although the graph in Figure 3.1.3 certainly resembles the graphs in Figures 3.1.1(b) and 3.1.1(c), and the graph in Figure 3.1.4 resembles the graphs in Figures 3.1.1(d)–3.11(f), the functions graphed in these two figures are *not* power functions $f(x) = x^n$, n odd, or $f(x) = x^n$, n even. We will not tell you what the specific functions are except to say that they were both graphed on the interval $[-1000, 1000]$. The point is this: the function whose graph is given in Figure 3.1.3 could be almost *any* polynomial function

$$f(x) = a_n x^n \boxed{+ \ a_{n-1}x^{n-1} + \cdots + a_1 x + a_0} \tag{3}$$

with $a_n > 0$, of *odd* degree n, $n = 3, 5, \ldots$ when graphed on $[-1000, 1000]$. Similarly, the graph in Figure 3.1.4 could be that of any polynomial function given in (1), with $a_n > 0$, of *even* degree n, $n = 2, 4, \ldots$ when graphed on a large interval around the origin. As the next theorem indicates, the terms enclosed in the blue rectangle in (3) are irrelevant when we look at a graph of a polynomial globally—that is, for large values of $|x|$. How $f(x)$ behaves when $|x|$ is very large is said to be the **end behavior** a polynomial function f.

THEOREM 3.1.1   End Behavior

For $|x|$ very large, that is, for $x \to -\infty$ and $x \to \infty$, the graph of the polynomial function $f(x) = a_n x^n + a_{n-1}x^{n-1} + \cdots + a_2 x^2 + a_1 x + a_0$, resembles the graph of $y = a_n x^n$.

To see why the graph of a polynomial function such as $f(x) = -2x^3 + 4x^2 + 5$ resembles the graph of the single-term polynomial $y = -2x^3$ when $|x|$ is large, let's factor out the highest power of x, that is, x^3:

both these terms become
negligible when $|x|$ is large
↓   ↓

$$f(x) = x^3\left(-2 + \frac{4}{x} + \frac{5}{x^3}\right). \tag{4}$$

By letting $|x|$ increase without bound, both $4/x$ and $5/x^3$ can be made as close to 0 as we want. Thus when $|x|$ is large, the values of the function f in (4) are closely approximated by the values of $y = -2x^3$. For example, for $x = 1000$ we see that

$$f(1000) = -2(1000)^3 + 4(1000)^2 + 5 = -1,995,999,995$$

whereas $y = -2(1000)^3 = -2,000,000,000.$

☐ **Types of End Behavior** There can be only four types of end behavior for a polynomial function f. Although two of the end behaviors are already illustrated in Figures 3.1.3 and 3.1.4, we include them in FIGURE 3.1.5. This pictorial summary shows that:

> *The end behavior of a polynomial function f depends on its degree n and on the sign of its leading coefficient a_n.*

For example, to interpret the red arrows in Figure 3.1.5 let's examine part (a) of this figure. The position and direction of the left arrow (left arrow points down) indicates that the graph is heading downward (or is falling) when x is negative and large in magnitude. Similarly, the position and direction of the right arrow (right arrow points up) indicates that the graph is heading upward (or is rising) as x increases without bound in the positive direction of the x-axis. In symbols, this end behavior is written, in turn,

$$f(x) \to -\infty \quad \text{as} \quad x \to -\infty,$$

The symbol $\to$ represents the word ▶
approaches. See Section 1.5.   and $f(x) \to \infty \quad \text{as} \quad x \to \infty.$

To describe the end behavior in, say, Figure 3.1.5(c), we write

$$f(x) \to \infty \quad \text{as} \quad x \to -\infty,$$

and $f(x) \to \infty \quad \text{as} \quad x \to \infty.$

$$f(x) = a_n x^n + a_{n-1} x^{n-1} + \cdots + a_1 x + a_0$$

Pictorial Summary of End Behavior of Polynomial Functions

n an odd positive integer

(a) $a_n > 0$ (b) $a_n < 0$

n an even positive integer

(c) $a_n > 0$ (d) $a_n < 0$

FIGURE 3.1.5 End behavior of a polynomial function f depends on its degree n and on the sign of its leading coefficient a_n

For the function $f(x) = -2x^3 + 4x^2 + 5$ in (4), the end behavior is $f(x) \to \infty$ as $x \to -\infty$, and $f(x) \to -\infty$ as $x \to \infty$.

☐ **Relative Extrema** The gaps between the arrows in Figure 3.1.5 correspond to some interval around the origin. In these gaps the graph of f exhibits **local behavior**, in other words, the graph of f shows the characteristics of a polynomial function of a particular degree. This local behavior includes the x- and y-intercepts of the graph, the behavior of the graph at an x-intercept, the turning points of the graph, and observable symmetry of the graph (if any). A **turning point** is a point $(c, f(c))$ at which the graph of a polynomial function f changes direction, that is, the function f changes from increasing to decreasing or vice versa. The graph of a polynomial function of degree n can have up to $n - 1$ turning points. In calculus a turning point corresponds to a **relative**, or **local**, **extremum** of a function f. A relative extremum of f is classified as either a **maximum** or a **minimum**. This leads to the following definition.

DEFINITION 3.1.2   Relative Extremum

(*i*) A number $f(c)$ is a **relative maximum** of a function f if $f(x) \leq f(c)$ for every x in some open interval (a, b) that contains c.

(*ii*) A number $f(c)$ is a **relative minimum** of a function f if $f(x) \geq f(c)$ for every x in some open interval (a, b) that contains c.

---

relative maximum $f(c_1)$

relative minimum $f(c_2)$

$y = f(x)$

FIGURE 3.1.6 Two relative extrema of a function f

If $(c, f(c))$ is a turning point of a polynomial function, then in some interval (a, b) containing c the function value $f(c)$ is either the *largest* (relative maximum) or the *smallest* (relative minimum) function value on the interval. If $f(c)$ is a relative maximum, then the graph of a polynomial function f must change from increasing immediately to the left of c to decreasing immediately to the right of c, whereas if $f(c)$ is a relative minimum the function f changes from decreasing to increasing at c. The graph in FIGURE 3.1.6 shows a graph of a function f with two relative extrema; $f(c_1)$ is a relative maximum on the interval (a_1, b_1) and $f(c_2)$ is a relative minimum on the interval (a_2, b_2).

☐ **Symmetry** It is easy to tell by inspection those polynomial functions whose graphs possess symmetry with respect to either the y-axis or the origin. The words "even" and "odd" functions have special meaning for polynomial functions. Recall that an even function is one for which $f(-x) = f(x)$ and an odd function is one for which $f(-x) = -f(x)$. These two conditions hold for polynomial functions in which all the powers of x are even integers and odd integers, respectively. For example,

even powers
$$f(x) = 5x^4 - 7x^2$$
even function

odd powers
$$f(x) = 10x^5 + 7x^3 + 4x$$
odd function

mixed powers
$$f(x) = -3x^7 + 2x^4 + x^3 + 2$$
neither even nor odd

A function such as $f(x) = 3x^6 - x^4 + 6$ is an even function because the obvious powers are even integers; the constant term 6 is actually $6x^0$, and 0 is an even nonnegative integer.

☐ **Intercepts** The graph of every polynomial function f passes through the y-axis since $x = 0$ is the domain of the function. The y-intercept is the point $(0, f(0))$. Recall that a number c is a **zero** of a function f if $f(c) = 0$. In this discussion we assume c is a real zero. If $x - c$ is a factor of a polynomial function f, that is, $f(x) = (x - c)q(x)$ where $q(x)$ is another polynomial, then clearly $f(c) = 0$ and the corresponding point on the

(a) Simple zero

(b) Zero of odd
multiplicity $m = 3, 5, \ldots$

(c) Zero of even
multiplicity $m = 2, 4, \ldots$

FIGURE 3.1.7 *x*-intercepts of a polynomial function f

graph is $(c, 0)$. Thus the real zeros of a polynomial function are the *x*-coordinates of the *x*-intercepts of its graph. If $(x - c)^m$ is a factor of f, where $m > 1$ is a positive integer, and $(x - c)^{m+1}$ is *not* a factor of f, then c is said to be a **repeated zero**, or more precisely, a **zero of multiplicity m**. For example, $f(x) = x^2 - 10x + 25$ is equivalent to $f(x) = (x - 5)^2$. Hence 5 is a repeated zero or a zero of multiplicity 2. When $m = 1$, c is called a **simple zero**. For example, $-\frac{1}{3}$ and $\frac{1}{2}$ are simple zeros of $f(x) = 6x^2 - x - 1$ since f can be written as $f(x) = 6\left(x + \frac{1}{3}\right)\left(x - \frac{1}{2}\right)$. The behavior of the graph of f at an *x*-intercept $(c, 0)$ depends on whether c is a simple zero or a zero of multiplicity $m > 1$, where m is either an even or an odd integer.

- If c is a simple zero, then the graph of f passes directly through, or crosses, the *x*-axis at $(c, 0)$. See FIGURE 3.1.7(a).
- If c is a zero of odd multiplicity $m = 3, 5, \ldots$, then the graph of f passes through the *x*-axis but is flattened at $(c, 0)$. See Figure 3.1.7(b).
- If c is a zero of even multiplicity $m = 2, 4, \ldots$, then the graph of f is tangent to, or touches, the *x*-axis at $(c, 0)$. See Figure 3.1.7(c).

In the case when c is either a simple zero or a zero of odd multiplicity $m = 3, 5, \ldots$, $f(x)$ changes sign *at* $(c, 0)$, whereas if c is a zero of even multiplicity $m = 2, 4, \ldots$, $f(x)$ does not change sign at $(c, 0)$. We note that depending on the sign of the leading coefficient of the polynomial, the graphs in Figure 3.1.7 could be reflected in the *x*-axis. For example, at a zero of even multiplicity the graph of f could be tangent to the *x*-axis from below that axis.

■ EXAMPLE 2   **Graphing a Polynomial Function**

Graph $f(x) = x^3 - 9x$.

Solution Here are some of the things we look at to sketch the graph of f:

End Behavior: By ignoring all terms but the first, we see that the graph of f resembles the graph of $y = x^3$ for large $|x|$. In other words, the end behavior of the graph is as shown in Figure 3.1.5(a): $f(x) \to -\infty$ as $x \to -\infty$, and $f(x) \to \infty$ as $x \to \infty$.

Symmetry: Since all the powers are odd integers, f is an odd function. The graph of f is symmetric with respect to the origin.

Intercepts: $f(0) = 0$, and so the *y*-intercept is $(0, 0)$. Setting $f(x) = 0$, we see that we must solve $x^3 - 9x = 0$. Factoring

$$\overset{\text{difference of two squares}}{\underset{\downarrow}{}}$$

$$x(x^2 - 9) = 0 \qquad \text{or} \qquad x(x - 3)(x + 3) = 0$$

shows that simple zeros of f are $x = 0$ and $x = \pm 3$. The *x*-intercepts are $(0, 0)$, $(-3, 0)$, and $(3, 0)$.

The Graph: From left to right, the graph rises (f is increasing) from the third quadrant and passes straight through $(-3, 0)$ since -3 is a simple zero. Although the graph is rising as it passes through this intercept it must turn back downward (f decreasing) at some point in the second quadrant to get through the intercept $(0, 0)$. Since the graph is symmetric with respect to the origin, its behavior is just the opposite in the first and fourth quadrants. See FIGURE 3.1.8. ■

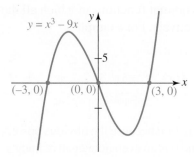

FIGURE 3.1.8 Graph of function in Example 2

In Example 2, the graph of f has two turning points. On the open interval $(-3, 0)$ there is a relative maximum of f and on the open interval $(0, 3)$ there is a relative minimum of f. We made no attempt to locate the corresponding turning points precisely; this is something that would, in general, require techniques from calculus. The best we can do using precalculus mathematics to refine the graph is to resort to plotting additional points on the intervals of interest. By the way, $f(x) = x^3 - 9x$ is the function whose graph on the interval $[-1000, 1000]$ is given in Figure 3.1.3.

■ EXAMPLE 3   Graphing a Polynomial Function

Graph $f(x) = (1 - x)(x + 1)^2$.

Solution Multiplying out, f is the same as $f(x) = -x^3 - x^2 + x + 1$.

End Behavior: From the preceding line we see that the graph of f resembles the graph of $y = -x^3$ for large values of $|x|$, just the opposite of the end behavior of the function in Example 2. See Figure 3.1.5(b).

Symmetry: As we see from $f(x) = -x^3 - x^2 + x + 1$, there are both even and odd powers of x present. Hence the function f is neither even nor odd; its graph possesses no y-axis or origin symmetry.

Intercepts: $f(0) = 1$ so the y-intercept is $(0, 1)$. From the given factored form of $f(x)$, we see that the points $(-1, 0)$ and $(1, 0)$ are the x-intercepts.

The Graph: From left to right, the graph falls (f decreasing) from the second quadrant and then, because -1 is a zero of multiplicity 2, the graph is tangent to the x-axis at $(-1, 0)$. The graph then rises (f increasing) as it passes through the y-intercept $(0, 1)$. At some point in the first quadrant the graph turns downward (f decreasing) and, since 1 is a simple zero, passes through the x-axis at $(1, 0)$, heading downward into the fourth quadrant. See FIGURE 3.1.9. ▬

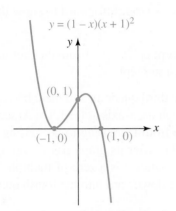

FIGURE 3.1.9 Graph of function in Example 3

In Example 3, there are again two turning points. It should be clear from Figure 3.1.9 that $(-1, 0)$ is a turning point; f changes from decreasing to increasing and so $f(-1) = 0$ is a relative minimum. At the turning point in the first quadrant f changes from increasing to decreasing. From the figure we see f has a relative maximum on the open interval $(0, 1)$.

■ EXAMPLE 4   Zeros of Multiplicity Two

Graph $f(x) = x^4 - 4x^2 + 4$.

Solution Before proceeding, note that the right-hand side of f is a perfect square. That is, $f(x) = (x^2 - 2)^2$. Since $x^2 - 2 = (x - \sqrt{2})(x + \sqrt{2})$, by the laws of exponents we can write

$$f(x) = (x - \sqrt{2})^2(x + \sqrt{2})^2. \tag{5}$$

End Behavior: Inspection of $f(x)$ shows that its graph resembles the graph of $y = x^4$ for large $|x|$. That is, the graph goes up to the left as $x \to -\infty$ and up to the right as $x \to \infty$, as shown in Figure 3.1.5(c).

Symmetry: Because $f(x)$ contains only even powers of x, it is an even function and so its graph is symmetric with respect to the y-axis.

Intercepts: $f(0) = 4$, so the y-intercept is $(0, 4)$. Inspection of (5) shows the x-intercepts are $(-\sqrt{2}, 0)$ and $(\sqrt{2}, 0)$.

The Graph: From left to right, the graph falls from the second quadrant and then, because $-\sqrt{2}$ is a zero of multiplicity 2, the graph touches the x-axis at $(-\sqrt{2}, 0)$. The graph then rises from this point to the y-intercept $(0, 4)$. We then use the y-axis symmetry to finish the graph in the first quadrant. See FIGURE 3.1.10. ▬

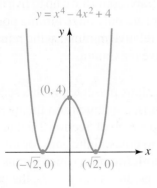

FIGURE 3.1.10 Graph of function in Example 4

3.1 Polynomial Functions   **147**

In Example 4, the graph of f has three turning points. From the even multiplicity of the zeros, along with the y-axis symmetry, it can be deduced that the x-intercepts $\left(-\sqrt{2}, 0\right)$ and $\left(\sqrt{2}, 0\right)$ are turning points and $f(-\sqrt{2}) = 0$ and $f(\sqrt{2}) = 0$ are relative minima, and that the y-intercept $(0, 4)$ is a turning point and $f(0) = 4$ is a relative maximum.

■ EXAMPLE 5   Zero of Multiplicity Three

Graph $f(x) = -(x + 4)(x - 2)^3$.

Solution *End Behavior*: Inspection of f shows that its graph resembles the graph of $y = -x^4$ for large values of $|x|$. This end behavior of f is shown in Figure 3.1.5(d).

Symmetry: The function f is neither even nor odd. It is straightforward to show that $f(-x) \neq f(x)$ and $f(-x) \neq -f(x)$.

Intercepts: $f(0) = (-4)(-2)^3 = 32$, so the y-intercept is $(0, 32)$. From the factored form of $f(x)$, we see that $(-4, 0)$ and $(2, 0)$ are the x-intercepts.

The Graph: From left to right, the graph rises from the third quadrant and then, because -4 is a simple zero, the graph of f passes directly through the x-axis at $(-4, 0)$. At some point in the second quadrant the function f must change from increasing to decreasing to enable its graph to pass through the y-intercept $(0, 32)$. After its graph passes through the y-intercept, the function f continues to decrease but, since 2 is a zero of multiplicity 3, its graph flattens as it passes through $(2, 0)$, heading downward into the fourth quadrant. See **FIGURE 3.1.11**. ■

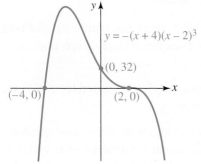

FIGURE 3.1.11 Graph of function in Example 5

Note in Example 5 that since f is of degree 4, its graph could have up to three turning points. But as can be seen from Figure 3.1.11, the graph of f possesses only one turning point and at this point the function value is a relative maximum.

■ EXAMPLE 6   Zeros of Multiplicity Two and Three

Graph $f(x) = (x - 3)(x - 1)^2(x + 2)^3$.

Solution The function f is of degree 6 and so its end behavior resembles the graph of $y = x^6$ for large $|x|$. See Figure 3.1.5(c). Also, the function f is neither even nor odd; its graph possesses no y-axis or origin symmetry. The y-intercept is $(0, f(0)) = (0, -24)$. From the factors of f we see that x-intercepts of the graph are $(-2, 0), (1, 0)$, and $(3, 0)$. Since -2 is a zero of multiplicity 3, the graph of f is flattened as it passes through $(-2, 0)$. Since 1 is a zero of multiplicity 2, the graph of f is tangent to the x-axis at $(1, 0)$. Since 3 is a simple zero, the graph of f passes directly through the x-axis at $(3, 0)$. Putting all these facts together we obtain the graph in **FIGURE 3.1.12**. ■

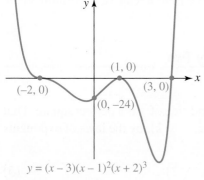

$y = (x - 3)(x - 1)^2(x + 2)^3$

FIGURE 3.1.12 Graph of function in Example 6

In Example 6, since the function f is of degree 6 its graph could have up to five turning points. But as the graph in Figure 3.1.12 shows, there are only three turning points. At two of these points the unknown function values are relative minima; at the remaining point $(1, 0)$, the function value $f(1) = 0$ is a relative maximum.

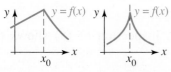

(a) Corner   (b) Cusp

FIGURE 3.1.13 The graph of a polynomial function cannot have corners or cusps

☐ **Continuous Function** As is apparent from the graphs presented in this section a polynomial function is **continuous everywhere**, that is, continuous on the interval $(-\infty, \infty)$. Recall from the discussion of continuity on page 90 of Section 2.5 that this means the graph of a polynomial function f can have no holes, finite gaps, or infinite breaks in it. Moreover, a polynomial function f is a **smooth function** which means that its graph does not contain any sharp corners or any cusps. In **FIGURE 3.1.13(a)** the point $(x_0, f(x_0))$ is a corner of the graph of f whereas $(x_0, f(x_0))$ is a cusp in Figure 3.1.13(b). For example, the functions $f(x) = |x|$ and $f(x) = x^{2/3}$ are continuous everywhere but are not smooth functions; the graph of $f(x) = |x|$ has a corner at the origin whereas the graph of $f(x) = x^{2/3}$ has a cusp at the origin. See Figures 2.5.6(a) and 2.2.1(i).

☐ **Intermediate Value Theorem** The following result is a direct consequence of continuity.

THEOREM 3.1.2 Intermediate Value Theorem

Suppose $y = f(x)$ is a continuous function on the closed interval $[a, b]$. If $f(a) \neq f(b)$ for $a < b$, and if N is any number between $f(a)$ and $f(b)$, then there exists a number c in the open interval (a, b) for which $f(c) = N$.

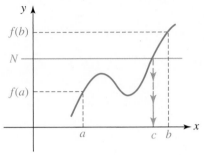

FIGURE 3.1.14 $f(x)$ takes on all values between $f(a)$ and $f(b)$

As we see in FIGURE 3.1.14, the Intermediate Value Theorem simply states that $f(x)$ takes on all values between the numbers $f(a)$ and $f(b)$. In particular, if the function values $f(a)$ and $f(b)$ have opposite signs, then by identifying $N = 0$, we can say that there is at least one number c in (a, b) for which $f(c) = 0$. In other words, if either $f(a) > 0$, $f(b) < 0$ or $f(a) < 0$, $f(b) > 0$, then $f(x)$ has at least one zero c in the interval (a, b). The plausibility of this conclusion is illustrated in FIGURE 3.1.15.

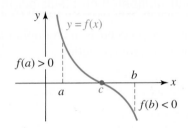

(a) One zero c in (a, b)

(b) Three zeros c_1, c_2, c_3 in (a, b)

FIGURE 3.1.15 Locating zeros using the Intermediate Value Theorem

---

EXAMPLE 7 **Using the Intermediate Value Theorem**

Consider the polynomial function $f(x) = x^3 + x - 1$. Because f is continuous on the interval $[-1, 1]$.

$$f(-1) = -3 < 0 \quad \text{and} \quad f(1) = 1 > 0,$$

and 0 satisfies $-3 < 0 < 1$, we know from the Intermediate Value Theorem that the graph of f must cross the line $y = 0$ (the x-axis). More precisely, there is a real number c in the open interval $(-1, 1)$ such that $f(c) = 0$, that is, c is a zero of the function f. ■

NOTES FROM THE CLASSROOM
••

In *(ii)* of the *Notes from the Classroom* at the end of Section 2.1 we noted that a function could depend on several variables. A generalization of the monomial (2) to two independent variables x and y is given by $f(x, y) = x^m y^n$ where m and n are nonnegative integers. The **degree** of f is defined to be $m + n$. For example, the degree of $f(x, y) = x^3 y^6$ is $m + n = 3 + 6 = 9$. Analogous to (1), **a polynomial in two variables** is finite sum of constant multiples of monomials. The degree of a polynomial of several variables is defined to be the highest degree of the monomials that comprise the function. For example, to find the degree of the polynomial function

$$f(x, y) = 3x^3y^2 - 10xy^6 + 2x^4y^2 + 2y^3 - 5x$$

we examine the degree of each of the five monomials:

$$\text{degree of } x^3y^2 : 3 + 2 = 5$$
$$\text{degree of } xy^6 : 1 + 6 = 7 \quad \leftarrow \text{highest degree}$$
$$\text{degree of } x^4y^2 : 4 + 2 = 6$$
$$\text{degree of } y^3 = x^0y^3 : 0 + 3 = 3$$
$$\text{degree of } x = x^1y^0 : 1 + 0 = 1.$$

Hence the degree of the function f is 7.

Exercises 3.1   Answers to selected odd-numbered problems begin on page ANS–11.

In Problems 1–8, proceed as in Example 1 and use transformations to sketch the graph of the given polynomial function.

1. $y = x^3 - 3$ **2.** $y = -(x + 2)^3$
3. $y = (x - 2)^3 + 2$ **4.** $y = 3 - (x + 2)^3$
5. $y = (x - 5)^4$ **6.** $y = x^4 - 1$
7. $y = 1 - (x - 1)^4$ **8.** $y = 4 + (x + 1)^4$

In Problems 9–12, determine whether the given polynomial function f is even, odd, or neither even nor odd. Do not graph.

9. $f(x) = -2x^3 + 4x$ **10.** $f(x) = x^6 - 5x^2 + 7$
11. $f(x) = x^5 + 4x^3 + 9x + 1$ **12.** $f(x) = x^3(x + 2)(x - 2)$

In Problems 13–18, match the given graph with one of the polynomial functions in (a)–(f).

(a) $f(x) = x^2(x - 1)^2$ **(b)** $f(x) = -x^3(x - 1)$
(c) $f(x) = x^3(x - 1)^3$ **(d)** $f(x) = -x(x - 1)^3$
(e) $f(x) = -x^2(x - 1)$ **(f)** $f(x) = x^3(x - 1)^2$

13.

FIGURE 3.1.16 Graph for Problem 13

14.

FIGURE 3.1.17 Graph for Problem 14

15.

FIGURE 3.1.18 Graph for Problem 15

CHAPTER 3 POLYNOMIAL AND RATIONAL FUNCTIONS

16.

FIGURE 3.1.19 Graph for
Problem 16

17.

FIGURE 3.1.20 Graph for
Problem 17

18.

FIGURE 3.1.21 Graph for
Problem 18

In Problems 19–22, construct a polynomial function f that has the given properties.
There is no unique answer.

19. f is of degree 4, its graph is symmetric with respect to the y-axis, y-intercept is $(0, -6)$

20. f is of degree 5, 0 is a zero of multiplicity 3, its graph is symmetric with respect to the origin

21. f has four real zeros, 1 is a simple zero, -3 is zero of multiplicity 2, behaves like $y = -7x^4$ for large values of $|x|$

22. f is of degree 6, has four real zeros, 2 is a zero of multiplicity 3, behaves like $y = 2x^6$ for large values of $|x|$, $f(0) = 8$

In Problems 23–44, proceed as in Example 2 and sketch the graph of the given polynomial function f.

23. $f(x) = x^3 - 4x$

24. $f(x) = 9x - x^3$

25. $f(x) = -x^3 + x^2 + 6x$

26. $f(x) = x^3 + 7x^2 + 12x$

27. $f(x) = (x + 1)(x - 2)(x - 4)$

28. $f(x) = (2 - x)(x + 2)(x + 1)$

29. $f(x) = x^4 - 4x^3 + 3x^2$

30. $f(x) = x^2(x - 2)^2$

31. $f(x) = (x^2 - x)(x^2 - 5x + 6)$

32. $f(x) = x^2(x^2 + 3x + 2)$

33. $f(x) = (x^2 - 1)(x^2 + 9)$

34. $f(x) = x^4 + 5x^2 - 6$

35. $f(x) = -x^4 + 2x^2 - 1$

36. $f(x) = x^4 - 6x^2 + 9$

37. $f(x) = x^4 + 3x^3$

38. $f(x) = x(x - 2)^3$

39. $f(x) = x^5 - 4x^3$

40. $f(x) = (x - 2)^5 - (x - 2)^3$

41. $f(x) = 3x(x + 1)^2(x - 1)^2$

42. $f(x) = (x + 1)^2(x - 1)^3$

43. $f(x) = -\frac{1}{2}x^2(x + 2)^3(x - 2)^2$

44. $f(x) = x(x + 1)^2(x - 2)(x - 3)$

FIGURE 3.1.22 Graph for Problem 45

45. The graph of $f(x) = x^3 - 3x$ is given in FIGURE 3.1.22.
   (a) Use the figure to obtain the graph of $g(x) = f(x) + 2$.
   (b) Using only the graph obtained in part (a) write an equation, in *factored* form, for $g(x)$. Then verify by multiplying out the factors that your equation for $g(x)$ is the same as $f(x) + 2 = x^3 - 3x + 2$.

46. Find a polynomial function f of lowest possible degree whose graph is consistent with the graph given in FIGURE 3.1.23.

47 Find the value of k such that $(2, 0)$ is an x-intercept for the graph of $f(x) = kx^5 - x^2 + 5x + 8$.

48. Find the values of k_1 and k_2 such that $(-1, 0)$ and $(1, 0)$ are x-intercepts for the graph of $f(x) = k_1x^4 - k_2x^3 + x - 4$.

49. Find the value of k such that $(0, 10)$ is the y-intercept for the graph of $f(x) = x^3 - 2x^2 + 14x - 3k$.

50. Consider the polynomial function $f(x) = (x - 2)^{n+1}(x + 5)$, where n is a positive integer. For what values of n does the graph of f touch, but not cross, the x-axis at $(2, 0)$?

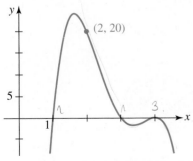

FIGURE 3.1.23 Graph for Problem 46

51. Consider the polynomial function $f(x) = (x - 1)^{n+2}(x + 1)$, where n is a positive integer. For what values of n does the graph of f cross the x-axis at $(1, 0)$?

52. Consider the polynomial function $f(x) = (x - 5)^{2m}(x + 1)^{2n-1}$, where m and n are positive integers.

    (a) For what values of m does the graph of f cross the x-axis at $(5, 0)$?

    (b) For what values of n does the graph of f cross the x-axis at $(-1, 0)$?

In Problems 53 and 54, use the Intermediate Value Theorem to determine whether there is a zero of the given function f in the indicated intervals. Do not graph.

53. $f(x) = 60x^3 - 13x^2 - 145x - 28; \quad [-2, -1], [-1, 0], [0, 1], [1, 2]$

54. $f(x) = 8x^4 - 23x^3 + 23x^2 - 26x + 15; \quad [-1, 0], [0, 1], [1, 2], [2, 3]$

Calculus-Related Problems

55. Constructing a Box An open box can be made from a rectangular piece of cardboard by cutting a square of length x from each corner and bending up the sides. See FIGURE 3.1.24. If the cardboard measures 30 cm by 40 cm, show that the volume of the resulting box is given by

$$V(x) = x(30 - 2x)(40 - 2x).$$

Sketch the graph of $V(x)$ for $x > 0$. What is the domain of the function V?

FIGURE 3.1.24 Box in Problem 55

56. Another Box In order to hold its shape, the box in Problem 55 will require tape or some other fastener at the corners. An open box that holds itself together can be made by cutting out a square of length x from each corner of a rectangular piece of cardboard, cutting on the solid line, and folding on the dashed lines, as shown in FIGURE 3.1.25. Find a polynomial function $V(x)$ that gives the volume of the resulting box if the original cardboard measures 30 cm by 40 cm. Sketch the graph of $V(x)$ for $x > 0$.

FIGURE 3.1.25 Box in Problem 56

57. **Making a Cup** A conical cup is made from a circular piece of paper of radius R by cutting out a circular sector and then joining the dashed edges as shown in FIGURE 3.1.26. Find a polynomial function $V(h)$ that gives the volume of the conical cup in terms of its height.

58. **Hourglass** Sand flows from the top half of the conical hourglass shown in FIGURE 3.1.27 to the bottom half at a constant rate. Find a polynomial function $V(h)$ that gives the volume of the bottom pile of sand in terms of the height of the sand. Assume that the top of the pile is level. [*Hint*: Use similar triangles.]

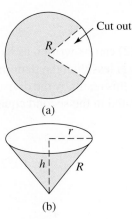

(a)

(b)

FIGURE 3.1.26 Cup in Problem 57

FIGURE 3.1.27 Hourglass in Problem 58

Calculator/Computer Problems

In Problems 59 and 60, use a graphing utility to examine the graph of the given polynomial function on the indicated intervals.

59. $f(x) = -(x - 8)(x + 10)^2$; $[-15, 15]$, $[-100, 100]$, $[-1000, 1000]$
60. $f(x) = (x - 5)^2(x + 5)^2$; $[-10, 10]$, $[-100, 100]$, $[-1000, 1000]$

For Discussion

61. Examine Figure 3.1.5. Then discuss whether there can exist cubic polynomial functions that have no real zeros.
62. Suppose a polynomial function f has three zeros, -3, 2, and 4, and has the end behavior that its graph goes down to the left as $x \to -\infty$ and down to the right as $x \to \infty$. Discuss possible equations for f.
63. Suppose f is a polynomial function of degree m and g is a polynomial function of degree n. What is the degree of $f + g$? Of fg?
64. Suppose f and g are polynomial functions. If c is a zero of f and of g, then show that c is a zero of $f + g$ and fg.

3.2 Division of Polynomial Functions

INTRODUCTION If $p > 0$ and $s > 0$ are integers such that $p \geq s$, then p/s is called an **improper fraction**. By dividing p by s, we obtain unique numbers q and r that satisfy

$$\frac{p}{s} = q + \frac{r}{s} \qquad \text{or} \qquad p = sq + r, \tag{1}$$

where $0 \leq r < s$. The number p is called the **dividend**, s is the **divisor**, q is the **quotient**, and r is the **remainder**. For example, consider the improper fraction $\frac{1052}{23}$. Performing long division gives

$$
\begin{array}{r}
45 \quad \leftarrow \text{quotient} \\
23 \overline{)1052} \quad \leftarrow \text{dividend} \\
92 \quad \leftarrow \text{subtract} \\
\hline
132 \\
115 \\
\hline
17. \quad \leftarrow \text{remainder}
\end{array}
\qquad (2)
$$

divisor $\rightarrow$

The result in (2) can be written as $\frac{1052}{23} = 45 + \frac{17}{23}$, where $\frac{17}{23}$ is a **proper fraction** since the numerator is less than the denominator; in other words, the fraction is less than 1. If we multiply this result by the divisor 23 we obtain the special way of writing the dividend p illustrated in the second equation in (1):

$$
1052 = 23 \cdot \overset{\text{quotient}}{45} + \overset{\text{remainder}}{17}.
\qquad (3)
$$

divisor

☐ **Division of Polynomials** The method for dividing two polynomial functions $f(x)$ and $d(x)$ is similar to division of positive integers. If the degree of a polynomial $f(x)$ is greater than or equal to the degree of the polynomial $d(x)$, then $f(x)/d(x)$ is also called an **improper fraction**. A result analogous to (1) is called the **Division Algorithm** for polynomials.

THEOREM 3.2.1 Division Algorithm

Let $f(x)$ and $d(x) \neq 0$ be polynomials where the degree of $f(x)$ is greater than or equal to the degree of $d(x)$. Then there exist unique polynomials $q(x)$ and $r(x)$ such that

$$
\frac{f(x)}{d(x)} = q(x) + \frac{r(x)}{d(x)} \qquad \text{or} \qquad f(x) = d(x)q(x) + r(x) \qquad (4)
$$

where $r(x)$ has degree less than the degree of $d(x)$.

The polynomial $f(x)$ is called the **dividend**, $d(x)$ the **divisor**, $q(x)$ the **quotient**, and $r(x)$ the **remainder**. Because $r(x)$ has degree less than the degree of $d(x)$, the rational expression $r(x)/d(x)$ is called a **proper fraction**.

Observe in (4) when $r(x) = 0$, then $f(x) = d(x)q(x)$, and so the divisor $d(x)$ is a factor of $f(x)$. In this case, we say that $f(x)$ is **divisible** by $d(x)$ or, in older terminology, $d(x)$ **divides evenly** into $f(x)$.

EXAMPLE 1   **Division of Two Polynomials**

Use long division to find the quotient $q(x)$ and remainder $r(x)$ when the polynomial $f(x) = 3x^3 - x^2 - 2x + 6$ is divided by the polynomial $d(x) = x^2 + 1$.

Solution By long division,

$$
\begin{array}{r}
3x \;-\; 1 \quad \leftarrow \text{quotient} \\
x^2 + 1 \overline{)3x^3 - x^2 - 2x + 6} \quad \leftarrow \text{dividend} \\
\underline{3x^3 + 0x^2 + 3x} \quad \leftarrow \text{subtract} \\
-x^2 - 5x + 6 \\
\underline{-x^2 + 0x - 1} \\
-5x + 7 \quad \leftarrow \text{remainder}
\end{array}
$$

(5)

The result of the division in (5) can be written

$$
\frac{3x^3 - x^2 - 2x + 6}{x^2 + 1} = \overset{q(x)}{\overbrace{3x - 1}} + \overset{r(x)}{\overbrace{\frac{-5x + 7}{x^2 + 1}}}.
$$

If we multiply both sides of the last equation by the divisor $x^2 + 1$, we get the second form given in (4):

$$
3x^3 - x^2 - 2x + 6 = (x^2 + 1)(3x - 1) + (-5x + 7). \qquad (6) \quad \blacksquare
$$

If the divisor $d(x)$ is a linear polynomial $x - c$, it follows from the Division Algorithm that the degree of the remainder r is 0, that is to say, r is a constant. Thus (4) becomes

$$
f(x) = (x - c)q(x) + r. \qquad (7)
$$

When the number $x = c$ is substituted into (7), we discover an alternative way of evaluating a polynomial function:

$$
f(c) = \overset{0}{\overbrace{(c - c)}}q(c) + r \quad \text{or} \quad f(c) = r.
$$

The foregoing result is called the **Remainder Theorem**.

THEOREM 3.2.2 Remainder Theorem

If a polynomial $f(x)$ is divided by a linear polynomial $x - c$, then the remainder r is the value of $f(x)$ at $x = c$, that is, $f(c) = r$.

EXAMPLE 2 **Finding the Remainder**

Use the Remainder Theorem to find r when $f(x) = 4x^3 - x^2 + 4$ is divided by $x - 2$.

Solution From the Remainder Theorem, the remainder r is the value of the function f evaluated at $x = 2$:

$$
r = f(2) = 4(2)^3 - (2)^2 + 4 = 32. \qquad (8) \quad \blacksquare
$$

Example 2, where a remainder r is determined by calculating a function value $f(c)$, is more interesting than important. What *is* important is the reverse problem: Determine the function value $f(c)$ by finding the remainder r by division of f by $x - c$. The next two examples illustrate this concept.

EXAMPLE 3 **Evaluation by Division**

Use the Remainder Theorem to find $f(c)$ for $f(x) = x^5 - 4x^3 + 2x - 10$ when $c = -3$.

Solution The value $f(-3)$ is the remainder when $f(x) = x^5 - 4x^3 + 2x - 10$ is divided by $x - (-3) = x + 3$. For the purposes of long division we must account for the missing x^4 and x^2 terms by rewriting the dividend as

$$f(x) = x^5 + 0x^4 - 4x^3 + 0x^2 + 2x - 10.$$

Then,

$$
\begin{array}{r}
x^4 - 3x^3 + 5x^2 - 15x + 47 \\
x + 3 \overline{) x^5 + 0x^4 - 4x^3 + 0x^2 + 2x - 10} \\
\underline{x^5 + 3x^4} \\
-3x^4 - 4x^3 + 0x^2 + 2x - 10 \\
\underline{-3x^4 - 9x^3} \\
5x^3 + 0x^2 + 2x - 10 \\
\underline{5x^3 + 15x^2} \\
-15x^2 + 2x - 10 \\
\underline{-15x^2 - 45x} \\
47x - 10 \\
\underline{47x + 141} \\
-151
\end{array}
\qquad (9)
$$

The remainder r in the division is the value of the function f at $x = -3$, that is, $f(-3) = -151$. ■

☐ **Synthetic Division** After working through Example 3 one could justifiably ask the question: Why would anyone want to calculate the value of a polynomial function f by division? The answer is: We would not bother to do this were it not for **synthetic division**. Synthetic division is a shorthand method of dividing a polynomial $f(x)$ by a *linear* polynomial $x - c$; it does not require writing down the various powers of the variable x but only the coefficients of these powers in the dividend $f(x)$ (which must include all 0 coefficients). It is also a very efficient and quick way of evaluating $f(c)$, since the process utilizes only the arithmetic operations of multiplication and addition. No exponentiations such as 2^3 and 2^2 in (8) are involved. Here is the same division in (9) done synthetically:

For a review of synthetic division please ▶ see the *Student Resource Manual* that accompanies this text.

$$
\begin{array}{r|rrrrrr}
-3 & 1 & 0 & -4 & 0 & 2 & -10 \\
   &   & -3 & 9 & -15 & 45 & -141 \\
\hline
   & 1 & -3 & 5 & -15 & 47 & \boxed{-151} = r = f(-3)
\end{array}
\qquad (10)
$$

Recall that the bottom line of numbers in (10) are the coefficients of the various powers of x in the quotient $q(x)$ when $f(x) = x^5 - 4x^3 + 2x - 10$ is divided by $x + 3$. You should compare this with the quotient obtained by the long division in (9).

▉ EXAMPLE 4   Using Synthetic Division to Evaluate a Function

Use the remainder theorem to find $f(c)$ for

$$f(x) = -3x^6 + 4x^5 + x^4 - 8x^3 - 6x^2 + 9$$

when $c = 2$.

Solution We will use synthetic division to find the remainder r in the division of f by $x - 2$. We begin by writing down all the coefficients in $f(x)$, including 0 as the coefficient of x. From

$$
\begin{array}{r|rrrrrrr}
2 & -3 & 4 & 1 & -8 & -6 & 0 & 9 \\
  &   & -6 & -4 & -6 & -28 & -68 & -136 \\
\hline
  & -3 & -2 & -3 & -14 & -34 & -68 & \boxed{-127} = r
\end{array}
$$

we see that $f(2) = -127$. ■

156

EXAMPLE 5   Using Synthetic Division to Evaluate a Function

Use synthetic division to evaluate $f(x) = x^3 - 7x^2 + 13x - 15$ at $x = 5$.

Solution From the synthetic division

$$
\begin{array}{r|rrrr}
5 & 1 & -7 & 13 & -15 \\
  &   & 5 & -10 & 15 \\
\hline
  & 1 & -2 & 3 & \boxed{0} = r
\end{array}
\tag{11}
$$

we see that $f(5) = 0$.

☐ **Zeros and Factors** The result $f(5) = 0$ in Example 5 shows that 5 is a zero of the given function f. Moreover, because $r = 0$ in (11) we have shown additionally that f is divisible by the linear polynomial $x - 5$, or put another way, $x - 5$ is a *factor* of f. The synthetic division (11) shows that $f(x) = x^3 - 7x^2 + 13x - 15$ is equivalent to

$$f(x) = (x - 5)(x^2 - 2x + 3).$$

In the next section we will further explore the use of the Division Algorithm and Remainder Theorem as a help in finding zeros and factors of a polynomial function. Problems 47–50 in Exercises 3.2 give a preview of this material.

Exercises 3.2   Answers to selected odd-numbered problems begin on page ANS-11.

In Problems 1–10, use long division to find the quotient $q(x)$ and remainder $r(x)$ when the polynomial $f(x)$ is divided by the given polynomial $d(x)$. In each case write your answer in the form $f(x) = d(x)q(x) + r(x)$.

1. $f(x) = 8x^2 + 4x - 7$; $d(x) = x^2$
2. $f(x) = x^2 + 2x - 3$; $d(x) = x^2 + 1$
3. $f(x) = 5x^3 - 7x^2 + 4x + 1$; $d(x) = x^2 + x - 1$
4. $f(x) = 14x^3 - 12x^2 + 6$; $d(x) = x^2 - 1$
5. $f(x) = 2x^3 + 4x^2 - 3x + 5$; $d(x) = (x + 2)^2$
6. $f(x) = x^3 + x^2 + x + 1$; $d(x) = (2x + 1)^2$
7. $f(x) = 27x^3 + x - 2$; $d(x) = 3x^2 - x$
8. $f(x) = x^4 + 8$; $d(x) = x^3 + 2x - 1$
9. $f(x) = 6x^5 + 4x^4 + x^3$; $d(x) = x^3 - 2$
10. $f(x) = 5x^6 - x^5 + 10x^4 + 3x^2 - 2x + 4$; $d(x) = x^2 + x - 1$

In Problems 11–16, proceed as in Example 2 and use the Remainder Theorem to find r when $f(x)$ is divided by the given linear polynomial.

11. $f(x) = 2x^2 - 4x + 6$; $x - 2$
12. $f(x) = 3x^2 + 7x - 1$; $x + 3$
13. $f(x) = x^3 - 4x^2 + 5x + 2$; $x - \frac{1}{2}$
14. $f(x) = 5x^3 + x^2 - 4x - 6$; $x + 1$
15. $f(x) = x^4 - x^3 + 2x^2 + 3x - 5$; $x - 3$
16. $f(x) = 2x^4 - 7x^2 + x - 1$; $x + \frac{3}{2}$

In Problems 17–22, proceed as in Example 3 and use the Remainder Theorem to find $f(c)$ for the given value of c.

17. $f(x) = 4x^2 - 10x + 6$; $c = 2$
18. $f(x) = 6x^2 + 4x - 2$; $c = \frac{1}{4}$

19. $f(x) = x^3 + 3x^2 + 6x + 6; \quad c = -5$

20. $f(x) = 15x^3 + 17x^2 - 30; \quad c = \frac{1}{5}$

21. $f(x) = 3x^4 - 5x^2 + 20; \quad c = \frac{1}{2}$

22. $f(x) = 14x^4 - 60x^3 + 49x^2 - 21x + 19; \quad c = 1$

In Problems 23–32, use synthetic division to find the quotient $q(x)$ and remainder r when $f(x)$ is divided by the given linear polynomial.

23. $f(x) = 2x^2 - x + 5; \quad x - 2$

24. $f(x) = 4x^2 - 8x + 6; \quad x - \frac{1}{2}$

25. $f(x) = x^3 - x^2 + 2; \quad x + 3$

26. $f(x) = 4x^3 - 3x^2 + 2x + 4; \quad x - 7$

27. $f(x) = x^4 + 16; \quad x - 2$

28. $f(x) = 4x^4 + 3x^3 - x^2 - 5x - 6; \quad x + 3$

29. $f(x) = x^5 + 56x^2 - 4; \quad x + 4$

30. $f(x) = 2x^6 + 3x^3 - 4x^2 - 1; \quad x + 1$

31. $f(x) = x^3 - (2 + \sqrt{3})x^2 + 3\sqrt{3}x - 3; \quad x - \sqrt{3}$

32. $f(x) = x^6 + x^4 - 3x^2 + 7; \quad x + \sqrt{2}$

In Problems 33–40, use synthetic division and the Remainder Theorem to find $f(c)$ for the given value of c.

33. $f(x) = 4x^2 - 2x + 9; \quad c = -3$

34. $f(x) = -9x^2 + 18x - 10; \quad c = \frac{1}{3}$

35. $f(x) = 2x^3 - 3x^2 - 8x + 6; \quad c = \sqrt{2}$

36. $f(x) = 3x^4 - 5x^2 + 27; \quad c = \frac{1}{2}$

37. $f(x) = 14x^4 - 60x^3 + 49x^2 - 21x + 19; \quad c = 1$

38. $f(x) = 3x^5 + x^2 - 16; \quad c = -2$

39. $f(x) = 2x^6 - 3x^5 + x^4 - 2x + 1; \quad c = 4$

40. $f(x) = x^7 - 3x^5 + 2x^3 - x + 10; \quad c = 5$

In Problems 41 and 42, use long division to find a value of k such that $f(x)$ is divisible by $d(x)$.

41. $f(x) = x^4 + x^3 + 3x^2 + kx - 4; \quad d(x) = x^2 - 1$

42. $f(x) = x^5 - 3x^4 + 7x^3 + kx^2 + 9x - 5; \quad d(x) = x^2 - x + 1$

In Problems 43 and 44, use synthetic division to find a value of k such that $f(x)$ is divisible by $d(x)$.

43. $f(x) = kx^4 + 2x^2 + 9k; \quad d(x) = x - 1$

44. $f(x) = x^3 + kx^2 - 2kx + 4; \quad d(x) = x + 2$

45. Find a value of k such that the remainder in the division of $f(x) = 3x^2 - 4kx + 1$ by $d(x) = x + 3$ is $r = -20$.

46. When $f(x) = x^2 - 3x - 1$ is divided by $x - c$, the remainder is $r = 3$. Determine c.

For Discussion

47. Reread the last paragraph of this section. Then use the fact that c is a zero of $f(x) = x^5 - c^5$ to show that

$$f(x) = (x - c)(x^4 + cx^3 + c^2x^2 + c^3x + c^4).$$

48. Use the result in Problem 47 to write $f(x) = x^5 - 32$ as a product of two factors.

49. Use synthetic division to show that

$$f(x) = x^3 + 3x^2 - 13x - 15$$

is divisible by the linear factor $x + 1$. Use this result to express f as a product of three linear factors.

50. If $n > 2$ is an even positive integer, explain why $f(x) = x^n - c^n$ can be written as a product of three factors.

51. Use the result in Problem 50 to write $f(x) = x^8 - 256$ as a product of three factors.

52. Suppose n is a positive integer. Show that $f(x) = x^{3n} + 1$ is divisible by $d(x) = x^n + 1$.

3.3 Zeros and Factors of Polynomial Functions

INTRODUCTION In Section 2.1 we saw that a zero of a function f is a number c for which $f(c) = 0$. A zero c of a function f can be a *real number* or a *complex number*. Recall from algebra that a **complex number** is a number of the form

$$z = a + bi, \qquad \text{where} \qquad i^2 = -1,$$

and a and b are real numbers. The number a is called the **real part** of z and b is called the **imaginary part** of z. The symbol i is called the **imaginary unit** and it is common practice to write it as $i = \sqrt{-1}$. If $z = a + bi$ is a complex number, then $\bar{z} = a - bi$ is called its **conjugate**. Thus the simple polynomial function $f(x) = x^2 + 1$ has two complex zeros since the solutions of $x^2 + 1 = 0$ are $\pm\sqrt{-1}$, that is, i and $-i$.

The operations of addition and subtraction of complex numbers $z_1 = a + bi$ and $z_2 = c + di$ are defined as

$$z_1 + z_2 = (a + bi) + (c + di) = (a + c) + (b + d)i$$

$$z_1 - z_2 = (a + bi) - (c + di) = (a - c) + (b - d)i.$$

Do not memorize the foregoing definitions. Because the basic properties of the real number system carry over to the complex number system, it suffices to know that:

> The **sum** (*difference*) *of two complex numbers is obtained by adding* (*subtracting*) *their corresponding real and imaginary parts.*

Also,

> The **product** *of two complex numbers can be obtained by using the associative, commutative, and distributive laws and the fact that* $i^2 = -1$.

For example, the sum $z_1 + z_2$, difference $z_1 - z_2$, and product $z_1 z_2$ of $z_1 = 5 - 6i$ and $z_2 = 2 + 4i$ are, in turn,

$$z_1 + z_2 = (5 - 6i) + (2 + 4i) = (5 + 2) + (-6 + 4)i = 7 - 2i$$
$$z_1 - z_2 = (5 - 6i) - (2 + 4i) = (5 - 2) + (-6 - 4)i = 3 - 10i$$
$$z_1 z_2 = (5 - 6i)(2 + 4i) = (5 - 6i)2 + (5 - 6i)4i \leftarrow \text{distributive law}$$

$$= 10 - 12i + 20i - 24i^2 \leftarrow \left\{ \begin{array}{l} \text{factor } i \text{ from the two middle} \\ \text{terms and then replace } i^2 \text{ with } -1 \end{array} \right.$$

$$= 10 - 24(-1) + (-12 + 20)i$$

$$= 34 + 8i.$$

In this section we explore the connection between the zeros of a polynomial function f, the operation of division, and the factors of f.

EXAMPLE 1 **A Real Zero**

Consider the polynomial function $f(x) = 2x^3 - 9x^2 + 6x - 1$. The real number $\frac{1}{2}$ is a zero of the function since

$$\begin{aligned} f(\tfrac{1}{2}) &= 2(\tfrac{1}{2})^3 - 9(\tfrac{1}{2})^2 + 6(\tfrac{1}{2}) - 1 \\ &= 2(\tfrac{1}{8}) - \tfrac{9}{4} + 3 - 1 \\ &= \tfrac{1}{4} - \tfrac{9}{4} + \tfrac{8}{4} = 0. \end{aligned}$$

EXAMPLE 2 **A Complex Zero**

See (7) in Section 1.5. ▶

Consider the polynomial function $f(x) = x^3 - 5x^2 + 8x - 6$. The complex number $1 + i$ is a zero of the function. To verify this we use the binomial expansion of $(a + b)^3$ and the fact that $i^2 = -1$ and $i^3 = i^2 \cdot i = (-1)i = -i$:

$$\begin{aligned} f(1 + i) &= (1 + i)^3 - 5(1 + i)^2 + 8(1 + i) - 6 \\ &= (1^3 + 3 \cdot 1^2 \cdot i + 3 \cdot 1 \cdot i^2 + i^3) - 5(1^2 + 2i + i^2) + 8(1 + i) - 6 \\ &= (-2 + 2i) - 5(2i) + (2 + 8i) \\ &= (-2 + 2) + (10 - 10)i = 0 + 0i = 0. \end{aligned}$$

☐ **Factor Theorem** We can now relate the notion of a zero of a polynomial function f with division of polynomials. From the Remainder Theorem we know that when $f(x)$ is divided by the linear polynomial $x - c$ the remainder is $r = f(c)$. If c is a zero of f, then $f(c) = 0$ implies $r = 0$. From the form of the Division Algorithm given in (4) of Section 3.2 we can then write f as

$$f(x) = (x - c)q(x). \tag{1}$$

Thus, if c is a zero of a polynomial function f, then $x - c$ is a factor of $f(x)$. Conversely, if $x - c$ is a factor of $f(x)$, then f has the form given in (1). In this case, we see immediately that $f(c) = (c - c)q(c) = 0$. These results are summarized in the **Factor Theorem** given next.

THEOREM 3.3.1 Factor Theorem

A number c is a zero of a polynomial function f if and only if $x - c$ is a factor of $f(x)$.

Recall, if a polynomial function f is of degree n and $(x - c)^m$, $m \leq n$, is a factor of $f(x)$, then c is said to be a **zero of multiplicity m**. When $m = 1$, c is a **simple zero**. Equivalently, we say that the number c is a **root of multiplicity m** of the equation $f(x) = 0$. We have already examined the graphical significance of repeated real zeros of a polynomial function f in Section 3.1. See Figure 3.1.7.

EXAMPLE 3 **Factors of a Polynomial**

Determine whether

(a) $x + 1$ is a factor of $f(x) = x^4 - 5x^2 + 6x - 1$,
(b) $x - 2$ is a factor of $f(x) = x^3 - 3x^2 + 4$.

Solution We use synthetic division to divide $f(x)$ by the given linear term.
(a) From the division

$$\begin{array}{r|rrrrr} -1 & 1 & 0 & -5 & 6 & -1 \\ & & -1 & 1 & 4 & -10 \\ \hline & 1 & -1 & -4 & 10 & \boxed{-11} = r = f(-1) \end{array}$$

we see that $f(-1) = -11$ and so -1 is not a zero of f. We conclude that $x - (-1) = x + 1$ is not a factor of $f(x)$.

(b) From the division

$$
\begin{array}{r|rrrr}
2 & 1 & -3 & 0 & 4 \\
  &   & 2 & -2 & -4 \\
\hline
  & 1 & -1 & -2 & \;\;0 = r = f(2) \\
\end{array}
$$

we see that $f(2) = 0$. This means that 2 is a zero and that $x - 2$ is a factor of $f(x)$. From the division we see that the quotient is $q(x) = x^2 - x - 2$ and so $f(x) = (x - 2)(x^2 - x - 2)$. ∎

□ **Number of Zeros** In Example 6 of Section 3.1 we graphed the polynomial function

$$f(x) = (x - 3)(x - 1)^2(x + 2)^3. \tag{2}$$

The number 3 is a simple zero of f; the number 1 is a zero of multiplicity 2; and -2 is a zero of multiplicity 3. Although the function f has three *distinct* zeros (different from one another), it is, nevertheless, standard practice to say that f has *six zeros* because we count the multiplicities of each zero. Hence for the function f in (2), the number of zeros is $1 + 2 + 3 = 6$. The question:

How many zeros does a polynomial function f have?

is answered next.

THEOREM 3.3.2   Fundamental Theorem of Algebra

A polynomial function f of degree $n > 0$ has at least one zero.

The foregoing theorem, first proved by the German mathematician **Carl Friedrich Gauss** (1777–1855) in 1799, is considered one of the major milestones in the history of mathematics. At first reading this theorem does not appear to say much, but when combined with the Factor Theorem, the Fundamental Theorem of Algebra shows:

Every polynomial function f of degree $n > 0$ has exactly n zeros. (3)

Of course if a zero is repeated—say, it has multiplicity k—we count that zero k times. To prove (3), we know from the Fundamental Theorem of Algebra that f has a zero (call it c_1). By the Factor Theorem we can write

$$f(x) = (x - c_1)q_1(x), \tag{4}$$

where q_1 is a polynomial function of degree $n - 1$. If $n - 1 \neq 0$, then in like manner we know that q_1 must have a zero (call it c_2) and so (4) becomes

$$f(x) = (x - c_1)(x - c_2)q_2(x),$$

where q_2 is a polynomial function of degree $n - 2$. If $n - 2 \neq 0$, we continue and arrive at

$$f(x) = (x - c_1)(x - c_2)(x - c_3)q_3(x), \tag{5}$$

and so on. Eventually we arrive at a factorization of $f(x)$ with n linear factors and the last factor $q_n(x)$ of degree 0. In other words, $q_n(x) = a_n$, where a_n is a constant. We have arrived at the **complete factorization** of $f(x)$.

THEOREM 3.3.3 Complete Factorization Theorem

Let $c_1, c_2, \ldots, c_n$ be the n (not necessarily distinct) zeros of the polynomial function of degree $n > 0$:

$$f(x) = a_n x^n + a_{n-1} x^{n-1} + \cdots + a_2 x^2 + a_1 x + a_0$$

Then $f(x)$ can be written as a product of n linear factors

$$f(x) = a_n(x - c_1)(x - c_2) \cdots (x - c_n) \tag{6}$$

Bear in mind that some or all the zeros $c_1, \ldots, c_n$ in (6) may be complex numbers $a + bi$, where $b \neq 0$.

In the case of a second-degree, or quadratic, polynomial function $f(x) = ax^2 + bx + c$, where the coefficients a, b, and c are real numbers, the zeros c_1 and c_2 of f can be found using the quadratic formula:

$$c_1 = \frac{-b - \sqrt{b^2 - 4ac}}{2a} \quad \text{and} \quad c_2 = \frac{-b + \sqrt{b^2 - 4ac}}{2a}. \tag{7}$$

The results in (7) tell the whole story about the zeros of the quadratic function: the zeros are real and distinct when $b^2 - 4ac > 0$, real with multiplicity two when $b^2 - 4ac = 0$, and complex and distinct when $b^2 - 4ac < 0$. It follows from (6) that the complete factorization of a quadratic polynomial function is

$$f(x) = a(x - c_1)(x - c_2). \tag{8}$$

> **EXAMPLE 4**   **Example 1 Revisited**

In Example 1 we demonstrated that $\frac{1}{2}$ is a zero of $f(x) = 2x^3 - 9x^2 + 6x - 1$. We now know that $x - \frac{1}{2}$ is a factor of $f(x)$ and that $f(x)$ has three zeros. The synthetic division

$$\begin{array}{r|rrrr}
\frac{1}{2} & 2 & -9 & 6 & -1 \\
  &   & 1 & -4 & 1 \\
\hline
  & 2 & -8 & 2 & \boxed{0} = r
\end{array}$$

again demonstrates that $\frac{1}{2}$ is a zero of $f(x)$ (the 0 remainder is the value of $f\left(\frac{1}{2}\right)$) and, additionally, gives us the quotient $q(x)$ obtained in the division of $f(x)$ by $x - \frac{1}{2}$, that is, $f(x) = \left(x - \frac{1}{2}\right)(2x^2 - 8x + 2)$. As shown in (8), we can now factor the quadratic quotient $q(x) = 2x^2 - 8x + 2$ by finding the roots of $2x^2 - 8x + 2 = 0$ by the quadratic formula:

$$\downarrow \sqrt{48} = \sqrt{16 \cdot 3} = \sqrt{16}\sqrt{3} = 4\sqrt{3}$$

$$x = \frac{-(-8) \pm \sqrt{(-8)^2 - 4(2)(2)}}{4} = \frac{8 \pm \sqrt{48}}{4} = \frac{8 \pm 4\sqrt{3}}{4}$$

$$= \frac{4(2 \pm \sqrt{3})}{4} = 2 \pm \sqrt{3}.$$

Thus the remaining zeros of $f(x)$ are the irrational numbers $2 + \sqrt{3}$ and $2 - \sqrt{3}$. With the identification of the leading coefficient as $a_3 = 2$, it follows from (6) that the complete factorization of $f(x)$ is then

$$f(x) = 2\left(x - \tfrac{1}{2}\right)\left(x - \left(2 + \sqrt{3}\right)\right)\left(x - \left(2 - \sqrt{3}\right)\right)$$

$$= 2\left(x - \tfrac{1}{2}\right)\left(x - 2 - \sqrt{3}\right)\left(x - 2 + \sqrt{3}\right). \qquad \blacksquare$$

EXAMPLE 5   **Using Synthetic Division**

Find the complete factorization of

$$f(x) = x^4 - 12x^3 + 47x^2 - 62x + 26$$

given that 1 is a zero of f of multiplicity 2.

Solution We know that $x - 1$ is a factor of $f(x)$, so by the division

$$
\begin{array}{r|rrrrr}
1| & 1 & -12 & 47 & -62 & 26 \\
& & 1 & -11 & 36 & -26 \\
\hline
& 1 & -11 & 36 & -26 & \boxed{0} = r
\end{array}
$$

we find

$$f(x) = (x - 1)(x^3 - 11x^2 + 36x - 26).$$

Since 1 is a zero of multiplicity 2, $x - 1$ must also be a factor of the quotient $q(x) = x^3 - 11x^2 + 36x - 26$. By the division,

$$
\begin{array}{r|rrrr}
1| & 1 & -11 & 36 & -26 \\
& & 1 & -10 & 26 \\
\hline
& 1 & -10 & 26 & \boxed{0} = r
\end{array}
$$

we conclude that $q(x)$ can be written $q(x) = (x - 1)(x^2 - 10x + 26)$. Therefore,

$$f(x) = (x - 1)^2(x^2 - 10x + 26).$$

The remaining two zeros, found by solving $x^2 - 10x + 26 = 0$ by the quadratic formula, are the complex numbers $5 + i$ and $5 - i$. Since the leading coefficient is $a_4 = 1$ the complete factorization of $f(x)$ is

$$
\begin{aligned}
f(x) &= (x - 1)^2(x - (5 + i))(x - (5 - i)) \\
&= (x - 1)^2(x - 5 - i)(x - 5 + i).
\end{aligned}
$$
■

EXAMPLE 6   **Complete Linear Factorization**

Find a polynomial function f of degree three, with zeros 1, -4, and 5 such that its graph possesses the y-intercept $(0, 5)$.

Solution Because we have three zeros 1, -4, and 5 we know $x - 1, x + 4$, and $x - 5$ are factors of f. However, the function we seek is *not*

$$f(x) = (x - 1)(x + 4)(x - 5). \tag{9}$$

The reason for this is that any nonzero constant multiple of f is a different polynomial with the same zeros. Notice, too, that the function in (9) gives $f(0) = 20$, but we want $f(0) = 5$. Hence by (6) we must assume that f has the form

$$f(x) = a_3(x - 1)(x + 4)(x - 5), \tag{10}$$

where a_3 is some real constant. Using (10), $f(0) = 5$ gives

$$f(0) = a_3(0 - 1)(0 + 4)(0 - 5) = 20a_3 = 5$$

and so $a_3 = \frac{5}{20} = \frac{1}{4}$. The desired function is then

$$f(x) = \tfrac{1}{4}(x - 1)(x + 4)(x - 5).$$
■

We have seen in the introduction that complex zeros of $f(x) = x^2 + 1$ are i and $-i$. In Example 5 the complex zeros are $5 + i$ and $5 - i$. In each case the complex zeros of the polynomial function are conjugate pairs. In other words, one complex zero is the conjugate of the other. This is no coincidence; complex zeros of polynomials with *real* coefficients *always* appear in conjugate pairs. In order to prove this, we use the following results concerning conjugates.

If z_1 and z_2 are complex numbers, then

$$\overline{z_1 + z_2} = \overline{z_1} + \overline{z_2} \quad \text{and} \quad \overline{z_1^n} = \overline{z_1}^n. \tag{11}$$

See Problem 65 in Exercises 3.3.

THEOREM 3.3.4 Conjugate Zeros Theorem

Let $f(x)$ be a polynomial function of degree $n > 1$ with real coefficients. If $z = a + bi$, $b \neq 0$, is a complex zero of $f(x)$, then the conjugate $\overline{z} = a - bi$ is also a zero of $f(x)$.

PROOF: Let $f(x) = a_n x^n + a_{n-1} x^{n-1} + \cdots + a_2 x^2 + a_1 x + a_0$, where the coefficients a_i, $i = 0, 1, 2, \ldots, n$ are real numbers. By assumption $f(z) = 0$ so

$$a_n z^n + a_{n-1} z^{n-1} + \cdots + a_2 z^2 + a_1 z + a_0 = 0.$$

Taking the conjugate of both sides of this equation gives

$$\overline{a_n z^n + a_{n-1} z^{n-1} + \cdots + a_2 z^2 + a_1 z + a_0} = \overline{0}.$$

Now using (11) along with the fact that the conjugate of any real number is itself, we obtain

$$a_n \overline{z}^n + a_{n-1} \overline{z}^{n-1} + \cdots + a_2 \overline{z}^2 + a_1 \overline{z} + a_0 = 0.$$

This means $f(\overline{z}) = 0$ and, therefore, $\overline{z}$ is a zero of $f(x)$ whenever z is a zero. ∎

EXAMPLE 7 **Example 2 Revisited**

In Example 2 we demonstrated that $1 + i$ is a complex zero of

$$f(x) = x^3 - 5x^2 + 8x - 6.$$

Since the coefficients of f are real numbers we conclude that another zero is the conjugate of $1 + i$, namely, $1 - i$. Thus we know two factors of $f(x)$, $x - (1 + i)$ and $x - (1 - i)$. Carrying out the multiplication, we find

$$(x - 1 - i)(x - 1 + i) = x^2 - 2x + 2.$$

Thus we can write

$$f(x) = (x - 1 - i)(x - 1 + i)q(x) = (x^2 - 2x + 2)q(x).$$

We determine $q(x)$ by performing the *long division* of $f(x)$ by $x^2 - 2x + 2$. (We can't do synthetic division because we are not dividing by a linear factor.) From

$$
\begin{array}{r}
x - 3 \\
x^2 - 2x + 2 \overline{)x^3 - 5x^2 + 8x - 6} \\
\underline{x^3 - 2x^2 + 2x} \\
-3x^2 + 6x - 6 \\
\underline{-3x^2 + 6x - 6} \\
0
\end{array}
$$

we see that the complete factorization of $f(x)$ is

$$f(x) = (x - 1 - i)(x - 1 + i)(x - 3).$$

The three zeros of $f(x)$ are $1 + i$, $1 - i$, and 3. ∎

NOTES FROM THE CLASSROOM

An alternative way of stating Theorem 3.3.4 is that complex zeros of a polynomial function $f(x)$ of degree $n > 1$ with real coefficients appear in conjugate pairs. As an immediate consequence we can also say that:

A polynomial function $f(x)$ of odd degree $n > 1$ with real coefficients must possess at least one real zero. (12)

For example, without knowing anything about the cubic (degree 3) polynomial function $f(x) = 4x^3 - 2x^2 - 5x + 1$ it follows from (12) that it has at least one real zero. We know from (3) that $f(x)$ must have three zeros, but all those zeros cannot be complex numbers $a + bi$, $b \neq 0$, because Theorem 3.3.4 implies $f(x)$ would necessarily have to have *four* complex zeros.

Exercises 3.3   Answers to selected odd-numbered problems begin on page ANS–12.

In Problems 1–10, find $z_1 + z_2$, $z_1 - z_2$, and $z_1 z_2$ for the given complex numbers.

1. $z_1 = 3 - 2i$, $z_2 = 1 - i$ **2.** $z_1 = 4 + 6i$, $z_2 = -3 + 4i$

3. $z_1 = 7 + 14i$, $z_2 = 2 + i$ **4.** $z_1 = 5 + 3i$, $z_2 = -10 + i$

5. $z_1 = 10 + 5i$, $z_2 = -4 - i$ **6.** $z_1 = 6 - 9i$, $z_2 = -7i$

7. $z_1 = -5 - 11i$, $z_2 = 1 + 3i$ **8.** $z_1 = -5 - \sqrt{3}i$, $z_2 = -5 + \sqrt{3}i$

9. $z_1 = \frac{1}{2} - \frac{3}{2}i$, $z_2 = 4 - 4i$ **10.** $z_1 = \frac{1}{2} + \frac{1}{3}i$, $z_2 = -\frac{1}{4} + \frac{5}{3}i$

In Problems 11–14, use the binomial expansions (6) and (7) of Section 1.5 to find z^2 and z^3 for the given complex number.

11. $z = 3 + i$ **12.** $z = 4 + 2i$

13. $z = 5 - 5i$ **14.** $z = \sqrt{2} - i$

In Problems 15–20, determine whether the indicated real number is a zero of the given polynomial function f. If yes, find all other zeros and then give the complete factorization of $f(x)$.

15. 1; $f(x) = 4x^3 - 9x^2 + 6x - 1$ **16.** $\frac{1}{2}$; $f(x) = 2x^3 - x^2 + 32x - 16$

17. 5; $f(x) = x^3 - 6x^2 + 6x + 5$ **18.** 3; $f(x) = x^3 - 3x^2 + 4x - 12$

19. $-\frac{2}{3}$; $f(x) = 3x^3 - 10x^2 - 2x + 4$ **20.** -2; $f(x) = x^3 - 4x^2 - 2x + 20$

In Problems 21–26, verify that each of the indicated numbers is a zero of the given polynomial function f. Find all other zeros and then give the complete factorization of $f(x)$.

21. $-3, 5$; $f(x) = 4x^4 - 8x^3 - 61x^2 + 2x + 15$

22. $\frac{1}{4}, \frac{3}{2}$; $f(x) = 8x^4 - 30x^3 + 23x^2 + 8x - 3$

23. $1, -\frac{1}{3}$ (multiplicity 2); $f(x) = 9x^4 + 69x^3 - 29x^2 - 41x - 8$

24. $-\sqrt{5}, \sqrt{5}$; $f(x) = 3x^4 + x^3 - 17x^2 - 5x + 10$

25. $1, 5$; $f(x) = x^5 - 3x^4 - 22x^3 + 74x^2 - 75x + 25$

26. $0, \frac{1}{2}$; $f(x) = 2x^5 - 17x^4 + 40x^3 - 16x^2$

In Problems 27–32, use synthetic division to determine whether the indicated linear polynomial is a factor of the given polynomial function f. If yes, find all other zeros and then give the complete factorization of $f(x)$.

27. $x - 5$; $\quad f(x) = 2x^2 + 6x - 25$
28. $x + \frac{1}{2}$; $\quad f(x) = 10x^2 - 27x + 11$
29. $x - 1$; $\quad f(x) = x^3 + x - 2$
30. $x + \frac{1}{2}$; $\quad f(x) = 2x^3 - x^2 + x + 1$
31. $x - \frac{1}{3}$; $\quad f(x) = 3x^3 - 3x^2 + 8x - 2$
32. $x - 2$; $\quad f(x) = x^3 - 6x^2 - 16x + 48$

In Problems 33–36, use division to show that the indicated polynomial is a factor of the given polynomial function f. Find all other zeros and then give the complete factorization of $f(x)$.

33. $(x - 1)(x - 2)$; $\quad f(x) = x^4 - 3x^3 + 6x^2 - 12x + 8$
34. $x(3x - 1)$; $\quad f(x) = 3x^4 - 7x^3 + 5x^2 - x$
35. $(x - 1)^2$; $\quad f(x) = 2x^4 + x^3 - 5x^2 - x + 3$
36. $(x + 3)^2$; $\quad f(x) = x^4 - 4x^3 - 22x^2 + 84x + 261$

In Problems 37–42, verify that the indicated complex number is a zero of the given polynomial function f. Proceed as in Example 7 to find all other zeros and then give the complete factorization of $f(x)$.

37. $2i$; $\quad f(x) = 3x^3 - 5x^2 + 12x - 20$
38. $\frac{1}{2}i$; $\quad f(x) = 12x^3 + 8x^2 + 3x + 2$
39. $-1 + i$; $\quad f(x) = 5x^3 + 12x^2 + 14x + 4$
40. $-i$; $\quad f(x) = 4x^4 - 8x^3 + 9x^2 - 8x + 5$
41. $1 + 2i$; $\quad f(x) = x^4 - 2x^3 - 4x^2 + 18x - 45$
42. $1 + i$; $\quad f(x) = 6x^4 - 11x^3 + 9x^2 + 4x - 2$

In Problems 43–48, find a polynomial function f with real coefficients of the indicated degree that possesses the given zeros.

43. degree 4; $\quad 2, 1, -3$ (multiplicity 2) $\qquad$ **44.** degree 5; $\quad -4i, -\frac{1}{3}, \frac{1}{2}$ (multiplicity 2)
45. degree 5; $\quad 3 + i, 0$ (multiplicity 3) $\qquad$ **46.** degree 4; $\quad 5i, 2 - 3i$
47. degree 2; $\quad 1 - 6i$ $\qquad\qquad\qquad$ **48.** degree 2; $\quad 4 + 3i$

In Problems 49–52, find a polynomial function f with real coefficients that satisfies the given conditions.

49. degree 2; zero $1 + i$; $f(1) = 5$
50. degree 3; zeros $1, 3i$; $f(0) = 27$
51. degree 4; zeros $-\frac{1}{2}, 2, -i$; $f(-1) = 252$
52. degree 4; zeros 0 (multiplicity 2), $2 - i$; $f(2) = 48$

In Problems 53–56, find the zeros of the given polynomial function f. State the multiplicity of each zero.

53. $f(x) = x(4x - 5)^2(2x - 1)^3$ $\qquad$ **54.** $f(x) = x^4 + 6x^3 + 9x^2$
55. $f(x) = (9x^2 - 4)^2$ $\qquad\qquad$ **56.** $f(x) = (x^2 + 25)(x^2 - 5x + 4)^2$

In Problems 57 and 58, solve the given equation if the indicated number c is a zero of the function f.

57. $2x^3 - 3x^2 - 8x - 3 = 0$; $\quad f(x) = 2x^3 - 3x^2 - 8x - 3$, $\quad c = -1$
58. $x^4 - 3x^3 - 8x^2 - 10x = 0$; $\quad f(x) = x^4 - 3x^3 - 8x^2 - 10x$, $\quad c = 5$

In Problems 59 and 60, find the value(s) of k such that the indicated number is a zero of $f(x)$. Then give the complete factorization of $f(x)$.

59. 3; $f(x) = 2x^3 - 2x^2 + k$ **60.** 1; $f(x) = x^3 + 5x^2 - k^2x + k$

In Problems 61 and 62, find a polynomial function f of the indicated degree whose graph is given in the figure.

61. degree 3

FIGURE 3.3.1 Graph for Problem 61

62. degree 5

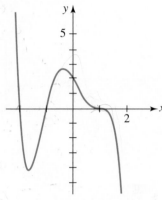

FIGURE 3.3.2 Graph for Problem 62

For Discussion

63. Let $z = a + bi$. Show that $z + \bar{z}$ and $z\bar{z}$ are real numbers.

64. Let $z = a + bi$. Use the results of Problem 63 to show that

$$f(x) = (x - z)(x - \bar{z})$$

is a polynomial function with real coefficients.

65. Let $z_1 = a + bi$ and $z_2 = c + di$. Show that

$$\overline{z_1 + z_2} = \bar{z}_1 + \bar{z}_2 \quad \text{and} \quad \overline{z_1^2} = \bar{z}_1^2.$$

FIGURE 3.3.3 Graph for Problem 66

66. Find an equation of the cubic polynomial function $y = f(x)$ whose graph is given in FIGURE 3.3.3 and has $2 + i$ as a zero.

67. Counting the multiplicity, the cubic polynomial function $f(x) = (x - i)^3$ has exactly three complex zeros. Explain why this does not contradict (12) and the comments in the *Notes from the Classroom* on page 165.

68. Suppose the complex numbers $1 - 2i$ and $3 + i$ are zeros of multiplicity 2 of a polynomial function f with real coefficients. Discuss: What is the degree of f?

69. Factor the polynomial function $f(x) = x^2 - i$. [*Hint:* Suppose $z = a + bi$ is a zero of the function f. Then use the fact that two complex numbers are equal when their real and imaginary parts are the same.]

70. A **square root** of a complex number $w = c + di$ is defined to be a complex number $z = a + bi$ that satisfies the polynomial equation $z^2 = c + di$. Find two square roots of i.

3.4   Real Zeros of Polynomial Functions

INTRODUCTION In the preceding section we saw that as a consequence of the Fundamental Theorem of Algebra, a polynomial function f of degree n has n zeros when the multiplicities of the zeros are counted. We also saw that a zero of a polynomial function could be either a real or a complex number. In this section we confine our attention to *real zeros* of polynomial functions with real coefficients.

☐ **Real Zeros** If a polynomial function f of degree $n > 0$ has m (not necessarily distinct) real zeros $c_1, c_2, \ldots, c_m$, then by the Factor Theorem each of the linear polynomials $x - c_1, x - c_2, \ldots, x - c_m$ are factors of $f(x)$. That is,

$$f(x) = (x - c_1)(x - c_2) \cdots (x - c_m)q(x),$$

where $q(x)$ is a polynomial. Thus n, the degree of f, must be greater than or possibly equal to m, the number of real zeros when each is counted according to its multiplicity. Using slightly different words, we restate the last sentence.

THEOREM 3.4.1 Number of Real Zeros

A polynomial function f of degree $n > 0$ has at most n real zeros (not necessarily distinct).

Let's summarize some facts about real zeros of a polynomial function f of degree n:

- f may not have any real zeros.

For example, the fourth degree polynomial function $f(x) = x^4 + 9$ has no real zeros, since there exists no real number x satisfying $x^4 + 9 = 0$ or $x^4 = -9$.

- f may have m real zeros where $m < n$.

For example, the third degree polynomial function $f(x) = (x - 1)(x^2 + 1)$ has one real zero.

- f may have n real zeros.

For example, by factoring the third degree polynomial function $f(x) = x^3 - x$ as $f(x) = x(x^2 - 1) = x(x + 1)(x - 1)$, we see that it has three real zeros.

- f has at least one real zero when its degree n is odd.

This is a consequence of the fact that complex zeros of a polynomial function f with real coefficients must appear in conjugate pairs. Thus if we write down an arbitrary cubic polynomial function such as $f(x) = x^3 + x + 1$, we know that f cannot have just one complex zero, nor can it have three complex zeros. Put another way, $f(x) = x^3 + x + 1$ either has exactly one real zero or it has exactly three real zeros.

See (12) in Section 3.3. ▶

- If the coefficients of $f(x)$ are positive and the constant term $a_0 \neq 0$, then any real zeros of f must be negative.

☐ **Finding Real Zeros** It is one thing to talk about the existence of real and complex zeros of a polynomial function; it is an entirely different problem to actually find these zeros. The problem of finding a *formula* that expresses the zeros of a general nth degree polynomial function f in terms of its coefficients perplexed mathematicians for centuries. We have seen in Sections 2.4 and 3.3 that in the case of a second-degree, or quadratic, polynomial function $f(x) = ax^2 + bx + c$, where the coefficients a, b, and c are real numbers, the zeros c_1 and c_2 of f can be found using the quadratic formula.

The problem of finding zeros of third-degree, or cubic, polynomial functions was solved in the sixteenth century through the pioneering work of the Italian mathematician

Niels Henrik Abel

Niccolò Fontana (1499–1557), also known as Tartaglia—"the stammerer." Around 1540 another Italian mathematician, **Lodovico Ferrari** (1522–1565) discovered an algebraic formula for determining the zeros for fourth degree, or quartic, polynomial functions. Since these formulas are complicated and difficult to use, they are seldom discussed in elementary courses.

For the next 284 years no one discovered any formulas for zeros for general polynomial functions of degrees five, six, ... For good reason! In 1824, at age 22, the Norwegian mathematician **Niels Henrik Abel** (1802–1829) proved it was impossible to find such formulas for the zeros of all general polynomials of degrees $n \geq 5$ in terms of their coefficients.

☐ **Rational Zeros** Real zeros of a polynomial function are either rational or irrational numbers. A rational number is a number of the form p/s, where p and s are integers and $s \neq 0$. An irrational number is one that is not rational. For example, $\frac{1}{4}$ and -9 are rational numbers, but $\sqrt{2}$ and π are irrational, that is, neither $\sqrt{2}$ nor π can be written as a fraction p/s where p and s are integers. So how do we find real zeros for polynomial functions of degree $n > 2$? The bad news: For irrational real zeros, we *may* have to be content to use an accurate graph to "eyeball" their location on the x-axis and then use one of the many sophisticated methods for *approximating* the zero that have been developed over the years. The good news: We can always find the rational real zeros of *any* polynomial function with rational coefficients. We have already seen that synthetic division is a useful method for determining whether a given number c is a zero of a polynomial function $f(x)$. When the remainder in the division of $f(x)$ by $x - c$ is $r = 0$, we have found a zero of the polynomial function f, since $r = f(c) = 0$. For example, $\frac{2}{3}$ is a zero of $f(x) = 18x^3 - 15x^2 + 14x - 8$, since

$$
\begin{array}{r|rrrr}
\frac{2}{3} & 18 & -15 & 14 & -8 \\
 & & 12 & -2 & 8 \\
\hline
 & 18 & -3 & 12 & \boxed{0} = r.
\end{array}
$$

Hence by the Factor Theorem, both $x - \frac{2}{3}$ and the quotient $18x^2 - 3x + 12$ are factors of f and so we can write the polynomial function as the product

$$
\begin{aligned}
f(x) &= \left(x - \tfrac{2}{3}\right)(18x^2 - 3x + 12) \quad \leftarrow \text{factor 3 from the quadratic polynomial} \\
&= \left(x - \tfrac{2}{3}\right)(3)(6x^2 - x + 4) \\
&= (3x - 2)(6x^2 - x + 4).
\end{aligned} \tag{1}
$$

As discussed in the preceding section, if we can factor the polynomial to the point where the remaining factor is a quadratic polynomial, we can then find the remaining two zeros by the quadratic formula. For this example, the factorization in (1) is as far as we can go using real numbers since the zeros of the quadratic factor $6x^2 - x + 4$ are complex (verify). But the indicated multiplication in (1) illustrates something important about rational zeros. The leading coefficient 18 and the constant term -8 of $f(x)$ are obtained from the products

$$
\overset{\displaystyle -8}{(3x - 2)(6x^2 - x + 4)}.
$$
$$
\underset{\displaystyle 18}{}
$$

Thus we see that the denominator 3 of the rational zero $\frac{2}{3}$ is a *factor* of the leading coefficient 18 of $f(x) = 18x^3 - 15x^2 + 14x - 8$, and the numerator 2 of the rational zero is a factor of the constant term -8.

This example illustrates the following general principle for determining the rational zeros of a polynomial function. Read the following theorem carefully; the coefficients of f are not only real numbers—they must be *integers*.

THEOREM 3.4.2 Rational Zeros Theorem

Let p/s be a rational number in lowest terms and a zero of the polynomial function

$$f(x) = a_n x^n + a_{n-1} x^{n-1} + \cdots + a_2 x^2 + a_1 x + a_0$$

where the coefficients $a_n, a_{n-1}, \ldots, a_2, a_1, a_0$ are integers with $a_n \neq 0$. Then p is an integer factor of the constant term a_0 and s is an integer factor of the leading coefficient a_n.

The Rational Zeros Theorem deserves to be read several times. Note that Theorem 3.4.2 *does not* assert that a polynomial function f with integer coefficients *must* have a rational zero; rather, it states that *if* a polynomial function f with integer coefficients has a rational zero p/s, then necessarily:

$$\frac{p}{s} = \frac{\text{integer factor of constant term } a_0}{\text{integer factor of leading coefficient } a_n}.$$

By forming all possible quotients of each integer factor of a_0 to each integer factor of a_n, we can construct a list of *potential* rational zeros of f.

EXAMPLE 1   **Rational Zeros**

Find all rational zeros of $f(x) = 3x^4 - 10x^3 - 3x^2 + 8x - 2$.

Solution We identify the constant term $a_0 = -2$ and leading coefficient $a_4 = 3$, and then list all the integer factors of a_0 and a_4, respectively:

$$p: \ \pm 1, \pm 2,$$
$$s: \ \pm 1, \pm 3.$$

Now we form a list of all possible rational zeros p/s by dividing all the factors of p by ± 1 and then by ± 3:

$$\frac{p}{s} = \frac{\text{integer factor of } -2}{\text{integer factor of } 3} = \pm 1, \pm 2, \pm \tfrac{1}{3}, \pm \tfrac{2}{3}. \tag{2}$$

We know that the given fourth-degree polynomial function f has four zeros; if any of these zeros is a real number and is rational, then it must appear in the list (2).

To determine which, if any, of the numbers in (2) are zeros, we could use direct substitution into $f(x)$. Synthetic division, however, is usually a more efficient means of evaluating $f(x)$. We begin by testing -1:

$$
\begin{array}{r|rrrrr}
-1 & 3 & -10 & -3 & 8 & -2 \\
   &   & -3 & 13 & -10 & 2 \\
\hline
   & 3 & -13 & 10 & -2 & 0 = r.
\end{array}
\tag{3}
$$

The zero remainder shows $r = f(-1) = 0$, and so -1 is a zero of f. Hence $x - (-1) = x + 1$ is a factor of f. Using the quotient found in (3) we can write

$$f(x) = (x + 1)(3x^3 - 13x^2 + 10x - 2). \tag{4}$$

From (4) we see that any other rational zero of f must be a zero of the quotient $3x^3 - 13x^2 + 10x - 2$. Since the latter polynomial is of lower degree, it will be easier to use synthetic division on it rather than on $f(x)$ to check the next rational zero. At this point in the process you should check to see whether the zero just found is a repeated zero. This is done by determining whether the found zero is also a zero of the quotient. A quick check, using synthetic division, shows that -1 is *not* a repeated zero of f since it is not a zero of $3x^3 - 13x^2 + 10x - 2$. So we move on and determine whether the number 1 is a rational zero of f. Indeed, it is *not* because the division

$$
\begin{array}{r|rrrr}
1 & 3 & -13 & 10 & -2 \qquad \leftarrow \text{coefficients of the quotient in (3)}\\
  &   & 3 & -10 & 0 \\
\hline
  & 3 & -10 & 0 & \boxed{-2} = r
\end{array}
\tag{5}
$$

shows that the remainder is $r = -2 \neq 0$. Checking $\frac{1}{3}$, we have

$$
\begin{array}{r|rrrr}
\frac{1}{3} & 3 & -13 & 10 & -2 \\
   &   & 1 & -4 & 2 \\
\hline
   & 3 & -12 & 6 & \boxed{0} = r.
\end{array}
\tag{6}
$$

Thus $\frac{1}{3}$ is a zero. At this point we can stop using synthetic division since (6) indicates that the remaining factor of f is the quadratic polynomial $3x^2 - 12x + 6$. From the quadratic formula we find that the remaining real zeros are $2 + \sqrt{2}$ and $2 - \sqrt{2}$. Therefore the given polynomial function f has two rational zeros, -1 and $\frac{1}{3}$, and two irrational zeros, $2 + \sqrt{2}$ and $2 - \sqrt{2}$. ■

If you have access to technology, your selection of rational numbers to test in Example 1 can be motivated by a graph of the function $f(x) = 3x^4 - 10x^3 - 3x^2 + 8x - 2$. With the aid of a graphing utility we obtain the graphs in FIGURE 3.4.1. In Figure 3.4.1(a) it would appear that f has at least three real zeros. But by "zooming-in" on the graph on the interval $[0, 1]$, Figure 3.4.1(b) reveals that f actually has four real zeros: one negative and three positive. Thus, once you have determined one negative rational zero of f you may disregard all other negative numbers as potential zeros.

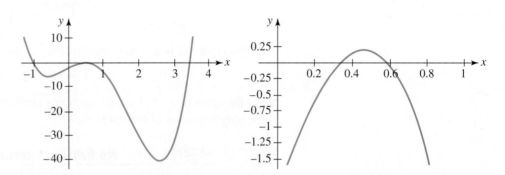

(a) Graph of f on the interval $[-1, 4]$ (b) Zoom-in of graph on the interval $[0, 1]$

FIGURE 3.4.1 Graph of function in Example 1

EXAMPLE 2   Complete Factorization

Since the function f in Example 1 is of degree 4 and we have found four real zeros, we can give its complete factorization. Using the leading coefficient $a_4 = 3$, it follows from (6) of Section 3.3 that

$$
\begin{aligned}
f(x) &= 3(x + 1)\left(x - \tfrac{1}{3}\right)\left(x - (2 - \sqrt{2})\right)\left(x - (2 + \sqrt{2})\right) \\
     &= 3(x + 1)\left(x - \tfrac{1}{3}\right)\left(x - 2 + \sqrt{2}\right)\left(x - 2 - \sqrt{2}\right).
\end{aligned}
$$
■

EXAMPLE 3 **Rational Zeros**

Find all rational zeros of $f(x) = x^4 + 4x^3 + 5x^2 + 4x + 4$.

Solution In this case the constant term is $a_0 = 4$ and the leading coefficient is $a_4 = 1$. The integer factors of a_0 and a_4 are, respectively:

$$p: \ \pm 1, \ \pm 2, \ \pm 4,$$
$$s: \ \pm 1.$$

The list of all possible rational zeros p/s is:

$$\frac{p}{s}: \ \pm 1, \ \pm 2, \ \pm 4.$$

Since all the coefficients of f are positive, substituting a positive number from the foregoing list into $f(x)$ can never result in $f(x) = 0$. Thus the only numbers that are potential rational zeros are -1, -2, and -4. From the synthetic division

$$
\begin{array}{r|rrrrr}
-1 & 1 & 4 & 5 & 4 & 4 \\
   &   & -1 & -3 & -2 & -2 \\
\hline
   & 1 & 3 & 2 & 2 & \boxed{2} = r
\end{array}
$$

we see that -1 is not a zero. However, from

$$
\begin{array}{r|rrrrr}
-2 & 1 & 4 & 5 & 4 & 4 \\
   &   & -2 & -4 & -2 & -4 \\
\hline
   & 1 & 2 & 1 & 2 & \boxed{0} = r
\end{array}
$$

we see -2 is a zero. We now test to see whether -2 is a repeated zero. Using the coefficients in the quotient,

$$
\begin{array}{r|rrrr}
-2 & 1 & 2 & 1 & 2 \\
   &   & -2 & 0 & -2 \\
\hline
   & 1 & 0 & 1 & \boxed{0} = r
\end{array}
$$

it follows that -2 is a zero of multiplicity 2. So far we have shown that

$$f(x) = (x + 2)^2(x^2 + 1).$$

Since the zeros of $x^2 + 1$ are the complex conjugates i and $-i$, we can conclude that -2 is the only rational real zero of $f(x)$. ■

EXAMPLE 4 **No Rational Zeros**

Consider the polynomial function $f(x) = x^5 - 4x - 1$. The only possible rational zeros are -1 and 1, and it is easy to see that neither $f(-1)$ nor $f(1)$ are 0. Thus f has no rational zeros. Since f is of odd degree we know that it has at least one real zero, and so that zero must be an irrational number. With the aid of a graphing utility we obtain the graph in FIGURE 3.4.2. Note in the figure that the graph to the right of $x = 2$ cannot turn back down *and* the graph to the left of $x = -2$ cannot turn back up, so that the graph crosses the x-axis five times because that shape of the graph would be inconsistent with the end behavior of f. Thus we can conclude that the function f possesses three irrational real zeros and two complex conjugate zeros. The best we can do here is to approximate these zeros. Using a computer algebra system such as *Mathematica* we can approximate both the real and the complex zeros. We find these approximations to be -1.34, -0.25, 1.47, $0.061 + 1.42i$, and $0.061 - 1.42i$. ■

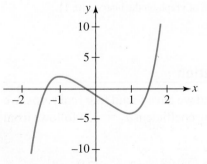

FIGURE 3.4.2 Graph of function in Example 4

Although the Rational Zeros Theorem requires that the coefficients of a polynomial function f be integers, in some circumstances we can apply the theorem to a polynomial function with some real *noninteger* coefficients. The next example illustrates the concept.

EXAMPLE 5　　Noninteger Coefficients

Find the rational zeros of $f(x) = \frac{5}{6}x^4 - \frac{23}{12}x^3 + \frac{10}{3}x^2 - 3x - \frac{3}{4}$.

Solution By multiplying f by the least common denominator 12 of all the rational coefficients, we obtain a new function g with integer coefficients:

$$g(x) = 10x^4 - 23x^3 + 40x^2 - 36x - 9.$$

In other words, $g(x) = 12f(x)$. If c is a zero of the function g, then c is also zero of f because $g(c) = 0 = 12f(c)$ implies $f(c) = 0$. After working through the numbers in the list of potential rational zeros

$$\frac{p}{s}: \pm 1, \pm 3, \pm 9, \pm \tfrac{1}{2}, \pm \tfrac{3}{2}, \pm \tfrac{9}{2}, \pm \tfrac{1}{5}, \pm \tfrac{3}{5}, \pm \tfrac{9}{5}, \pm \tfrac{1}{10}, \pm \tfrac{3}{10}, \pm \tfrac{9}{10},$$

we find that $-\frac{1}{5}$ and $\frac{3}{2}$ are zeros of g, and hence are the rational zeros of f. ∎

Exercises 3.4　Answers to selected odd-numbered problems begin on page ANS-12.

In Problems 1–20, find all rational zeros of the given polynomial function f.

1. $f(x) = 5x^3 - 3x^2 + 8x + 4$ 　　　　**2.** $f(x) = 2x^3 + 3x^2 - x + 2$

3. $f(x) = x^3 - 8x - 3$ 　　　　　　　　**4.** $f(x) = 2x^3 - 7x^2 - 17x + 10$

5. $f(x) = 4x^4 - 7x^2 + 5x - 1$ 　　　　**6.** $f(x) = 8x^4 - 2x^3 + 15x^2 - 4x - 2$

7. $f(x) = x^4 + 2x^3 + 10x^2 + 14x + 21$ 　　**8.** $f(x) = 3x^4 + 5x^2 + 1$

9. $f(x) = 6x^4 - 5x^3 - 2x^2 - 8x + 3$ 　　**10.** $f(x) = x^4 + 2x^3 - 2x^2 - 6x - 3$

11. $f(x) = x^4 + 6x^3 - 7x$ 　　　　　　　**12.** $f(x) = x^5 - 2x^2 - 12x$

13. $f(x) = x^5 + x^4 - 5x^3 + x^2 - 6x$ 　　**14.** $f(x) = 128x^6 - 2$

15. $f(x) = \frac{1}{2}x^3 - \frac{9}{4}x^2 + \frac{17}{4}x - 3$ 　　　**16.** $f(x) = 0.2x^3 - x + 0.8$

17. $f(x) = 2.5x^3 + x^2 + 0.6x + 0.1$ 　　**18.** $f(x) = \frac{3}{4}x^3 + \frac{9}{4}x^2 + \frac{5}{3}x + \frac{1}{3}$

19. $f(x) = 6x^4 + 2x^3 - \frac{11}{6}x^2 - \frac{1}{3}x + \frac{1}{6}$ 　　**20.** $f(x) = x^4 + \frac{5}{2}x^3 + \frac{3}{2}x^2 - \frac{1}{2}x - \frac{1}{2}$

In Problems 21–30, find all real zeros of the given polynomial function f. Then factor $f(x)$ using only real numbers.

21. $f(x) = 8x^3 + 5x^2 - 11x + 3$

22. $f(x) = 6x^3 + 23x^2 + 3x - 14$

23. $f(x) = 10x^4 - 33x^3 + 66x - 40$

24. $f(x) = x^4 - 2x^3 - 23x^2 + 24x + 144$

25. $f(x) = x^5 + 4x^4 - 6x^3 - 24x^2 + 5x + 20$

26. $f(x) = 18x^5 + 75x^4 + 47x^3 - 52x^2 - 11x + 3$

27. $f(x) = 4x^5 - 8x^4 - 24x^3 + 40x^2 - 12x$

28. $f(x) = 6x^5 + 11x^4 - 3x^3 - 2x^2$

29. $f(x) = 16x^5 - 24x^4 + 25x^3 + 39x^2 - 23x + 3$

30. $f(x) = x^6 - 12x^4 + 48x^2 - 64$

In Problems 31–36, find all real solutions of the given equation.

31. $2x^3 + 3x^2 + 5x + 2 = 0$ 　　　　　**32.** $x^3 - 3x^2 = -4$

33. $2x^4 + 7x^3 - 8x^2 - 25x - 6 = 0$ 　　**34.** $9x^4 + 21x^3 + 22x^2 + 2x - 4 = 0$

35. $x^5 - 2x^4 + 2x^3 - 4x^2 + 5x - 2 = 0$ 　　**36.** $8x^4 - 6x^3 - 7x^2 + 6x - 1 = 0$

In Problems 37 and 38, find a polynomial function f of the indicated degree with integer coefficients that possesses the given rational zeros.

37. degree 4; $-4, \frac{1}{3}, 1, 3$

38. degree 5; $-2, -\frac{2}{3}, \frac{1}{2}, 1$ (multiplicity 2)

39. If $f(x) = 4x^3 - 11x^2 + 14x - 6$, then show the values $f(0)$ and $f(1)$ have different algebric signs. Explain why this information along with the fact that a polynomial function is a continuous function (see page 148) enables us to conclude that f has a zero in the interval $[0, 1]$. Find the zero.

40. List, but do not test, all possible rational zeros of

$$f(x) = 24x^3 - 14x^2 + 36x + 105.$$

In Problems 41 and 42, find a cubic polynomial function f that satisfies the given conditions.

41. rational zeros 1 and 2, $f(0) = 1$ and $f(-1) = 4$

42. rational zero $\frac{1}{2}$, irrational zeros $1 + \sqrt{3}$ and $1 - \sqrt{3}$, coefficient of x is 2

Cut out

Fold

(a)

(b)

FIGURE 3.4.3 Box in Problem 43

Calculus-Related Problems

43. Construction of a Box A box with no top is made from a square piece of cardboard by cutting square pieces from each corner and then folding up the sides. See FIGURE 3.4.3. The length of one side of the cardboard is 10 inches. Find the length of one side of the squares that were cut from the corners if the volume of the box is 48 in³.

44. Deflection of a Beam A cantilever beam 20 ft long with a load of 600 lb at its right end is deflected by an amount $d(x) = \frac{1}{16,000}(60x^2 - x^3)$, where d is measured in inches and x in feet. See FIGURE 3.4.4. Find x when the deflection is 0.1215 in. When the deflection is 1 in.

x

d

600 lb

20 ft

FIGURE 3.4.4 Cantilever beam in Problem 44

For Discussion

45. Discuss: What is the maximum number of times the graphs of the given polynomial functions can intersect?
(a) $f(x) = a_3x^3 + a_2x^2 + a_1x + a_0$, $g(x) = b_2x^2 + b_1x + b_0$
(b) $f(x) = x^3 + a_2x^2 + a_1x + a_0$, $g(x) = x^3 + b_2x^2 + b_1x + b_0$

46. Consider the polynomial function $f(x) = x^n + a_{n-1}x^{n-1} + \cdots + a_1x + a_0$, where the coefficients $a_{n-1}, \ldots, a_1, a_0$ are nonzero even integers. Discuss why -1 and 1 cannot be zeros of f.

47. If the leading coefficient of a polynomial function f with integer coefficients is 1, then what can be said about the possible real zeros of f?

48. If k is a prime number (a positive integer greater than 1 whose only positive integer factors are itself and 1) such that $k > 2$, then what are the possible rational zeros of $f(x) = 6x^4 - 9x^2 + k$?

49. (a) The real number $\sqrt{2}$ is a zero of the polynomial function $f(x) = x^2 - 2$. How does the discussion in this section prove that $\sqrt{2}$ is irrational?
(b) Use the idea implied in part (a) to prove that the real number $1 + \sqrt{2}$ is irrational.

50. Without doing any work, explain why the polynomial function

$$f(x) = 4x^{10} + 9x^6 + 5x^4 + 13x^2 + 3$$

has no real zeros.

3.5 Rational Functions

INTRODUCTION Many functions are built up out of polynomial functions by means of arithmetic operations and function composition (see Section 2.6). In this section we construct a class of functions by forming the quotient of two polynomial functions.

DEFINITION 3.5.1 Rational Function

A **rational function** $y = f(x)$ is a function of the form

$$f(x) = \frac{P(x)}{Q(x)} \qquad (1)$$

where P and Q are polynomial functions.

For example, the following functions are rational functions:

$$y = \frac{x}{x^2 + 5}, \qquad y = \frac{\overset{\text{polynomial}}{\overset{\downarrow}{x^3 - x + 7}}}{\underset{\uparrow}{\underset{\text{polynomial}}{x + 3}}}, \qquad y = \frac{1}{x}.$$

The function

$$y = \frac{\sqrt{x}}{x^2 - 1} \qquad \leftarrow \text{not a polynomial}$$

is not a rational function. In (1) we cannot allow the denominator to be zero. So the **domain** of a rational function $f(x) = P(x)/Q(x)$ is the set of all real numbers *except* those numbers for which the denominator $Q(x)$ is zero. For example, the domain of the rational function $f(x) = (2x^3 - 1)/(x^2 - 9)$ is $\{x \mid x \neq -3, x \neq 3\}$ or $(-\infty, -3) \cup (-3, 3) \cup (3, \infty)$. It goes without saying that we also disallow the zero polynomial $Q(x) = 0$ as a denominator.

☐ **Graphs** Graphing a rational function f is a little more complicated than graphing a polynomial function because in addition to paying attention to

- intercepts,
- symmetry, and
- shifting/reflecting/stretching of known graphs,

you should also keep an eye on

- the domain of f, and
- the degrees of $P(x)$ and $Q(x)$.

The latter two topics are important in determining whether a graph of a rational function possesses *asymptotes*.

The y-intercept is the point $(0, f(0))$, provided the number 0 is in the domain of f. For example, the graph of the rational function $f(x) = (1 - x)/x$ does not cross the y-axis since $f(0)$ is not defined. If the polynomials $P(x)$ and $Q(x)$ have no common factors, then the x-intercepts of the graph of the rational function $f(x) = P(x)/Q(x)$ are the points whose x-coordinates are the real zeros of the numerator $P(x)$. In other words, the only way we can have $f(x) = P(x)/Q(x) = 0$ is to have $P(x) = 0$. The graph of a rational function f is symmetric with respect to the y-axis if $f(-x) = f(x)$, and symmetric with respect to the origin if $f(-x) = -f(x)$. Since it is easy to spot an even or an odd polynomial function (see page 145), here is an easy way to determine symmetry of the graph of a rational function. We again assume $P(x)$ and $Q(x)$ have no common factors.

- The quotient of two even functions is even. (2)
- The quotient of two odd functions is even. (3)
- The quotient of an even and an odd function is odd. (4)

See Problem 48 in Exercises 3.5.

We have already seen the graphs of two simple rational functions, $y = 1/x$ and $y = 1/x^2$, in Figures 2.2.1(e) and 2.2.1(f). You are encouraged to review those graphs at this time. Note that $P(x) = 1$ is an even function and $Q(x) = x$ is an odd function, so $y = 1/x$ is an odd function by (4). On the other hand, $P(x) = 1$ is an even function and $Q(x) = x^2$ is an even function, so $y = 1/x^2$ is an even function by (2).

EXAMPLE 1   Shifted Reciprocal Function

Graph the function $f(x) = \dfrac{2}{x - 1}$.

Solution The graph possesses no symmetry since $Q(x) = x - 1$ is neither even nor odd. Since $f(0) = -2$, the y-intercept is $(0, -2)$. Because $P(x) = 2$ is never 0, there are no x-intercepts. You might also recognize that the graph of this rational function is the graph of the reciprocal function $y = 1/x$ stretched vertically by a factor of 2 and shifted 1 unit to the right. The point $(1, 1)$ is on the graph of $y = 1/x$; in FIGURE 3.5.1, after the vertical stretch and horizontal shift, the corresponding point on the graph of $y = 2/(x - 1)$ is $(2, 2)$.

The vertical line $x = 1$ and the horizontal line $y = 0$ (the equation of the x-axis) are of special importance for this graph.

The vertical dashed line $x = 1$ in Figure 3.5.1 is the y-axis in Figure 2.2.1(e) shifted 1 unit to the right. Although the number 1 is not in the domain of the given function, we can evaluate f at values of x that are *near* 1. For example, you should verify that

x	0.999	1.001
$f(x)$	-2000	2000

(5)

The table in (5) shows that for values of x close to 1, the corresponding function values $f(x)$ are large in absolute value. On the other hand, for values of x for which $|x|$ is large, the corresponding function values $f(x)$ are near 0. For example, you should verify that

x	-999	1001
$f(x)$	-0.002	0.002

(6)

Geometrically, as x approaches 1, the graph of the function approaches the vertical line $x = 1$, and as $|x|$ increases without bound the graph of the function approaches the horizontal line $y = 0$. ■

☐ **Asymptotes** To indicate that x is approaching a number a we use the arrow notation:

- $x \to a^-$ to mean that x is approaching a from the *left*, that is, through numbers that are less than a;

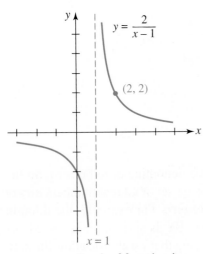

FIGURE 3.5.1 Graph of function in Example 1

- $x \to a^+$ to mean that x is approaching a from the *right*, that is, through numbers that are greater than a; and
- $x \to a$ to mean that x is approaching a from both the *left* and the *right*.

We also use the infinity symbols and the arrow notation:

- $x \to -\infty$ to mean that x becomes *unbounded in the negative direction*, and
- $x \to \infty$ to mean that x becomes *unbounded in the positive direction*.

Similar interpretations are given to the symbols $f(x) \to -\infty$ and $f(x) \to \infty$. These notational devices are a convenient way of describing the behavior of a function either near a number $x = a$ or as x increases to the right or decreases to the left. Thus, in Example 1 it is apparent from (5) and Figure 3.5.1 that

$$f(x) \to -\infty \text{ as } x \to 1^- \qquad \text{and} \qquad f(x) \to \infty \text{ as } x \to 1^+.$$

In words, the notation in the preceding line signifies that the function values are decreasing without bound as x approaches 1 from the left, and the function values are increasing without bound as x approaches 1 from the right. From (6) and Figure 3.5.1 it should also be apparent that

$$f(x) \to 0 \text{ as } x \to -\infty \qquad \text{and} \qquad f(x) \to 0 \text{ as } x \to \infty.$$

In Figure 3.5.1, the vertical line whose equation is $x = 1$ is a called a **vertical asymptote** for the graph of f, and the horizontal line whose equation is $y = 0$ is called a **horizontal asymptote** for the graph of f.

In this section we will examine three types of asymptotes, which correspond to the three types of lines studied in Section 2.3: *vertical lines*, *horizontal lines*, and *slant* (or *oblique*) *lines*. The characteristic of any asymptote is that the graph of a function f must get close to, or approach, the line.

DEFINITION 3.5.2 Vertical Asymptote

A line $x = a$ is said to be a **vertical asymptote** for the graph of a function f if at least one of the following six statements is true:

$$
\begin{array}{ll}
f(x) \to -\infty \text{ as } x \to a^- & f(x) \to \infty \text{ as } x \to a^- \\
f(x) \to -\infty \text{ as } x \to a^+ & f(x) \to \infty \text{ as } x \to a^+ \\
f(x) \to -\infty \text{ as } x \to a & f(x) \to \infty \text{ as } x \to a
\end{array}
\qquad (7)
$$

FIGURE 3.5.2 illustrates four of the possibilities listed in (7) of Definition 3.5.2 for the unbounded behavior of a function f near a vertical asymptote $x = a$. If the function exhibits the *same kind of unbounded behavior from both sides of* $x = a$, then we write either

$$f(x) \to \infty \text{ as } x \to a, \qquad (8)$$

or

$$f(x) \to -\infty \text{ as } x \to a. \qquad (9)$$

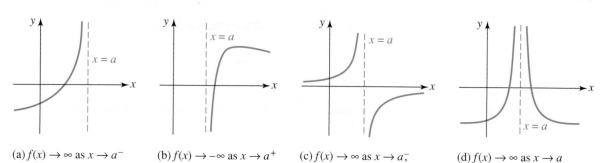

(a) $f(x) \to \infty$ as $x \to a^-$ (b) $f(x) \to -\infty$ as $x \to a^+$ (c) $f(x) \to \infty$ as $x \to a^-$ (d) $f(x) \to \infty$ as $x \to a$

$f(x) \to -\infty$ as $x \to a^+$

FIGURE 3.5.2 The line $x = a$ is a vertical asymptote

In Figure 3.5.2(d) we see that $f(x) \to \infty$ as $x \to a^{-}$ and $f(x) \to \infty$ as $x \to a^{+}$, and so we write $f(x) \to \infty$ as $x \to a$.

If $x = a$ is a vertical asymptote for the graph of a *rational function* $f(x) = P(x)/Q(x)$, then the function values $f(x)$ become unbounded as x approaches a from *both sides*, that is, from the right $(x \to a^{+})$ *and* from the left $(x \to a^{-})$. The graphs in Figures 3.5.2(c) and 3.5.2(d) (or the reflection of these graphs in the x-axis) are typical graphs of a rational function with a single vertical asymptote. As can be seen from these figures, a rational function with a vertical asymptote is a **discontinuous function**. There is an infinite break in each graph at $x = a$. As seen in Figures 3.5.2(c) and 3.5.2(d), a single vertical asymptote divides the xy-plane into two regions, and within each region there is a single piece or **branch** of the graph of the rational function f.

DEFINITION 3.5.3   Horizontal Asymptote

A line $y = c$ is said to be a **horizontal asymptote** for the graph of a function f if

$$f(x) \to c \text{ as } x \to -\infty \qquad \text{or} \qquad f(x) \to c \text{ as } x \to \infty \qquad (10)$$

In FIGURE 3.5.3 we have illustrated some typical horizontal asymptotes. We note, in conjunction with Figure 3.5.3(d) that, in general, the graph of a function can have at most

Remember this. ▶ *two* horizontal asymptotes, but the graph of a *rational function* $f(x) = P(x)/Q(x)$ can have at most *one*. If the graph of a rational function f possesses a horizontal asymptote $y = c$, then as shown in Figure 3.5.3(c),

$$f(x) \to c \text{ as } x \to -\infty \qquad \text{and} \qquad f(x) \to c \text{ as } x \to \infty.$$

The last line is a mathematical description of the end behavior of the graph of a rational function with a horizontal asymptote. Also, the graph of a function can *never* cross a vertical asymptote but, as suggested in Figure 3.5.3(a), a graph can cross a horizontal asymptote.

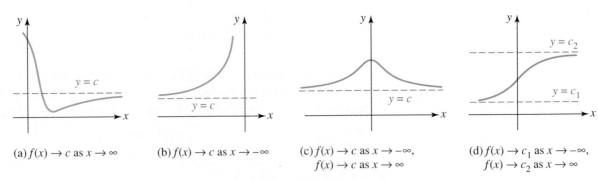

(a) $f(x) \to c$ as $x \to \infty$

(b) $f(x) \to c$ as $x \to -\infty$

(c) $f(x) \to c$ as $x \to -\infty$, $f(x) \to c$ as $x \to \infty$

(d) $f(x) \to c_1$ as $x \to -\infty$, $f(x) \to c_2$ as $x \to \infty$

FIGURE 3.5.3 The line $y = c$ is a horizontal asymptote in (a), (b), and (c)

DEFINITION 3.5.4   Slant Asymptote

A slant asymptote is also called an ▶ oblique asymptote.

A line $y = mx + b, m \neq 0$, is said to be a **slant asymptote** for the graph of a function f if

$$f(x) \to mx + b \text{ as } x \to -\infty$$

or

$$f(x) \to mx + b \text{ as } x \to \infty \qquad (11)$$

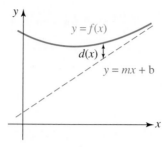

FIGURE 3.5.4 Slant asymptote is $y = mx + b$

The notation in (11) of Definition 3.5.4 means that the graph of f possesses a slant asymptote whenever the function values $f(x)$ become closer and closer to the values of y on the line $y = mx + b$ as x becomes large in absolute value. Another way of stating (11) is: A line $y = mx + b$ is a slant asymptote for the graph of f if the vertical distance $d(x)$ between points with the same x-coordinate on the two graphs satisfies

$$d(x) = f(x) - (mx + b) \to 0 \text{ as } x \to -\infty \text{ or as } x \to \infty.$$

See FIGURE 3.5.4. We note that if a graph of a rational function $f(x) = P(x)/Q(x)$ possesses a slant asymptote it can have vertical asymptotes, but the graph *cannot* have a horizontal asymptote.

On a practical level, vertical and horizontal asymptotes of the graph of a rational function f can be determined by inspection. So for the sake of discussion let us suppose that

$$f(x) = \frac{P(x)}{Q(x)} = \frac{a_n x^n + a_{n-1} x^{n-1} + \cdots + a_1 x + a_0}{b_m x^m + b_{m-1} x^{m-1} + \cdots + b_1 x + b_0}, \quad a_n \neq 0, b_m \neq 0, \quad (12)$$

represents a general rational function.

☐ **Finding a Vertical Asymptote** Let us assume that the polynomial functions $P(x)$ and $Q(x)$ in (12) have no common factors. In that case:

- If a is a real number such that $Q(a) = 0$, then the line $x = a$ is a vertical asymptote for the graph of f.

Since $Q(x)$ is a polynomial function of degree m, it can have up to m real zeros, and so the graph of a rational function f can have up to m vertical asymptotes. If the graph of a rational function f has, say, k ($k \leq m$) vertical asymptotes, then the k vertical lines divide the xy-plane into $k + 1$ regions. Thus the graph of this rational function would have $k + 1$ branches.

EXAMPLE 2　　　　**Vertical Asymptotes**

(a) Inspection of the rational function $f(x) = \dfrac{2x + 1}{x^2 - 4}$ shows that the denominator $Q(x) = x^2 - 4 = (x + 2)(x - 2) = 0$ at $x = -2$ and $x = 2$. These are equations of vertical asymptotes for the graph of f. The graph of f has three branches: one to the left of the line $x = -2$, one between the lines $x = -2$ and $x = 2$, and one to the right of the line $x = 2$.

(b) The graph of the rational function $f(x) = \dfrac{1}{x^2 + x + 4}$ has no vertical asymptotes, since $Q(x) = x^2 + x + 4 \neq 0$ for all real numbers. ■

☐ **Finding a Horizontal Asymptote** When we discussed end behavior of a polynomial function $P(x)$ of degree n, we pointed out that $P(x)$ behaves like $y = a_n x^n$, that is, $P(x) \approx a_n x^n$, for values of x large in absolute value. As a consequence, we see from

lower powers of x are
irrelevant as $x \to \pm\infty$
↓

$$f(x) = \frac{\boxed{a_n x^n} + a_{n-1} x^{n-1} + \cdots + a_1 x + a_0}{\boxed{b_m x^m} + b_{m-1} x^{m-1} + \cdots + b_1 x + b_0}$$

that $f(x)$ behaves like $y = \dfrac{a_n}{b_m}x^{n-m}$ because $f(x) \approx \dfrac{a_n x^n}{b_m x^m} = \dfrac{a_n}{b_m}x^{n-m}$ for $x \to \pm\infty$.

Therefore:

$$\text{If } n = m, \qquad f(x) \approx \dfrac{a_n}{b_m}x^{\overset{\overset{0}{\downarrow}}{n-n}} \to \dfrac{a_n}{b_m} \qquad \text{as } x \to \pm\infty. \tag{13}$$

$$\text{If } n < m, \qquad f(x) \approx \dfrac{a_n}{b_m}x^{\overset{\overset{\text{negative}}{\downarrow}}{n-m}} = \dfrac{a_n}{b_m}\dfrac{1}{x^{m-n}} \to 0 \qquad \text{as } x \to \pm\infty. \tag{14}$$

$$\text{If } n > m, \qquad f(x) \approx \dfrac{a_n}{b_m}x^{\overset{\overset{\text{positive}}{\downarrow}}{n-m}} \to \pm\infty \qquad \text{as } x \to \pm\infty. \tag{15}$$

From (13), (14), and (15) we glean the following three facts about horizontal asymptotes for the graph of $f(x) = P(x)/Q(x)$:

- If degree of $P(x) =$ degree of $Q(x)$, then $y = a_n/b_m$ (the quotient of the leading coefficients) is a horizontal asymptote. (16)
- If degree of $P(x) <$ degree of $Q(x)$, then $y = 0$ is a horizontal asymptote. (17)
- If degree of $P(x) >$ degree of $Q(x)$, then the graph of f has *no* horizontal asymptote. (18)

EXAMPLE 3   Horizontal Asymptotes

Determine whether the graph of each of the following rational functions possesses a horizontal asymptote.

(a) $f(x) = \dfrac{3x^2 + 4x - 1}{8x^2 + x}$ **(b)** $f(x) = \dfrac{4x^3 + 7x + 8}{2x^4 + 3x^2 - x + 6}$ **(c)** $f(x) = \dfrac{5x^3 + x^2 + 1}{2x + 3}$

Solution **(a)** Since the degree of the numerator $3x^2 + 4x - 1$ is the same as the degree of the denominator $8x^2 + x$ (both degrees are 2), we see from (13) that

$$f(x) \approx \dfrac{3}{8}x^{2-2} = \dfrac{3}{8} \qquad \text{as} \qquad x \to \pm\infty.$$

As summarized in (16), $y = \frac{3}{8}$ is a horizontal asymptote for the graph of f.

(b) Since the degree of the numerator $4x^3 + 7x + 8$ is 3 and the degree of the denominator $2x^4 + 3x^2 - x + 6$ is 4 (and $3 < 4$), we see from (14) that

$$f(x) \approx \dfrac{4}{2}x^{3-4} = \dfrac{2}{x} \to 0 \qquad \text{as} \qquad x \to \pm\infty.$$

As summarized in (17), $y = 0$ (the x-axis) is a horizontal asymptote for the graph of f.

(c) Since the degree of the numerator $5x^3 + x^2 - 1$ is 3 and the degree of the denominator $2x + 3$ is 1 (and $3 > 1$), we see from (15) that

$$f(x) \approx \dfrac{5}{2}x^{3-1} = \dfrac{5}{2}x^2 \to \infty \qquad \text{as} \qquad x \to \pm\infty.$$

As summarized in (18), the graph of f has no horizontal asymptote.   ■

CHAPTER 3 POLYNOMIAL AND RATIONAL FUNCTIONS

In the graphing examples that follow we will assume again that $P(x)$ and $Q(x)$ in (12) have no common factors.

EXAMPLE 4   **Graph of a Rational Function**

Graph the function $f(x) = \dfrac{3 - x}{x + 2}$.

Solution Here are some things we look at to sketch the graph of f.

Symmetry: No symmetry. $P(x) = 3 - x$ and $Q(x) = x + 2$ are neither even nor odd.

Intercepts: $f(0) = \frac{3}{2}$ and so the y-intercept is $\left(0, \frac{3}{2}\right)$. Setting $P(x) = 0$ or $3 - x = 0$ implies 3 is a zero of P. The single x-intercept is $(3, 0)$.

Vertical Asymptotes: Setting $Q(x) = 0$ or $x + 2 = 0$ gives $x = -2$. The line $x = -2$ is a vertical asymptote.

Branches: Because there is only a single vertical asymptote, the graph of f consists of two distinct branches, one to the left of $x = -2$ and one to the right of $x = -2$.

Horizontal Asymptote: The degree of P and the degree of Q are the same (namely, 1), and so the graph of f has a horizontal asymptote. By rewriting f as

$f(x) = \dfrac{-x + 3}{x + 2}$ we see that the quotient of leading coefficients is $-1/1 = -1$. From (16) we see that the line $y = -1$ is a horizontal asymptote.

The Graph: We draw the vertical and horizontal asymptotes using dashed lines. The right branch of the graph of f is drawn through the intercepts $\left(0, \frac{3}{2}\right)$ and $(3, 0)$ in such a manner that it approaches both asymptotes. The left branch is drawn *below* the horizontal asymptote $y = -1$. Were we to draw this branch above the horizontal asymptote it would have to be near the horizontal asymptote from above and near the vertical asymptote from the left. In order to do this the branch of the graph would have to cross the x-axis, but since there are no more x-intercepts this is impossible. See FIGURE 3.5.5.

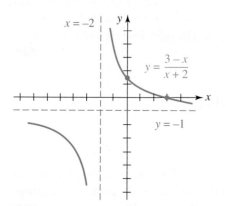

FIGURE 3.5.5 Graph of function in Example 4

EXAMPLE 5   **Example 4 Using Transformations**

Long division and rigid transformations can sometimes be aids in graphing rational functions. Note that if we carry out the long division for the function f in Example 4, we see that

$$f(x) = \frac{3 - x}{x + 2} \quad \text{is the same as} \quad f(x) = -1 + \frac{5}{x + 2}.$$

Thus, starting with the graph of $y = 1/x$, we stretch it vertically by a factor of 5. Next, shift the graph of $y = 5/x$ to the left 2 units. Finally, shift $y = 5/(x + 2)$ vertically downward 1 unit. You should verify that the net result is the graph in Figure 3.5.5.

EXAMPLE 6   **Graph of a Rational Function**

Graph the function $f(x) = \dfrac{x}{1 - x^2}$.

Solution *Symmetry*: Since $P(x) = x$ is odd and $Q(x) = 1 - x^2$ is even, the quotient $P(x)/Q(x)$ is odd. The graph of f is symmetric with respect to the origin.

Intercepts: $f(0) = 0$, and so the *y*-intercept is $(0, 0)$. Setting $P(x) = x = 0$ gives $x = 0$. Thus the only intercept is $(0, 0)$.

Vertical Asymptotes: Setting $Q(x) = 0$ or $1 - x^2 = 0 = 0$ gives $x = -1$ and $x = 1$. The lines $x = -1$ and $x = 1$ are vertical asymptotes.

Branches: Because there are two vertical asymptotes, the graph of *f* consists of three distinct branches, one to the left of the line $x = -1$, one between the lines $x = -1$ and $x = 1$, and one to the right of the line $x = 1$.

Horizontal Asymptote: Since the degree of the numerator x is 1 and the degree of the denominator $1 - x^2$ is 2 (and $1 < 2$), it follows from (14) and (17) that $y = 0$ is a horizontal asymptote for the graph of *f*.

The Graph: We can plot the graph for $x \geq 0$ and then use symmetry to obtain the remaining part of the graph of $x < 0$. We begin by drawing the vertical asymptotes using dashed lines. The half-branch of the graph of *f* on the interval $[0, 1)$ is drawn starting at $(0, 0)$. The function *f* must then increase because $P(x) = x > 0$, and $Q(x) = 1 - x^2 > 0$ indicates that $f(x) > 0$ for $0 < x < 1$. This implies that near the vertical asymptote $x = 1, f(x) \to \infty$ as $x \to 1^-$. The branch of the graph for $x > 1$ is drawn below the horizontal asymptote $y = 0$, since $P(x) = x > 0$ and $Q(x) = 1 - x^2 < 0$ imply $f(x) < 0$. Thus $f(x) \to -\infty$ as $x \to 1^+$ and $f(x) \to 0$ as $x \to \infty$. The remainder of the graph for $x < 0$ is obtained by reflecting the graph for $x > 0$ through the origin. See FIGURE 3.5.6. ■

FIGURE 3.5.6 Graph of function in Example 6

EXAMPLE 7   Graph of a Rational Function

Graph the function $f(x) = \dfrac{x}{1 + x^2}$.

Solution The given function *f* is similar to the function in Example 6 in that *f* is an odd function, $(0, 0)$ is the only intercept of its graph, and its graph has the horizontal asymptote $y = 0$. However, note that since $1 + x^2 > 0$ for all real numbers, there are no vertical asymptotes. Thus there are no branches; the graph is one continuous curve. For $x \geq 0$, the graph passes through $(0, 0)$ and then must increase since $f(x) > 0$ for $x > 0$. Also, *f* must attain a relative maximum and then decrease in order to satisfy the condition $f(x) \to 0$ as $x \to \infty$. As mentioned in Section 3.1, the exact value of this relative maximum can be obtained through calculus techniques. Finally, we reflect the portion of the graph for $x > 0$ through the origin. The graph must look something like that shown in FIGURE 3.5.7. ■

FIGURE 3.5.7 Graph of function in Example 7

☐ **Finding a Slant Asymptote** Let us again assume that the polynomials $P(x)$ and $Q(x)$ in (12) have no common factors. In that case we can recognize the existence of a slant asymptote in the following manner:

- If the degree of $P(x)$ is precisely one greater than the degree of $Q(x)$, that is, if the degree of $Q(x)$ is *m* and the degree of $P(x)$ is $m + 1$, then the graph of *f* possesses a slant asymptote.

We find the slant asymptote by division. Using long division to divide $P(x)$ by $Q(x)$ yields a quotient that is a linear polynomial $mx + b$ and a polynomial remainder $R(x)$:

$$f(x) = \frac{P(x)}{Q(x)} = mx + b + \frac{R(x)}{Q(x)}. \qquad (19)$$

with labels: quotient (pointing to $mx + b$) and remainder (pointing to $\dfrac{R(x)}{Q(x)}$).

Because the degree of $R(x)$ must be less than the degree of the divisor $Q(x)$, we have $R(x)/Q(x) \to 0$ as $x \to -\infty$ and as $x \to \infty$, and consequently

$$f(x) \to mx + b \text{ as } x \to -\infty \qquad \text{and} \qquad f(x) \to mx + b \text{ as } x \to \infty.$$

In other words, an equation of the slant asymptote is $y = mx + b$, where $mx + b$ is the quotient in (19).

If the denominator $Q(x)$ is a *linear* polynomial, we can then use synthetic division to carry out the long division.

▌EXAMPLE 8   Graph with a Slant Asymptote

Graph the function $f(x) = \dfrac{x^2 - x - 6}{x - 5}$.

Solution *Symmetry*: No symmetry. $P(x) = x^2 - x - 6$ and $Q(x) = x - 5$ are neither even nor odd.

Intercepts: $f(0) = \frac{6}{5}$, and so the y-intercept is $\left(0, \frac{6}{5}\right)$. Setting $P(x) = 0$ or $x^2 - x - 6 = 0$ or $(x + 2)(x - 3) = 0$ shows that -2 and 3 are zeros of $P(x)$. The x-intercepts are $(-2, 0)$ and $(3, 0)$.

Vertical Asymptotes: Setting $Q(x) = 0$ or $x - 5 = 0$ gives $x = 5$. The line $x = 5$ is a vertical asymptote.

Branches: The graph of f consists of two branches, one to the left of $x = 5$ and one to the right of $x = 5$.

Horizontal Asymptote: None.

Slant Asymptote: Since the degree of $P(x) = x^2 - x - 6$ (which is 2) is exactly one greater than the degree of $Q(x) = x - 5$ (which is 1), the graph of $f(x)$ has a slant asymptote. To find it, we divide $P(x)$ by $Q(x)$. Because $Q(x)$ is a linear polynomial we can use synthetic division:

$$\begin{array}{r|rrr} 5 & 1 & -1 & -6 \\ & & 5 & 20 \\ \hline & 1 & 4 & \underline{|14} \end{array}$$

Recall that the latter notation means that

$$\overset{\displaystyle y = x + 4 \text{ is the slant asymptote}}{\underset{\displaystyle \downarrow}{}}$$

$$\frac{x^2 - x - 6}{x - 5} = x + 4 + \frac{14}{x - 5}.$$

Note again that $14/(x - 5) \to 0$ as $x \to \pm\infty$. Hence the line $y = x + 4$ is a slant asymptote.

The Graph: Using the foregoing information we obtain the graph in FIGURE 3.5.8. The asymptotes are the red dashed lines in the figure.   ▬

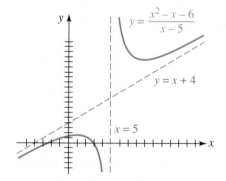

FIGURE 3.5.8 Graph of function in Example 8

▌EXAMPLE 9   Graph with a Slant Asymptote

By inspection it should be apparent that the graph of the rational function $f(x) = \dfrac{x^3 - 8x + 12}{x^2 + 1}$ possesses a slant asymptote but no vertical asymptotes. Since the denominator is a quadratic polynomial we resort to long division to obtain

$$\frac{x^3 - 8x + 12}{x^2 + 1} = x + \frac{-9x + 12}{x^2 + 1}.$$

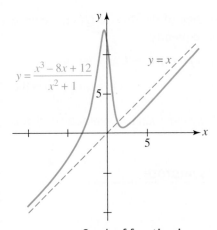

$$y = \frac{x^3 - 8x + 12}{x^2 + 1}$$

FIGURE 3.5.9 Graph of function in Example 9

The slant asymptote is the line $y = x$. The graph has no symmetry. The y-intercept is $(0, 12)$. The lack of vertical asymptotes indicates that the function f is continuous; its graph consists of an unbroken curve. Because the numerator is a polynomial of odd degree, we know that it has at least one real zero. Since $x^3 - 8x + 12 = 0$ has no rational roots, we use approximation or graphical techniques to show that the equation possesses only one real irrational root. Thus the x-intercept is approximately $(-3.4, 0)$. The graph of f is given in FIGURE 3.5.9. Notice in the figure that the graph of f crosses the slant asymptote. ▬

☐ **Graph with a Hole** We assumed throughout the preceding discussion of asymptotes of rational functions that the polynomial functions $P(x)$ and $Q(x)$ in (1) have no common factors. We now know that if a is a real number such that $Q(a) = 0$, and $P(x)$ and $Q(x)$ have no common factors, then the line $x = a$ is a vertical asymptote for the graph of f. Because Q is a polynomial function, it follows from the Factor Theorem that $Q(x) = (x - a)q(x)$. The assumption that the numerator P and denominator Q have no common factors tells us that $x - a$ is not a factor of P and so $P(a) \neq 0$. When $P(a) = 0$ *and* $Q(a) = 0$, then $x = a$ *may not* be a vertical asymptote. For example, when a is a *simple zero* of both P and Q, then $x = a$ is *not* a vertical asymptote for the graph of $f(x) = P(x)/Q(x)$. To see this, we know from the Factor Theorem that if $P(a) = 0$ and $Q(a) = 0$, then $x - a$ is a common factor of P and Q:

$$P(x) = (x - a)p(x) \qquad \text{and} \qquad Q(x) = (x - a)q(x),$$

where p and q are polynomials such that $p(a) \neq 0$ and $q(a) \neq 0$. After canceling

$$f(x) = \frac{P(x)}{Q(x)} = \frac{(x - a)p(x)}{(x - a)q(x)} = \frac{p(x)}{q(x)}, \quad x \neq a,$$

we see that $f(x)$ is undefined at a, but the function values $f(x)$ do not become unbounded as $x \to a^-$ or as $x \to a^+$ because $q(x)$ is not approaching 0. As an example, we saw in Section 2.5 that the graph of the rational function

$$f(x) = \frac{x^2 - 4}{x - 2} = \frac{(x - 2)(x + 2)}{x - 2} = x + 2, \quad x \neq 2,$$

is basically a straight line. But since $f(2)$ is undefined there is no point $(2, f(2))$ on the line. Instead, there is a **hole** in the graph corresponding to the point $(2, 4)$. See Figure 2.5.5(a) on page 90.

▐ EXAMPLE 10   **Graph with a Hole**

Graph the function $f(x) = \dfrac{x^2 - 2x - 3}{x^2 - 1}$.

Solution Although $x^2 - 1 = 0$ for $x = -1$ and $x = 1$, only $x = 1$ is a vertical asymptote. Note that the numerator $P(x)$ and denominator $Q(x)$ have the common factor $x + 1$, which we cancel provided $x \neq -1$:

equality is true for $x \neq -1$
↓

$$f(x) = \frac{(x + 1)(x - 3)}{(x + 1)(x - 1)} = \frac{x - 3}{x - 1}. \tag{20}$$

Thus we see from (20) that there is no infinite break in the graph at $x = -1$. We graph $y = \dfrac{x - 3}{x - 1}, x \neq -1$, by observing that the y-intercept is $(0, 3)$, an x-intercept is $(3, 0)$, a vertical asymptote is $x = 1$, and a horizontal asymptote is $y = 1$. The graph of this function has two branches, but the branch to the left of the vertical asymptote $x = 1$ has a hole in it corresponding to the point $(-1, 2)$. See FIGURE 3.5.10. ▬

$$f(x) = \frac{x - 3}{x - 1}, x \neq -1$$

$(-1, 2)$

$y = 1$

$x = 1$

FIGURE 3.5.10 Graph of function in Example 10

CHAPTER 3 POLYNOMIAL AND RATIONAL FUNCTIONS

When asked whether they have ever heard the statement "*An asymptote is a line that the graph approaches but does not cross*," a surprising number of students will raise their hands. First, let's make it clear that the statement is false; a graph *can* cross a horizontal asymptote and *can* cross a slant asymptote. A graph can never cross a *vertical* asymptote $x = a$, since the function is inherently undefined at $x = a$. We can even find the points where a graph crosses a horizontal or slant asymptote. For example, the rational function $f(x) = \dfrac{x^2 + 2x}{x^2 - 1}$ has the horizontal asymptote $y = 1$. Determining whether the graph of f crosses the horizontal line $y = 1$ is equivalent to asking whether $y = 1$ is in the range of the function f. Setting $f(x)$ equal to 1, that is,

$$\frac{x^2 + 2x}{x^2 - 1} = 1$$

implies $\qquad x^2 + 2x = x^2 - 1 \quad$ and $\quad x = -\tfrac{1}{2}$.

Since $x = -\tfrac{1}{2}$ is in the domain of f, the graph of f crosses the horizontal asymptote at $\left(-\tfrac{1}{2}, f\left(-\tfrac{1}{2}\right)\right) = \left(-\tfrac{1}{2}, 1\right)$. Observe in Example 9 we can find the point where the slant asymptote crosses the graph of $y = x$ by solving $f(x) = x$. You should verify that the point of intersection is $\left(\tfrac{4}{3}, \tfrac{4}{3}\right)$. See Problems 31–36 in Exercises 3.5.

Exercises 3.5 Answers to selected odd-numbered problems begin on page ANS–12.

In Problems 1 and 2, use a calculator to fill out the given table for the rational function $f(x) = \dfrac{2x}{x - 3}$.

1. $x = 3$ is a vertical asymptote for the graph of f

x	3.1	3.01	3.001	3.0001	3.00001
$f(x)$					
x	2.9	2.99	2.999	2.9999	2.99999
$f(x)$					

2. $y = 2$ is a horizontal asymptote for the graph of f

x	10	100	1000	10,000	100,000
$f(x)$					
x	-10	-100	-1000	$-10,000$	$-100,000$
$f(x)$					

In Problems 3–22, find the vertical and horizontal asymptotes for the graph of the given rational function. Find x- and y-intercepts of the graph. Sketch the graph of f.

3. $f(x) = \dfrac{1}{x-2}$

4. $f(x) = \dfrac{4}{x+3}$

5. $f(x) = \dfrac{x}{x+1}$

6. $f(x) = \dfrac{x}{2x-5}$

7. $f(x) = \dfrac{4x-9}{2x+3}$

8. $f(x) = \dfrac{2x+4}{x-2}$

9. $f(x) = \dfrac{1-x}{x+1}$

10. $f(x) = \dfrac{2x-3}{x}$

11. $f(x) = \dfrac{1}{(x-1)^2}$

12. $f(x) = \dfrac{4}{(x+2)^3}$

13. $f(x) = \dfrac{1}{x^3}$

14. $f(x) = \dfrac{8}{x^4}$

15. $f(x) = \dfrac{x}{x^2-1}$

16. $f(x) = \dfrac{x^2}{x^2-4}$

17. $f(x) = \dfrac{1}{x(x-2)}$

18. $f(x) = \dfrac{1}{x^2-2x-8}$

19. $f(x) = \dfrac{1-x^2}{x^2}$

20. $f(x) = \dfrac{16}{x^2+4}$

21. $f(x) = \dfrac{-2x^2+8}{(x-1)^2}$

22. $f(x) = \dfrac{x(x-5)}{x^2-9}$

In Problems 23–30, find the vertical and slant asymptotes for the graph of the given rational function. Find x- and y-intercepts of the graph. Sketch the graph f.

23. $f(x) = \dfrac{x^2-9}{x}$

24. $f(x) = \dfrac{x^2-3x-10}{x}$

25. $f(x) = \dfrac{x^2}{x+2}$

26. $f(x) = \dfrac{x^2-2x}{x+2}$

27. $f(x) = \dfrac{x^2-2x-3}{x-1}$

28. $f(x) = \dfrac{-(x-1)^2}{x+2}$

29. $f(x) = \dfrac{x^3-8}{x^2-x}$

30. $f(x) = \dfrac{5x(x+1)(x-4)}{x^2+1}$

In Problems 31–34, find the point where the graph of f crosses its horizontal asymptote. Sketch the graph of f.

31. $f(x) = \dfrac{x-3}{x^2+3}$

32. $f(x) = \dfrac{(x-3)^2}{x^2-5x}$

33. $f(x) = \dfrac{4x(x-2)}{(x-3)(x+4)}$

34. $f(x) = \dfrac{2x^2}{x^2+x+1}$

In Problems 35 and 36, find the point where the graph of f crosses its slant asymptote. Use a graphing utility to obtain the graph of f and the slant asymptote in the same coordinate plane.

35. $f(x) = \dfrac{x^3-3x^2+2x}{x^2+1}$

36. $f(x) = \dfrac{x^3+2x-4}{x^2}$

In Problems 37–40, find a rational function that satisfies the given conditions. There is no unique answer.

37. vertical asymptote: $x = 2$
horizontal asymptote: $y = 1$
x-intercept: $(5, 0)$

38. vertical asymptote: $x = 1$
horizontal asymptote: $y = -2$
y-intercept: $(0, -1)$

39. vertical asymptotes: $x = -1, x = 2$
horizontal asymptote: $y = 3$
x-intercept: $(3, 0)$

40. vertical asymptote: $x = 4$
slant asymptote: $y = x + 2$

In Problems 41–44, find the asymptotes and any holes in the graph of the given rational function. Find x- and y-intercepts of the graph. Sketch the graph f.

41. $f(x) = \dfrac{x^2 - 1}{x - 1}$

42. $f(x) = \dfrac{x - 1}{x^2 - 1}$

43. $f(x) = \dfrac{x + 1}{x(x^2 + 4x + 3)}$

44. $f(x) = \dfrac{x^3 + 8}{x + 2}$

Applications

5 ohms

r ohms

FIGURE 3.5.11 Parallel resistors in Problem 45

45. Parallel Resistors A 5-ohm resistor and a variable resistor are placed in parallel as shown in FIGURE 3.5.11. The resulting resistance R (in ohms) is related to the resistance r (in ohms) of the variable resistor by the equation

$$R = \frac{5r}{5 + r}.$$

Sketch the graph of R as a function of r for $r > 0$. What is the resulting resistance R as r becomes very large?

46. Power The electrical power P produced by a certain source is given by

$$P = \frac{E^2 r}{R^2 + 2Rr + r^2},$$

where E is the voltage of the source, R is the resistance of the source, and r is the resistance in the circuit. Sketch the graph of P as a function of r using the values $E = 5$ volts and $R = 1$ ohm.

47. Illumination Intensity The intensity of illumination from a light source at any point is directly proportional to the strength of the source and inversely proportional to the square of the distance from the source. Given two sources of strengths 16 units and 2 units that are 100 cm apart, as shown in FIGURE 3.5.12, the intensity I at any point P between them is given by

$$I(x) = \frac{16}{x^2} + \frac{2}{(100 - x)^2},$$

16-unit source

2-unit source

P

x

100 cm

FIGURE 3.5.12 Two light sources in Problem 47

where x is the distance from the 16-unit source. Sketch the graph of $I(x)$ on the interval $(0, 100)$. Describe the behavior of $I(x)$ as $x \to 0^+$. As $x \to 100^-$.

For Discussion

48. Suppose $f(x) = P(x)/Q(x)$. Prove the symmetry rules (2), (3), and (4) for rational functions. Assume $P(x)$ and $Q(x)$ have no common factors.

49. Construct a rational function $f(x) = P(x)/Q(x)$ whose graph crosses its slant asymptote twice.

50. If you have studied Section 1.5, then discuss how topics in this section and Section 3.2 can be used to help find the limit: $\lim\limits_{x \to 1} \dfrac{x^5 - 1}{x - 1}$.

3.6 Partial Fractions

INTRODUCTION When two rational functions, say, $f(x) = \dfrac{2}{x+5}$ and $g(x) = \dfrac{1}{x+1}$, are added, the terms are combined by means of a common denominator:

$$\frac{2}{x+5} + \frac{1}{x+1} = \frac{2}{x+5}\left(\frac{x+1}{x+1}\right) + \frac{1}{x+1}\left(\frac{x+5}{x+5}\right). \tag{1}$$

Adding numerators on the right-hand side of (1) yields the single rational expression

$$\frac{3x+7}{(x+5)(x+1)}. \tag{2}$$

An important procedure in the study of integral calculus requires that we be able to reverse the process. In other words, we would start with a rational expression such as (2) and then break it down, or *decompose* it, into simpler component fractions $2/(x+5)$ and $1/(x+1)$, called **partial fractions**.

☐ **Partial Fractions** The algebraic process for breaking down a rational expression such as (2) into partial fractions is known as **partial fraction decomposition**. For convenience we will assume that the rational function $P(x)/Q(x)$, $Q(x) \neq 0$, is a **proper fraction** or **proper rational expression**, that is, the degree of $P(x)$ is less than the degree of $Q(x)$. We will also assume once again that the polynomials $P(x)$ and $Q(x)$ have no common factors.

In the discussion that follows we consider four cases of partial fraction decomposition of $P(x)/Q(x)$. The cases depend on the factors in the denominator $Q(x)$. When the polynomial $Q(x)$ is factored as a product of $(ax+b)^n$ and $(ax^2+bx+c)^m$, $n = 1, 2, \ldots, m = 1, 2, \ldots$, where the coefficients a, b, and c are real numbers and the quadratic polynomial ax^2+bx+c is **irreducible** over the real numbers (that is, does not factor using real numbers), the rational expression $P(x)/Q(x)$ can be decomposed into a sum of partial fractions of the form

$$\frac{C_k}{(ax+b)^k} \quad \text{and} \quad \frac{A_k x + B_k}{(ax^2+bx+c)^k},$$

where k is a positive integer.

CASE 1: $Q(x)$ **Contains Only Nonrepeated Linear Factors**
We state the following fact from algebra without proof. If the denominator can be factored completely into linear factors,

$$Q(x) = (a_1 x + b_1)(a_2 x + b_2) \cdots (a_n x + b_n),$$

where all the $a_i x + b_i$, $i = 1, 2, \ldots, n$ are distinct (that is, no two factors are the same), then unique real constants $C_1, C_2, \ldots, C_n$ can be found such that

$$\frac{P(x)}{Q(x)} = \frac{C_1}{a_1 x + b_1} + \frac{C_2}{a_2 x + b_2} + \cdots + \frac{C_n}{a_n x + b_n}. \tag{3}$$

In practice we will use the letters A, B, C, $\ldots$, in place of the subscripted coefficients $C_1, C_2, C_3, \ldots$. The next example illustrates this first case.

EXAMPLE 1   **Distinct Linear Factors**

To decompose $\dfrac{2x + 1}{(x - 1)(x + 3)}$ into individual partial fractions we make the assumption, based on the form given in (3), that the rational function can be written as

$$\frac{2x + 1}{(x - 1)(x + 3)} = \frac{A}{x - 1} + \frac{B}{x + 3}. \qquad (4)$$

We now clear (4) of fractions; this can be done by either combining the terms on the right-hand side of the equality over a least common denominator and equating numerators or by simply multiplying both sides of the equality by the denominator $(x - 1)(x + 3)$ on the left-hand side. Either way, we arrive at

$$2x + 1 = A(x + 3) + B(x - 1). \qquad (5)$$

Multiplying out the right-hand side of (5) and grouping by powers of x gives

$$2x + 1 = A(x + 3) + B(x - 1) = (A + B)x + (3A - B). \qquad (6)$$

Each of the equations (5) and (6) is an identity, which means that the equality is true for *all* real values of x. As a consequence, the coefficients of x on the left-hand side of (6) must be the same as the coefficients of the corresponding powers of x on the right-hand side, that is,

$$2x + 1x^0 = (A + B)x + (3A - B)x^0.$$

The result is a system of two equations in two unknowns A and B:

$$\begin{aligned} 2 &= A + B \\ 1 &= 3A - B. \end{aligned} \qquad (7)$$

By adding the two equations we get $3 = 4A$, and so we find that $A = \frac{3}{4}$. Substituting this value into either equation in (7) then yields $B = \frac{5}{4}$. Hence the desired decomposition is

$$\frac{2x + 1}{(x - 1)(x + 3)} = \frac{\frac{3}{4}}{x - 1} + \frac{\frac{5}{4}}{x + 3}.$$

You are encouraged to verify the foregoing result by combining the terms on the right-hand side of the last equation by means of a common denominator. ∎

☐ **A Shortcut Worth Knowing** If the denominator contains, say, three linear factors such as in $\dfrac{4x^2 - x + 1}{(x - 1)(x + 3)(x - 6)}$, then the partial fraction decomposition looks like this

$$\frac{4x^2 - x + 1}{(x - 1)(x + 3)(x - 6)} = \frac{A}{x - 1} + \frac{B}{x + 3} + \frac{C}{x - 6}.$$

By following the same steps as in Example 1, we would find that the analogue of (7) is now three equations in the three unknowns A, B, and C. The point is this: the more linear factors in the denominator, the larger the system of equations we must solve. There is an algebraic procedure worth learning that can cut down on some of the algebra. To illustrate, let's return to the identity (5). Since the equality is true for every value of x, it holds for $x = 1$ and $x = -3$, *the zeros of the denominator*. Setting $x = 1$ in (5) gives

$3 = 4A$, from which it follows immediately that $A = \frac{3}{4}$. Similarly, by setting $x = -3$ in (5), we obtain $-5 = (-4)B$ or $B = \frac{5}{4}$.

CASE 2: $Q(x)$ Contains Repeated Linear Factors

If the denominator $Q(x)$ contains a repeated linear factor $(ax + b)^n, n > 1$, then unique real constants $C_1, C_2, \ldots, C_n$ can be found such that the partial fraction decomposition of $P(x)/Q(x)$ contains the terms

$$\frac{C_1}{ax + b} + \frac{C_2}{(ax + b)^2} + \cdots + \frac{C_n}{(ax + b)^n}. \tag{8}$$

▌ EXAMPLE 2   Repeated Linear Factors

To decompose $\dfrac{6x - 1}{x^3(2x - 1)}$ into partial fractions, we first observe that the denominator consists of the repeated linear factor x and the nonrepeated linear factor $2x - 1$. Based on the forms in (3) and (8), we assume that

$$\frac{6x - 1}{x^3(2x - 1)} = \overbrace{\frac{A}{x} + \frac{B}{x^2} + \frac{C}{x^3}}^{\text{according to Case 2}} + \overbrace{\frac{D}{2x - 1}}^{\text{according to Case 1}}. \tag{9}$$

Multiplying (9) by $x^3(2x - 1)$ clears it of fractions and yields

$$6x - 1 = Ax^2(2x - 1) + Bx(2x - 1) + C(2x - 1) + Dx^3 \tag{10}$$

$$\text{or} \qquad 6x - 1 = (2A + D)x^3 + (-A + 2B)x^2 + (-B + 2C)x - C. \tag{11}$$

Now the zeros of the denominator in the original expression are $x = 0$ and $x = \frac{1}{2}$. If we then set $x = 0$ and $x = \frac{1}{2}$ in (10), we find, in turn, that $C = 1$ and $D = 16$. Because the denominator of the original expression has only two distinct zeros, we can find A and B by equating the corresponding coefficients of x^3 and x^2 in (11):

▶ The coefficients of x^3 and x^2 on the left-hand side of (11) are both 0.

$$0 = 2A + D, \qquad 0 = -A + 2B.$$

Using the known value of D, the first equation yields $A = -D/2$ or $A = -8$. The second then gives $B = A/2$ or $B = -4$. The partial fraction decomposition is

$$\frac{6x - 1}{x^3(2x - 1)} = -\frac{8}{x} - \frac{4}{x^2} + \frac{1}{x^3} + \frac{16}{2x - 1}. \qquad ▬$$

CASE 3: $Q(x)$ Contains Nonrepeated Irreducible Quadratic Factors

If the denominator $Q(x)$ contains nonrepeated irreducible quadratic factors $a_i x^2 + b_i x + c_i$, then unique real constants $A_1, A_2, \ldots, A_n, B_1, B_2, \ldots, B_n$ can be found such that the partial fraction decomposition of $P(x)/Q(x)$ contains the terms

$$\frac{A_1 x + B_1}{a_1 x^2 + b_1 x + c_1} + \frac{A_2 x + B_2}{a_2 x^2 + b_2 x + c_2} + \cdots + \frac{A_n x + B_n}{a_n x^2 + b_n x + c_n}. \tag{12}$$

▌ EXAMPLE 3   Irreducible Quadratic Factors

To decompose $\dfrac{4x}{(x^2 + 1)(x^2 + 2x + 3)}$ into partial fractions, we first observe that the quadratic polynomials $x^2 + 1$ and $x^2 + 2x + 3$ are irreducible over the real numbers. Hence by (12) we assume that

▶ Use the quadratic formula. For either factor you will find that $b^2 - 4ac < 0$.

$$\frac{4x}{(x^2 + 1)(x^2 + 2x + 3)} = \frac{Ax + B}{x^2 + 1} + \frac{Cx + D}{x^2 + 2x + 3}.$$

After clearing fractions in the preceding line, we find

$$4x = (Ax + B)(x^2 + 2x + 3) + (Cx + D)(x^2 + 1)$$
$$= (A + B)x^3 + (2A + B + D)x^2 + (3A + 2B + C)x + (3B + D).$$

Because the denominator of the original fraction has no real zeros, we have no recourse except to form a system of equations by comparing coefficients of all powers of x:

$$0 = A + C$$
$$0 = 2A + B + D$$
$$4 = 3A + 2B + C$$
$$0 = 3B + D.$$

Using $C = -A$ and $D = -3B$ from the first and fourth equations, we can eliminate C and D in the second and third equations:

$$0 = A - B$$
$$2 = A + B.$$

Solving this simpler system of equations yields $A = 1$ and $B = 1$. Hence, $C = -1$ and $D = -3$. The partial fraction decomposition is

$$\frac{4x}{(x^2 + 1)(x^2 + 2x + 3)} = \frac{x + 1}{x^2 + 1} - \frac{x + 3}{x^2 + 2x + 3}.$$ ∎

CASE 4: $Q(x)$ Contains Repeated Irreducible Quadratic Factors

If the denominator $Q(x)$ contains a repeated irreducible quadratic factor $(ax^2 + bx + c)^n$, $n > 1$ a positive integer, then unique real constants $A_1, A_2, \ldots, A_n, B_1, B_2, \ldots, B_n$ can be found such that the partial fraction decomposition of $P(x)/Q(x)$ contains the terms

$$\frac{A_1 x + B}{ax^2 + bx + c} + \frac{A_2 x + B_2}{(ax^2 + bx + c)^2} + \cdots + \frac{A_n x + B_n}{(ax^2 + bx + c)^n}. \tag{13}$$

■ EXAMPLE 4   Repeated Quadratic Factor

Decompose $\dfrac{x^2}{(x^2 + 4)^2}$ into partial fractions.

Solution The denominator contains only the repeated irreducible quadratic factor $x^2 + 4$. As indicated in (13), we assume a decomposition of the form

$$\frac{x^2}{(x^2 + 4)^2} = \frac{Ax + B}{x^2 + 4} + \frac{Cx + D}{(x^2 + 4)^2}.$$

Clearing fractions by multiplying both sides of the preceding equality by $(x^2 + 4)^2$ gives

$$x^2 = (Ax + B)(x^2 + 4) + Cx + D. \tag{14}$$

As in Example 3, the denominator of the original has no real zeros, and so we must solve a system of four equations for $A, B, C,$ and D. To that end we rewrite (14) as

$$0x^3 + 1x^2 + 0x + 0x^0 = Ax^3 + Bx^2 + (4A + C)x + (4B + D)x^0$$

and compare coefficients of like powers (match the colors) to obtain

$$0 = A$$
$$1 = B$$
$$0 = 4A + C$$
$$0 = 4B + D.$$

From this system we find that $A = 0$, $B = 1$, $C = 0$, and $D = -4$. The required partial fraction decomposition is then

$$\frac{x^2}{(x^2 + 4)^2} = \frac{1}{x^2 + 4} - \frac{4}{(x^2 + 4)^2}.$$

EXAMPLE 5   Combination of Cases

Determine the form of the decomposition of $\dfrac{x^2 + 3x + 5}{(x - 5)(x + 2)^2(x^2 + 1)^2}$.

Solution The denominator contains a single linear factor $x - 5$, a repeated linear factor $x + 2$, and a repeated irreducible quadratic factor $x^2 + 1$. By Cases 1, 2, and 4 the assumed form of the partial fraction decomposition is

$$\frac{x + 3}{(x - 5)(x + 2)^2(x^2 + 1)^2} = \overbrace{\frac{A}{x - 5}}^{\text{Case 1}} + \overbrace{\frac{B}{x + 2} + \frac{C}{(x + 2)^2}}^{\text{Case 2}} + \overbrace{\frac{Dx + E}{x^2 + 1} + \frac{Fx + G}{(x^2 + 1)^2}}^{\text{Case 4}}$$

NOTES FROM THE CLASSROOM

We assumed throughout the foregoing discussion that the degree of the numerator $P(x)$ was less than the degree of the denominator $Q(x)$. If, however, the degree of $P(x)$ is greater than or equal to the degree of $Q(x)$, then $P(x)/Q(x)$ is an **improper fraction**. We can still do partial fraction decomposition, but the process starts with long division until a polynomial quotient and a proper fraction is attained. For example, long division gives

$$\underset{\uparrow}{\overset{\text{improper fraction}}{}} \qquad \underset{\uparrow}{\overset{\text{proper fraction}}{}}$$
$$\frac{x^3 + x - 1}{x^2 - 3x} = x + 3 + \frac{10x - 1}{x(x - 3)}.$$

Then by using Case 1 we finish the problem with the decomposition of the proper-fraction term in the last equality:

$$\frac{x^3 + x - 1}{x^2 - 3x} = x + 3 + \frac{10x - 1}{x(x - 3)} = x + 3 + \frac{\frac{1}{3}}{x} + \frac{\frac{29}{3}}{x - 3}.$$

See Problems 35–40 in Exercises 3.6.

In Problems 1–8, write out the appropriate form of the partial fraction decomposition of the given expression. Do not evaluate the coefficients.

1. $\dfrac{x - 1}{x^2 + x}$

2. $\dfrac{9x - 8}{x^2 - 1}$

3. $\dfrac{2x^2 - 3}{x^3 + x^2}$

4. $\dfrac{x^3}{(x^2 - 1)(x + 1)^2}$

5. $\dfrac{3x^2 - x + 4}{x^4 + 2x^3 + x^2}$

6. $\dfrac{4}{x^3(x^2 + 1)}$

7. $\dfrac{2x^3 - x}{(x^2 + 1)^2}$

8. $\dfrac{-x^2 + 3x + 7}{(x^2 + x - 2)(x^2 + x + 1)^3}$

In Problems 9–34, find the partial fraction decomposition of the given rational expression.

9. $\dfrac{1}{x(x + 2)}$

10. $\dfrac{2}{x(4x - 1)}$

11. $\dfrac{-9x + 27}{x^2 - 4x - 5}$

12. $\dfrac{-5x + 18}{x^2 + 2x - 63}$

13. $\dfrac{2x^2 - x}{(x + 1)(x + 2)(x + 3)}$

14. $\dfrac{1}{x(x - 2)(2x - 1)}$

15. $\dfrac{3x}{x^2 - 16}$

16. $\dfrac{10x - 5}{25x^2 - 1}$

17. $\dfrac{5x - 6}{(x - 3)^2}$

18. $\dfrac{5x^2 - 25x + 28}{x^2(x - 7)}$

19. $\dfrac{1}{x^2(x + 2)^2}$

20. $\dfrac{-4x + 6}{(x - 2)^2(x - 1)^2}$

21. $\dfrac{3x - 1}{x^3(x - 1)(x + 3)}$

22. $\dfrac{x^2 - x}{x(x + 4)^3}$

23. $\dfrac{6x^2 - 7x + 11}{(x - 1)(x^2 + 9)}$

24. $\dfrac{2x + 10}{2x^3 + x}$

25. $\dfrac{4x^2 + 4x - 6}{(2x - 3)(x^2 - x + 1)}$

26. $\dfrac{2x^2 - x + 7}{(x - 6)(x^2 + x + 5)}$

27. $\dfrac{t + 8}{t^4 - 1}$

28. $\dfrac{y^2 + 1}{y^3 - 1}$

29. $\dfrac{x^3}{(x^2 + 2)(x^2 + 1)}$

30. $\dfrac{x - 15}{(x^2 + 2x + 5)(x^2 + 6x + 10)}$

31. $\dfrac{(x + 1)^2}{(x^2 + 1)^2}$

32. $\dfrac{2x^2}{(x - 2)(x^2 + 4)^2}$

33. $\dfrac{40}{x^3 + 3x^2 - 4x - 12}$

34. $\dfrac{-5x^2 - 31x + 10}{2x^3 - 9x^2 - 6x + 5}$

In Problems 35–40, first use long division followed by partial fraction decomposition.

35. $\dfrac{x^5}{x^2 - 1}$

36. $\dfrac{(x + 2)^2}{x(x + 3)}$

37. $\dfrac{x^2 - 4x + 1}{2x^2 + 5x + 2}$

38. $\dfrac{x^4 + 3x}{x^2 + 2x + 1}$

39. $\dfrac{x^6}{x^3 - 2x^2 + x - 2}$

40. $\dfrac{x^3 + x^2 - x + 1}{x^3 + 3x^2 + 3x + 1}$

3.7 The Area Problem

INTRODUCTION As we saw in Section 2.10, the fundamental motivating problem of differential calculus, *Find a tangent line to the graph of a function f*, is answered by the notion of the **derivative** of the function. Differential calculus is the study of the properties and applications of the derivative of a function $y = f(x)$. Integral calculus, on the other hand, is the study of the properties and the applications of the **definite integral** of a function $y = f(x)$. As mentioned in Section 2.10, the historical problem that leads to the concept of the definite integral is, *Find the area under the graph of a function f*. We examine the area problem in this section.

☐ **Area Under a Graph** Throughout the discussion that follows we will assume that $y = f(x)$ is a function that is continuous and nonnegative on an interval $[a, b]$. Recall that the concept of continuity has been mentioned several times in previous sections; in this case, the graph of f has no breaks, gaps, or holes in it anywhere on the interval $[a, b]$. The requirement that f be nonnegative, that is, $f(x) \geq 0$ for all x in $[a, b]$, means that no portion of its graph on the interval is below the x-axis. Specifically, then,

> By the **area under a graph** we mean the area A of the region in the plane bounded by the graph of f, the lines $x = a$ and $x = b$, and the x-axis.

See **FIGURE 3.7.1**.

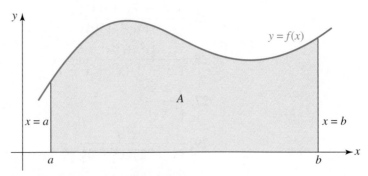

FIGURE 3.7.1 Area A under a graph

To get to the answer of the question, What is the *exact* value of A? we begin with a method for systematically *approximating A*. The basic idea is simply this: build rectangles across the interval $[a, b]$ and use the sum of the areas of the rectangles as an approximation for A.

☐ **Approximating the Area** One possible systematic procedure for approximating the value of the area A under a graph is summarized next.

(*i*) Subdivide the interval $[a, b]$ into n subintervals $[x_{k-1}, x_k]$, where

$$a = x_0 < x_1 < x_2 < \cdots < x_{n-1} < x_n = b,$$

so that each subinterval has the same width $\Delta x = \dfrac{b - a}{n}$. This is called a **regular partition** of the interval $[a, b]$.

(*ii*) Choose a number x_k^* in each of the n subintervals $[x_{k-1}, x_k]$ and form the n products $f(x_k^*)\,\Delta x$. Since the area of a rectangle is length × width, $f(x_k^*)\,\Delta x$ is the area of the rectangle of length $f(x_k^*)$ and width Δx built up on the kth subinterval $[x_{k-1}, x_k]$. The n numbers $x_1^*, x_2^*, x_3^*, \ldots, x_n^*$ are called **sample points**. See FIGURE 3.7.2.

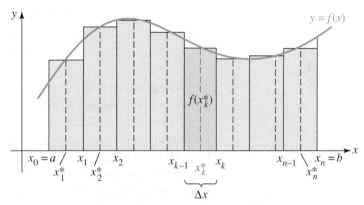

FIGURE 3.7.2 n rectangles of width Δx and length $f(x_k^*)$

(*iii*) The sum of the areas of the n rectangles represents an approximation to the value of the area of the region:

$$A \approx f(x_1^*)\Delta x + f(x_2^*)\Delta x + f(x_3^*)\Delta x + \cdots + f(x_n^*)\Delta x. \tag{1}$$

To simplify the hand calculations, the sample points x_k^*, $k = 1, 2, \ldots, n$, are generally chosen to be either the left-hand endpoint or the right-hand endpoint of each subinterval $[x_{k-1}, x_k]$.

EXAMPLE 1   Area of a Triangular Region

Approximate the area A under the graph of $f(x) = x$ on the interval $[0, 1]$ using four subintervals of equal width and choosing

(a) x_k^* as the left-hand endpoint of each subinterval, and
(b) x_k^* as the right-hand endpoint of each subinterval.

See FIGURE 3.7.3.

Solution By dividing $[0, 1]$ into four subintervals, the width of each subinterval is $\Delta x = \frac{1-0}{4} = \frac{1}{4}$.

(a) If x_k^* is the left-hand endpoint of each of the four subintervals, then $x_1^* = 0$, $x_2^* = \frac{1}{4}, x_3^* = \frac{2}{4} = \frac{1}{2}$, and $x_4^* = \frac{3}{4}$. See FIGURE 3.7.4(a). We have from (1),

$$\begin{aligned} A &\approx f(0)\tfrac{1}{4} + f\!\left(\tfrac{1}{4}\right)\tfrac{1}{4} + f\!\left(\tfrac{1}{2}\right)\tfrac{1}{4} + f\!\left(\tfrac{3}{4}\right)\tfrac{1}{4} \\ &= 0 \cdot \tfrac{1}{4} + \tfrac{1}{4} \cdot \tfrac{1}{4} + \tfrac{1}{2} \cdot \tfrac{1}{4} + \tfrac{3}{4} \cdot \tfrac{1}{4} \\ &= \tfrac{3}{8} = 0.375. \end{aligned}$$

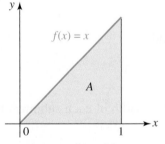

FIGURE 3.7.3 Area A in Example 1

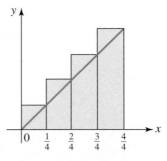

(a) Using left-hand endpoints   (b) Using right-hand endpoints

FIGURE 3.7.4 Approximating the area A in Example 1

(b) If x_k^* is the right-hand endpoint of each of the four subintervals, then $x_1^* = \frac{1}{4}$, $x_2^* = \frac{2}{4} = \frac{1}{2}$, $x_3^* = \frac{3}{4}$, and $x_4^* = \frac{4}{4} = 1$. See Figure 3.7.4(b). We have from (1),

$$A \approx f\left(\tfrac{1}{4}\right)\tfrac{1}{4} + f\left(\tfrac{1}{2}\right)\tfrac{1}{4} + f\left(\tfrac{3}{4}\right)\tfrac{1}{4} + f(1)\tfrac{1}{4}$$
$$= \tfrac{1}{4}\cdot\tfrac{1}{4} + \tfrac{1}{2}\cdot\tfrac{1}{4} + \tfrac{3}{4}\cdot\tfrac{1}{4} + 1\cdot\tfrac{1}{4}$$
$$= \tfrac{5}{8} = 0.625.$$

As can be seen in Figures 3.7.4(a) and 3.7.4(b), the value obtained in part (a) of Example 1 underestimates the area A, whereas the value in part (b) overestimates A, that is, $0.375 < A < 0.625$. We can compare these approximations with the actual area. Since the area under the graph of $f(x) = x$ on the interval $[0, 1]$ is the area of a right triangle of base $= 1$ and height $= 1$, the exact area is $A = \frac{1}{2}\cdot$ base $\cdot$ height $= \frac{1}{2}\cdot 1 \cdot 1 = \frac{1}{2} = 0.5$.

There is no special reason that we chose the sample points x_k^*, $k = 1, 2, \ldots, n$, to be the left-hand and then the right-hand endpoints of the subintervals $[x_{k-1}, x_k]$, other than *convenience*. We could pick x_k^* randomly in each subinterval. In Problem 3 of Exercises 3.7 you are asked to approximate the area in Example 1 using the midpoint of each subinterval.

Intuitively, the more rectangles we use the better (1) approximates the area A under a graph. The trade-off, of course, is that we must do more calculations.

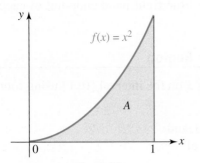

FIGURE 3.7.5 Area A in Example 2

■ EXAMPLE 2   **Area Under a Parabola**

Approximate the area A under the graph of $f(x) = x^2$ on the interval $[0, 1]$ using eight subintervals of equal width and choosing

(a) x_k^* as the left-hand endpoint of each subinterval, and
(b) x_k^* as the right-hand endpoint of each subinterval.

See FIGURE 3.7.5.

Solution By dividing $[0, 1]$ into eight subintervals, the width of each subinterval is $\Delta x = \frac{1-0}{8} = \frac{1}{8}$.

(a) If x_k^* is the left-hand endpoint of each of the four subintervals, then $x_1^* = 0$, $x_2^* = \frac{1}{8}$, $x_3^* = \frac{2}{8} = \frac{1}{4}$, $x_4^* = \frac{3}{8}$, $x_5^* = \frac{4}{8} = \frac{1}{2}$, $x_6^* = \frac{5}{8}$, $x_7^* = \frac{6}{8} = \frac{3}{4}$, $x_8^* = \frac{7}{8}$. See FIGURE 3.7.6(a). We have from (1),

$$A \approx f(0)\cdot\tfrac{1}{8} + f\left(\tfrac{1}{8}\right)\cdot\tfrac{1}{8} + f\left(\tfrac{1}{4}\right)\cdot\tfrac{1}{8} + f\left(\tfrac{3}{8}\right)\cdot\tfrac{1}{8} + f\left(\tfrac{1}{2}\right)\cdot\tfrac{1}{8} + f\left(\tfrac{5}{8}\right)\cdot\tfrac{1}{8} + f\left(\tfrac{3}{4}\right)\cdot\tfrac{1}{8} + f\left(\tfrac{7}{8}\right)\cdot\tfrac{1}{8}$$
$$= 0\cdot\tfrac{1}{8} + \tfrac{1}{64}\cdot\tfrac{1}{8} + \tfrac{1}{16}\cdot\tfrac{1}{8} + \tfrac{9}{64}\cdot\tfrac{1}{8} + \tfrac{1}{4}\cdot\tfrac{1}{8} + \tfrac{25}{64}\cdot\tfrac{1}{8} + \tfrac{9}{16}\cdot\tfrac{1}{8} + \tfrac{49}{64}\cdot\tfrac{1}{8}$$
$$= \tfrac{35}{128} = 0.2734375.$$

(a) Using left-hand endpoints

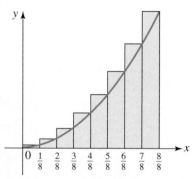

(b) Using right-hand endpoints

FIGURE 3.7.6 Approximating the area A in Example 2

(b) If x_k^* is the right-hand endpoint of each of the four subintervals, then $x_1^* = \frac{1}{8}$, $x_2^* = \frac{2}{8} = \frac{1}{4}$, $x_3^* = \frac{3}{8}$, $x_4^* = \frac{4}{8} = \frac{1}{2}$, $x_5^* = \frac{5}{8}$, $x_6^* = \frac{6}{8} = \frac{3}{4}$, $x_7^* = \frac{7}{8}$, $x_8^* = \frac{8}{8} = 1$. See Figure 3.7.6(b). We have from (1),

$$A \approx f\left(\tfrac{1}{8}\right) \cdot \tfrac{1}{8} + f\left(\tfrac{1}{4}\right) \cdot \tfrac{1}{8} + f\left(\tfrac{3}{8}\right) \cdot \tfrac{1}{8} + f\left(\tfrac{1}{2}\right) \cdot \tfrac{1}{8} + f\left(\tfrac{5}{8}\right) \cdot \tfrac{1}{8} + f\left(\tfrac{3}{4}\right) \cdot \tfrac{1}{8} + f\left(\tfrac{7}{8}\right) \cdot \tfrac{1}{8} + f(1) \cdot \tfrac{1}{8}$$

$$= \tfrac{1}{64} \cdot \tfrac{1}{8} + \tfrac{1}{16} \cdot \tfrac{1}{8} + \tfrac{9}{64} \cdot \tfrac{1}{8} + \tfrac{1}{4} \cdot \tfrac{1}{8} + \tfrac{25}{64} \cdot \tfrac{1}{8} + \tfrac{9}{16} \cdot \tfrac{1}{8} + \tfrac{49}{64} \cdot \tfrac{1}{8} + 1 \cdot \tfrac{1}{8}$$

$$= \tfrac{51}{128} = 0.3984375.$$

From Figure 3.7.6(a) we see that the area of the seven rectangles underestimates A in Example 2, whereas the eight rectangles in Figure 3.7.6(b) overestimates A. From the calculations in Example 2 we can write $0.2734375 < A < 0.3984375$. But an observation is in order at this point. Don't assume that by using left-hand endpoints followed by the right-hand endpoints of the subintervals for x_k^* that we *always* get, in turn, a lower estimate followed by an upper estimate of the area A under the graph of f on $[a, b]$. This occurred in Examples 1 and 2 simply because, in both cases, the function f was increasing on the interval $[0, 1]$.

☐ **Summation Notation** Writing out sums such as (1) can become very tedious. To facilitate the discussion of the area problem, a special notation for summation is used in calculus. Suppose a_k denotes a real number that depends on an integer k. The sum of n such real numbers a_k, $a_1 + a_2 + a_3 + \cdots + a_n$, is denoted by the symbol $\sum_{k=1}^{n} a_k$, that is,

$$\sum_{k=1}^{n} a_k = a_1 + a_2 + a_3 + \cdots + a_n. \tag{2}$$

Because Σ is the capital Greek letter *sigma*, (2) is called **sigma notation**, but more commonly, (2) is referred to as **summation notation**. The integer k is called the **index of summation** and takes on consecutive integer values starting with $k = 1$ and ending with $k = n$. For example, the sum of the first 100 squared positive integers,

$$1^2 + 2^2 + 3^2 + 4^2 + \cdots + 98^2 + 99^2 + 100^2,$$

can be written compactly as

sum ends with this number
↓
$$\sum_{k=1}^{100} k^2.$$
↑
sum starts with this number

Using summation notation, the sum of the areas in (1) can be written as

$$A \approx \sum_{k=1}^{n} f(x_k^*) \Delta x.$$

☐ **Area** It should seem believable that we can reduce the error inherent in the method of approximating an area A under a graph by summing areas of rectangles by using more and more rectangles ($n \to \infty$) of decreasing width $\left(\Delta x = \dfrac{b-a}{n} \to 0 \right)$. Thus the 32 rectangles in FIGURE 3.7.7 should give us a better approximation to area A in Figure 3.7.1 than the eight rectangles shown in Figure 3.7.2. Indeed that is the case. It can be proved that when f is continuous on $[a, b]$ and $f(x) \geq 0$ for all x in the interval, the area A under the graph of the function $y = f(x)$ on the interval is given by the limit

$$A = \lim_{n \to \infty} \sum_{k=1}^{n} f(x_k^*) \Delta x. \tag{3}$$

FIGURE 3.7.7 Using more rectangles improves the approximation to area A

The limit (3) exists regardless of how the sample points $x_1^*, x_2^*, x_3^*, \ldots, x_n^*$ are chosen in the subintervals $[x_0, x_1], [x_1, x_2], [x_2, x_3], \ldots, [x_{n-1}, x_n]$. Thus in (3), each sample point x_k^* could always be chosen, say, to be the right-hand endpoint of each subinterval. Since we are in no position to deal, in general terms, with limits of the kind given in (3), we leave that aspect of the area problem to a course in calculus. But if you are willing to put in the time to work Problems 15–22 in Exercises 3.7 then Problems 23 and 24 will give you a small taste of what is involved in computing area A by the limiting process given in (3).

We said at the start that the area problem is the motivating problem for the definite integral. You ask: So what is a definite integral? It is now just a small jump from (3) to the concept of the definite integral.

DEFINITION 3.7.1   Definite Integral

Let the function f be continuous on $[a, b]$. The **definite integral** of f from $x = a$ to $x = b$, denoted by $\int_a^b f(x)dx$, is

$$\int_a^b f(x)dx = \lim_{n \to \infty} \sum_{k=1}^{n} f(x_k^*) \Delta x \tag{4}$$

The integral symbol $\int$ in (4), as used by **Wilhelm Gottfried Leibniz** (1646–1716) who is considered the co-inventor of calculus, along with **Isaac Newton** (1643–1727), is simply an elongated S for the word "sum."

NOTES FROM THE CLASSROOM

(*i*) If read quickly, you might conclude that formula (4) is the same as (3). In a way this is correct; however, (4) is a more general concept (notice that we are not requiring f to be nonnegative on the interval $[a, b]$). Thus, a *definite integral need not be area*. Also, in its most general setting, even the conditions of continuity of f and the use of a regular partition are dropped in the definition of the definite integral. What, then, is a definite integral? For now, accept the fact that a definite integral is simply a real number that can be negative, zero, or positive. When the conditions of continuity and nonnegativity are imposed on $y = f(x)$ on the interval $[a, b]$, then the area under the graph is $A = \int_a^b f(x)\,dx$. Also, you should be aware that the interpretations of derivative and the definite integral are much broader than just slopes of tangent lines and areas under graphs. As you progress through courses in mathematics, sciences, and engineering you will see many diverse applications of the derivative and the definite integral.

(*ii*) In this chapter we worked principally with polynomial functions. Polynomial functions are the fundamental building blocks of a class known as **algebraic functions**. In Section 3.5 we saw that a rational function is the quotient of two polynomial functions. In general, an algebraic function f involves a finite number of additions, subtractions, multiplications, divisions, and roots of polynomial functions. Thus

$$y = 2x^2 - 5x, \quad y = \sqrt[3]{x^2}, \quad y = x^4 + \sqrt{x^2 + 5}, \text{ and } y = \frac{\sqrt{x}}{x^3 - 2x^2 + 7}$$

are algebraic functions. Indeed, all the functions in Chapters 2 and 3 are algebraic functions. Starting with the next chapter we consider functions that belong to a different class known as **transcendental functions**. A transcendental function f is defined to be one that is *not algebraic*. The six trigonometric functions (Chapter 4) and the exponential and logarithmic functions (Chapter 5) are examples of transcendental functions.

Exercises 3.7   Answers to selected odd-numbered problems begin on page ANS–13.

In Problems 1–4, the function f and the interval are given in Example 1.

1. Approximate the area A, this time using eight subintervals of equal width and choosing x_k^* as the left-hand endpoint of each subinterval. Draw the eight rectangles.

2. Approximate the area A, this time using eight subintervals of equal width and choosing x_k^* as the right-hand endpoint of each subinterval. Draw the eight rectangles.

3. Approximate the area A, this time using four subintervals of equal width but choosing x_k^* as the midpoint of each subinterval. Draw the four rectangles.

4. Compare the approximation obtained in Problem 3 of the exact area $A = 0.5$. Explain why your answer in Problem 3 is not surprising.

5. Approximate the area under the graph of $f(x) = x + 2$ on the interval $[-1, 2]$ using six subintervals of equal width and choosing:
   (a) x_k^* as the left-hand endpoint of each subinterval, and
   (b) x_k^* as the right-hand endpoint of each subinterval.

FIGURE 3.7.8 Graph for Problem 11

FIGURE 3.7.9 Graph for Problem 12

FIGURE 3.7.10 Land in Problem 13

FIGURE 3.7.11 Fish pond in Problem 14

6. Repeat Problem 5 using twelve subintervals of equal width.
7. Approximate the area under the graph of $f(x) = -x^2 + 5x$ on the interval $[0, 5]$ using five subintervals of equal width and choosing:
   (a) x_k^* as the left-hand endpoint of each subinterval, and
   (b) x_k^* as the right-hand endpoint of each subinterval.
8. Repeat Problem 7 using ten subintervals of equal width.
9. Approximate the area under the graph of $f(x) = -x^2 + 5x$ on the interval $[0, 5]$ using five subintervals of equal width and choosing x_k^* as the midpoints of each subinterval.
10. Approximate the area under the graph of $f(x) = -x^3 + 2x^2$ on the interval $[0, 2]$ using ten subintervals of equal width and choosing x_k^* as the right-hand endpoint of each subinterval.
11. Find two different approximations for the area A under the graph $y = f(x)$ on the interval $[1, 4]$ shown in FIGURE 3.7.8.
12. Find two different approximations for the area A under the graph $y = f(x)$ on the interval shown in FIGURE 3.7.9.

Applications

13. **Lakefront Property** Suppose a realtor wants to find the area of an irregularly shaped piece of land that is bounded between a 1 mile-long segment of a straight road and the shore of a lake. Measurements (in feet) of the perpendicular distances from the road to the lake are taken at equally spaced intervals along the road as shown in FIGURE 3.7.10. Find two different approximations of the area of the land. Express your answer in acres using the fact that 1 acre $= 43,560$ ft^2.
14. **For the Fish** The large irregularly shaped fish pond shown in FIGURE 3.7.11 is filled with water to a uniform depth of 4 ft. Find an approximation to the number of gallons of water in the pond. Measurements are in feet and the vertical spacing between the horizontal measurements is 1.86 ft. There are 7.48 gallons in 1 cubic foot of water. [*Hint*: The volume of water is surface area × depth.]

For Discussion

If c denotes a constant—that is, independent of the summation index k—then $\sum_{k=1}^{n} c$ means $c + c + c + \cdots + c$. Since there are n c's in this sum, we have

$$\sum_{k=1}^{n} c = nc. \tag{5}$$

In Problems 15 and 16, use (5) to find the numerical value of the given sum.

15. $\displaystyle\sum_{k=1}^{75} 6$ 16. $\displaystyle\sum_{k=1}^{25} 10$

The sum of the first n positive integers can be written $\sum_{k=1}^{n} k$. If this sum is denoted by S, then

$$S = 1 + 2 + 3 + \cdots + (n - 1) + n \tag{6}$$

can also be written as

$$S = n + (n - 1) + \cdots + 3 + 2 + 1. \tag{7}$$

If we add (6) and (7), then

$$2S = \underbrace{(n + 1) + (n + 1) + (n + 1) + \cdots + (n + 1)}_{n \text{ terms of } n + 1} = n(n + 1).$$

Solving for S gives $S = \frac{1}{2}n(n + 1)$, or

$$\sum_{k=1}^{n} k = \tfrac{1}{2}n(n + 1). \qquad (8)$$

In Problems 17 and 18, use (8) to find the numerical value of the given sum.

17. $\displaystyle\sum_{k=1}^{50} k$ **18.** $\displaystyle\sum_{k=1}^{1000} k$

Here are two properties of summation notation:

$$\sum_{k=1}^{n} ca_k = c \sum_{k=1}^{n} a_k, \quad c \text{ a constant}, \qquad (9)$$

$$\sum_{k=1}^{n} (a_k \pm b_k) = \sum_{k=1}^{n} a_k \pm \sum_{k=1}^{n} b_k. \qquad (10)$$

In Problems 19–22, use (5) and (8)–(10) to find the numerical value of the given sum.

19. $\displaystyle\sum_{k=1}^{20} 2k$ **20.** $\displaystyle\sum_{k=1}^{15} (-6k)$

21. $\displaystyle\sum_{k=1}^{10} (4k + 5)$ **22.** $\displaystyle\sum_{k=1}^{20} (4k - 3)$

In Problems 23 and 24, use the results in (5) and (8)–(10) and the limit definition of area given in (3) to find the *exact* value of the area A. In each case, partition the given interval into n subintervals of width $\Delta x = (b - a)/n$ and use x_k^* as the right-hand end-point of each subinterval.

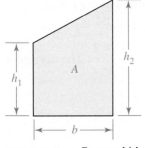

FIGURE 3.7.12 Trapezoid in Problem 25

23. A is the area under the graph of $f(x) = 2x + 1$ on the interval $[0, 4]$

24. A is the area under the graph of $f(x) = -3x + 12$ on the interval $[1, 3]$

25. Consider the trapezoid given in FIGURE 3.7.12.
    (a) Discuss how the area A can be approximated using (1) of this section.
    (b) Using well-known area formulas, find a formula that expresses A in terms of h_1, h_2, and b.

Chapter 3 Review Exercises   Answers to selected odd-numbered Problems begin on page ANS-14.

A. Fill in the Blanks

In Problems 1–22, fill in the blanks.

1. The graph of the polynomial function $f(x) = x^3(x - 1)^2(x - 5)$ is tangent to the x-axis at _____ and passes through the x-axis at _____.

2. A third-degree polynomial function with zeros 1 and $3i$ is _____.

3. The end behavior of the graph of $f(x) = x^2(x + 3)(x - 5)$ resembles the graph of the power function $f(x) =$ _____.

4. The polynomial function $f(x) = x^4 - 3x^3 + 17x^2 - 2x + 2$ has _____ (how many) possible rational zeros.

5. For $f(x) = kx^2(x - 2)(x - 3), f(-1) = 8$ if $k =$ _____.

6. The y-intercept of the graph of the rational function $f(x) = \dfrac{2x + 8}{x^2 - 5x + 4}$ is _____.

7. The vertical asymptotes for the graph of the rational function

$$f(x) = \frac{2x + 8}{x^2 - 5x + 4} \text{ are } \underline{\hspace{1cm}}.$$

8. The x-intercepts of the graph of the rational function $f(x) = \dfrac{x^3 - x}{4 - 2x^3}$ are $\underline{\hspace{1cm}}$.

9. The horizontal asymptote for the graph of the rational function $f(x) = \dfrac{x^3 - x}{4 - 2x^3}$

is $\underline{\hspace{1cm}}$.

10. A rational function whose graph has the horizontal asymptote $y = 1$ and x-intercept $(3, 0)$ is $\underline{\hspace{1cm}}$.

11. The graph of the rational function $f(x) = \dfrac{x^n}{x^3 + 1}$, where n is a nonnegative

integer, has the horizontal asymptote $y = 0$ when $n = \underline{\hspace{1cm}}$.

12. The graph of the polynomial function $f(x) = 3x^5 - 4x^2 + 5x - 2$ has at most $\underline{\hspace{1cm}}$ turning points.

13. If i is a zero of $f(x) = x^4 + 2x^3 + 3x^2 + 2x + 2$ then three other zeros are $\underline{\hspace{1cm}}$.

14. The graph of the rational function $f(x) = \dfrac{x^3}{x^2 + 1}$ has $\underline{\hspace{1cm}}$ (how many)

asymptotes.

15. Suppose that when a polynomial function f of degree 5 is divided by $x - 3$

we get $\dfrac{f(x)}{x - 3} = q(x) + \dfrac{7}{x - 3}$. The degree of the quotient $q(x)$ is $\underline{\hspace{1cm}}$.

16. If f is the polynomial function in Problem 15, then $f(3) = \underline{\hspace{1cm}}$.

17. A rational function f can be written $f(x) = 2x + 4 + \dfrac{x - 3}{x^2 - x}$. The asymptotes

for the graph of f are $\underline{\hspace{1cm}}$.

18. If $f(x) = \dfrac{x + 10}{2 - x}$, then $f(x) \to \underline{\hspace{1cm}}$ as $x \to 2^+$.

19. The graph of the rational function $f(x) = \dfrac{x^3}{x^2 - 9}$ is symmetric with respect

to $\underline{\hspace{1cm}}$.

20. If $f(-2) > 1$ and $f(3) < -1$, then the polynomial function f has at least one zero in the interval $\underline{\hspace{1cm}}$.

21. If n is a positive integer, then $x + 1$ a factor of $f(x) = x^n - 1$ for $n = \underline{\hspace{1cm}}$.

22. If n is a positive integer, then $x + 1$ a factor of $f(x) = x^n + 1$ for $n = \underline{\hspace{1cm}}$.

B. True/False

In Problems 1–24, answer true or false.

1. $f(x) = 2x^3 - 8x^{-2} + 5$ is not a polynomial function. $\underline{\hspace{1cm}}$

2. $f(x) = x + \dfrac{1}{x}$ is a rational function. $\underline{\hspace{1cm}}$

3. The graph of a polynomial function f can have no holes in it. $\underline{\hspace{1cm}}$

4. A polynomial function f of degree 4 has exactly four real zeros. $\underline{\hspace{1cm}}$

5. When a polynomial of degree greater than one is divided by $x - 1$, the remainder is always a constant. $\underline{\hspace{1cm}}$

6. If the coefficients $a, b, c,$ and d of the polynomial function $f(x) = ax^3 + bx^2 + cx + d$ are positive integers, then f has no positive real zeros. $\underline{\hspace{1cm}}$

7. The polynomial equation $2x^7 = 1 - x$ has a solution in the interval $[0, 1]$. $\underline{\hspace{1cm}}$

8. The graph of the rational function $f(x) = (x^2 + 1)/x$ has a slant asymptote. _____

9. The graph of the polynomial function $f(x) = 4x^6 + 3x^2$ is symmetric with respect to the y-axis. _____

10. The graph of a polynomial function f that is an odd function, must pass through the origin. _____

11. An asymptote is a line that the graph of a function approaches but never crosses. _____

12. The point $\left(\frac{1}{3}, \frac{7}{4}\right)$ is on the graph of $f(x) = \dfrac{2x + 4}{3 - x}$. _____

13. The graph of a rational function $f(x) = P(x)/Q(x)$ has a slant asymptote when the degree of P is greater than the degree of Q. _____

14. If $3 - 4i$ is a zero of a polynomial function $f(x)$ with real coefficients, then $3 + 4i$ is also a zero of f. _____

15. A polynomial function f must have at least one rational zero. _____

16. The graph of $f(x) = x^4 + 5x^2 + 2$ does not cross the x-axis. _____

17. If $(-1, 6)$ and $(4, -2)$ are two points on the graph of a polynomial function f, then f has at least one zero in the open interval $(-1, 4)$. _____

18. If the end behavior of a polynomial function f is

$$f(x) \to \infty \text{ as } x \to -\infty \text{ and } f(x) \to \infty \text{ as } x \to \infty,$$

then the degree of f must be an even positive integer. _____

19. The graph of a rational function f can have at most one horizontal asymptote. _____

20. If a polynomial function f is an odd function and 3 is a zero of f, then -3 is also a zero of f. _____

21. The graph of a polynomial function f always has a y-intercept. _____

22. If the leading coefficient of a polynomial function f is 1, then any rational zeros of the function must be integers. _____

23. If 5 and $1 + 2i$ are zeros of a polynomial function f of degree 4 with real coefficients, then one of the remaining two zeros of must be real. _____

24. The graph of a polynomial function f of degree 4 that has four real zeros must pass through the x-axis four times. _____

C. Review Exercises

In Problems 1 and 2, use long division to divide $f(x)$ by $d(x)$.

1. $f(x) = 6x^5 - 4x^3 + 2x^2 + 4, \quad d(x) = 2x^2 - 1$

2. $f(x) = 15x^4 - 2x^3 + 8x + 6, \quad d(x) = 5x^3 + x + 2$

In Problems 3 and 4, use synthetic division to divide $f(x)$ by $d(x)$.

3. $f(x) = 7x^4 - 6x^2 + 9x + 3, \quad d(x) = x - 2$

4. $f(x) = 4x^3 + 7x^2 - 8x, \quad d(x) = x + 1$

5. Without actually performing the division, determine the remainder when $f(x) = 5x^3 - 4x^2 + 6x - 9$ is divided by $d(x) = x + 3$.

6. Use synthetic division and the Remainder Theorem to find $f(c)$ for

$$f(x) = x^6 - 3x^5 + 2x^4 + 3x^3 - x^2 + 5x - 1$$

when $c = 2$.

7. Without graphing, does the rational function

$$f(x) = \frac{x^3 + 6x^2 - 7x}{(x - 1)^2}$$

possess a hole in the graph at $x = 1$ or is $x = 1$ a vertical asymptote?

8. Suppose that
$$f(x) = 36x^{98} - 40x^{25} + 18x^{14} - 3x^7 + 40x^4 + 5x^2 - x + 2$$
is divided by $d(x) = x - 1$. What is the remainder?

9. List, but do not test, all possible rational zeros of
$$f(x) = 8x^4 + 19x^3 + 31x^2 + 38x - 15.$$

10. Find the complete factorization of $f(x) = 12x^3 + 16x^2 + 7x + 1$.

In Problems 11 and 12, verify that each of the indicated numbers is a zero of the given polynomial function $f(x)$. Find all other zeros and then give the complete factorization of $f(x)$.

11. 2; $f(x) = (x - 3)^3 + 1$ **12.** -1; $f(x) = (x + 2)^4 - 1$

In Problems 13–16, find the real value of k so that the given condition is satisfied.

13. the remainder in the division of $f(x) = x^4 - 3x^3 - x^2 + kx - 1$ by $g(x) = x - 4$ is $r = 5$

14. $x + \frac{1}{2}$ is a factor of $f(x) = 8x^2 - 4kx + 9$

15. $x - k$ is a factor of $f(x) = 2x^3 + x^2 + 2x - 12$

16. the graph of $f(x) = \dfrac{x - k}{x^2 + 5x + 6}$ has a hole at $x = k$

In Problems 17 and 18, find a polynomial function f of indicated degree whose graph is given in the figure.

17. fifth degree   **18.** sixth degree

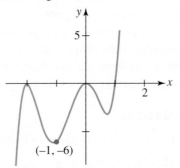

FIGURE 3.R.1 Graph for Problem 17

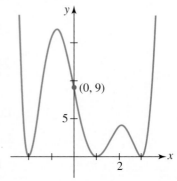

FIGURE 3.R.2 Graph for Problem 18

In Problems 19 and 20, find a rational function f whose graph is given in the figure.

19.   **20.**

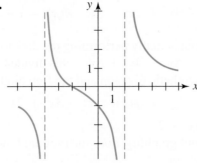

FIGURE 3.R.3 Graph for Problem 19

FIGURE 3.R.4 Graph for Problem 20

In Problems 21–30, match the given rational function with one of the graphs (a)–(j).

(a)

FIGURE 3.R.5

(b)

FIGURE 3.R.6

(c)

FIGURE 3.R.7

(d)

FIGURE 3.R.8

(e)

FIGURE 3.R.9

(f)

FIGURE 3.R.10

(g)

FIGURE 3.R.11

(h)

FIGURE 3.R.12

(i)

FIGURE 3.R.13

(j)

FIGURE 3.R.14

21. $f(x) = \dfrac{2x}{x^2 + 1}$

22. $f(x) = \dfrac{x^2 - 1}{x^2 + 1}$

23. $f(x) = \dfrac{2x}{x - 2}$

24. $f(x) = 2 - \dfrac{1}{x^2}$

25. $f(x) = \dfrac{x}{(x - 2)^2}$

26. $f(x) = \dfrac{(x - 1)^2}{x - 2}$

27. $f(x) = \dfrac{x^2 - 10}{2x - 4}$

28. $f(x) = \dfrac{-x^2 + 5x - 5}{x - 2}$

29. $f(x) = \dfrac{2x}{x^3 + 1}$

30. $f(x) = \dfrac{3}{x^2 + 1}$

In Problems 31 and 32, find the asymptotes for the graph of the given rational function. Find x- and y-intercepts of the graph. Sketch the graph of f.

31. $f(x) = \dfrac{x + 2}{x^2 + 2x - 8}$

32. $f(x) = \dfrac{-x^3 + 2x^2 + 9}{x^2}$

In Problems 33–36, find the partial fraction decomposition of the given rational expression.

33. $\dfrac{2x - 1}{x(x^2 + 2x - 3)}$

34. $\dfrac{1}{x^4(x^2 + 5)}$

35. $\dfrac{x^2}{(x^2 + 9)^2}$

36. $\dfrac{x^5 - x^4 + 2x^3 + 5x - 1}{(x - 1)^2}$

4 Trigonometric Functions

Chapter Contents

4.1 Angles and Their Measurement

Terminal side

Vertex

Initial side

(a) Two half-rays

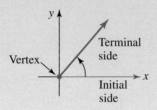

(b) Standard position

FIGURE 4.1.1 Initial and terminal sides of an angle

INTRODUCTION We begin our study of trigonometry by discussing angles and two methods of measuring them: degrees and radians. As we will see it is the radian measure of an angle that enables us to define trigonometric functions on sets of real numbers.

☐ **Angles** An **angle** is formed by two half-rays, or half-lines, which have a common endpoint, called the **vertex**. We designate one ray the **initial side** of the angle and the other the **terminal side**. It is useful to consider the angle as having been formed by a rotation from the initial side to the terminal side as shown in FIGURE 4.1.1(a). An angle is said to be in **standard position** if its vertex is placed at the origin of a rectangular coordinate system with its initial side coinciding with the positive x-axis, as shown in Figure 4.1.1(b).

☐ **Degree Measure** The **degree measure** of an angle is based on the assignment of 360 degrees (written 360°) to the angle formed by one complete counterclockwise rotation, as shown in FIGURE 4.1.2. Other angles are then measured in terms of a 360° angle, with a 1° angle being formed by $\frac{1}{360}$ of a complete rotation. If the rotation is counterclockwise, the measure will be *positive*; if clockwise, the measure is *negative*. For example, the angle in FIGURE 4.1.3(a) obtained by one-fourth of a complete counterclockwise rotation will be

$$\tfrac{1}{4}(360°) = 90°.$$

Shown in Figure 4.1.3(b) is the angle formed by three-fourths of a complete clockwise rotation. This angle has measure

$$\tfrac{3}{4}(-360°) = -270°.$$

FIGURE 4.1.2 Angle of 360 degrees

(a) 90° angle

(b) −270° angle

FIGURE 4.1.3 Positive measure in (a); negative measure in (b)

FIGURE 4.1.4 Three coterminal angles

☐ **Coterminal Angles** Comparison of Figure 4.1.3(a) with Figure 4.1.3(b) shows that the terminal side of a 90° angle coincides with the terminal side of a −270° angle. When two angles in standard position have the same terminal sides we say they are **coterminal**. For example, the angles θ, $\theta + 360°$, and $\theta - 360°$ shown in FIGURE 4.1.4 are coterminal. In fact, the addition of any integer multiple of 360° to a given angle results in a coterminal angle. Conversely, any two coterminal angles have degree measures that differ by an integer multiple of 360°.

◼EXAMPLE 1　　**Angles and Coterminal Angles**

For a 960° angle:
(a) Locate the terminal side and sketch the angle.
(b) Find a coterminal angle between 0° and 360°.
(c) Find a coterminal angle between −360° and 0°.

Solution (a) We first determine how many full rotations are made in forming this angle. Dividing 960 by 360 we obtain a quotient of 2 and a remainder of 240. Equivalently we can write

$$960 = 2(360) + 240.$$

Thus, this angle is formed by making two counterclockwise rotations before completing $\frac{240}{360} = \frac{2}{3}$ of another rotation. As illustrated in FIGURE 4.1.5(a), the terminal side of 960° lies in the third quadrant.

(b) Figure 4.1.5(b) shows that the angle 240° is coterminal with a 960° angle.

(c) Figure 4.1.5(c) shows that the angle $-120°$ is coterminal with a 960° angle.

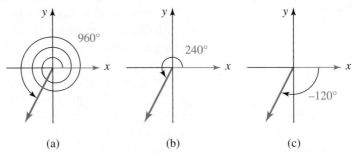

(a)   (b)   (c)

FIGURE 4.1.5 Angles in (b) and (c) are coterminal with the angle in (a)

☐ **Minutes and Seconds** With calculators it is convenient to represent fractions of degrees by decimals, such as 42.23°. Traditionally, however, fractions of degrees were expressed in **minutes** and **seconds**, where

$$1° = 60 \text{ minutes (written } 60') \text{*} \qquad (1)$$

and $\qquad\qquad\qquad 1' = 60 \text{ seconds (written } 60''). \qquad\qquad (2)$

For example, an angle of 7 degrees, 30 minutes, and 5 seconds is expressed as $7°30'5''$. Some calculators have a special DMS key for converting an angle given in decimal degrees to Degrees, Minutes, and Seconds (DMS notation), and vice versa. The following example shows how to perform these conversions by hand.

■ **EXAMPLE 2**   **Using (1) and (2)**

Convert:

(a) 86.23° to degrees, minutes, and seconds,
(b) $17°47'13''$ to decimal notation.

Solution In each case we will use (1) and (2).

(a) Since 0.23° represents $\frac{23}{100}$ of 1° and $1° = 60'$, we have

$$86.23° = 86° + 0.23°$$
$$= 86° + (0.23)(60')$$
$$= 86° + 13.8'.$$

Now $13.8' = 13' + 0.8'$, so we must convert 0.8' to seconds. Since 0.8' represents $\frac{8}{10}$ of 1' and $1' = 60''$, we have

$$86° + 13' + 0.8' = 86° + 13' + (0.8)(60'')$$
$$= 86° + 13' + 48''.$$

Hence, $86.23° = 86°13'48''$.

---

*The use of the number 60 as a base dates back to the Babylonians. Another example of the use of this base in our culture in the measurement of time (1 hour = 60 minutes and 1 minute = 60 seconds).

(b) Since $1° = 60'$, it follows that $1' = \left(\frac{1}{60}\right)°$. Similarly, $1'' = \left(\frac{1}{60}\right)' = \left(\frac{1}{3600}\right)°$. Thus we have

$$17°47'13'' = 17° + 47' + 13''$$
$$= 17° + 47\left(\tfrac{1}{60}\right)° + 13\left(\tfrac{1}{3600}\right)°$$
$$\approx 17° + 0.7833° + 0.0036°.$$

Thus we see that $17°47'13'' \approx 17.7869°$. ▬

☐ **Radian Measure** Another measure for angles is **radian measure**, which is generally used in almost all applications of trigonometry that involve calculus. The radian measure of an angle θ is based on the length of an arc on a circle. If we place the vertex of the angle θ at the center of a circle of radius r, then θ is called a **central angle**. As we know, an angle θ in standard position can be viewed as having been formed by the initial side rotating from the positive x-axis to the terminal side. The region formed by the initial and terminal sides with a central angle θ is called a **circular sector**. As shown in FIGURE 4.1.6(a), if the initial side of θ traverses a distance s along the circumference of the circle, then the **radian measure of θ** is defined by

$$\theta = \frac{s}{r}. \tag{3}$$

In the case when the terminal side of θ traverses an arc of length s along the circumference of the circle equal to the radius r of the circle, then we see from (3) that the measure of the angle θ is **1 radian**. See Figure 4.1.6(b).

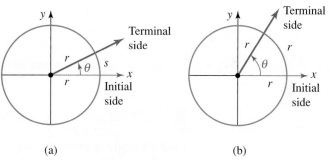

(a)   (b)

FIGURE 4.1.6 Central angle in (a); angle of 1 radian in (b)

The definition given in (3) does not depend on the size of the circle. To see this, all we need do is to draw another circle centered at the vertex of θ of radius r' and subtended arc length s'. See FIGURE 4.1.7. Because the two circular sectors are similar the ratios s/r and s'/r' are equal. Therefore, regardless of which circle we use, we obtain the same radian measure for θ.

In equation (3) any convenient unit of length may be used for s and r, but the same unit must be used for *both* s and r. Thus,

$$\theta(\text{in radians}) = \frac{s\,(\text{units of length})}{r\,(\text{units of length})}$$

appears to be a "dimensionless" quantity. For example, if $s = 4$ in. and $r = 2$ in., then the radian measure of the angle is

$$\theta = \frac{4 \text{ in.}}{2 \text{ in.}} = 2,$$

where 2 is simply a real number. This is the reason why sometimes the word *radians* is omitted when an angle is measured in radians. We will come back to this idea in Section 4.2.

FIGURE 4.1.7 Concentric circles

One complete rotation of the initial side of θ will traverse an arc equal in length to the circumference of the circle $2\pi r$. It follows from (3) that

$$\text{one rotation} = \frac{s}{r} = \frac{2\pi r}{r} = 2\pi \text{ radians.}$$

We have the same convention as before: An angle formed by a counterclockwise rotation is considered positive, whereas an angle formed by a clockwise rotation is negative. In FIGURE 4.1.8 we illustrate angles in standard position of $\pi/2$, $-\pi/2$, π, and 3π radians, respectively. Note that the angle of $\pi/2$ radians shown in 4.1.8(a) is obtained by one-fourth of a complete counterclockwise rotation; that is

$$\frac{1}{4}(2\pi \text{ radians}) = \frac{\pi}{2} \text{ radians.}$$

The angle shown in Figure 4.1.8(b), obtained by one-fourth of a complete clockwise rotation, is $-\pi/2$ radians. The angle shown in Figure 4.1.8(c) is coterminal with the angle shown in Figure 4.1.8(d). In general, the addition of any integer multiple of 2π radians to an angle measured in radians results in a coterminal angle. Conversely, any two coterminal angles measured in radians will differ by an integer multiple of 2π.

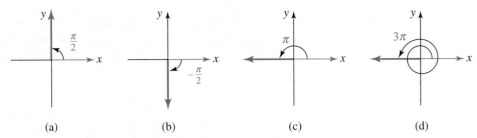

(a) (b) (c) (d)

FIGURE 4.1.8 Angles measured in radians

EXAMPLE 3   A Coterminal Angle

Find an angle between 0 and 2π radians that is coterminal with $\theta = 11\pi/4$ radians. Sketch the angle.

Solution Since $2\pi < 11\pi/4 < 3\pi$, we subtract the equivalent of one rotation, or 2π radians, to obtain

$$\frac{11\pi}{4} - 2\pi = \frac{11\pi}{4} - \frac{8\pi}{4} = \frac{3\pi}{4}.$$

Alternatively, we can proceed as in part (a) of Example 1 and divide: $11\pi/4 = 2\pi + 3\pi/4$. Thus, an angle of $3\pi/4$ radians is coterminal with θ, as illustrated in FIGURE 4.1.9.

FIGURE 4.1.9 Coterminal angles in Example 3

☐ **Conversion Formulas** Although many scientific calculators have keys that convert between degree and radian measure, there is an easy way to remember the relationship between the two measures. Since the circumference of a unit circle is 2π, one complete rotation has measure 2π radians as well as $360°$. It follows that $360° = 2\pi$ radians or

$$180° = \pi \text{ radians.} \tag{4}$$

If we interpret (4) as $180(1°) = \pi(1 \text{ radian})$, then we obtain the following two formulas for converting between degree and radian measure.

Conversion Between Degrees and Radians

$$1° = \frac{\pi}{180} \text{ radian} \qquad (5)$$

$$1 \text{ radian} = \left(\frac{180}{\pi}\right)° \qquad (6)$$

Using a calculator to carry out the divisions in (5) and (6), we find that

$$1° \approx 0.0174533 \text{ radian} \quad \text{and} \quad 1 \text{ radian} \approx 57.29578°.$$

Although we will continue to use the terms *radian* and *radians* it is common to use the abbreviation *rad* for both words.

▮ EXAMPLE 4   Conversion Between Degrees and Radians

Convert:

(a) 20° to radians   **(b)** $7\pi/6$ radians to degrees   **(c)** 2 radians to degrees.

Solution **(a)** To convert from degrees to radians we use (5):

$$20° = 20(1°) = 20 \cdot \left(\frac{\pi}{180} \text{ radian}\right) = \frac{\pi}{9} \text{ radian}.$$

(b) To convert from radians to degrees we use (6):

$$\frac{7\pi}{6} \text{ radians} = \frac{7\pi}{6} \cdot (1 \text{ radian}) = \frac{7\pi}{6}\left(\frac{180}{\pi}\right)° = 210°.$$

(c) We again use (6):

approximate answer
rounded to two
decimal places

$$2 \text{ radians} = 2 \cdot (1 \text{ radian}) = 2 \cdot \left(\frac{180}{\pi}\right)° = \left(\frac{360}{\pi}\right)° \approx \overbrace{114.59°}. \quad ▬$$

TABLE 4.1.1

Degrees	0	30	45	60	90	180
Radians	0	$\frac{\pi}{6}$	$\frac{\pi}{4}$	$\frac{\pi}{3}$	$\frac{\pi}{2}$	π

Table 4.1.1 provides the radian and degree measure of the most commonly used angles.

☐ **Terminology** You may recall from geometry that a 90° angle is called a **right angle** and a 180° angle is called a **straight angle**. In radian measure, $\pi/2$ is a right angle and π is a straight angle. An **acute angle** has measure between 0° and 90° (or between 0 and $\pi/2$ radians); and an **obtuse angle** has measure between 90° and 180° (or between $\pi/2$ and π radians). Two acute angles are said to be **complementary** if their sum is 90° (or $\pi/2$ radians). Two positive angles are **supplementary** if their sum is 180° (or π radians). The angle 180° (or π radians) is a **straight angle**. An angle whose terminal side coincides with a coordinate axis is called a **quadrantal angle**. For example, 90° (or $\pi/2$ radians) is a quadrantal angle. A triangle that contains a right angle is called a **right triangle**. The lengths a, b, and c of the sides of a right triangle satisfy the Pythagorean theorem $a^2 + b^2 = c^2$, where c is the length of the side opposite the right angle (the hypotenuse).

EXAMPLE 5 **Complementary and Supplementary Angles**

(a) Find the angle that is complementary to $\theta = 74.23°$.
(b) Find the angle that is supplementary to $\phi = \pi/3$ radians.

Solution (a) Since two acute angles are complementary if their sum is $90°$, we find the angle that is complementary to $\theta = 74.23°$ is

$$90° - \theta = 90° - 74.23° = 15.77°.$$

(b) Since two positive angles are supplementary if their sum is π radians, we find the angle that is supplementary to $\phi = \pi/3$ radians is

$$\pi - \phi = \pi - \frac{\pi}{3} = \frac{3\pi}{3} - \frac{\pi}{3} = \frac{2\pi}{3} \text{ radians.} \qquad \blacksquare$$

☐ **Arc Length** In many applications it is necessary to find the length s of the arc subtended by a central angle θ in a circle of radius r. See FIGURE 4.1.10. From the definition of radian measure given in (3),

$$\theta \text{ (in radians) } = \frac{s}{r}.$$

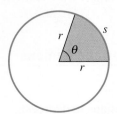

FIGURE 4.1.10 Length of arc s determined by a central angle θ

By multiplying both sides of the last equation by r we obtain the **arc length formula** $s = r\theta$. We summarize the result.

THEOREM 4.1.1 Arc Length Formula

For a circle of radius r, a central angle of θ radians subtends an **arc of length**

$$s = r\theta \qquad (7)$$

EXAMPLE 6 **Finding Arc Length**

Find the arc length subtended by a central angle of (a) 2 radians in a circle of radius 6 inches, (b) $30°$ in a circle of radius 12 feet.

Solution (a) From the arc length formula (7) with $\theta = 2$ radians and $r = 6$ inches, we have $s = r\theta = 2 \cdot 6 = 12$. So the arc length is 12 inches.

Students often apply the arc length formula ▶ incorrectly by using degree measure. Remember $s = r\theta$ is valid only if θ is measured in radians.

(b) We must first express $30°$ in radians. Recall that $30° = \pi/6$ radians. Then from the arc length formula (7) we have $s = r\theta = (12)(\pi/6) = 2\pi$. So the arc length is $2\pi \approx 6.28$ feet. $\qquad \blacksquare$

☐ **Area of a Circular Sector** The area of, say, a *quarter* circle is the fractional amount *one-fourth* of the total area πr^2, that is, the area is $\frac{1}{4}\pi r^2$. This reasoning carries over in finding the area of any circular sector. If θ in radians is the central angle of the circular sector shown in Figure 4.1.10, then the area A of the sector is simply the fractional amount $\theta/2\pi$ of the total area of the circle. Thus

$$A = \frac{\theta}{2\pi}(\pi r^2) = \frac{1}{2}r^2\theta.$$

Note for the quarter circle example, $\theta = \pi/2$ and the fractional amount of the total area is $(\pi/2)/2\pi = \frac{1}{4}$. We summarize this result in the next theorem.

THEOREM 4.1.2   Area of a Circular Sector

For a circle of radius r, the area A of a circular sector with central angle θ measured in radians is given by

$$A = \frac{1}{2}r^2\theta \qquad (8)$$

EXAMPLE 7   **Area of a Circular Sector**

A 14-inch pizza is cut into 8 slices. Let us assume that the pizza is a perfect circle and that the slices are exactly the same size. Find the area of one slice.

Solution One slice of pizza is a circular sector with radius $r = 7$ inches. The central angle of the sector is $360°/8 = 45°$. We convert this central angle from degree measure to radian measure:

$$45° = 45 \cdot \frac{\pi}{180} = \frac{\pi}{4} \text{ radian.}$$

Then by (8) the area of the sector, or slice, is

$$A = \frac{1}{2} \cdot 7^2 \cdot \frac{\pi}{4} = \frac{49}{2} \cdot \frac{\pi}{4} = \frac{49\pi}{8} \text{ in}^2 \approx 19.24 \text{ in}^2. \qquad \blacksquare$$

A 14-inch pizza is usually cut into eight slices (or pieces)

© Tobik/ShutterStock, Inc.

Exercises 4.1   Answers to selected odd-numbered problems begin on page ANS–14.

In Problems 1–16, draw the given angle in standard position. Bear in mind that the lack of a degree symbol (°) in an angular measurement indicates that the angle is measured in radians.

1. 60°	**2.** −120°	**3.** 135°	**4.** 150°
5. 1140°	**6.** −315°	**7.** −240°	**8.** −210°
9. $\dfrac{\pi}{3}$	**10.** $\dfrac{5\pi}{4}$	**11.** $\dfrac{7\pi}{6}$	**12.** $-\dfrac{2\pi}{3}$
13. $-\dfrac{\pi}{6}$	**14.** -3π	**15.** 3	**16.** 4

In Problems 17–20, express the given angle in decimal notation.

17. 10°39′17″   **18.** 143°7′2″   **19.** 5°10′   **20.** 10°25′

In Problems 21–24, express the given angle in terms of degrees, minutes, and seconds.

21. 210.78°   **22.** 15.45°   **23.** 30.81°   **24.** 110.5°

In Problems 25–32, convert the given angle from degrees to radians.

25. 10°	**26.** 15°	**27.** 75°	**28.** 215°
29. 270°	**30.** −120°	**31.** −230°	**32.** 540°

In Problems 33–40, convert the given angle from radians to degrees.

33. $\dfrac{2\pi}{9}$	**34.** $\dfrac{11\pi}{6}$	**35.** $\dfrac{2\pi}{3}$	**36.** $\dfrac{7\pi}{12}$
37. $\dfrac{5\pi}{4}$	**38.** 7π	**39.** 3.1	**40.** 12

In Problems 41–44, find the angle between $0°$ and $360°$ that is coterminal with the given angle.

41. $875°$ **42.** $400°$ **43.** $-610°$ **44.** $-150°$

45. Find the angle between $-360°$ and $0°$ that is coterminal with the angle in Problem 41.

46. Find the angle between $-360°$ and $0°$ that is coterminal with the angle in Problem 43.

In Problems 47–52, find the angle between 0 and 2π that is coterminal with the given angle.

47. $-\dfrac{9\pi}{4}$ **48.** $\dfrac{17\pi}{2}$ **49.** 5.3π

50. $-\dfrac{9\pi}{5}$ **51.** -4 **52.** 7.5

53. Find the angle between -2π and 0 radians that is coterminal with the angle in Problem 47.

54. Find the angle between -2π and 0 radians that is coterminal with the angle in Problem 49.

In Problems 55–62, find an angle that is **(a)** complementary and **(b)** supplementary to the given angle, or state why no such angle can be found.

55. $48.25°$ **56.** $93°$ **57.** $98.4°$ **58.** $63.08°$

59. $\dfrac{\pi}{4}$ **60.** $\dfrac{\pi}{6}$ **61.** $\dfrac{2\pi}{3}$ **62.** $\dfrac{5\pi}{6}$

In Problems 63 and 64, find both the degree and the radian measures of the angle formed by the given rotation. Refer to Figures 4.1.2 and 4.1.3.

63. three-fifths of a counterclockwise rotation
64. five and one-eighth clockwise rotations

In Problems 65–68, find the measure of a central angle θ in a circle of radius r that subtends an arc length s. Give θ in **(a)** radians and **(b)** degrees.

65. $r = 5$ ft, $s = 7.5$ ft   **66.** $r = 10$ in, $s = 36$ in
67. $r = 9$ m, $s = 15$ m   **68.** $r = 20$ cm, $s = 90$ cm

In Problems 69–72, find the arc length s subtended by a central angle θ in a circle of radius r.

69. $\theta = 3$ radians, $r = 5$ in   **70.** $\theta = 1.5$ radians, $r = 4$ cm
71. $\theta = 30°$, $r = 2$ m   **72.** $\theta = 15°$, $r = 6$ ft

In Problems 73–76, find the area of the circular sector having the given radius r and central angle θ.

73. $r = 3$ ft, $\theta = 7.2$ radians   **74.** $r = 18$ in, $\theta = \dfrac{2\pi}{3}$ radians

75. $r = 6$ m, $\theta = 30°$ **76.** $r = 12$ cm, $\theta = 75°$

In Problems 77 and 78, the light red region in the given figure is portion of a circle. Find the area A of the region if θ is measured in **(a)** radians, and **(b)** degrees.

77.

FIGURE 4.1.11 Region in Problem 77

78.

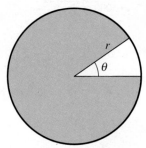

FIGURE 4.1.12 Region in Problem 78

Applications

FIGURE 4.1.13 Analog clock in Problems 79–82

Planet Mercury in Poblem 84

Satellite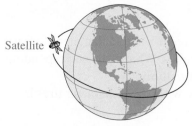

FIGURE 4.1.14 Satellite in Problem 85

79. Analog Clock Consider the analog clock shown in FIGURE 4.1.13. What are the degree and the radian measures of the angle between two adjacent hour tick marks on the clock face?

80. What are the degree and the radian measures of the angle between two adjacent minute tick marks on the analog clock face in Figure 4.1.13?

81. What are the degree and the radian measures of the smallest positive angle formed by the hands of the analog clock in Figure 4.1.13 at **(a)** 8:00, **(b)** 2:00, and **(c)** 7:30?

82. What are degree and the radian measures of the angle through which the minute hand on the analog clock in Figure 4.1.13 rotates in **(a)** $\frac{3}{4}$ hour, and **(b)** 3.5 hours?

83. Planet Earth The Earth rotates on its axis once every 24 hours. How long does it take the Earth to rotate through an angle of **(a)** $240°$ and **(b)** $\pi/6$ radians?

84. Planet Mercury The planet Mercury completes one rotation on its axis every 59 days. Through what angle (measured in degrees) does it rotate in **(a)** 1 day, **(b)** 1 hour, and **(c)** 1 minute?

85. Angular and Linear Speed If we divide (7) by time t we get the relationship $v = r\omega$, where $v = s/t$ is called the **linear speed** of a point on the circumference of a circle and $\omega = \theta/t$ is called the **angular speed** of the point. A communications satellite is placed in a circular geosynchronous orbit 35,786 km above the surface of the Earth. The time it takes the satellite to make one full revolution around the Earth is 23 hours, 56 minutes, 4 seconds and the radius of the Earth is 6378 km. See FIGURE 4.1.14.
(a) What is the angular speed of the satellite in rad/s?
(b) What is the linear speed of the satellite in km/s?

86. Pendulum Clock A clock pendulum is 1.3 m long and swings back and forth along a 15-cm arc. Find **(a)** the central angle and **(b)** the area of the sector through which the pendulum sweeps in one swing.

87. Sailing at Sea A nautical mile is defined as the arc length subtended on the surface of the Earth by an angle of measure 1 minute. If the diameter of the Earth is 7927 miles, find how many statute (land) miles there are in a nautical mile.

88. Circumference of the Earth Around 230 B.C.E. **Eratosthenes** calculated the circumference of the Earth from the following observations. At noon on the longest day of the year, the Sun was directly overhead in Syene, while it was inclined $7.2°$ from the vertical in Alexandria. He believed the two cities to be on

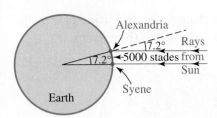

FIGURE 4.1.15 Earth in Problem 88

Yo-Yo in
Problems 89 and 90

the same longitudinal line and assumed that the rays of the Sun are parallel. Thus he concluded that the arc from Syene to Alexandria was subtended by a central angle of 7.2° at the center of the Earth. See FIGURE 4.1.15. At that time the distance from Syene to Alexandria was measured as 5000 stades. If one stade = 559 feet, find the circumference of the Earth in (a) stades and (b) miles. Show that Eratosthenes' data gives a result that is within 7% of the correct value if the polar diameter of the Earth is 7900 miles (to the nearest mile).

89. **Circular Motion of a Yo-Yo** A yo-yo is whirled around in a circle at the end of its 100-cm string.
   (a) If it makes six revolutions in 4 seconds, find its rate of turning, or angular speed, in radians per second.
   (b) Find the speed at which the yo-yo travels in centimeters per second; that is, its linear speed.

90. **More Yo-Yos** If there is a knot in the yo-yo string described in Problem 79 at a point 40 cm from the yo-yo, find (a) the angular speed of the knot and (b) the linear speed.

91. **Locating a Cell Phone** As shown in the accompanying photo, many cell phone antennas have a triangular shape. The reception sectors α, β, and γ corresponding to the three sides of the antenna are shown in FIGURE 4.1.16. A cell tower with a triangular antenna is located at the common center of the circles in the figure; the circles have a radius 1 mile, 2 miles, 3 miles, and so on. Suppose a cell phone signal is detected in the β sector approximately 5.3 miles from the antenna. Determine the area of the circular band where the cell phone is located bounded between radius 5 miles and radius 6 miles. (A more precise location of a cell phone can be obtained using either triangulation between three cell towers or GPS.)

Triangular cell phone
antenna

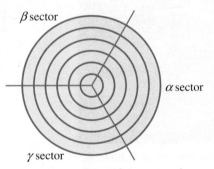

FIGURE 4.1.16 Reception sectors in
Problem 91

92. **Circular Motion of a Car Tire** As shown in FIGURE 4.1.17 the diameter of a car tire is 26 inches. Suppose the car is driven 1.5 miles. What is the corresponding radian measure of the angle through which the tire turns?

93. **Circular Motion of a Car Tire** An automobile with 26-in. diameter tires is traveling at a rate of 55 mi/h.
   (a) Find the number of revolutions per minute that its tires are making.
   (b) Find the angular speed of its tires in radians per minute.

94. **Diameter of the Moon** The average distance from the Earth to the Moon as given by NASA is 238,855 miles. If the angle subtended by the Moon at the eye of an observer on Earth is 0.52°, then what is the approximate diameter of the Moon? FIGURE 4.1.18 is not to scale.

FIGURE 4.1.17 Car tire in
Problems 92 and 93

CHAPTER 4 TRIGONOMETRIC FUNCTIONS

FIGURE 4.1.18 The curved red arc represents the approximate diameter of the Moon

For Discussion

95. Intersection of Circles Each of the circles in FIGURE 4.1.19 has its center on an axis, passes through the origin, and has radius r.

(a) Construct a right triangle in the figure and then use that triangle to express the area A of the intersection of the circles, the yellow region in the figure, as a function of r.

(b) Find the area of the intersection of the circles

$$x^2 + (y - 5)^2 = 25 \quad \text{and} \quad (x - 5)^2 + y^2 = 25.$$

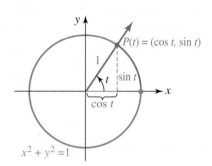

FIGURE 4.1.19 Intersecting circles in Problem 95

4.2 The Sine and Cosine Functions

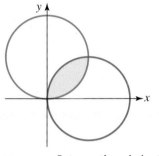

FIGURE 4.2.1 Coordinates of $P(t)$ are $(\cos t, \sin t)$

INTRODUCTION Originally, the trigonometric functions were defined using angles in right triangles. A more modern approach, and one that is used in calculus, is to define the trigonometric functions on sets of real numbers. As we will see, the radian measure for angles is key in making these definitions.

☐ **Trigonometric Functions** For each real number t there corresponds an angle of t radians in standard position. As shown in FIGURE 4.2.1 we denote the point of intersection of the terminal side of the angle t with the **unit circle** by $P(t)$. The x and y coordinates of this point give us the values of the six basic trigonometric functions. The y-coordinate of $P(t)$ is called the **sine of** t, while the x-coordinate of $P(t)$ is called the **cosine of** t.

DEFINITION 4.2.1 Sine and Cosine Functions

Let t be any real number and $P(t) = (x, y)$ be the point of intersection of the unit circle with the terminal side of the angle of t radians in standard position. Then, the **sine of** t, denoted $\sin t$, and the **cosine of** t, denoted $\cos t$, are

$$\sin t = y \tag{1}$$

and
$$\cos t = x \tag{2}$$

Since to each real number t there corresponds a unique point $P(t) = (\cos t, \sin t)$, we have just defined two functions – the sine function and the cosine function – each with domain the set R of real numbers. Four additional trigonometric functions are defined in terms of the coordinates of $P(t) = (x, y)$.

DEFINITION 4.2.2 Tangent, Cotangent, Secant, and Cosecant Functions

The **tangent, cotangent, secant,** and **cosecant functions** of the real number t are

$$\tan t = \frac{y}{x}, \quad x \neq 0 \tag{3}$$

$$\cot t = \frac{x}{y}, \quad y \neq 0 \tag{4}$$

$$\sec t = \frac{1}{x}, \quad x \neq 0 \tag{5}$$

and

$$\csc t = \frac{1}{y}, \quad y \neq 0 \tag{6}$$

Using $\sin t = y$ and $\cos t = x$ in (3)–(6) of Definition 4.2.2 we obtain the important identities:

$$\tan t = \frac{\sin t}{\cos t} \qquad \cot t = \frac{\cos t}{\sin t} \tag{7}$$

$$\sec t = \frac{1}{\cos t} \qquad \csc t = \frac{1}{\sin t}. \tag{8}$$

Because of the role played by the unit circle in Definitions 4.2.1 and 4.2.2, the six trigonometric functions are referred to as the **circular functions**.

For the remainder of this section and the next we are going to examine the sine and cosine functions in detail. We will come back to the tangent, cotangent, secant, and cosecant functions in Section 4.4.

A number of properties of the sine and cosine functions follow from the fact that $P(t) = (\cos t, \sin t)$ lies on the unit circle. For instance, the coordinates of $P(t)$ must satisfy the equation of the circle:

$$x^2 + y^2 = 1. \tag{9}$$

Substituting $x = \cos t$ and $y = \sin t$ gives an important relationship between the sine and the cosine called the **Pythagorean identity**:

$$(\cos t)^2 + (\sin t)^2 = 1.$$

From now on we will follow two standard practices in writing this identity: $(\cos t)^2$ and $(\sin t)^2$ will be written as $\cos^2 t$ and $\sin^2 t$, respectively, and the $\sin^2 t$ term will be written first.

THEOREM 4.2.1 Pythagorean Identity

For all real numbers t,

$$\sin^2 t + \cos^2 t = 1 \tag{10}$$

Again, if $P(x, y)$ denotes a point on the unit circle (9), it follows that the coordinates of P must satisfy the inequalities $-1 \leq x \leq 1$ and $-1 \leq y \leq 1$. Because $x = \cos t$ and $y = \sin t$ we have the following bounds on the values of the sine and cosine functions.

THEOREM 4.2.2 Bounds on the Values of Sine and Cosine

For all real numbers t,

$$-1 \leq \sin t \leq 1 \qquad \text{and} \qquad -1 \leq \cos t \leq 1$$

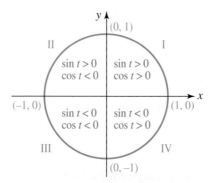

FIGURE 4.2.2 Algebraic signs of $\sin t$ and $\cos t$ in the four quadrants

The foregoing inequalities can also be expressed as $|\sin t| \leq 1$ and $|\cos t| \leq 1$. Thus, for example, there is no real number t such that $\sin t = \frac{3}{2}$.

☐ **Domain and Range** From the preceding observations we have the sine and cosine functions $f(t) = \sin t$ and $g(t) = \cos t$ each with **domain** the set R of real numbers and **range** the interval $[-1, 1]$.

☐ **Signs of the Circular Functions** The signs of the function values $\sin t$ and $\cos t$ are determined by the quadrant in which the point $P(t)$ lies, and conversely. For example, if $\sin t$ and $\cos t$ are both negative, then the point $P(t)$ and terminal side of the corresponding angle of t radians must lie in quadrant III. FIGURE 4.2.2 displays the signs of the cosine and sine functions in each of the four quadrants.

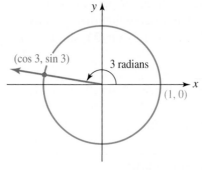

FIGURE 4.2.3 The point $P(3)$

EXAMPLE 1 Using the Pythagorean Identity

Given that $\cos t = \frac{1}{3}$ and that $P(t)$ is a point in the fourth quadrant, find $\sin t$.

Solution Substitution of $\cos t = \frac{1}{3}$ into the Pythagorean identity (10) gives $\sin^2 t + \left(\frac{1}{3}\right)^2 = 1$ or $\sin^2 t = \frac{8}{9}$. Since $\sin t$ is the y-coordinate of $P(t)$, a point in the fourth quadrant, we must take the negative square root for $\sin t$:

$$\sin t = -\sqrt{\frac{8}{9}} = -\frac{2\sqrt{2}}{3}. \qquad \blacksquare$$

EXAMPLE 2 Sine and Cosine of a Real Number

Use a calculator to approximate $\sin 3$ and $\cos 3$ and give a geometric interpretation of these values.

Solution From a calculator set in *radian mode*, we obtain $\cos 3 \approx -0.9899925$ and $\sin 3 \approx 0.1411200$. These values represent the x- and y-coordinates, respectively, of the point of intersection of the terminal side of the angle of 3 radians in standard position with the unit circle. As shown in FIGURE 4.2.3, this point lies in the second quadrant because $\pi/2 < 3 < \pi$. This would also be expected in view of Figure 4.2.2 since $\cos 3$, the x-coordinate, is *negative* and $\sin 3$, the y-coordinate, is *positive*. $\qquad \blacksquare$

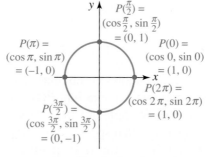

FIGURE 4.2.4 Sine and cosine values for quadrantal angles

☐ **Values Corresponding to Unit Circle Intercepts** As shown in FIGURE 4.2.4, the x- and y-intercepts of the unit circle give us the values of the sine and cosine functions for the real numbers corresponding to **quadrantal angles** listed next.

Values of the Sine and Cosine

For $t = 0$:	$\sin 0 = 0$	and	$\cos 0 = 1$
For $t = \dfrac{\pi}{2}$:	$\sin \dfrac{\pi}{2} = 1$	and	$\cos \dfrac{\pi}{2} = 0$
For $t = \pi$:	$\sin \pi = 0$	and	$\cos \pi = -1$
For $t = \dfrac{3\pi}{2}$:	$\sin \dfrac{3\pi}{2} = -1$	and	$\cos \dfrac{3\pi}{2} = 0$

4.2 The Sine and Cosine Functions

Periodicity In Section 4.1 we saw that for any real number t, the angles of t radians and $t \pm 2\pi$ radians are coterminal. Thus they determine the same point (x, y) on the unit circle. Therefore

$$\sin t = \sin(t \pm 2\pi) \quad \text{and} \quad \cos t = \cos(t \pm 2\pi). \tag{11}$$

In other words, the sine and cosine functions repeat their values every 2π units. It also follows that for any integer n:

$$\sin(t + 2n\pi) = \sin t \quad \text{and} \quad \cos(t + 2n\pi) = \cos t. \tag{12}$$

DEFINITION 4.2.3 Periodic Functions

A nonconstant function f is said to be **periodic** if there is a positive number p such that

$$f(t) = f(t + p) \tag{13}$$

for every t in the domain of f. If p is the smallest positive number for which (13) is true, then p is called the **period** of the function f.

The equations in (11) imply that the sine and the cosine functions are periodic with period $p \le 2\pi$. To see that the period of $\sin t$ is actually 2π, we observe that there is only one point on the unit circle with y-coordinate 1, namely, $P(\pi/2) = (\cos(\pi/2), \sin(\pi/2)) = (0, 1)$. Therefore,

$$\sin t = 1 \quad \text{only for} \quad t = \frac{\pi}{2}, \frac{\pi}{2} \pm 2\pi, \frac{\pi}{2} \pm 4\pi,$$

and so on. Thus the smallest possible positive value of p is 2π.

THEOREM 4.2.3 Period of the Sine and Cosine

The sine and cosine functions are periodic with **period 2π**. Therefore,

$$\sin(t + 2\pi) = \sin t \quad \text{and} \quad \cos(t + 2\pi) = \cos t \tag{14}$$

for every real number t.

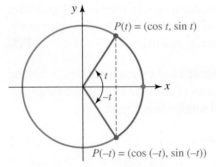

FIGURE 4.2.5 Coordinates of $P(t)$ and $P(-t)$

□ **Even–Odd Properties** The symmetry of the unit circle endows the circular functions with several additional properties. For any real number t, the points $P(t)$ and $P(-t)$ on the unit circle are located on the terminal side of an angle of t and $-t$ radians, respectively. These two points will always be symmetric with respect to the x-axis. FIGURE 4.2.5 illustrates the situation for a point $P(t)$ lying in the first quadrant: The x-coordinates of the two points are identical; however, the y-coordinates have equal magnitudes but opposite signs. The same symmetries will hold regardless of which quadrant contains $P(t)$. Thus, for any real number t, $\cos(-t) = \cos t$ and $\sin(-t) = -\sin t$. Applying the definitions of **even** and **odd functions** from Section 2.2 we have the following result.

THEOREM 4.2.4 Even and Odd Functions

The cosine function is **even** and the sine function is **odd**. That is, for every real number t,

$$\cos(-t) = \cos t \quad \text{and} \quad \sin(-t) = -\sin t \tag{15}$$

The following additional properties of the sine and cosine functions can be verified by considering the symmetries of appropriately chosen points on the unit circle.

THEOREM 4.2.5 Additional Properties

$$\cos\left(\frac{\pi}{2} - t\right) = \sin t \qquad \text{and} \qquad \sin\left(\frac{\pi}{2} - t\right) = \cos t \qquad (16)$$

$$\cos(t + \pi) = -\cos t \qquad \text{and} \qquad \sin(t + \pi) = -\sin t \qquad (17)$$

$$\cos(\pi - t) = -\cos t \qquad \text{and} \qquad \sin(\pi - t) = \sin t \qquad (18)$$

For example, to justify the properties in (16) of Theorem 4.2.5 for $0 < t < \pi/2$, consider FIGURE 4.2.6. Since the points $P(t)$ and $P(\pi/2 - t)$ are symmetric with respect to the line $y = x$, we can obtain the coordinates of $P(\pi/2 - t)$, by interchanging the coordinates of $P(t)$. Thus,

$$\cos t = \sin\left(\frac{\pi}{2} - t\right) \quad \text{and} \quad \sin t = \cos\left(\frac{\pi}{2} - t\right).$$

The special properties of the sine and cosine functions in Theorem 4.2.5 will become quite useful as soon as we determine additional values for $\sin t$ and $\cos t$ in the interval $[0, 2\pi)$. Using results from plane geometry we will now find the values of the sine and cosine functions for $t = \pi/6$, $t = \pi/4$, and $t = \pi/3$.

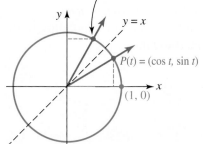

FIGURE 4.2.6 Geometric justification of (16) in Theorem 4.2.5

☐ **Finding $\sin(\pi/4)$ and $\cos(\pi/4)$** We draw an angle of $\pi/4$ radians (45°) in standard position and locate and label $P(\pi/4) = (\cos(\pi/4), \sin(\pi/4))$ on the unit circle. As shown in FIGURE 4.2.7, we form a right triangle by dropping a perpendicular from $P(\pi/4)$ to the x-axis. Since the sum of the angles in any triangle is π radians (180°), the third angle of this triangle is also $\pi/4$ radians, hence the triangle is isosceles. Therefore the coordinates of $P(\pi/4)$ are equal; that is, $\cos(\pi/4) = \sin(\pi/4)$. It follows from the Pythagorean identity (10)

$$\sin^2\frac{\pi}{4} + \cos^2\frac{\pi}{4} = 1 \qquad \text{that} \qquad 2\cos^2\frac{\pi}{4} = 1.$$

Dividing by 2 and taking the square root, we obtain $\cos(\pi/4) = \pm\sqrt{2}/2$. Since $P(\pi/4)$ lies in the first quadrant, both coordinates must be positive. So we have found the (equal) coordinates of $P(\pi/4)$:

$$\cos\frac{\pi}{4} = \frac{\sqrt{2}}{2} \qquad \text{and} \qquad \sin\frac{\pi}{4} = \frac{\sqrt{2}}{2}.$$

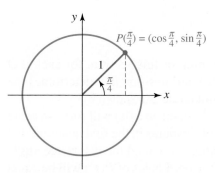

FIGURE 4.2.7 The point $P(\pi/4)$

☐ **Finding $\sin(\pi/6)$ and $\cos(\pi/6)$** We construct two angles of $\pi/6$ radians (30°) in the first and fourth quadrants, as shown in FIGURE 4.2.8, and label the points of intersection with the unit circle $P(\pi/6)$ and Q, respectively. By drawing perpendicular line segments from P and Q to the x-axis, we obtain two *congruent* right triangles because each triangle has a hypotenuse of length 1 and angles of 30°, 60°, and 90°. Since the 90° angles form a straight angle, these two right triangles form an *equilateral* triangle $\triangle POQ$ with sides of length 1. Since $\sin(\pi/6)$ is equal to half of the vertical side of $\triangle POQ$, we have

$$\sin\frac{\pi}{6} = \frac{1}{2}.$$

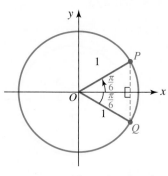

FIGURE 4.2.8 The point $P(\pi/6)$

From this result and the Pythagorean identity (10) we find the value of $\cos(\pi/6)$:

$$\left(\frac{1}{2}\right)^2 + \cos^2\frac{\pi}{6} = 1 \qquad \text{implies} \qquad \cos^2\frac{\pi}{6} = \frac{3}{4}$$

We take the positive square root here ▶
because $P(\pi/6)$ lies in the first quadrant.

or

$$\cos\frac{\pi}{6} = \frac{\sqrt{3}}{2}.$$

FIGURE 4.2.9 The point $P(\pi/3)$

☐ **Finding $\sin(\pi/3)$ and $\cos(\pi/3)$** We draw angles of $\pi/6$ and $\pi/3$ in standard position and locate and label the points $P(\pi/6)$ and $P(\pi/3)$, as shown in **FIGURE 4.2.9**. We then construct two congruent 30°-60°-90° triangles by dropping perpendiculars to the x- and y-axes, respectively. It follows from the congruence of these triangles that

$$\cos\frac{\pi}{3} = \sin\frac{\pi}{6} = \frac{1}{2} \qquad \text{and} \qquad \sin\frac{\pi}{3} = \cos\frac{\pi}{6} = \frac{\sqrt{3}}{2}.$$

The foregoing results also follow from (16) of Theorem 4.2.5 with $t = \pi/6$.

We summarize the values of the sine and cosine functions corresponding to the basic fractional multiples of π that we have determined so far.

Values of the Sine and Cosine (Continued)

$$\text{For } t = \frac{\pi}{6}: \quad \sin\frac{\pi}{6} = \frac{1}{2} \qquad \text{and} \qquad \cos\frac{\pi}{6} = \frac{\sqrt{3}}{2}$$

$$\text{For } t = \frac{\pi}{4}: \quad \sin\frac{\pi}{4} = \frac{\sqrt{2}}{2} \qquad \text{and} \qquad \cos\frac{\pi}{4} = \frac{\sqrt{2}}{2}$$

$$\text{For } t = \frac{\pi}{3}: \quad \sin\frac{\pi}{3} = \frac{\sqrt{3}}{2} \qquad \text{and} \qquad \cos\frac{\pi}{3} = \frac{1}{2}$$

☐ **Reference Angle** As we noted at the beginning of this section, for each real number t there is a unique angle of t radians in standard position that determines the point $P(t)$, with coordinates $(\cos t, \sin t)$, on the unit circle. As shown in **FIGURE 4.2.10**, the terminal side of any angle of t radians (with $P(t)$ not on an axis) will form an acute angle with the x-axis. We can then locate an angle of t' radians in the first quadrant that is congruent to this acute angle. The angle of t' radians is called the **reference angle** for t. Because of the symmetry of the unit circle, the coordinates of $P(t')$ will be equal *in absolute value* to the respective coordinates of $P(t)$. Hence

$$\sin t = \pm \sin t' \qquad \text{and} \qquad \cos t = \pm \cos t'.$$

As the following examples will show, reference angles can be used to find the trigonometric function values of any integer multiple of $\pi/6$, $\pi/4$, and $\pi/3$.

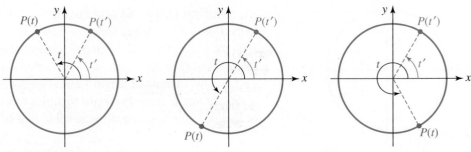

FIGURE 4.2.10 Reference angle t' is an acute angle

CHAPTER 4 TRIGONOMETRIC FUNCTIONS

FIGURE 4.2.11 Reference angle in part (a) of Example 3

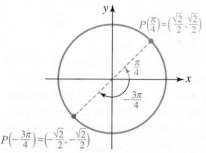

FIGURE 4.2.12 Reference angle in part (b) of Example 3

Find exact values of $\sin t$ and $\cos t$ for the given real number t.

(a) $t = 5\pi/3$ **(b)** $t = -3\pi/4$

Solution In each part we begin by finding the reference angle corresponding to the given value of t.

(a) From FIGURE 4.2.11 we find that an angle of $t = 5\pi/3$ radians determines a point $P(5\pi/3)$ in the fourth quadrant and has the reference angle $t' = \pi/3$ radians. After adjusting the signs of the coordinates of $P(\pi/3) = (1/2, \sqrt{3}/2)$ to obtain the fourth quadrant point $P(5\pi/3) = (1/2, -\sqrt{3}/2)$, we find that

$$\sin\frac{5\pi}{3} = -\sin\frac{\pi}{3} = -\frac{\sqrt{3}}{2} \quad \text{and} \quad \cos\frac{5\pi}{3} = \cos\frac{\pi}{3} = \frac{1}{2}.$$

(b) The point $P(-3\pi/4)$ lies in the third quadrant and has reference angle $\pi/4$, as shown in FIGURE 4.2.12. Therefore,

$$\sin\left(-\frac{3\pi}{4}\right) = -\sin\frac{\pi}{4} = -\frac{\sqrt{2}}{2} \quad \text{and} \quad \cos\left(-\frac{3\pi}{4}\right) = -\cos\frac{\pi}{4} = -\frac{\sqrt{2}}{2}. \quad ■$$

Sometimes, in order to find the trigonometric values of multiples of our basic fractions of π we must use periodicity or the even-odd function properties in addition to reference numbers.

Find exact values of $\sin t$ and $\cos t$ for $t = 29\pi/6$.

Solution Since $29\pi/6$ is greater than 2π, we rewrite $29\pi/6$ as an integer multiple of 2π plus a number less than 2π:

$$\frac{29\pi}{6} = 4\pi + \frac{5\pi}{6} = 2(2\pi) + \frac{5\pi}{6}.$$

FIGURE 4.2.13 Reference angle in Example 4

From the periodicity equations (12) with $n = 2$ and $t = 5\pi/6$ we know that $\sin(29\pi/6) = \sin(5\pi/6)$ and $\cos(29\pi/6) = \cos(5\pi/6)$. Next we see from FIGURE 4.2.13 that the reference angle for $5\pi/6$ is $\pi/6$. Since $P(5\pi/6)$ is a second quadrant point, we have

$$\sin\frac{29\pi}{6} = \sin\frac{5\pi}{6} = \sin\frac{\pi}{6} = \frac{1}{2}$$

and

$$\cos\frac{29\pi}{6} = \cos\frac{5\pi}{6} = -\cos\frac{\pi}{6} = -\frac{\sqrt{3}}{2}. \quad ■$$

4.2 The Sine and Cosine Functions

EXAMPLE 5 Using the Even–Odd Properties

Find exact values of $\sin t$ and $\cos t$ for $t = -\pi/6$.

Solution Since sine is an odd function and cosine is an even function,

See (15) in Theorem 4.2.4. ▶

$$\sin\left(-\frac{\pi}{6}\right) = -\sin\frac{\pi}{6} = -\frac{1}{2} \quad \text{and} \quad \cos\left(-\frac{\pi}{6}\right) = \cos\frac{\pi}{6} = \frac{\sqrt{3}}{2}.$$

This problem could also have been solved by using a reference angle. ◼

☐ **Trigonometric Functions of Angles** In this section we have defined sine and cosine functions of the real number t by using the coordinates of a point $P(t)$ on the unit circle. It is now possible to define the **trigonometric functions of any angle θ**. For any angle θ, we simply let

$$\sin\theta = \sin t \quad \text{and} \quad \cos\theta = \cos t,$$

where the real number t is the radian measure of θ. As mentioned in Section 4.1, it is common to omit the word radians when measuring an angle. So we write $\sin(\pi/6)$ for both the sine of the real number $\pi/6$ and for the sine of the angle of $\pi/6$ radians. Furthermore, since the values of the trigonometric functions are determined by the coordinates of the point $P(t)$ on the unit circle, it really does not matter whether θ is measured in radians or in degrees. For example, regardless of whether we are given $\theta = \pi/6$ radians or $\theta = 30°$, the point on the unit circle corresponding to this angle in standard position is $\left(\sqrt{3}/2, 1/2\right)$. Thus,

$$\sin\frac{\pi}{6} = \sin 30° = \frac{1}{2} \quad \text{and} \quad \cos\frac{\pi}{6} = \cos 30° = \frac{\sqrt{3}}{2}.$$

Exercises 4.2 Answers to selected odd-numbered problems begin on page ANS–14.

1. Given that $\cos t = -\frac{2}{5}$ and that $P(t)$ is a point in the second quadrant, find $\sin t$.
2. Given that $\sin t = \frac{1}{4}$ and that $P(t)$ is a point in the second quadrant, find $\cos t$.
3. Given that $\sin t = -\frac{2}{3}$ and that $P(t)$ is a point in the third quadrant, find $\cos t$.
4. Given that $\cos t = \frac{3}{4}$ and that $P(t)$ is a point in the fourth quadrant, find $\sin t$.
5. If $\sin t = -\frac{2}{7}$, find all possible values of $\cos t$.
6. If $\cos t = \frac{3}{10}$, find all possible values of $\sin t$.
7. If $\cos t = -0.2$, find all possible values of $\sin t$.
8. If $\sin t = 0.4$, find all possible values of $\cos t$.
9. If $2\sin t - \cos t = 0$, find all possible values of $\sin t$ and $\cos t$.
10. If $3\sin t - 2\cos t = 0$, find all possible values of $\sin t$ and $\cos t$.

In Problems 11–14, find the exact value of **(a)** $\sin t$ and **(b)** $\cos t$ for the given value of t. Do not use a calculator.

11. $t = -\pi/2$ **12.** $t = 3\pi$
13. $t = 8\pi$ **14.** $t = -3\pi/2$

In Problems 15–26, for the given value of t determine the reference angle t' and the exact values of $\sin t$ and $\cos t$. Do not use a calculator.

15. $t = 2\pi/3$ **16.** $t = 4\pi/3$ **17.** $t = 5\pi/4$
18. $t = 3\pi/4$ **19.** $t = 11\pi/6$ **20.** $t = 7\pi/6$

21. $t = -\pi/4$
22. $t = -7\pi/4$
23. $t = -5\pi/6$
24. $t = -11\pi/6$
25. $t = -5\pi/3$
26. $t = -2\pi/3$

In Problems 27–32, find the given trigonometric function value. Do not use a calculator.

27. $\sin(-11\pi/3)$
28. $\cos(17\pi/6)$
29. $\cos(-7\pi/4)$
30. $\sin(-19\pi/2)$
31. $\cos 5\pi$
32. $\sin(23\pi/3)$

In Problems 33–38, justify the given statement with one of the properties of the trigonometric functions.

33. $\sin \pi = \sin 3\pi$
34. $\cos(\pi/4) = \sin(\pi/4)$
35. $\sin(-3 - \pi) = -\sin(3 + \pi)$
36. $\cos 16.8\pi = \cos 14.8\pi$
37. $\cos 0.43 = \cos(-0.43)$
38. $\sin(2\pi/3) = \sin(\pi/3)$

In Problems 39–46, find the given trigonometric function value. Do not use a calculator.

39. $\sin 135°$
40. $\cos 150°$
41. $\cos 210°$
42. $\sin 270°$
43. $\cos 330°$
44. $\sin(-180°)$
45. $\sin(-60°)$
46. $\cos(-300°)$

In Problems 47–50, find all angles t, where $0 \le t < 2\pi$, that satisfy the given condition.

47. $\sin t = 0$
48. $\cos t = -1$
49. $\cos t = \sqrt{2}/2$
50. $\sin t = \frac{1}{2}$

In Problems 51–54, find all angles θ, where $0° \le \theta < 360°$, that satisfy the given condition.

51. $\cos \theta = \sqrt{3}/2$
52. $\sin \theta = -\frac{1}{2}$
53. $\sin \theta = -\sqrt{2}/2$
54. $\cos \theta = 1$

Applications

55. Free Throw Under certain conditions the maximum height y attained by a basketball released from a height h at an angle θ measured from the horizontal with an initial velocity v_0 is given by $y = h + (v_0^2\sin^2\theta)/2g$, where g is the acceleration due to gravity. Compute the maximum height reached by a free throw if $h = 2.15$ m, $v_0 = 8$ m/s, $\theta = 64.47°$, and $g = 9.81$ m/s^2.

56. Putting the Shot The range of a shot put released from a height h above the ground with an initial velocity v_0 at an angle θ to the horizontal can be approximated by

$$R = \frac{v_0\cos\theta}{g}\left(v_0\sin\theta + \sqrt{v_0^2\sin^2\theta + 2gh}\right),$$

where g is the acceleration due to gravity. If $v_0 = 13.7$ m/s, $\theta = 40°$, and $g = 9.81$ m/s^2, compare the ranges achieved for the release heights **(a)** $h = 2.0$ m and **(b)** $h = 2.4$ m. **(c)** Explain why an increase in h yields an increase in R if the other parameters are held fixed. **(d)** What does this imply about the advantage that height gives a shot putter?

57. Acceleration Due to Gravity Because of its rotation the Earth bulges at the equator and is flattened at the poles. As a consequence, the acceleration due to gravity is

Free throw

© Pétur Ásgeirsson/ShutterStock, Inc.

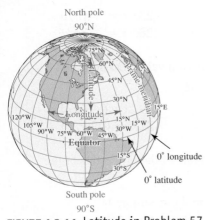

North pole
90°N

Equator

0° longitude

0° latitude

South pole
90°S

FIGURE 4.2.14 Latitude in Problem 57

not a constant 980 cm/s², but varies with latitude. As shown in FIGURE 4.2.14, the latitude of a point on the Earth is an angle ϕ measured (usually in degrees, minutes, seconds) north (N) or south (S) from the equatorial plane. Based on satellite studies, a mathematical model for the acceleration due to gravity g_{lat} is given by

$$g_{lat} = 978.0309 + 5.18552\sin^2\phi - 0.00570\sin^2 2\phi.$$

(a) Find g_{lat} at Mexico City, Mexico (19.42° N), Los Angeles, CA (34.05° N), New York City, NY (40.70° N), and Fairbanks, AK (64.83° N).
(b) At what latitude is g_{lat} a minimum? A maximum?

For Discussion

58. Discuss how it is possible to determine without a calculator that the point $P(6) = (\cos 6, \sin 6)$ lies in the fourth quadrant.
59. Discuss how it is possible to determine without the aid of a calculator that both $\sin 4$ and $\cos 4$ are negative.
60. Is there a real number t satisfying $3\sin t = 5$? Explain why or why not.
61. Is there an angle θ satisfying $\cos\theta = -2$? Explain why or why not.
62. Suppose f is a periodic function with period p. Show that $F(x) = f(ax)$, $a > 0$, is periodic with period p/a.

4.3 Graphs of Sine and Cosine Functions

INTRODUCTION One way to further your understanding of the trigonometric functions is to examine their graphs. In this section we consider the graphs of the sine and cosine functions.

☐ **Graphs of Sine and Cosine** In Section 4.2 we saw that the domain of the sine function $f(t) = \sin t$ is the set of real numbers $(-\infty, \infty)$ and the interval $[-1, 1]$ is its range. Since the sine function has period 2π, we begin by sketching its graph on the interval $[0, 2\pi]$. We obtain a rough sketch of the graph given in FIGURE 4.3.1(b) by considering various positions of the point $P(t)$ on the unit circle, as shown in Figure 4.3.1(a). As t varies from 0 to $\pi/2$, the value $\sin t$ increases from 0 to its maximum value 1. But as t varies from $\pi/2$ to $3\pi/2$, the value $\sin t$ decreases from 1 to its minimum value -1. We note that $\sin t$ changes from positive to negative at $t = \pi$. For t between $3\pi/2$ and 2π, we see that the corresponding values of $\sin t$ increase from -1 to 0. The graph of *any* periodic function over an interval of length equal

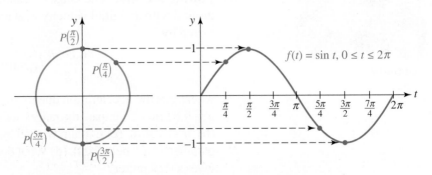

(a) Unit circle   (b) One cycle of sine graph

FIGURE 4.3.1 Points $P(t)$ on a circle corresponding to points on the graph to its period is said to be one **cycle of its graph**. In the case of the sine function, the graph over the interval $[0, 2\pi]$ in Figure 4.3.1(b) is one cycle of the graph of $f(t) = \sin t$.

Note: Change of symbols ▶ From this point on we will revert to the traditional symbols x and y when graphing trigonometric functions. Thus, $f(t) = \sin t$ will either be written $f(x) = \sin x$ or simply $y = \sin x$.

The graph of a periodic function is easily obtained by repeatedly drawing one cycle of its graph. In other words, the graph of $y = \sin x$ on, say, the intervals $[-2\pi, 0]$ and $[2\pi, 4\pi]$ is the same as that given in Figure 4.3.1(b). Recall from Theorem 4.2.4 of Section 4.2 that the sine function is an odd function since $f(-x) = \sin(-x) = -\sin x = -f(x)$. In other words, if (x, y) is a point on the graph of f, then so is $(-x, -y)$. Thus, from Theorem 2.2.1 it follows that the graph of $y = \sin x$ shown in FIGURE 4.3.2 is symmetric with respect to the origin.

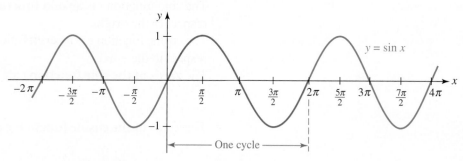

FIGURE 4.3.2 Graph of $y = \sin x$

By working again with the unit circle we can obtain one cycle of the graph of the cosine function $g(x) = \cos x$ on the interval $[0, 2\pi]$. In contrast to the graph of $f(x) = \sin x$ where $f(0) = f(2\pi) = 0$, for the cosine function we have $g(0) = g(2\pi) = 1$. FIGURE 4.3.3 shows one cycle (in red) of $y = \cos x$ on $[0, 2\pi]$ along with the extension of that cycle (in blue) to the adjacent intervals $[-2\pi, 0]$ and $[2\pi, 4\pi]$. We see from this figure that the graph of the cosine function is symmetric with respect to the y-axis. This is a consequence of g being an even function:

$$g(-x) = \cos(-x) = \cos x = g(x).$$

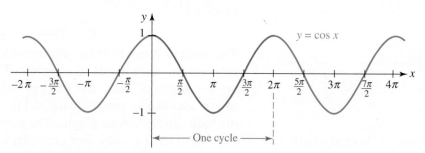

FIGURE 4.3.3 Graph of $y = \cos x$

☐ **Intercepts** In this and subsequent courses in mathematics it is important that you know the x-coordinates of the x-intercepts of the sine and cosine graphs—in other words, the zeros of $f(x) = \sin x$ and $g(x) = \cos x$. From the sine graph in Figure 4.3.2 we see that the zeros of the sine function, or the numbers for which $\sin x = 0$, are

$$x = 0, \pm\pi, \pm 2\pi, \pm 3\pi, \ldots.$$

4.3 Graphs of Sine and Cosine Functions

These numbers are integer multiples of π. From the cosine graph in Figure 4.3.3 we see that $\cos x = 0$ when

$$x = \pm\pi/2, \pm3\pi/2, \pm5\pi/2, \ldots.$$

These numbers are odd-integer multiples of $\pi/2$.

We summarize the foregoing discussion.

Properties of the Sine and Cosine Functions

- The **domain** of $f(x) = \sin x$ and of $g(x) = \cos x$ is the set of all real numbers $(-\infty, \infty)$.
- The **range** of $f(x) = \sin x$ and of $g(x) = \cos x$ is the interval $[-1, 1]$ on the y-axis.
- The **period** of $f(x) = \sin x$ and of $g(x) = \cos x$ is the number 2π.
- The sine function f is an **odd function**, and so its graph is symmetric with respect to the origin.
- The cosine function g is an **even function**, and so its graph is symmetric with respect to the y-axis.
- The **zeros of the sine function** f are the numbers

$$x = n\pi, n = 0, \pm1, \pm2, \ldots. \tag{1}$$

- The **zeros of the cosine function** g are the numbers

$$x = (2n + 1)\pi/2, n = 0, \pm1, \pm2, \ldots. \tag{2}$$

Note in (2) that if n is an integer, then $2n + 1$ is an odd integer. Using the distributive law, the zeros of the cosine function are often written as $x = \pi/2 + n\pi$.

☐ **Variation of the Graphs** As we did in Chapters 2 and 3 we can obtain variations of the basic sine and cosine graphs through rigid and nonrigid transformations. For the remainder of the discussion in this section we will consider graphs of functions of the form

$$y = A\sin(Bx + C) + D \quad \text{or} \quad y = A\cos(Bx + C) + D, \tag{3}$$

where $A, B, C,$ and D are real constants.

☐ **Graphs of $y = A\sin x$ and $y = A\cos x$** We begin by considering the special cases of (3):

$$y = A\sin x \quad \text{and} \quad y = A\cos x.$$

The multiple A can be either positive or negative, but does not affect the period of the function; in other words, the **period** of both $y = A\sin x$ and $y = A\cos x$ is 2π. For $|A| > 1$ graphs of these functions can be interpreted as a **vertical stretch** of the graphs of $y = \sin x$ or $y = \cos x$; when $0 < |A| < 1$ the graphs are a **vertical compression** of the basic sine or cosine graphs. The number $|A|$ is called the **amplitude** of the functions or of their graphs. For example, as **FIGURE 4.3.4** shows we obtain the graph of $y = 2\sin x$ by stretching the graph of $y = \sin x$ vertically by a factor of 2. Note that the maximum and minimum values of $y = 2\sin x$ occur at the same x-values as the maximum and minimum values of $y = \sin x$. In general, the maximum distance from any point on the graph of $y = A\sin x$ or $y = A\cos x$ to the x-axis is $|A|$. The amplitude of the basic functions $y = \sin x$ and $y = \cos x$ is $|A| = 1$. In general, if a periodic function f is continuous, then over a closed interval of length equal to its period, f has both a maximum value M and a minimum value m. The amplitude is defined by

$$\text{amplitude} = \tfrac{1}{2}[M - m]. \tag{4}$$

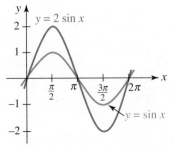

FIGURE 4.3.4 Vertical stretch of the graph of $y = \sin x$

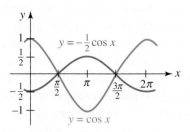

FIGURE 4.3.5 Graph of function in Example 1

As the next example shows, when $A < 0$ the graphs are also reflected in the x-axis.

EXAMPLE 1 **Vertically Compressed/Reflected Graph**

Graph $y = -\frac{1}{2}\cos x$.

Solution With the identification $A = -\frac{1}{2}$ we see that the amplitude of the function is $|A| = |-\frac{1}{2}| = \frac{1}{2}$. Because $|A| < 1$ and A is negative, the graph of $y = -\frac{1}{2}\cos x$ is the graph of $y = \cos x$ compressed vertically by a factor of $\frac{1}{2}$ and then reflected in the x-axis. The graph of $y = -\frac{1}{2}\cos x$ on the interval $[0, 2\pi]$ is shown in red in FIGURE 4.3.5. ▆

Vertical Stretch/Compression

- The graphs of the functions

$$y = A\sin x \quad \text{and} \quad y = A\cos x$$

have amplitude $|A|$ and period 2π. The graphs of $y = A\sin x$ and $y = A\cos x$ are the graphs of $y = \sin x$ and $y = \cos x$ stretched vertically if $|A| > 1$, and compressed vertically if $0 < |A| < 1$.

☐ **Graphs of $y = A\sin x + D$ and $y = A\cos x + D$** The graphs of

$$y = A\sin x + D \quad \text{and} \quad y = A\cos x + D$$

are the graphs of $y = A\sin x$ and $y = A\cos x$ shifted vertically, up for $D > 0$ and down for $D < 0$. The amplitude of the graph of either $y = A\sin x + D$ or $y = A\cos x + D$ is still $|A|$.

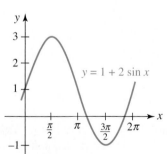

FIGURE 4.3.6 Graph of function in Example 2

EXAMPLE 2 **Vertically Shifted Sine Graph**

Graph $y = 1 + 2\sin x$.

Solution The graph of $y = 1 + 2\sin x$ is the graph of $y = 2\sin x$ given in Figure 4.3.4 shifted up 1 unit. From the identification $A = 2$, the amplitude is $|A| = 2$. Because the maximum 2 and minimum -2 of $y = 2\sin x$ occur, respectively, at $x = \pi/2$ and $x = 3\pi/2$, a rigid upward translation of the graph does not change the latter numbers but increases the maximum and minimum by 1 unit. We see in FIGURE 4.3.6 that the maximum of $y = 1 + 2\sin x$ is 3 at $x = \pi/2$ and the minimum of $y = 1 + 2\sin x$ is -1 at $x = 3\pi/2$. Using the function values $M = 3$ and $m = -1$, we can verify the amplitude of $y = 1 + 2\sin x$ using (4):

$$\tfrac{1}{2}[M - m] = \tfrac{1}{2}[3 - (-1)] = 2.$$

Note that the range $[-1, 3]$ of the function $y = 1 + 2\sin x$ is the range $[-2, 2]$ of $y = 2\sin x$ shifted up 1 unit on the y-axis. ▆

☐ **Graphs of $y = A\sin Bx$ and $y = A\cos Bx$** We now consider the graph of $y = A\sin Bx$ and $y = A\cos Bx$. Throughout the discussion we may assume that $B > 0$. Because 2π is the period of both $y = A\sin x$ and $y = A\cos x$ a cycle of the graphs of $y = A\sin Bx$ and $y = A\cos Bx$ begins at $x = 0$ and will start to repeat its values when $Bx = 2\pi$. Dividing the last equality by B shows that the **period** of each of the functions

$y = \sin Bx$ and $y = \cos Bx$ is $2\pi/B$ and so the graph of each function over the interval $[0, 2\pi/B]$ is one **cycle** of its graph. If $0 < B < 1$, then the period $2\pi/B$ is greater than 2π, and we can we can characterize the cycle of either $y = \sin Bx$ or $y = \cos Bx$ on $[0, 2\pi/B]$ as a **horizontal stretch** of the graphs of $y = \sin x$ and $y = \cos x$ on the interval $[0, 2\pi]$. On the other hand, if $B > 1$, then the period $2\pi/B$ is less than 2π, and so the graphs on $[0, 2\pi/B]$ can be interpreted as a **horizontal compression** of the graphs of the functions $y = \sin x$ and $y = \cos x$ on $[0, 2\pi]$. Finally, we can easily find the x-coordinates of the x-intercepts of the graphs of $y = A\sin Bx$ and $y = A\cos Bx$ by replacing the symbol x by Bx in (1) and (2) and solving for x.

The next example illustrates these concepts.

■EXAMPLE 3   Horizontally Compressed Sine Graph

Careful here: $\sin 2x \neq 2\sin x$ ▶

With the identification $B = 2$ the period of the sine function $y = \sin 2x$ is $2\pi/B = 2\pi/2 = \pi$. Therefore one cycle of the graph is completed on the interval $[0, \pi]$. FIGURE 4.3.7 shows that two cycles of the graph of $y = \sin 2x$ (in red) are completed on the interval $[0, 2\pi]$ whereas the graph of $y = \sin x$ (in blue) has completed only one cycle. Interpreted in terms of transformations, the graph of $y = \sin 2x$ on $[0, \pi]$ is a horizontal compression of the graph of $y = \sin x$ on $[0, 2\pi]$.

To find the x-intercepts of the graph of $y = \sin 2x$ we solve the equation $\sin 2x = 0$. From (1) with x replaced by $2x$, we get

$$2x = n\pi \quad \text{or} \quad x = \tfrac{1}{2}n\pi.$$

By letting $n = 0, \pm 1, \pm 2, \pm 3, \pm 4. \ldots$, the zeros of $y = \sin 2x$ are $x = 0, \pm\tfrac{1}{2}\pi,$ $\pm\pi, \pm\tfrac{3}{2}\pi, \pm 2\pi$, and so on. As we see in Figure 4.3.7, the x-intercepts on the nonnegative x-axis are the points

$$(0, 0), (\pi/2, 0), (\pi, 0), (3\pi/2, 0), (2\pi, 0), \ldots.$$

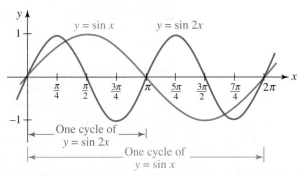

FIGURE 4.3.7 Graph of function in Example 3

Horizontal Stretch/Compression

• The graphs of the functions

$$y = A\sin Bx \quad \text{and} \quad y = A\cos Bx$$

for $B > 0$ have amplitude $|A|$ and period $2\pi/B$. The graphs of $y = A\sin Bx$ and $y = A\cos Bx$ are the graphs of $y = A\sin x$ and $y = A\cos x$ stretched horizontally if $0 < B < 1$, and compressed horizontally if $B > 1$.

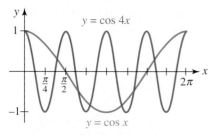

FIGURE 4.3.8 Graph of function in Example 4

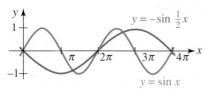

FIGURE 4.3.9 Graph of function in Example 5

EXAMPLE 4 **Horizontally Compressed Cosine Graph**

Find the period of $y = \cos 4x$ and graph the function.

Solution Since $B = 4$, we see that the period of $y = \cos 4x$ is $2\pi/4 = \pi/2$. We conclude that the graph of $y = \cos 4x$ is the graph of $y = \cos x$ compressed horizontally. To graph the function, we draw one cycle of the cosine graph with amplitude 1 on the interval $[0, \pi/2]$ and then use periodicity to extend the graph. FIGURE 4.3.8 shows four complete cycles of $y = \cos 4x$ (in red) and one cycle of $y = \cos x$ (in blue) on $[0, 2\pi]$. Notice that $y = \cos 4x$ attains its minimum at $x = \pi/4$ since $\cos 4(\pi/4) = \cos \pi = -1$ and its maximum at $x = \pi/2$ since $\cos 4(\pi/2) = \cos 2\pi = 1$. ■

In the case when $B < 0$ in either $y = A \sin Bx$ or $y = A \cos Bx$, we can use the even/odd properties, (15) of Section 4.2, to rewrite the function with positive B. This is illustrated in the next example.

EXAMPLE 5 **Horizontally Stretched Sine Graph**

Find the amplitude and period of $y = \sin(-\frac{1}{2}x)$. Graph the function.

Solution Since we require $B > 0$, we use $\sin(-x) = -\sin x$ to rewrite the function as

$$y = \sin(-\tfrac{1}{2}x) = -\sin\tfrac{1}{2}x.$$

With the identification $A = -1$, the amplitude is seen to be $|A| = |-1| = 1$. Now with $B = \frac{1}{2}$ we find that the period is $2\pi/\frac{1}{2} = 4\pi$. Hence we can interpret the cycle of $y = -\sin\frac{1}{2}x$ on $[0, 4\pi]$ as a horizontal stretch and a reflection in the x-axis (because $A < 0$) of the cycle of $y = \sin x$ on $[0, 2\pi]$. FIGURE 4.3.9 shows that on the interval $[0, 4\pi]$ the graph of $y = -\sin\frac{1}{2}x$ (in red) completes one cycle whereas the graph of $y = \sin x$ (in blue) completes two cycles. ■

☐ **Graphs of $y = A \sin(Bx + C)$ and $y = A \cos(Bx + C)$** We have seen that the basic graphs of $y = \sin x$ and $y = \cos x$ can be stretched or compressed vertically:

$$y = A \sin x \quad \text{and} \quad y = A \cos x,$$

shifted vertically:

$$y = A \sin x + D \quad \text{and} \quad y = A \cos x + D,$$

and stretched or compressed horizontally:

$$y = A \sin Bx + D \quad \text{and} \quad y = A \cos Bx + D.$$

The graphs of

$$y = A \sin(Bx + C) \quad \text{and} \quad y = A \cos(Bx + C)$$

are the graphs of $y = A \sin Bx$ and $y = A \cos Bx$ shifted horizontally. And finally, the graphs of

$$y = A \sin(Bx + C) + D \quad \text{and} \quad y = A \cos(Bx + C) + D$$

are the graphs of $y = A \sin(Bx + C)$ and $y = A \cos(Bx + C)$ shifted vertically.

Since the last case is straightforward, we are going to focus on the graphs of $y = A \sin(Bx + C)$ and $y = A \cos(Bx + C)$ in the remaining discussion. For example, we know from Section 2.2 that the graph of $y = \cos(x - \pi/2)$ is the basic cosine graph shifted $\pi/2$ units to the right. In FIGURE 4.3.10 the graph of $y = \cos(x - \pi/2)$ (in red) on the interval $[0, 2\pi]$ is one cycle of $y = \cos x$ on the interval $[-\pi/2, 3\pi/2]$

(in blue) shifted horizontally $\pi/2$ units to the right.   Similarly, the graphs of $y = \sin(x + \pi/2)$ and $y = \sin(x - \pi/2)$ are the basic sine graphs shifted $\pi/2$ units to the left and to the right, respectively. See FIGURES 4.3.11 and 4.3.12.

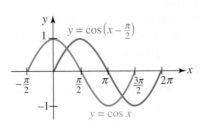

FIGURE 4.3.10 Horizontally shifted cosine graph

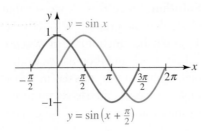

FIGURE 4.3.11 Horizontally shifted sine graph

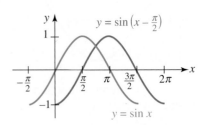

FIGURE 4.3.12 Horizontally shifted sine graph

By comparing the red graphs in Figures 4.3.10–4.3.12 with the graphs in Figures 4.3.2 and 4.3.3 we see that:

- the cosine graph shifted $\pi/2$ units to the right is the sine graph,
- the sine graph shifted $\pi/2$ units to the left is the cosine graph, and
- the sine graph shifted $\pi/2$ units to the right is the cosine graph reflected in the x-axis.

In other words, we have graphically verified the identities

$$\cos\left(x - \frac{\pi}{2}\right) = \sin x, \quad \sin\left(x + \frac{\pi}{2}\right) = \cos x, \quad \text{and} \quad \sin\left(x - \frac{\pi}{2}\right) = -\cos x. \quad (5)$$

◀ Everything said in this paragraph also holds for $y = A \cos(Bx + C)$.

For convenience we now consider the graph of $y = A \sin(Bx + C)$, for $B > 0$. Since the values of $\sin(Bx + C)$ range from -1 to 1, it follows that $A \sin(Bx + C)$ varies between $-A$ and A. That is, the **amplitude** of $y = A \sin(Bx + C)$ is $|A|$. Also, as $Bx + C$ varies from 0 to 2π, the graph will complete one cycle. By solving $Bx + C = 0$ and $Bx + C = 2\pi$, we find that one cycle is completed as x varies from $-C/B$ to $(2\pi - C)/B$. Therefore, the function $y = A \sin(Bx + C)$ has the **period**

$$\frac{2\pi - C}{B} - \left(-\frac{C}{B}\right) = \frac{2\pi}{B}.$$

Moreover, if $f(x) = A \sin Bx$, then

$$f\left(x + \frac{C}{B}\right) = A \sin B\left(x + \frac{C}{B}\right) = A \sin(Bx + C). \quad (6)$$

The result in (6) shows that the graph of $y = A \sin(Bx + C)$ can be obtained by shifting the graph of $f(x) = A \sin Bx$ horizontally a distance $|C|/B$. The number C/B is called the **phase shift** of the graph of $y = A \sin(Bx + C)$. If $C/B < 0$ the shift is to the right whereas if $C/B > 0$, the shift is to the left.

▮ EXAMPLE 6   Equation of Shifted Cosine Graph

The graph of $y = 10\cos 4x$ is shifted $\pi/12$ units to the right. Find its equation.

Solution We first identify $f(x) = 10\cos 4x$ and $B = 4$. Because we want to shift the graph of f to the right the phase shift is $C/B = -\pi/12$. Then the analogue of (6) for the cosine function is:

$$f\left(x - \frac{\pi}{12}\right) = 10 \cos 4\left(x - \frac{\pi}{12}\right) \quad \text{or} \quad y = 10 \cos\left(4x - \frac{\pi}{3}\right). \quad ▬$$

As a practical matter, if we are given the equation $y = A\sin(Bx + C)$ (or $y = A\cos(Bx + C)$), then the phase shift C/B of the graph of can be obtained by factoring the number B from $Bx + C$:

$$y = A\sin(Bx + C) = A\sin B\left(x + \frac{C}{B}\right).$$

The foregoing information is summarized next.

Horizontally Shifted Sine and Cosine Graphs

- The graphs of the functions

$$y = A\sin(Bx + C) \text{ and } y = A\cos(Bx + C), B > 0, \qquad (7)$$

  are the graphs of $y = A\sin Bx$ and $y = A\cos Bx$ shifted horizontally by $|C|/B$ units. The number C/B is called the **phase shift** of the graph.
- The horizontal shift is to the right if $C/B < 0$ and to left if $C/B > 0$.
- The **amplitude** of each function in (7) is $|A|$ and the **period** of each function is $2\pi/B$.
- The **range** of each function in (7) is the interval $[-|A|, |A|]$ on the y-axis.

The numbers $|A|$ and $2\pi/B$ are also referred to as the amplitude and period *of the graphs* of the functions in (7).

■ EXAMPLE 7   Horizontally Shifted Sine Graph

Graph $y = 3\sin(2x - \pi/3)$.

Solution For purposes of comparison we will first graph $y = 3\sin 2x$. The amplitude of $y = 3\sin 2x$ is $|A| = 3$ and its period is $2\pi/B = 2\pi/2 = \pi$. Thus one cycle of $y = 3\sin 2x$ is completed on the interval $[0, \pi]$. We extend this graph to the adjacent interval $[\pi, 2\pi]$ as shown in blue in FIGURE 4.3.13. Next, we rewrite $y = 3\sin(2x - \pi/3)$ by factoring 2 from $2x - \pi/3$:

$$y = 3\sin\left(2x - \frac{\pi}{3}\right) = 3\sin 2\left(x - \frac{\pi}{6}\right).$$

FIGURE 4.3.13 Graph of function in Example 7

From the last form we see that the phase shift is $-\pi/6 < 0$. This means that the graph of the given function, shown in red in Figure 4.3.13, is obtained by shifting the graph of $y = 3\sin 2x$ to the right $|C|/B = |(-\pi/3)|/2 = \pi/6$ units. Remember, this means that if (x, y) is a point on the blue graph, then $(x + \pi/6, y)$ is the corresponding point on the red graph. For example, $x = 0$ and $x = \pi$ are the x-coordinates of two x-intercepts of the blue graph. Thus $x = 0 + \pi/6 = \pi/6$ and $x = \pi + \pi/6 = 7\pi/6$ are x-coordinates of the x-intercepts of the red or shifted graph. These numbers are indicated by the arrows in Figure 4.3.13. Note that one cycle of the red graph is completed on the interval $[\pi/6, 7\pi/6]$. ■

■ EXAMPLE 8   Horizontally Shifted Graphs

Determine the amplitude, the period, the phase shift, and the direction of horizontal shift for each of the following functions.

(a) $y = 15\cos\left(5x - \dfrac{3\pi}{2}\right)$ **(b)** $y = -8\sin\left(2x + \dfrac{\pi}{4}\right)$

Solution (a) We first make the identifications $A = 15$, $B = 5$, and $C = -3\pi/2$. Thus the amplitude is $|A| = 15$ and the period is $2\pi/B = 2\pi/5$. The phase shift can be computed either by $C/B = (-3\pi/2)/5 = -3\pi/10$ or by rewriting the function as

$$y = 15\cos 5\left(x - \frac{3\pi}{10}\right).$$

The last form indicates that the graph of $y = 15\cos(5x - 3\pi/2)$ is the graph of $y = 15\cos 5x$ shifted $3\pi/10$ units to the right.

(b) Since $A = -8$ the amplitude is $|A| = |-8| = 8$. With $B = 2$ the period is $2\pi/B = 2\pi/2 = \pi$. By factoring 2 from $2x + \pi/4$, we see from

$$y = -8\sin\left(2x + \frac{\pi}{4}\right) = -8\sin 2\left(x + \frac{\pi}{8}\right)$$

that the phase shift is $\pi/8 > 0$ and so the graph of $y = -8\sin(2x + \pi/4)$ is the graph of $y = -8\sin 2x$ shifted $\pi/8$ units to the left. ∎

▌EXAMPLE 9   Horizontally Shifted Cosine Graph

Graph $y = 2\cos(\pi x + \pi)$.

Solution The amplitude of $y = 2\cos\pi x$ is $|A| = 2$ and the period is $2\pi/\pi = 2$. Thus one cycle of $y = 2\cos\pi x$ is completed on the interval $[0, 2]$. In FIGURE 4.3.14 two cycles of the graph of $y = 2\cos\pi x$ (in blue) are shown. The x-intercepts of this graph correspond to the values of x for which $\cos\pi x = 0$. By (2), this implies $\pi x = (2n + 1)\pi/2$ or $x = (2n + 1)/2$, n an integer. In other words, for $n = 0, -1, 1, -2, 2, -3, \ldots$ we get $x = \pm\frac{1}{2}, \pm\frac{3}{2}, \pm\frac{5}{2}$, and so on. Now by rewriting the given function as

$$y = 2\cos\pi(x + 1)$$

we see the phase shift is $1 > 0$. Thus the graph of $y = 2\cos(\pi x + \pi)$ (in red) in Figure 4.3.14, is obtained by shifting the graph of $y = 2\cos\pi x$ to the left 1 unit. This means that the x-intercepts are the same for both graphs. ∎

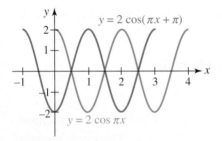

FIGURE 4.3.14 Graph of function in Example 9

▌EXAMPLE 10   Alternating Current

The current I (in amperes) in a wire of an alternating-current circuit is given by $I(t) = 30\sin 120\pi t$, where t is time measured in seconds. Sketch one cycle of the graph. What is the maximum value of the current?

Solution The graph has amplitude 30 and period $2\pi/120\pi = \frac{1}{60}$. Therefore, we sketch one cycle of the basic sine curve on the interval $[0, \frac{1}{60}]$, as shown in FIGURE 4.3.15. From the figure it is evident that the maximum value of the current is $I = 30$ amperes and occurs at $t = \frac{1}{240}$ since

$$I\left(\frac{1}{240}\right) = 30\sin\left(120\pi \cdot \frac{1}{240}\right) = 30\sin\frac{\pi}{2} = 30.$$ ∎

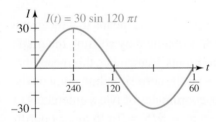

FIGURE 4.3.15 Graph of current in Example 10

Exercises 4.3   Answers to selected odd-numbered problems begin on page ANS–15.

In Problems 1–6, use the techniques of shifting, stretching, compressing, and reflecting to sketch at least one cycle of the graph of the given function.

1. $y = \frac{1}{2} + \cos x$ **2.** $y = -1 + \cos x$ **3.** $y = 2 - \sin x$

4. $y = 3 + 3\sin x$ **5.** $y = -2 + 4\cos x$ **6.** $y = 1 - 2\sin x$

In Problems 7–10, the given figure shows one cycle of a sine or cosine graph. From the figure, determine A and D and write an equation of the form $y = A\sin x + D$ or $y = A\cos x + D$ for the graph.

7.

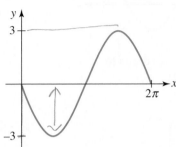

FIGURE 4.3.16 Graph for Problem 7

8.

FIGURE 4.3.17 Graph for Problem 8

9.

FIGURE 4.3.18 Graph for Problem 9

10.

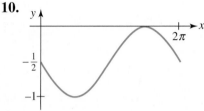

FIGURE 4.3.19 Graph for Problem 10

In Problems 11–16, use (1) and (2) of this section to find the x-intercepts for the graph of the given function. Do not graph.

11. $y = \sin \pi x$

12. $y = -\cos 2x$

13. $y = 10\cos\dfrac{x}{2}$

14. $y = 3\sin(-5x)$

15. $y = \sin\left(x - \dfrac{\pi}{4}\right)$

16. $y = \cos(2x - \pi)$

In Problems 17 and 18, find the x-intercepts of the graph of the given function on the interval $[0, 2\pi]$. Then find all intercepts using periodicity.

17. $y = -1 + \sin x$

18. $y = 1 - 2\cos x$

In Problems 19–24, the given figure shows one cycle of a sine or cosine graph. From the figure, determine A and B and write an equation of the form $y = A\sin Bx$ or $y = A\cos Bx$ for the graph.

19.

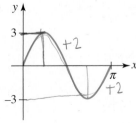

FIGURE 4.3.20 Graph for Problem 19

20.

FIGURE 4.3.21 Graph for Problem 20

21.

FIGURE 4.3.22 Graph for
Problem 21

22.

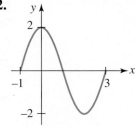

FIGURE 4.3.23 Graph for
Problem 22

23.

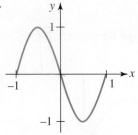

FIGURE 4.3.24 Graph for
Problem 23

24.

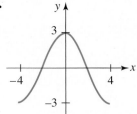

FIGURE 4.3.25 Graph for
Problem 24

In Problems 25–30, find the amplitude and period of the given function. Sketch at least one cycle of the graph.

25. $y = 4 \sin \pi x$

26. $y = -5 \sin \dfrac{x}{2}$

27. $y = -3 \cos 2\pi x$

28. $y = \dfrac{5}{2} \cos 4x$

29. $y = -2 \sin(-2x)$

30. $y = 5 \cos\left(-\dfrac{\pi}{2}x\right)$

In Problems 31–36, **(a)** sketch one cycle of the graph of the given function. **(b)** Find the amplitude $|A|$ by inspection of the function. **(c)** Find the maximum value M and the minimum value m of the function on the interval in part (a). **(d)** Then use (4) to verify the amplitude $|A|$ of the function. **(e)** Give the range of each function.

31. $y = 3 - 4\cos x$

32. $y = -3 + 3\sin x$

33. $y = -1 + \sin \dfrac{\pi}{2}x$

34. $y = -1 - 5\cos \pi x$

35. $y = 3 - 2\sin \pi x$

36. $y = 1 - 4\sin \dfrac{2}{3}x$

In Problems 37 and 38, from the given figure determine A, B, and D and write an equation of the form $y = A \sin Bx + D$ or $y = A \cos Bx + D$ for the graph.

37.

FIGURE 4.3.26 Graph for
Problem 37

38.

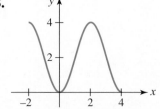

FIGURE 4.3.27 Graph for
Problem 38

In Problems 39–48, find the amplitude, period, and phase shift of the given function. Sketch at least one cycle of the graph.

39. $y = \sin\left(x - \dfrac{\pi}{6}\right)$

40. $y = \sin\left(3x - \dfrac{\pi}{4}\right)$

41. $y = \cos\left(x + \dfrac{\pi}{4}\right)$

42. $y = -2\cos\left(2x - \dfrac{\pi}{6}\right)$

43. $y = 4\cos\left(2x - \dfrac{3\pi}{2}\right)$

44. $y = 3\sin\left(2x + \dfrac{\pi}{4}\right)$

45. $y = 3\sin\left(\dfrac{x}{2} - \dfrac{\pi}{3}\right)$

46. $y = -\cos\left(\dfrac{x}{2} - \pi\right)$

47. $y = -4\sin\left(\dfrac{\pi}{3}x - \dfrac{\pi}{3}\right)$

48. $y = 2\cos\left(-2\pi x - \dfrac{4\pi}{3}\right)$

In Problems 49 and 50, write an equation of the function whose graph is described in words.

49. The graph of $y = \cos x$ is vertically stretched up by a factor of 3 and shifted down by 5 units. One cycle of $y = \cos x$ on $[0, 2\pi]$ is compressed to $[0, \pi/3]$ and then the compressed cycle is shifted horizontally $\pi/4$ units to the left.

50. One cycle of $y = \sin x$ on $[0, 2\pi]$ is stretched to $[0, 8\pi]$ and then the stretched cycle is shifted horizontally $\pi/12$ units to the right. The graph is also compressed vertically by a factor of $\frac{3}{4}$ and then reflected in the x-axis.

In Problems 51–54, find horizontally shifted sine and cosine functions so that each function satisfies the given conditions. Graph the functions.

51. Amplitude 3, period $2\pi/3$, shifted by $\pi/3$ units to the right
52. Amplitude $\frac{2}{3}$, period π, shifted by $\pi/4$ units to the left
53. Amplitude 0.7, period 0.5, shifted by 4 units to the right
54. Amplitude $\frac{5}{4}$, period 4, shifted by $1/2\pi$ units to the left

In Problems 55 and 56, graphically verify the given identity.

55. $\cos(x + \pi) = -\cos x$

56. $\sin(x + \pi) = -\sin x$

Applications

57. Pendulum The angular displacement θ of a pendulum from the vertical at time t seconds is given by $\theta(t) = \theta_0 \cos \omega t$, where θ_0 is the initial displacement at $t = 0$ seconds. See FIGURE 4.3.28. For $\omega = 2$ rad/s and $\theta_0 = \pi/10$, sketch two cycles of the resulting function.

58. Fahrenheit Temperature Suppose that

$$T(t) = 50 + 10\sin\frac{\pi}{12}(t - 8),$$

$0 \le t \le 24$, is a mathematical model of the Fahrenheit temperature at t hours after midnight on a certain day of the week.
(a) What is the temperature at 8 A.M.?
(b) At what time(s) does $T(t) = 60$?
(c) Sketch the graph of T.
(d) Find the maximum and minimum temperatures and the times at which they occur.

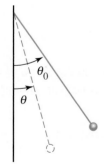

FIGURE 4.3.28 Pendulum in Problem 57

59. Depth of Water The depth d of water at the entrance to a small harbor at time t is modeled by a function of the form

$$d(t) = A \sin B\left(t - \frac{\pi}{2}\right) + C,$$

where A is one-half the difference between the high- and low-tide depths; $2\pi/B$, $B > 0$, is the tidal period; and C is the average depth. Assume that the tidal period is 12 hours, the depth at high tide is 18 feet, and the depth at low tide is 6 feet. Sketch two cycles of the graph of d.

60. Hours of Daylight The number H of daylight hours per day in various locations in the world can be modeled by a function of the form

$$H(t) = A \sin B(t - C) + D,$$

where the variable t represents the number of days in a year corresponding to a specific calendar date (for example, February 1 corresponds to $t = 32$ days), and $A, B, C,$ and D are positive constants. In this problem we construct a model for Los Angeles, CA for the year 2017 (not a leap year) using data obtained from the U.S. Naval Observatory, Washington, D.C.

(a) Find the amplitude A if 14.43 is the maximum number of daylight hours at the summer solstice and if 9.88 is the minimum number of daylight hours at the winter solstice.

(b) Find B if the function $H(t)$ is to have the period 365 days.

(c) For Los Angeles in the year 2017, we choose $C = 79$. Explain the significance of this number. [*Hint*: C has the same units as t.]

(d) Find D if the number of daylight hours at the vernal equinox for 2017 is 12.14 and occurs on March 20.

(e) What does the model $H(t)$ predict to be the number of daylight hours on January 1? On June 21? On August 1? On December 21?

(f) Using a graphing utility to obtain the graph of $H(t)$ on the interval $[0, 365]$.

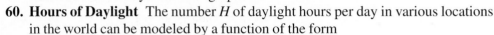

Sunset in Los Angeles

Calculator/Computer Problems

In Problems 61–64, use a graphing utility to investigate whether the given function is periodic.

61. $f(x) = \sin\left(\frac{1}{x}\right)$

62. $f(x) = \dfrac{1}{\sin 2x}$

63. $f(x) = 1 + (\cos x)^2$

64. $f(x) = x \sin x$

For Discussion

In Problems 65 and 66, describe in words how you would obtain the graph of the given function by starting with the graph of $y = \sin x$ (Problem 65) and the graph of $y = \cos x$ (Problem 66).

65. $y = 5 + 3 \sin(2x - \pi)$

66. $y = -6 + \frac{1}{4}\cos(\frac{1}{2}x + \pi)$

In Problems 67 and 68, find the period of the given function.

67. $f(x) = \sin\frac{1}{2}x \sin 2x$

68. $f(x) = \sin\frac{3}{2}x + \cos\frac{5}{2}x$

In Problems 69 and 70, discuss and then sketch the graph of the given function.

69. $f(x) = |\sin x|$

70. $f(x) = |\cos x|$

4.4 Other Trigonometric Functions

INTRODUCTION Recall that the remaining four trigonometric functions are the **tangent**, **cotangent**, **secant**, and **cosecant functions** and are denoted, in turn, as $\tan x$, $\cot x$, $\sec x$, and $\csc x$. We saw in Section 4.2 that by using (1) and (2) of Definition 4.2.1 in (3)–(6) of Definition 4.2.2 we can express these four new functions in terms of $\sin x$ and $\cos x$:

$$\tan x = \frac{\sin x}{\cos x} \qquad \cot x = \frac{\cos x}{\sin x} \tag{1}$$

$$\sec x = \frac{1}{\cos x} \qquad \csc x = \frac{1}{\sin x}. \tag{2}$$

☐ **Domain and Range** Because the functions in (1) and (2) are quotients, we know from Definition 2.6.1 that the **domain** of each function consists of the set of real numbers *except* those numbers for which the denominator is zero. We have seen in (2) of Section 4.3 that $\cos x = 0$ for $x = (2n + 1)\pi/2$, $n = 0, \pm1, \pm2, \ldots$, and so

- the domain of $\tan x$ and of $\sec x$ is $\{x \mid x \neq (2n + 1)\pi/2, n = 0, \pm1, \pm2, \ldots\}$.

Similarly, from (1) of Section 4.3, $\sin x = 0$ for $x = n\pi$, $n = 0, \pm1, \pm2, \ldots$, and so it follows that

- the domain of $\cot x$ and of $\csc x$ is $\{x \mid x \neq n\pi, n = 0, \pm1, \pm2, \ldots\}$.

We know that the values of the sine and cosine are bounded, that is, $|\sin x| \leq 1$ and $|\cos x| \leq 1$. From these last inequalities we have

$$|\sec x| = \left|\frac{1}{\cos x}\right| = \frac{1}{|\cos x|} \geq 1 \tag{3}$$

and

$$|\csc x| = \left|\frac{1}{\sin x}\right| = \frac{1}{|\sin x|} \geq 1. \tag{4}$$

Recall that an inequality such as (3) means that $\sec x \geq 1$ or $\sec x \leq -1$. Hence the range of the secant function is $(-\infty, -1] \cup [1, \infty)$. The inequality in (4) implies that the cosecant function has the same range $(-\infty, -1] \cup [1, \infty)$. When we consider the graphs of the tangent and cotangent functions we will see that they have the same range: $(-\infty, \infty)$.

If we interpret x as an angle, then **FIGURE 4.4.1** illustrates the algebraic signs of the tangent, cotangent, secant, and cosecant functions in each of the four quadrants. This is easily verified using the signs of the sine and cosine functions displayed in Figure 4.2.2.

II $\tan x < 0$ $\cot x < 0$ $\sec x < 0$ $\csc x > 0$	I $\tan x > 0$ $\cot x > 0$ $\sec x > 0$ $\csc x > 0$
III $\tan x > 0$ $\cot x > 0$ $\sec x < 0$ $\csc x < 0$	IV $\tan x < 0$ $\cot x < 0$ $\sec x > 0$ $\csc x < 0$

FIGURE 4.4.1 Signs of $\tan x$, $\cot x$, $\sec x$, and $\csc x$, in the four quadrants

▮ EXAMPLE 1   Example 5 of Section 4.2 Revisited

Find $\tan x$, $\cot x$, $\sec x$, and $\csc x$ for $x = -\pi/6$.

Solution In Example 5 of Section 4.2 we saw that

$$\sin\left(-\frac{\pi}{6}\right) = -\sin\frac{\pi}{6} = -\frac{1}{2} \qquad \text{and} \qquad \cos\left(-\frac{\pi}{6}\right) = \cos\frac{\pi}{6} = \frac{\sqrt{3}}{2}.$$

Therefore, by the definitions in (1) and (2):

$$\tan\left(-\frac{\pi}{6}\right) = \frac{-1/2}{\sqrt{3}/2} = -\frac{1}{\sqrt{3}}, \quad \cot\left(-\frac{\pi}{6}\right) = \frac{\sqrt{3}/2}{-1/2} = -\sqrt{3},$$

$$\sec\left(-\frac{\pi}{6}\right) = \frac{1}{\sqrt{3}/2} = \frac{2}{\sqrt{3}}, \quad \csc\left(-\frac{\pi}{6}\right) = \frac{1}{-1/2} = -2.$$

TABLE 4.4.1

x	0	$\frac{\pi}{6}$	$\frac{\pi}{4}$	$\frac{\pi}{3}$	$\frac{\pi}{2}$
$\tan x$	0	$\frac{1}{\sqrt{3}}$	1	$\sqrt{3}$	$-$
$\cot x$	$-$	$\sqrt{3}$	1	$\frac{1}{\sqrt{3}}$	0
$\sec x$	1	$\frac{2}{\sqrt{3}}$	$\sqrt{2}$	2	$-$
$\csc x$	$-$	2	$\sqrt{2}$	$\frac{2}{\sqrt{3}}$	1

Table 4.4.1 summarizes some important values of the tangent, cotangent, secant, and cosecant and was constructed using values of the sine and cosine from Section 4.2. A dash in the table indicates that the trigonometric function is not defined at that particular value of x.

☐ **Identities** The tangent is related to the secant by a useful identity. If we divide the Pythagorean identity

$$\sin^2 x + \cos^2 x = 1 \tag{5}$$

by $\cos^2 x$, we see that

$$\frac{\sin^2 x}{\cos^2 x} + \frac{\cos^2 x}{\cos^2 x} = \frac{1}{\cos^2 x}. \tag{6}$$

Similarly, dividing (5) by $\sin^2 x$ yields an identity relating the cotangent with the cosecant:

$$\frac{\sin^2 x}{\sin^2 x} + \frac{\cos^2 x}{\sin^2 x} = \frac{1}{\sin^2 x}. \tag{7}$$

Using the laws of exponents,

$$\frac{\sin^2 x}{\cos^2 x} = \left(\frac{\sin x}{\cos x}\right)^2 = \tan^2 x, \qquad \frac{1}{\cos^2 x} = \left(\frac{1}{\cos x}\right)^2 = \sec^2 x,$$

$$\frac{\cos^2 x}{\sin^2 x} = \left(\frac{\cos x}{\sin x}\right)^2 = \cot^2 x, \qquad \frac{1}{\sin^2 x} = \left(\frac{1}{\sin x}\right)^2 = \csc^2 x,$$

we see that (6) and (7) can be written in a simpler manner:

$$1 + \tan^2 x = \sec^2 x$$
$$1 + \cot^2 x = \csc^2 x.$$

Since the last two identities are direct consequences of $\sin^2 x + \cos^2 x = 1$ they too are called **Pythagorean identities**.

Finally, note that the tangent and cotangent function are related by the **reciprocal identity**

$$\cot x = \frac{\cos x}{\sin x} = \frac{1}{\dfrac{\sin x}{\cos x}} = \frac{1}{\tan x}.$$

☐ **Summary** For future reference, especially for the work in the next section, we pause here to summarize a small collection of identities that are so basic to the study of trigonometry that they are known collectively as the **fundamental identities**. You should firmly commit these identities to memory.

CHAPTER 4 TRIGONOMETRIC FUNCTIONS

Fundamental Trigonometric Identities

Pythagorean identities:

$$\sin^2 x + \cos^2 x = 1 \qquad (8)$$
$$1 + \tan^2 x = \sec^2 x \qquad (9)$$
$$1 + \cot^2 x = \csc^2 x \qquad (10)$$

Quotient identities:

$$\tan x = \frac{\sin x}{\cos x} \qquad \cot x = \frac{\cos x}{\sin x} \qquad (11)$$

Reciprocal identities:

$$\sec x = \frac{1}{\cos x} \qquad \csc x = \frac{1}{\sin x} \qquad \cot x = \frac{1}{\tan x} \qquad (12)$$

EXAMPLE 2   Using a Pythagorean Identity

Given that $\csc x = -5$ and $3\pi/2 < x < 2\pi$, determine the values of $\tan x$ and $\cot x$.

Solution We first compute $\cot x$. It follows from (10) that

$$\cot^2 x = \csc^2 x - 1.$$

For $3\pi/2 < x < 2\pi$, we see from Figure 4.4.1 that $\cot x$ must be negative and so we take the negative square root:

$$\cot x = -\sqrt{\csc^2 x - 1} = -\sqrt{(-5)^2 - 1} = -\sqrt{24} = -2\sqrt{6}.$$

Using $\cot x = 1/\tan x$, we have

$$\overset{\text{rationalizing the denominator}}{\underset{\downarrow}{}}$$

$$\tan x = \frac{1}{\cot x} = \frac{1}{-2\sqrt{6}} = -\frac{\sqrt{6}}{12}.$$

In Example 2, given the information $\csc x = -5$ and $3\pi/2 < x < 2\pi$, we could easily find the values of the remaining five trigonometric functions. One way of proceeding would be to use $\csc x = 1/\sin x$ to find $\sin x = 1/\csc x = -\frac{1}{5}$. Then we use $\sin^2 x + \cos^2 x = 1$ to find $\cos x$. After we have found $\cos x$, the remaining three trigonometric functions can be obtained from (1) and (2).

☐ **Periodicity** Because the sine and cosine functions are 2π periodic, each of the functions in (1) and (2) have a period 2π. But from (17) of Theorem 4.2.5 we have

Also see Problems 55 and 56 in ▶
Exercises 4.3.

$$\tan(x + \pi) = \frac{\sin(x + \pi)}{\cos(x + \pi)} = \frac{-\sin x}{-\cos x} = \tan x. \qquad (13)$$

Thus (13) implies that $\tan x$ and $\cot x$ are periodic with a period $p \leq \pi$. In the case of the tangent function, $\tan x = 0$ only if $\sin x = 0$, that is, only if $x = 0, \pm\pi, \pm 2\pi$, and so on. Therefore, the smallest positive number p for which $\tan(x + p) = \tan x$ is $p = \pi$. The cotangent function has the same period since it is the reciprocal of the tangent function.

THEOREM 4.4.1 Period of the Tangent and Cotangent

The tangent and cotangent functions are periodic with **period** π. Therefore,

$$\tan(x + \pi) = \tan x \qquad \text{and} \qquad \cot(x + \pi) = \cot x \qquad (14)$$

for every real number x for which the functions are defined.

THEOREM 4.4.2 Period of the Secant and Cosecant

The secant and cosecant functions are periodic with **period** 2π. Therefore,

$$\sec(x + 2\pi) = \sec x \qquad \text{and} \qquad \csc(x + 2\pi) = \csc x \qquad (15)$$

for every real number x for which the functions are defined.

☐ **Even-Odd Properties** Because the cosine function is even and the sine function is odd, each of the remaining four trigonometric functions is either even or odd.

THEOREM 4.4.3 Even and Odd Functions

The tangent, cotangent, and cosecant functions are **odd functions**, whereas the secant function is an **even function**. That is,

$$\tan(-x) = -\tan x \qquad \text{and} \qquad \cot(-x) = -\cot x \qquad (16)$$
$$\sec(-x) = \sec x \qquad \text{and} \qquad \csc(-x) = -\csc x \qquad (17)$$

for every real number x for which the functions are defined,

PROOF: We prove the first entries in (16) and (17). Because $\cos(-x) = \cos x$ and $\sin(-x) = -\sin x$,

$$\tan(-x) = \frac{\sin(-x)}{\cos(-x)} = \frac{-\sin x}{\cos x} = -\frac{\sin x}{\cos x} = -\tan x,$$
$$\sec(-x) = \frac{1}{\cos(-x)} = \frac{1}{\cos x} = \sec x$$

proves, in turn, that $\tan x$ is an odd function and $\sec x$ is an even function. ∎

This is a good time to review (7) of Section 3.5. ▶

FIGURE 4.4.2 One cycle of the graph of $y = \tan x$

☐ **Graphs of $y = \tan x$ and $y = \cot x$** The numbers that make the denominators of $\tan x$, $\cot x$, $\sec x$, and $\csc x$ equal to zero correspond to vertical asymptotes of their graphs. For example, we encourage you to verify, using a calculator, that

$$\tan x \to -\infty \text{ as } x \to -\frac{\pi}{2}^{+} \qquad \text{and} \qquad \tan x \to \infty \text{ as } x \to \frac{\pi}{2}^{-}.$$

In other words, $x = -\pi/2$ and $x = \pi/2$ are vertical asymptotes. The graph of $y = \tan x$ on the interval $(-\pi/2, \pi/2)$ given in FIGURE 4.4.2 is one **cycle** of the graph of $y = \tan x$. Using periodicity we extend the cycle in Figure 4.4.2 to adjacent intervals of length π, as shown in FIGURE 4.4.3. The x-intercepts of the graph of the tangent function are $(0, 0)$, $(\pm\pi, 0)$, $(\pm 2\pi, 0), \ldots$, and the vertical asymptotes of the graph are $x = \pm\pi/2$, $\pm 3\pi/2, \pm 5\pi/2, \ldots$.

CHAPTER 4 TRIGONOMETRIC FUNCTIONS

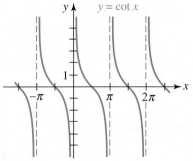

FIGURE 4.4.3 Graph of $y = \tan x$

FIGURE 4.4.4 Graph of $y = \cot x$

The graph of $y = \cot x$ is similar to the graph of the tangent function and is given in FIGURE 4.4.4. In this case, the graph of $y = \cot x$ on the interval $(0, \pi)$ is one **cycle** of the graph of $y = \cot x$. The x-intercepts of the graph of the cotangent function are $(\pm\pi/2, 0)$, $(\pm 3\pi/2, 0)$, $(\pm 5\pi/2, 0)$, . . . , and the vertical asymptotes of the graph are $x = 0$, $\pm\pi$, $\pm 2\pi$, $\pm 3\pi$, ... Because $y = \tan x$ and $y = \cot x$ are odd functions, their graphs are symmetric with respect to the origin.

☐ **Graphs of $y = \sec x$ and $y = \csc x$** For both $y = \sec x$ and $y = \csc x$ we know that $|y| \geq 1$, and so no portion of their graphs can appear in the horizontal strip $-1 < y < 1$ of the Cartesian plane. Hence the graphs of $y = \sec x$ and $y = \csc x$ have no x-intercepts. Both $y = \sec x$ and $y = \csc x$ have period 2π. The vertical asymptotes for the graph of $y = \sec x$ are the same as $y = \tan x$, namely, $x = \pm\pi/2$, $\pm 3\pi/2$, $\pm 5\pi/2$, ... Because $y = \sec x$ is an even function, its graph is symmetric with respect to the y-axis. On the other hand, the vertical asymptotes for the graph of $y = \csc x$ are the same as $y = \cot x$, namely, $x = 0$, $\pm\pi$, $\pm 2\pi$, $\pm 3\pi$, ... Because $y = \csc x$ is an odd function, its graph is symmetric with respect to the origin. One cycle of the graph of $y = \sec x$ on $[0, 2\pi]$ is extended to the interval $[-2\pi, 0]$ by periodicity (or y-axis symmetry) in FIGURE 4.4.5. Similarly, in FIGURE 4.4.6 we extend one cycle of $y = \csc x$ on $(0, 2\pi)$ to the interval $(-2\pi, 0)$ by periodicity (or origin symmetry).

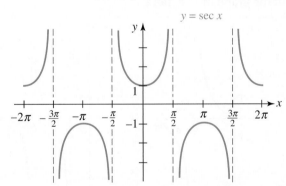

FIGURE 4.4.5 Graph of $y = \sec x$

FIGURE 4.4.6 Graph of $y = \csc x$

☐ **Transformations and Graphs** Similar to the sine and cosine graphs, rigid and nonrigid transformations can be applied to the graphs of $y = \tan x$, $y = \cot x$, $y = \sec x$, and $y = \csc x$. For example, a function such as $y = A\tan(Bx + C) + D$ can be analyzed in the following manner:

vertical stretch/compression/reflection $\qquad$ vertical shift
$\downarrow \qquad\qquad\qquad\qquad\qquad\qquad\qquad\qquad \downarrow$

$$y = A\tan(Bx + C) + D \qquad (18)$$

$\qquad\qquad\qquad \uparrow \qquad\quad \uparrow$
horizontal stretch/compression by changing period $\qquad$ horizontal shift

If $B > 0$, then the period of

$$y = A \tan(Bx + C) \qquad \text{and} \qquad y = A \cot(Bx + C) \text{ is } \pi/B, \qquad (19)$$

whereas the period of

$$y = A \sec(Bx + C) \qquad \text{and} \qquad y = A \csc(Bx + C) \text{ is } 2\pi/B. \qquad (20)$$

As we see in (18), the number A in each case can be interpreted as either a vertical stretch or a compression of a graph. However, you should be aware of the fact that the functions in (19) and (20) have no amplitude, because none of the functions has a maximum *and* a minimum value.

Of the six trigonometric functions, only the ▶ sine and cosine functions have an amplitude.

EXAMPLE 3   Comparison of Graphs

Find the period, x-intercepts, and vertical asymptotes for the graph of $y = \tan 2x$. Graph the function on $[0, \pi]$.

Solution With the identification $B = 2$, we see from (19) that the period is $\pi/2$. Since $\tan 2x = \sin 2x / \cos 2x$, the x-intercepts of the graph occur at the zeros of $\sin 2x$. From (1) of Section 4.3, $\sin 2x = 0$ for

$$2x = n\pi \qquad \text{so that} \qquad x = \tfrac{1}{2} n\pi, n = 0, \pm 1, \pm 2, \ldots.$$

That is, $x = 0, \pm \pi/2, \pm 2\pi/2 = \pi, \pm 3\pi/2, \pm 4\pi/2 = 2\pi$, and so on. The x-intercepts are $(0, 0), (\pm \pi/2, 0), (\pm \pi, 0), (\pm 3\pi/2, 0), \ldots$. The vertical asymptotes of the graph occur at zeros of $\cos 2x$. From (2) of Section 4.3, the numbers for which $\cos 2x = 0$ are found in the following manner:

$$2x = (2n + 1)\frac{\pi}{2} \qquad \text{so that} \qquad x = (2n + 1)\frac{\pi}{4}, n = 0, \pm 1, \pm 2, \ldots.$$

That is, the vertical asymptotes are $x = \pm \pi/4, \pm 3\pi/4, \pm 5\pi/4, \ldots$. On the interval $[0, \pi]$, the graph of $y = \tan 2x$ has three intercepts $(0, 0), (\pi/2, 0)$, and $(\pi, 0)$ and two vertical asymptotes $x = \pi/4$ and $x = 3\pi/4$. In FIGURE 4.4.7, we have compared the graphs of $y = \tan x$ and $y = \tan 2x$ on the interval. The graph of $y = \tan 2x$ is a horizontal compression of the graph of $y = \tan x$.

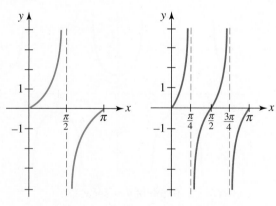

(a) $y = \tan x$ on $[0, \pi]$ (b) $y = \tan 2x$ on $[0, \pi]$

FIGURE 4.4.7 Graph of function in Example 3

EXAMPLE 4   Comparison of Graphs

Compare one cycle of the graphs of $y = \tan x$ and $y = \tan(x - \pi/4)$.

Solution The graph of $y = \tan(x - \pi/4)$ is the graph of $y = \tan x$ shifted horizontally $\pi/4$ units to the right. The intercept $(0, 0)$ for the graph of $y = \tan x$ is shifted to $(\pi/4, 0)$ on the graph of $y = \tan(x - \pi/4)$. The vertical asymptotes $x = -\pi/2$ and

$x = \pi/2$ for the graph of $y = \tan x$ are shifted to $x = -\pi/4$ and $x = 3\pi/4$ for the graph of $y = \tan(x - \pi/4)$. In FIGURES 4.4.8(a) and 4.4.8(b) we see, respectively, that a cycle of the graph of $y = \tan x$ on the interval $(-\pi/2, \pi/2)$ is shifted to the right to yield a cycle of the graph of $y = \tan(x - \pi/4)$ on the interval $(-\pi/4, 3\pi/4)$.

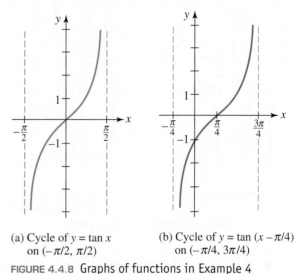

(a) Cycle of $y = \tan x$ (b) Cycle of $y = \tan(x - \pi/4)$
on $(-\pi/2, \pi/2)$ on $(-\pi/4, 3\pi/4)$

FIGURE 4.4.8 Graphs of functions in Example 4

As we did in the analysis of the graphs of $y = A \sin(Bx + C)$ and $y = A \cos(Bx + C)$, we can determine the amount of horizontal shift for graphs of functions such as $y = A \tan(Bx + C)$ and $y = A \sec(Bx + C)$ by factoring the number $B > 0$ from $Bx + C$.

EXAMPLE 5   Two Shifts and Two Compressions

Graph $y = 2 - \frac{1}{2}\sec(3x - \pi/2)$.

Solution Let's break down the analysis of the graph into four parts, namely, by transformations.

(*i*) One cycle of the graph of $y = \sec x$ occurs on $[0, 2\pi]$. Since the period of $y = \sec 3x$ is $2\pi/3$, one cycle of its graph occurs on the interval $[0, 2\pi/3]$. In other words, the graph of $y = \sec 3x$ is a horizontal compression of the graph of $y = \sec x$. Since $\sec 3x = 1/\cos 3x$, the vertical asymptotes occur at the zeros of $\cos 3x$. Using (2) of Section 4.3, we find

$$3x = (2n + 1)\frac{\pi}{2} \quad \text{or} \quad x = (2n + 1)\frac{\pi}{6}, n = 0, \pm 1, \pm 2, \ldots.$$

FIGURE 4.4.9(a) shows two cycles of the graph $y = \sec 3x$; one cycle on $[-2\pi/3, 0]$ and another on $[0, 2\pi/3]$. Within those intervals the vertical asymptotes are $x = -\pi/2$, $x = -\pi/6$, $x = \pi/6$, and $x = \pi/2$.

(*ii*) The graph of $y = -\frac{1}{2}\sec 3x$ is the graph of $y = \sec 3x$ compressed vertically by a factor of $\frac{1}{2}$ and then reflected in the x-axis. See Figure 4.4.9(b).

(*iii*) By factoring 3 from $3x - \pi/2$, we see from

$$y = -\tfrac{1}{2}\sec\left(3x - \frac{\pi}{2}\right) = -\tfrac{1}{2}\sec 3\left(x - \frac{\pi}{6}\right)$$

that the graph of $y = -\frac{1}{2}\sec(3x - \pi/2)$ is the graph of $y = -\frac{1}{2}\sec 3x$ shifted $\pi/6$ units to the right. By shifting the two intervals $[-2\pi/3, 0]$ and $[0, 2\pi/3]$ in Figure 4.4.9(b)

(a) Horizontal compression (b) Vertical compression (c) Horizontal shift (d) Vertical shift
and reflection in x-axis

FIGURE 4.4.9 Graph of function in Example 5

to the right $\pi/6$ units, we see in Figure 4.4.9(c) two cycles of $y = -\frac{1}{2}\sec(3x - \pi/2)$ on the intervals $[-\pi/2, \pi/6]$ and $[\pi/6, 5\pi/6]$. The vertical asymptotes $x = -\pi/2$, $x = -\pi/6, x = \pi/6$, and $x = \pi/2$ shown in Figure 4.4.9(b) are shifted to $x = -\pi/3$, $x = 0, x = \pi/3$, and $x = 2\pi/3$. Observe that the y-intercept $(0, -\frac{1}{2})$ in Figure 4.4.9(b) is now moved to $(\pi/6, -\frac{1}{2})$ in Figure 4.4.9(c).

(*iv*) Finally, we obtain the graph $y = 2 - \frac{1}{2}\sec(3x - \pi/2)$ in Figure 4.4.9(d) by shifting the graph of $y = -\frac{1}{2}\sec(3x - \pi/2)$ in Figure 4.4.9(c) upward 2 units. ▌

Exercises 4.4 Answers to selected odd-numbered problems begin on page ANS–16.

In Problems 1 and 2, complete the given table.

1.

x	$\frac{2\pi}{3}$	$\frac{3\pi}{4}$	$\frac{5\pi}{6}$	π	$\frac{7\pi}{6}$	$\frac{5\pi}{4}$	$\frac{4\pi}{3}$	$\frac{3\pi}{2}$	$\frac{5\pi}{3}$	$\frac{7\pi}{4}$	$\frac{11\pi}{6}$	2π
$\tan x$												
$\cot x$												

2.

x	$\frac{2\pi}{3}$	$\frac{3\pi}{4}$	$\frac{5\pi}{6}$	π	$\frac{7\pi}{6}$	$\frac{5\pi}{4}$	$\frac{4\pi}{3}$	$\frac{3\pi}{2}$	$\frac{5\pi}{3}$	$\frac{7\pi}{4}$	$\frac{11\pi}{6}$	2π
$\sec x$												
$\csc x$												

In Problems 3–18, find the indicated value without the use of a calculator.

3. $\cot\dfrac{13\pi}{6}$ **4.** $\csc\left(-\dfrac{3\pi}{2}\right)$ **5.** $\tan\dfrac{9\pi}{2}$ **6.** $\sec 7\pi$

7. $\csc\left(-\dfrac{\pi}{3}\right)$ **8.** $\cot\left(-\dfrac{13\pi}{3}\right)$ **9.** $\tan\dfrac{23\pi}{4}$ **10.** $\tan\left(-\dfrac{5\pi}{6}\right)$

11. $\sec\dfrac{10\pi}{3}$ **12.** $\cot\dfrac{17\pi}{6}$ **13.** $\csc 5\pi$ **14.** $\sec\dfrac{29\pi}{4}$

15. $\sec(-120°)$ **16.** $\tan 405°$ **17.** $\csc 495°$ **18.** $\cot(-720°)$

In Problems 19–26, use the given information to find the values of the remaining five trigonometric functions.

19. $\tan x = -2$, $\pi/2 < x < \pi$ **20.** $\cot x = \frac{1}{2}$, $\pi < x < 3\pi/2$

21. $\csc x = \frac{4}{3}$, $0 < x < \pi/2$ **22.** $\sec x = -5$, $\pi/2 < x < \pi$

23. $\sin x = \frac{1}{3}$, $\pi/2 < x < \pi$ **24.** $\cos x = -1/\sqrt{5}$, $\pi < x < 3\pi/2$

25. $\cos x = \frac{12}{13}$, $3\pi/2 < x < 2\pi$ **26.** $\sin x = \frac{4}{5}$, $0 < x < \pi/2$

27. If $3\cos x = \sin x$, find all values of $\tan x$, $\cot x$, $\sec x$, and $\csc x$.

28. If $\csc x = \sec x$, find all values of $\tan x$, $\cot x$, $\sin x$, and $\cos x$.

In Problems 29–36, find the period, x-intercepts, and the vertical asymptotes of the given function. Sketch at least one cycle of the graph.

29. $y = \tan \pi x$ **30.** $y = \tan \dfrac{x}{2}$

31. $y = \cot 2x$ **32.** $y = -\cot \dfrac{\pi x}{3}$

33. $y = \tan\left(\dfrac{x}{2} - \dfrac{\pi}{4}\right)$ **34.** $y = \dfrac{1}{4}\cot\left(x - \dfrac{\pi}{2}\right)$

35. $y = -1 + \cot \pi x$ **36.** $y = \tan\left(x + \dfrac{5\pi}{6}\right)$

In Problems 37–44, find the period and the vertical asymptotes of the given function. Sketch at least one cycle of the graph.

37. $y = -\sec x$ **38.** $y = 2\sec\dfrac{\pi x}{2}$

39. $y = 3\csc \pi x$ **40.** $y = -2\csc\dfrac{x}{3}$

41. $y = \sec\left(3x - \dfrac{\pi}{2}\right)$ **42.** $y = \csc(4x + \pi)$

43. $y = 3 + \csc\left(2x + \dfrac{\pi}{2}\right)$ **44.** $y = -1 + \sec(x - 2\pi)$

In Problems 45 and 46, use the graphs of $y = \tan x$ and $y = \sec x$ to find numbers A and C for which the given equality is true.

45. $\cot x = A\tan(x + C)$ **46.** $\csc x = A\sec(x + C)$

Calculus-Related Problems

In Problems 47–56, use the Pythagorean identities (8)–(10) and the indicated substitution to rewrite the given algebraic expression as a trigonometric expression without radicals. Assume that $a > 0$.

47. $\sqrt{a^2 - x^2}$; $x = a\cos\theta$, $0 \le \theta \le \pi$

48. $\sqrt{a^2 + x^2}$; $x = a\tan\theta$, $-\pi/2 < \theta < \pi/2$

49. $\sqrt{x^2 - a^2}$; $x = a\sec\theta$, $0 \le \theta < \pi/2$

50. $\sqrt{16 - 25x^2}$; $x = \frac{4}{5}\sin\theta$, $-\pi/2 \le \theta \le \pi/2$

51. $\dfrac{x}{\sqrt{9 - x^2}}$; $x = 3\sin\theta$, $-\pi/2 < \theta < \pi/2$

52. $x^2\sqrt{x^2 - 4}$; $x = 2\sec\theta$, $0 \le \theta < \pi/2$

53. $\dfrac{\sqrt{x^2 - 3}}{x^2}$; $\quad x = \sqrt{3}\sec\theta,\ 0 \le \theta < \pi/2$

54. $(36 + x^2)^{3/2}$; $\quad x = 6\tan\theta,\ -\pi/2 < \theta < \pi/2$

55. $\dfrac{1}{\sqrt{7 + x^2}}$; $\quad x = \sqrt{7}\tan\theta,\ -\pi/2 < \theta < \pi/2$

56. $\dfrac{\sqrt{5 - x^2}}{x}$; $\quad x = \sqrt{5}\cos\theta,\ 0 \le \theta \le \pi$

For Discussion

57. Use a calculator in radian mode to compare the values of $\tan(1.57)$ and $\tan(1.58)$. Explain the difference in these values.

58. Use a calculator in radian mode to compare the values of $\cot(3.14)$ and $\cot(3.15)$.

59. Explain why there are no real numbers x satisfying the equation $9\csc x = 1$.

60. For which real numbers x is **(a)** $\sin x \le \csc x$? **(b)** $\sin x < \csc x$?

61. For which real numbers x is **(a)** $\sec x \le \cos x$? **(b)** $\sec x < \cos x$?

62. Discuss and then sketch the graphs of $y = |\sec x|$ and $y = |\csc x|$.

4.5 Sum and Difference Formulas

INTRODUCTION There are *many* identities involving trigonometric functions. A **trigonometric identity** is an equation or formula involving only trigonometric functions that is valid for all angles measured in degrees or radians or for real numbers for which both sides of the equality are defined. The most basic of all these identities—the Pythagorean identities, the Quotient identities, and Reciprocal identities—were discussed in Section 4.4. In this section we develop some additional identities that are of particular importance in courses in mathematics and science.

☐ **Sum and Difference Formulas** The **sum** and **difference formulas** for the cosine and sine functions are identities that reduce $\cos(x_1 + x_2)$, $\cos(x_1 - x_2)$, $\sin(x_1 + x_2)$, and $\sin(x_1 - x_2)$ to expressions that involve $\cos x_1$, $\cos x_2$, $\sin x_1$, and $\sin x_2$. We will derive the formula for $\cos(x_1 - x_2)$ first, and then we will use that result to obtain the others.

For convenience, let us suppose that x_1 and x_2 represent angles measured in radians. As shown in FIGURE 4.5.1(a), let d denote the distance between $P(x_1)$ and $P(x_2)$. If we place the angle $x_1 - x_2$ in standard position as shown in Figure 4.5.1(b), then d is

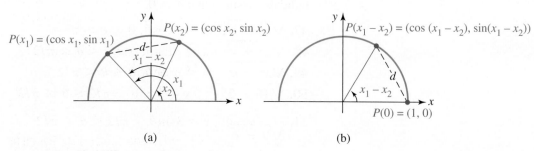

(a)   (b)

FIGURE 4.5.1 The difference of two angles

also the distance between $P(x_1 - x_2)$ and $P(0)$. Equating the squares of these distances gives

$$(\cos x_1 - \cos x_2)^2 + (\sin x_1 - \sin x_2)^2 = (\cos(x_1 - x_2) - 1)^2 + \sin^2(x_1 - x_2)$$

or $\cos^2 x_1 - 2\cos x_1 \cos x_2 + \cos^2 x_2 + \sin^2 x_1 - 2\sin x_1 \sin x_2 + \sin^2 x_2$

$$= \cos^2(x_1 - x_2) - 2\cos(x_1 - x_2) + 1 + \sin^2(x_1 - x_2).$$

In view of the Pythagorean identity (8) of Section 4.4,

$$\cos^2 x_1 + \sin^2 x_1 = 1, \quad \cos^2 x_2 + \sin^2 x_2 = 1, \quad \cos^2(x_1 - x_2) + \sin^2(x_1 - x_2) = 1,$$

and so the preceding equation simplifies to

$$\cos(x_1 - x_2) = \cos x_1 \cos x_2 + \sin x_1 \sin x_2.$$

This last result can be put to work immediately to find the cosine of the sum of two angles. Since $x_1 + x_2$ can be rewritten as the difference $x_1 - (-x_2)$,

$$\cos(x_1 + x_2) = \cos(x_1 - (-x_2))$$
$$= \cos x_1 \cos(-x_2) + \sin x_1 \sin(-x_2).$$

By the even–odd identities, $\cos(-x_2) = \cos x_2$ and $\sin(-x_2) = -\sin x_2$, it follows that the last line is the same as

$$\cos(x_1 + x_2) = \cos x_1 \cos x_2 - \sin x_1 \sin x_2.$$

The two results just obtained are summarized in the next theorem.

THEOREM 4.5.1 Sum and Difference Formulas for the Cosine

For all real numbers x_1 and x_2,

$$\cos(x_1 + x_2) = \cos x_1 \cos x_2 - \sin x_1 \sin x_2 \qquad (1)$$
$$\cos(x_1 - x_2) = \cos x_1 \cos x_2 + \sin x_1 \sin x_2 \qquad (2)$$

EXAMPLE 1　　　**Cosine of a Sum**

Evaluate $\cos(7\pi/12)$.

Solution We have no way of evaluating $\cos(7\pi/12)$ directly. However, observe that

$$\frac{7\pi}{12} \text{ radians} = 105° = 60° + 45° = \frac{\pi}{3} + \frac{\pi}{4}.$$

Because $7\pi/12$ radians is a second-quadrant angle, we know that the value of $\cos(7\pi/12)$ is negative. Proceeding, the sum formula (1) of the Theorem 4.5.1 gives

$$\cos\frac{7\pi}{12} = \cos\left(\frac{\pi}{3} + \frac{\pi}{4}\right) = \cos\frac{\pi}{3}\cos\frac{\pi}{4} - \sin\frac{\pi}{3}\sin\frac{\pi}{4}$$

$$= \frac{1}{2}\frac{\sqrt{2}}{2} - \frac{\sqrt{3}}{2}\frac{\sqrt{2}}{2} = \frac{\sqrt{2}}{4}(1 - \sqrt{3}).$$

Using $\sqrt{2}\sqrt{3} = \sqrt{6}$, this result can also be written as $\cos(7\pi/12) = (\sqrt{2} - \sqrt{6})/4$. Since $\sqrt{6} > \sqrt{2}$, we see that $\cos(7\pi/12) < 0$, as expected. ∎

To obtain the corresponding sum/difference identities for the sine function we will make use of two identities:

See (5) in Section 4.3. ▶

$$\cos\left(x - \frac{\pi}{2}\right) = \sin x \qquad \text{and} \qquad \sin\left(x - \frac{\pi}{2}\right) = -\cos x. \tag{3}$$

These identities were first discovered in Section 4.3 by shifting the graphs of the cosine and sine. However, both results in (3) can now be proved using (2):

$$\cos\left(x - \frac{\pi}{2}\right) = \cos x \cos\frac{\pi}{2} + \sin x \sin\frac{\pi}{2} = \cos x \cdot 0 + \sin x \cdot 1 = \sin x,$$

$$\underset{\substack{\uparrow \\ \text{zero}}}{}\qquad\qquad \underset{\substack{\uparrow \\ \text{by (2) of Theorem 4.5.1}}}{}$$

$$\cos x = \cos\left(\frac{\pi}{2} - \frac{\pi}{2} + x\right) = \cos\left(\frac{\pi}{2} - \left(\frac{\pi}{2} - x\right)\right) = \sin\left(\frac{\pi}{2} - x\right) = -\sin\left(x - \frac{\pi}{2}\right).$$

Now from the first equation in (3), the sine of the sum $x_1 + x_2$ can be written

$$\begin{aligned}
\sin(x_1 + x_2) &= \cos\left((x_1 + x_2) - \frac{\pi}{2}\right) \\
&= \cos\left(x_1 + \left(x_2 - \frac{\pi}{2}\right)\right) \\
&= \cos x_1 \cos\left(x_2 - \frac{\pi}{2}\right) - \sin x_1 \sin\left(x_2 - \frac{\pi}{2}\right) \qquad \leftarrow \text{by (1) of Theorem 4.5.1} \\
&= \cos x_1 \sin x_2 - \sin x_1 (-\cos x_2). \qquad\qquad\qquad \leftarrow \text{by (3)}
\end{aligned}$$

The last line is traditionally written as

$$\sin(x_1 + x_2) = \sin x_1 \cos x_2 + \cos x_1 \sin x_2.$$

To obtain the sine of the difference $x_1 - x_2$, we use again $\cos(-x_2) = \cos x_2$ and $\sin(-x_2) = -\sin x_2$:

$$\begin{aligned}
\sin(x_1 - x_2) = \sin(x_1 + (-x_2)) &= \sin x_1 \cos(-x_2) + \cos x_1 \sin(-x_2) \\
&= \sin x_1 \cos x_2 - \cos x_1 \sin x_2.
\end{aligned}$$

THEOREM 4.5.2   Sum and Difference Formulas for the Sine

For all real numbers x_1 and x_2,

$$\sin(x_1 + x_2) = \sin x_1 \cos x_2 + \cos x_1 \sin x_2 \tag{4}$$
$$\sin(x_1 - x_2) = \sin x_1 \cos x_2 - \cos x_1 \sin x_2 \tag{5}$$

■ **EXAMPLE 2**   **Sine of a Sum**

Evaluate $\sin(7\pi/12)$.

Solution We proceed as in Example 1, except we use the sum formula (4) of Theorem 4.5.2:

$$\sin\frac{7\pi}{12} = \sin\left(\frac{\pi}{3} + \frac{\pi}{4}\right) = \sin\frac{\pi}{3}\cos\frac{\pi}{4} + \cos\frac{\pi}{3}\sin\frac{\pi}{4}$$

$$= \frac{\sqrt{3}}{2}\frac{\sqrt{2}}{2} + \frac{1}{2}\frac{\sqrt{2}}{2} = \frac{\sqrt{2}}{4}(1 + \sqrt{3}).$$

As in Example 1, the result can be rewritten as $\sin(7\pi/12) = (\sqrt{2} + \sqrt{6})/4$.   ■

Since we know the value of $\cos(7\pi/12)$ from Example 1, we can also compute the value of $\sin(7\pi/12)$ using the Pythagorean identity (8) of Section 4.4:

$$\sin^2\frac{7\pi}{12} + \cos^2\frac{7\pi}{12} = 1.$$

We solve for $\sin(7\pi/12)$ and take the positive square root:

$$\sin\frac{7\pi}{12} = \sqrt{1 - \cos^2\frac{7\pi}{12}} = \sqrt{1 - \left[\frac{\sqrt{2}}{4}\left(1 - \sqrt{3}\right)\right]^2}$$

$$= \sqrt{\frac{4 + 2\sqrt{3}}{8}} = \frac{\sqrt{2 + \sqrt{3}}}{2}. \tag{6}$$

Although the number in (6) does not look like the result obtained in Example 2, the values are the same. See Problem 77 in Exercises 4.5.

There are sum and difference formulas for the tangent function as well. We can derive the sum formula using the sum formulas for the sine and cosine as follows:

$$\tan(x_1 + x_2) = \frac{\sin(x_1 + x_2)}{\cos(x_1 + x_2)} = \frac{\sin x_1\cos x_2 + \cos x_1\sin x_2}{\cos x_1\cos x_2 - \sin x_1\sin x_2}. \tag{7}$$

We now divide the numerator and denominator of (7) by $\cos x_1\cos x_2$ (assuming that x_1 and x_2 are such that $\cos x_1\cos x_2 \neq 0$),

$$\tan(x_1 + x_2) = \frac{\dfrac{\sin x_1}{\cos x_1}\dfrac{\cos x_2}{\cos x_2} + \dfrac{\cos x_1}{\cos x_1}\dfrac{\sin x_2}{\cos x_2}}{\dfrac{\cos x_1}{\cos x_1}\dfrac{\cos x_2}{\cos x_2} - \dfrac{\sin x_1}{\cos x_1}\dfrac{\sin x_2}{\cos x_2}} = \frac{\tan x_1 + \tan x_2}{1 - \tan x_1\tan x_2}. \tag{8}$$

The derivation of the difference formula for $\tan(x_1 - x_2)$ is obtained in a similar manner. We summarize the two results.

THEOREM 4.5.3   Sum and Difference Formulas for the Tangent

For real numbers x_1 and x_2 for which the functions are defined,

$$\tan(x_1 + x_2) = \frac{\tan x_1 + \tan x_2}{1 - \tan x_1\tan x_2} \tag{9}$$

$$\tan(x_1 - x_2) = \frac{\tan x_1 - \tan x_2}{1 + \tan x_1\tan x_2} \tag{10}$$

EXAMPLE 3   **Tangent of a Difference**

Evaluate $\tan(\pi/12)$.

Solution If we think of $\pi/12$ as an angle in radians, then

$$\frac{\pi}{12}\text{ radians} = 15° = 45° - 30° = \frac{\pi}{4} - \frac{\pi}{6}\text{ radians.}$$

It follows from formula (10) of Theorem 4.5.3:

You should rework this example using ▶

$\pi/12 = \pi/3 - \pi/4$

to see that the result is the same.

$$\tan\frac{\pi}{12} = \tan\left(\frac{\pi}{4} - \frac{\pi}{6}\right) = \frac{\tan\dfrac{\pi}{4} - \tan\dfrac{\pi}{6}}{1 + \tan\dfrac{\pi}{4}\tan\dfrac{\pi}{6}}$$

$$= \frac{1 - \frac{1}{\sqrt{3}}}{1 + 1\cdot\frac{1}{\sqrt{3}}} = \frac{\sqrt{3} - 1}{\sqrt{3} + 1}$$

$$= \frac{\sqrt{3} - 1}{\sqrt{3} + 1} \cdot \frac{\sqrt{3} - 1}{\sqrt{3} - 1} \quad \leftarrow \text{rationalizing the denominator}$$

$$= \frac{(\sqrt{3} - 1)^2}{2} = \frac{4 - 2\sqrt{3}}{2} = 2 - \sqrt{3}.$$

Strictly speaking, we really do not need the identities for $\tan(x_1 \pm x_2)$, since we can always compute $\sin(x_1 \pm x_2)$ and $\cos(x_1 \pm x_2)$ using (1), (2), (4), (5) and then proceed as in (7), that is, form the quotient $\sin(x_1 \pm x_2)/\cos(x_1 \pm x_2)$.

☐ **Double-Angle Formulas** Many useful trigonometric formulas can be derived from the sum and difference formulas. The **double-angle formulas** for the cosine and sine functions express the cosine and sine of $2x$ in terms of the cosine and sine of x.

If we set $x_1 = x_2 = x$ in (1) and use $\cos(x + x) = \cos 2x$, then

$$\cos 2x = \cos x\cos x - \sin x\sin x = \cos^2 x - \sin^2 x.$$

Similarly, by setting $x_1 = x_2 = x$ in (4) and using $\sin(x + x) = \sin 2x$, then

these two terms are equal
$$\downarrow \qquad\qquad \downarrow$$
$$\sin 2x = \sin x\cos x + \cos x\sin x = 2\sin x\cos x.$$

We summarize the last two results along with the double-angle formula for the tangent function.

THEOREM 4.5.4   Double-Angle Formulas

For a real number x for which the functions are defined,

$$\cos 2x = \cos^2 x - \sin^2 x \tag{11}$$
$$\sin 2x = 2\sin x\cos x \tag{12}$$
$$\tan 2x = \frac{2\tan x}{1 - \tan^2 x} \tag{13}$$

EXAMPLE 4   **Using the Double-Angle Formulas**

If $\sin x = -\frac{1}{4}$ and $\pi < x < 3\pi/2$, find the exact values of $\cos 2x$ and $\sin 2x$.

Solution First, we compute $\cos x$ using $\sin^2 x + \cos^2 x = 1$. Since $\pi < x < 3\pi/2$, $\cos x < 0$, and so we choose the negative square root:

$$\cos x = -\sqrt{1 - \sin^2 x} = -\sqrt{1 - \left(-\frac{1}{4}\right)^2} = -\frac{\sqrt{15}}{4}.$$

From the double-angle formula (11) of Theorem 4.5.4,

$$cos 2x = cos^2 x - sin^2 x$$

$$= \left(-\frac{\sqrt{15}}{4}\right)^2 - \left(-\frac{1}{4}\right)^2$$

$$= \frac{15}{16} - \frac{1}{16} = \frac{14}{16} = \frac{7}{8}.$$

Finally, from the double-angle formula (12),

$$sin 2x = 2 sin x cos x = 2\left(-\frac{1}{4}\right)\left(-\frac{\sqrt{15}}{4}\right) = \frac{\sqrt{15}}{8}.$$ ▬

The formula in (11) has two useful alternative forms. By (8) of Section 4.4, we know that $sin^2 x = 1 - cos^2 x$. Substituting the last expression into (11) yields $cos 2x = cos^2 x - (1 - cos^2 x)$ or

$$cos 2x = 2 cos^2 x - 1. \tag{14}$$

On the other hand, if we substitute $cos^2 x = 1 - sin^2 x$ into (11) we get

$$cos 2x = 1 - 2 sin^2 x. \tag{15}$$

☐ **Half-Angle Formulas** The alternative forms (14) and (15) of the double-angle formula (11) are the source of two **half-angle formulas**. Solving (14) and (15) for $cos^2 x$ and $sin^2 x$ gives, respectively,

$$cos^2 x = \frac{1}{2}(1 + cos 2x) \quad \text{and} \quad sin^2 x = \frac{1}{2}(1 - cos 2x). \tag{16}$$

By replacing the symbol x in (16) by $x/2$ and using $2(x/2) = x$, we obtain the half-angle formulas for the cosine and sine functions.

THEOREM 4.5.5   Half-Angle Formulas

For a real number x for which the functions are defined,

$$cos^2 \frac{x}{2} = \frac{1}{2}(1 + cos x) \tag{17}$$

$$sin^2 \frac{x}{2} = \frac{1}{2}(1 - cos x) \tag{18}$$

$$tan^2 \frac{x}{2} = \frac{1 - cos x}{1 + cos x} \tag{19}$$

EXAMPLE 5   **Using the Half-Angle Formulas**

Find the exact values of $cos(5\pi/8)$ and $sin(5\pi/8)$.

Solution If we let $x = 5\pi/4$, then $x/2 = 5\pi/8$ and formulas (17) and (18) of Theorem 4.5.5 yield, respectively,

$$cos^2 \frac{5\pi}{8} = \frac{1}{2}\left(1 + cos\frac{5\pi}{4}\right) = \frac{1}{2}\left[1 + \left(-\frac{\sqrt{2}}{2}\right)\right] = \frac{2 - \sqrt{2}}{4},$$

and

$$sin^2 \frac{5\pi}{8} = \frac{1}{2}\left(1 - cos\frac{5\pi}{4}\right) = \frac{1}{2}\left[1 - \left(-\frac{\sqrt{2}}{2}\right)\right] = \frac{2 + \sqrt{2}}{4}.$$

Because $5\pi/8$ radians is a second-quadrant angle, $\cos(5\pi/8) < 0$ and $\sin(5\pi/8) > 0$. Therefore, we take the negative square root for the value of the cosine,

$$\cos\frac{5\pi}{8} = -\sqrt{\frac{2-\sqrt{2}}{4}} = -\frac{\sqrt{2-\sqrt{2}}}{2},$$

and the positive square root for the value of the sine,

$$\sin\frac{5\pi}{8} = \sqrt{\frac{2+\sqrt{2}}{4}} = \frac{\sqrt{2+\sqrt{2}}}{2}.$$

If want the exact value of, say, $\tan(5\pi/8)$ we can use the results of Example 5 or formula (19) with $x = 5\pi/4$. Either way, the result is the same

$$\tan\frac{5\pi}{8} = -\frac{\sqrt{2+\sqrt{2}}}{\sqrt{2-\sqrt{2}}} = -1 - \sqrt{2}. \qquad \leftarrow\begin{cases}\text{rationalizing}\\\text{the denominator}\end{cases}$$

☐ **Reducing Powers of Sine and Cosine** As discussed in Section 3.7, the principal topics of study in calculus are *derivatives* and *integrals* of functions. Trigonometric identities are especially useful in integral calculus. To evaluate integrals of powers of the sine and cosine, specifically $\cos^n x$ and $\sin^n x$, where $n \geq 2$ is an even positive integer, you would use the half-angle formulas in the form given in (16). The idea is to express $\cos^n x$ and $\sin^n x$ in terms of one or more of cosine functions raised to the first power. The next example illustrates the idea.

EXAMPLE 6 Reducing a Power

Express $\sin^4 x$ in terms of first-powers of cosine functions.

Solution We first use the laws of exponents to rewrite $\sin^4 x$ as $(\sin^2 x)^2$ and then use the second identity in (16) to replace $\sin^2 x$ in terms of $\cos 2x$:

$$\sin^4 x = (\sin^2 x)^2 \qquad \leftarrow\begin{cases}\text{replace } \sin^2 x \text{ using}\\\text{second formula in (16)}\end{cases}$$

$$= \left[\frac{1}{2}(1 - \cos 2x)\right]^2 \qquad \leftarrow \text{expand}$$

$$= \frac{1}{4} - \frac{1}{2}\cos 2x + \frac{1}{4}\cos^2 2x \qquad \leftarrow\begin{cases}\text{now replace } \cos^2 2x \text{ using}\\\text{the first formula in (16)}\\\text{with } x \text{ replaced by } 2x\end{cases}$$

$$= \frac{1}{4} - \frac{1}{2}\cos 2x + \frac{1}{4}\left[\frac{1}{2}(1 + \cos 4x)\right]$$

$$= \frac{3}{8} - \frac{1}{2}\cos 2x + \frac{1}{8}\cos 4x.$$

As required, the last line contains only first-powers of cosine functions of multiples of the original independent variable x.

NOTES FROM THE CLASSROOM

(*i*) Should you memorize all the identities presented in this section? You should consult your instructor about this, but in the opinion of the authors, you should at the very least memorize formulas (1), (2), (4), (5), (11), and (12).

(*ii*) When you enroll in a calculus course, check the title of your text. If it has the words *Early Transcendentals* in its title, then your knowledge of the graphs and properties of the trigonometric functions will come into play almost immediately.

(*iii*) At some point in your study of integral calculus you may be required to evaluate integrals of products such as

$$\sin 2x \sin 5x \qquad \text{and} \qquad \sin 10x \cos 4x.$$

One way of doing this is to use the sum/difference formulas to devise an identity that converts these products into either a sum of sines or a sum of cosines. This will be the topic of discussion in the next section.

Exercises 4.5   Answers to selected odd-numbered problems begin on page ANS–16.

In Problems 1–22, use a sum or difference formula to find the exact value of the given trigonometric function. Do not use a calculator.

1. $\cos \dfrac{\pi}{12}$

2. $\sin \dfrac{\pi}{12}$

3. $\sin 75°$

4. $\cos 75°$

5. $\sin \dfrac{17\pi}{12}$

6. $\cos \dfrac{11\pi}{12}$

7. $\tan \dfrac{5\pi}{12}$

8. $\cos\left(-\dfrac{5\pi}{12}\right)$

9. $\sin\left(-\dfrac{\pi}{12}\right)$

10. $\tan \dfrac{11\pi}{12}$

11. $\sin \dfrac{11\pi}{12}$

12. $\tan \dfrac{7\pi}{12}$

13. $\cos 165°$

14. $\sin 165°$

15. $\tan 165°$

16. $\cos 195°$

17. $\sin 195°$

18. $\tan 195°$

19. $\cos 345°$

20. $\sin 345°$

21. $\cos \dfrac{13\pi}{12}$

22. $\tan \dfrac{17\pi}{12}$

In Problems 23–28, use a double-angle formula to write the given expression as a single trigonometric function of twice the angle.

23. $2\cos\beta \sin\beta$

24. $\cos^2 2t - \sin^2 2t$

25. $1 - 2\sin^2 \dfrac{\pi}{5}$

26. $2\cos^2 \dfrac{19}{2}x - 1$

27. $\dfrac{\tan 3t}{1 - \tan^2 3t}$

28. $2\sin \dfrac{y}{2}\cos \dfrac{y}{2}$

In Problems 29–34, use the given information to find (**a**) $\cos 2x$, (**b**) $\sin 2x$, and (**c**) $\tan 2x$.

29. $\sin x = \sqrt{2}/3, \quad \pi/2 < x < \pi$

30. $\cos x = \sqrt{3}/5, \quad 3\pi/2 < x < 2\pi$

31. $\tan x = \frac{1}{2}, \quad \pi < x < 3\pi/2$

32. $\csc x = -3, \quad \pi < x < 3\pi/2$

33. $\sec x = -\frac{13}{5}, \quad \pi/2 < x < \pi$

34. $\cot x = \frac{4}{3}, \quad 0 < x < \pi/2$

In Problems 35–44, use a half-angle formula to find the exact value of the given trigonometric function. Do not use a calculator.

35. $\cos(\pi/12)$

36. $\sin(\pi/8)$

37. $\sin(3\pi/8)$

38. $\tan(\pi/12)$

39. $\cos 67.5°$

40. $\sin 15°$

4.5 Sum and Difference Formulas

41. $\tan 105°$

42. $\cot 157.5°$

43. $\csc(13\pi/12)$

44. $\sec(-3\pi/8)$

In Problems 45–50, use the given information to find (**a**) $\cos(x/2)$, (**b**) $\sin(x/2)$, and (**c**) $\tan(x/2)$.

45. $\sin x = \frac{12}{13}$, $\pi/2 < x < \pi$

46. $\cos x = \frac{4}{5}$, $3\pi/2 < x < 2\pi$

47. $\tan x = 2$, $\pi < x < 3\pi/2$

48. $\csc x = 9$, $0 < x < \pi/2$

49. $\sec x = \frac{3}{2}$, $0 < x < 90°$

50. $\cot x = -\frac{1}{4}$, $90° < x < 180°$

In Problems 51–60, verify the given identity.

51. $\sin 4x = 4\cos x(\sin x - 2\sin^3 x)$

52. $\cos 3x = 4\cos^3 x - 3\cos x$

53. $(\sin x + \cos x)^2 = 1 + \sin 2x$

54. $\cos 2x = \cos^4 x - \sin^4 x$

55. $\cot 2x = \frac{1}{2}(\cot x - \tan x)$

56. $\sec 2x = \dfrac{1}{2\cos^2 x - 1}$

57. $\dfrac{2\tan x}{1 + \tan^2 x} = \sin 2x$

58. $\dfrac{\cot x - \tan x}{\cot x + \tan x} = \cos 2x$

59. $\tan\dfrac{x}{2} = \dfrac{1 - \cos x}{\sin x}$

60. $\tan\dfrac{x}{2} = \dfrac{\sin x}{1 + \cos x}$

In Problems 61–64, proceed as in Example 6 and use (16) to rewrite the given function in terms of first powers of cosine functions.

61. $\sin^2 5x$

62. $\cos^2 2x$

63. $\cos^4 x$

64. $\sin^4 3x$

In Problems 65 and 66, proceed as in Problems 61–64 to rewrite the given function in terms of first powers of cosine functions. You will also need (7) of Section 1.5 and the identity in Problem 52 written in the form

$$\cos^3 x = \frac{3}{4}\cos x + \frac{1}{4}\cos 3x.$$

65. $\sin^6 x$

66. $\cos^6 \frac{1}{2}x$

In Problems 67–70, rewrite the given function as a single trigonometric function involving no products or squares. Give the amplitude and the period of the function.

67. $y = 4\cos^2 x - 2$

68. $y = \sin(x/2)\cos(x/2)$

69. $y = 2\sin 2x\cos 2x$

70. $y = 5\cos^2 4x - 5\sin^2 4x$

71. If $P(x_1)$ and $P(x_2)$ are points in quadrant II on the terminal side of the angles x_1 and x_2, respectively, with $\cos x_1 = -\frac{1}{3}$ and $\sin x_2 = \frac{2}{3}$, find (**a**) $\sin(x_1 + x_2)$, (**b**) $\cos(x_1 + x_2)$, (**c**) $\sin(x_1 - x_2)$, and (**d**) $\cos(x_1 - x_2)$.

72. If x_1 is a quadrant II angle, x_2 is a quadrant III angle, $\sin x_1 = \frac{8}{17}$, and $\tan x_2 = \frac{3}{4}$, find (**a**) $\sin(x_1 + x_2)$, (**b**) $\sin(x_1 - x_2)$, (**c**) $\cos(x_1 + x_2)$, and (**d**) $\cos(x_1 - x_2)$.

Applications

73. Mach Number The ratio of the speed of an airplane to the speed of sound is called the Mach number M of the plane. If $M > 1$, the plane makes sound waves that form a (moving) cone, as shown in FIGURE 4.5.2. A sonic boom is heard at the intersection of the cone with the ground. If the vertex angle of the cone is θ, then

$$\sin\frac{\theta}{2} = \frac{1}{M}.$$

If $\theta = \pi/6$, find the exact value of the Mach number.

FIGURE 4.5.2 Airplane in Problem 73

CHAPTER 4 TRIGONOMETRIC FUNCTIONS

74. **Cardiovascular Branching** A mathematical model for blood flow in a large blood vessel predicts that the optimal values of the angles θ_1 and θ_2, which represent the (positive) angles of the smaller daughter branches (vessels) with respect to the axis of the parent branch, are given by

$$\cos\theta_1 = \frac{A_0^2 + A_1^2 - A_2^2}{2A_0A_1} \quad \text{and} \quad \cos\theta_2 = \frac{A_0^2 - A_1^2 + A_2^2}{2A_0A_2},$$

where A_0 is the cross-sectional area of the parent branch and A_1 and A_2 are the cross-sectional areas of the daughter branches. See FIGURE 4.5.3. Let $\psi = \theta_1 + \theta_2$ be the junction angle, as shown in the figure.

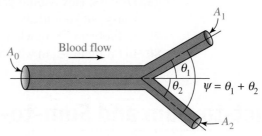

FIGURE 4.5.3 Branching of a large blood vessel in Problem 74

(a) Show that

$$\cos\psi = \frac{A_0^2 - A_1^2 - A_2^2}{2A_1A_2}.$$

(b) Show that for the optimal values of θ_1 and θ_2, the cross-sectional area of the daughter branches, $A_1 + A_2$, is greater than or equal to that of the parent branch. Therefore, the blood must slow down in the daughter branches.

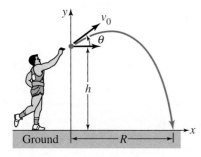

FIGURE 4.5.4 Projectile in Problem 75

75. **Range of a Projectile** We saw in Problem 56 of Exercises 4.2 that if a projectile, such as a shot put, is released from a height h, upward at an angle θ with velocity v_0, the range R at which it strikes the ground is given by

$$R = \frac{v_0\cos\theta}{g}\left(v_0\sin\theta + \sqrt{v_0^2\sin^2\theta + 2gh}\right),$$

where g is the acceleration due to gravity. See FIGURE 4.5.4.

(a) Show that when $h = 0$ the range of the projectile is

$$R = \frac{v_0^2\sin 2\theta}{g}.$$

(b) It can be shown that the maximum range $R_{\max}$ is achieved when the angle θ satisfies the equation

$$\cos 2\theta = \frac{gh}{v_0^2 + gh}.$$

Show that maximum range is

$$R_{\max} = \frac{v_0\sqrt{v_0^2 + 2gh}}{g}$$

by using the expressions for R and $\cos 2\theta$ and the half-angle formulas for the sine and the cosine with $t = 2\theta$.

76. Discuss: Why would you expect to get an error message from your calculator when you try to evaluate

$$\frac{\tan 35° + \tan 55°}{1 - \tan 35°\tan 55°}?$$

77. In Example 2 we showed that $\sin(7\pi/12) = \frac{1}{4}(\sqrt{2} + \sqrt{6})$. Following the example, we then showed that $\sin(7\pi/12) = \frac{1}{2}\sqrt{2 + \sqrt{3}}$. Demonstrate that these two answers are equivalent.

78. Discuss: How would you find a formula that expresses $\sin 3\theta$ in terms of $\sin\theta$? Carry out your ideas.

79. In Problem 71, in what quadrants do $P(x_1 + x_2)$ and $P(x_1 - x_2)$ lie?

80. In Problem 72, in which quadrant does the terminal side of $x_1 + x_2$ lie? The terminal side of $x_1 - x_2$?

4.6 Product-to-Sum and Sum-to-Product Formulas

INTRODUCTION There are instances, especially in integral calculus, where it is necessary to convert a product of sine and cosine functions to a sum of these functions. Moreover, in solving trigonometric equations we may find it convenient to convert a sum of sine and cosine functions into a product of these functions. In the discussion that follows we establish trigonometric identities or formulas that do the job.

☐ **Reduction of a Product to a Sum** The **product-to-sum formulas** given in the next theorem are direct consequences of the sum and difference formulas for the cosine and sine functions in Section 4.5.

THEOREM 4.6.1 Product-to-Sum Formulas

For all real numbers x_1 and x_2,

$$\sin x_1 \sin x_2 = \tfrac{1}{2}[\cos(x_1 - x_2) - \cos(x_1 + x_2)] \qquad (1)$$
$$\cos x_1 \cos x_2 = \tfrac{1}{2}[\cos(x_1 - x_2) + \cos(x_1 + x_2)] \qquad (2)$$
$$\sin x_1 \cos x_2 = \tfrac{1}{2}[\sin(x_1 + x_2) + \sin(x_1 - x_2)] \qquad (3)$$

PROOF: To prove (1), we use (1) and (2) of Theorem 4.5.1:

$$\cos(x_1 - x_2) = \cos x_1 \cos x_2 + \sin x_1 \sin x_2 \qquad (4)$$
$$\cos(x_1 + x_2) = \cos x_1 \cos x_2 - \sin x_1 \sin x_2. \qquad (5)$$

Subtracting (5) from (4) yields

$$\cos(x_1 - x_2) - \cos(x_1 + x_2) = 2\sin x_1 \sin x_2.$$

And so,

$$\sin x_1 \sin x_2 = \tfrac{1}{2}[\cos(x_1 - x_2) - \cos(x_1 + x_2)]$$

which is (1). Similarly, by adding (4) and (5) we get

$$\cos(x_1 - x_2) + \cos(x_1 + x_2) = 2\cos x_1 \cos x_2,$$

which in turn, yields formula (2):

$$\cos x_1 \cos x_2 = \tfrac{1}{2}[\cos(x_1 - x_2) + \cos(x_1 + x_2)].$$

Formula (3) follows analogously by adding the sum and difference formulas for the sine, (4) and (5) in Theorem 4.5.2 of Section 4.5. ▄

Although we do not feel that it is necessary to memorize (1)–(3) in Theorem 4.6.1, you should listen to what your instructor requires. By remembering the *procedure* just illustrated in the proof of Theorem 4.6.1 each of these formulas can be derived on the spot.

EXAMPLE 1 Using (2) of Theorem 4.6.1

Use a product-to-sum formula to rewrite the product $\cos 2\theta \cos 3\theta$ as a sum.

Solution From formula (2) of Theorem 4.6.1 with the identifications $x_1 = 2\theta$ and $x_2 = 3\theta$, we obtain

$$\begin{aligned}
\cos 2\theta \cos 3\theta &= \tfrac{1}{2}[\cos(2\theta - 3\theta) + \cos(2\theta + 3\theta)] \\
&= \tfrac{1}{2}[\cos(-\theta) + \cos 5\theta] \quad \leftarrow \cos(-\theta) = \cos\theta \\
&= \tfrac{1}{2}[\cos\theta + \cos 5\theta].
\end{aligned}$$

EXAMPLE 2 Using (3) of Theorem 4.6.1

Use a product-to-sum formula to find the exact value of the product $\sin 45° \cos 15°$.

Solution Using formula (3) of Theorem 4.6.1 with $x_1 = 45°$ and $x_2 = 15°$, we have

$$\begin{aligned}
\sin 45° \cos 15° &= \tfrac{1}{2}[\sin(45° + 15°) + \sin(45° - 15°)] \\
&= \tfrac{1}{2}[\sin 60° + \sin 30°].
\end{aligned}$$

Because $\sin 60° = \tfrac{1}{2}\sqrt{3}$ and $\sin 30° = \tfrac{1}{2}$ we observe that the exact value of the given product is

$$\sin 45° \cos 15° = \tfrac{1}{2}[\sin 60° + \sin 30°] = \tfrac{1}{2}\left(\tfrac{1}{2}\sqrt{3} + \tfrac{1}{2}\right) = \tfrac{1}{4}\left(\sqrt{3} + 1\right).$$

☐ **Reduction of a Sum to a Product** The results in Theorem 4.6.1 can now be used to derive the **sum-to-product formulas**.

THEOREM 4.6.2 Sum-to-Product Formulas

For all real numbers x_1 and x_2,

$$\sin x_1 + \sin x_2 = 2\sin\frac{x_1 + x_2}{2}\cos\frac{x_1 - x_2}{2} \qquad (6)$$

$$\sin x_1 - \sin x_2 = 2\cos\frac{x_1 + x_2}{2}\sin\frac{x_1 - x_2}{2} \qquad (7)$$

$$\cos x_1 + \cos x_2 = 2\cos\frac{x_1 + x_2}{2}\cos\frac{x_1 - x_2}{2} \qquad (8)$$

$$\cos x_1 - \cos x_2 = -2\sin\frac{x_1 + x_2}{2}\sin\frac{x_1 - x_2}{2} \qquad (9)$$

PROOF: By replacing, in turn, the symbols x_1 and x_2 in (1) of Theorem 4.6.1 by

$$\frac{x_1 + x_2}{2} \qquad \text{and} \qquad \frac{x_1 - x_2}{2}, \qquad (10)$$

we get

$$\sin\frac{x_1 + x_2}{2}\sin\frac{x_1 - x_2}{2} = \frac{1}{2}\left[\cos\left(\frac{x_1 + x_2}{2} - \frac{x_1 - x_2}{2}\right) - \cos\left(\frac{x_1 + x_2}{2} + \frac{x_1 - x_2}{2}\right)\right]$$

$$= \frac{1}{2}[\cos x_2 - \cos x_1].$$

Multiplying the last expression by -2 yields

$$-2\sin\frac{x_1 + x_2}{2}\sin\frac{x_1 - x_2}{2} = \cos x_1 - \cos x_2,$$

which is formula (9). In a similar manner, each of the remaining product-to-sum formulas together with the substitutions in (10) yields one of the sum-to-product formulas. ▄

EXAMPLE 3   Using (9) of Theorem 4.6.2

Use a sum-to-product formula to rewrite the sum $\cos t - \cos 5t$ as a product.

Solution We use formula (9) of Theorem 4.6.2 with $x_1 = t$ and $x_2 = 5t$:

$$\cos t - \cos 5t = -2\sin\left(\frac{t + 5t}{2}\right)\sin\left(\frac{t - 5t}{2}\right)$$

$$= -2\sin 3t \sin(-2t) \quad \leftarrow \sin(-2t) = -\sin 2t$$

$$= 2\sin 3t \sin 2t$$

EXAMPLE 4   Using (6) of Theorem 4.6.2

Use a sum-to-product formula to find the exact value of the sum $\sin 75° + \sin 15°$.

Solution In this case we use formula (6) of Theorem 4.6.2 with $x_1 = 75°$ and $x_2 = 15°$:

$$\sin 75° + \sin 15° = 2\sin\left(\frac{75° + 15°}{2}\right)\cos\left(\frac{75° - 15°}{2}\right)$$

$$= 2\sin 45° \cos 30°.$$

Because $\sin 45° = \frac{1}{2}\sqrt{2}$ and $\cos 30° = \frac{1}{2}\sqrt{3}$ the exact value of the given sum is

$$\sin 75° + \sin 15° = 2\sin 45° \cos 30° = 2\left(\tfrac{1}{2}\sqrt{2}\right)\left(\tfrac{1}{2}\sqrt{3}\right) = \tfrac{1}{2}\sqrt{6}. \quad ▄$$

Exercises 4.6   Answers to selected odd-numbered problems begin on page ANS–17.

In Problems 1–12, use a product-to-sum formula in Theorem 4.6.1 to write the given product as a sum of cosines or a sum of sines.

1. $\cos 4\theta \cos 3\theta$

2. $\sin\dfrac{3t}{2}\cos\dfrac{t}{2}$

3. $\sin 2x \sin 5x$

4. $\sin 10x \cos 4x$

5. $\cos\dfrac{4x}{3}\cos\dfrac{x}{3}$

6. $-\sin t \sin 2t$

7. $\sin 8x \cos 12x$

8. $\sin \pi\theta \cos 7\pi\theta$

9. $2\cos 3\beta \sin \beta$

10. $6\sin \alpha \sin 4\alpha$

11. $2\sin\left(x + \dfrac{\pi}{4}\right)\sin\left(x - \dfrac{\pi}{4}\right)$

12. $2\sin\left(t + \dfrac{\pi}{2}\right)\cos\left(t - \dfrac{\pi}{2}\right)$

In Problems 13–18, use a product-to-sum formula in Theorem 4.6.1 to find the exact value of the given expression. Do not use a calculator.

13. $\cos\dfrac{5\pi}{12}\sin\dfrac{\pi}{12}$

14. $\sin\dfrac{5\pi}{8}\cos\dfrac{\pi}{8}$

15. $\sin 75° \sin 15°$

16. $\cos 15° \cos 45°$

17. $\sin 97.5° \sin 52.5°$

18. $\sin 105° \cos 195°$

In Problems 19–30, use a sum-to-product-formula in Theorem 4.6.2 to write the given sum as a product of cosines, a product of sines, or a product of a sine and a cosine.

19. $\sin y - \sin 5y$

20. $\cos 3\theta - \cos\theta$

21. $\cos\dfrac{9x}{2} - \cos\dfrac{x}{2}$

22. $\sin\dfrac{x}{2} - \sin\dfrac{3x}{2}$

23. $\cos 2x + \cos 6x$

24. $\sin 5t + \sin 3t$

25. $\sin \omega_1 t + \sin \omega_2 t$

26. $\frac{1}{2}(\cos 2\alpha + \cos 2\beta)$

27. $-\frac{1}{2}\cos t + \frac{1}{2}\cos 5t$

28. $\sin(\theta + \pi) + \sin(\theta - \pi)$

29. $\sin\left(t + \dfrac{\pi}{2}\right) + \sin\left(t - \dfrac{\pi}{2}\right)$

30. $\cos\left(t + \dfrac{\pi}{2}\right) - \cos\left(t - \dfrac{\pi}{2}\right)$

In Problems 31–36, use a sum-to-product-formula in Theorem 4.6.2 to find the exact value of the given expression. Do not use a calculator.

31. $\sqrt{2}\sin\dfrac{13\pi}{12} + \sqrt{2}\sin\dfrac{5\pi}{12}$

32. $\sin\dfrac{\pi}{12} - \sin\dfrac{5\pi}{12}$

33. $\cos 105° - \cos 15°$

34. $\cos 15° + \cos 75°$

35. $\sin 195° + \sin 105°$

36. $2\cos 195° - 2\cos 105°$

Applications

37. Sound Wave A note produced by a certain musical instrument results in a sound wave described by

$$f(t) = 0.03\sin 500\pi t + 0.03\sin 1000\pi t,$$

where $f(t)$ is the difference between atmospheric pressure and air pressure in dynes per square centimeter at the eardrum after t seconds. Express f as the product of a sine and a cosine function.

38. Beats If two piano wires struck by the same key are slightly out of tune, the difference between the atmospheric pressure and air pressure at the eardrum can be represented by the function

$$f(t) = a\cos 2\pi b_1 t + a\cos 2\pi b_2 t,$$

where the value of the constant b_1 is close to the value of constant b_2. The variations in loudness that occur are called **beats**. See FIGURE 4.6.1. The two strings can be tuned to the same frequency by tightening one of them while sounding both until the beats disappear.

(a) Use a sum formula to write $f(t)$ as a product.

(b) Show that $f(t)$ can be considered a cosine function with period $2/(b_1 + b_2)$ and variable amplitude $2a\cos\pi(b_1 - b_2)t$.

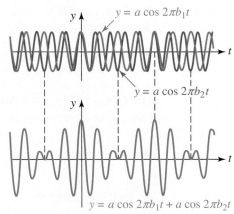

FIGURE 4.6.1 Graph for Problem 38

   (c) Use a graphing utility to obtain the graph of f in the case $2\pi b_1 = 5$, $2\pi b_2 = 4$, and $a = \frac{1}{2}$.
39. Alternating Current The term $\sin\omega t\sin(\omega t + \phi)$ is encountered in the derivation of an expression for the power in an alternating-current circuit. Show that this term can be written as $\frac{1}{2}[\cos\phi - \cos(2\omega t + \phi)]$.

For Discussion

40. If $x_1 + x_2 + x_3 = \pi$, then show that

$$\sin 2x_1 + \sin 2x_2 + \sin 2x_3 = 4\sin x_1 \sin x_2 \sin x_3.$$

41. Write as a product of cosines: $1 + \cos 2t + \cos 4t + \cos 6t$.
42. Simplify: $2\cos 2t\cos t - \cos 3t$.

4.7 Inverse Trigonometric Functions

INTRODUCTION Although we can find the values of the trigonometric functions of real numbers or angles, in many applications we must do the reverse: Given the value of a trigonometric function, find a corresponding angle or number. This suggests we consider inverse trigonometric functions. Before we define the inverse trigonometric functions, let's recall from Section 2.8 some of the properties of a one-to-one function f and its inverse f^{-1}.

Recall, a function f is one-to-one if every y ▶ in its range corresponds to exactly one x in its domain.

☐ **Properties of Inverse Functions** If $y = f(x)$ is a one-to-one function, then there is a unique inverse function f^{-1} with the following properties:

> ## Properties of Inverse Functions
>
> - The domain of $f^{-1} = $ range of f.
> - The range of $f^{-1} = $ domain of f.
> - $y = f(x)$ is equivalent to $x = f^{-1}(y)$.
> - The graphs of f and f^{-1} are reflections in the line $y = x$.
> - $f(f^{-1}(x)) = x$ for x in the domain of f^{-1}.
> - $f^{-1}(f(x)) = x$ for x in the domain of f.

Inspection of the graphs of the various trigonometric functions clearly shows that *none* of these functions are one-to-one. In Section 2.8 we discussed the fact that if a function f is not one-to-one, it may be possible to restrict the function to a portion of its domain where it is

See Example 8 in Section 2.8. ▶

one-to-one. Then we can define an inverse for *f* on that restricted domain. Normally, when we restrict the domain, we make sure to preserve the entire range of the original function.

☐ **Arcsine Function** From FIGURE 4.7.1(a) we see that the function $y = \sin x$ on the closed interval $[-\pi/2, \pi/2]$ takes on all values in its range $[-1, 1]$. Notice that any horizontal line drawn to intersect the red portion of the graph can do so at most once. Thus the sine function on this restricted domain is one-to-one and has an inverse. There are two commonly used notations to denote the inverse of the function shown in Figure 4.7.1(b):

$$\arcsin x \quad \text{or} \quad \sin^{-1}x,$$

and are read **arcsine of *x*** and **inverse sine of *x***, respectively.

(a) Not a one-to-one function   (b) A one-to-one function

FIGURE 4.7.1 Restricting the domain of $y = \sin x$ to produce a one-to-one function

In FIGURE 4.7.2(a) we have reflected the portion of the graph of $y = \sin x$ on the interval $[-\pi/2, \pi/2]$ (the red graph in Figure 4.7.1(b)) about the line $y = x$ to obtain the graph of $y = \arcsin x$ (in blue). For clarity, we have reproduced this blue graph in Figure 4.7.2(b). As this curve shows, the domain of the arcsine function is $[-1, 1]$ and the range is $[-\pi/2, \pi/2]$.

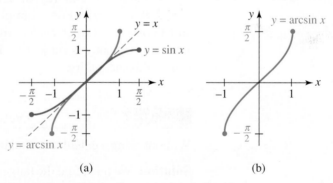

(a)   (b)

FIGURE 4.7.2 Graph of $y = \arcsin x$ is the blue curve

Proceeding as in (6) of Section 2.8, the inverse of $y = \sin x$, $-\pi/2 \le x \le \pi/2$, is obtained by interchanging the symbols *x* and *y*, that is, $y = \arcsin x$ is defined implicitly by

$$x = \sin y, \quad -\frac{\pi}{2} \le y \le \frac{\pi}{2}.$$

The following definition summarizes the discussion.

DEFINITION 4.7.1 Arcsine Function

The **arcsine function**, or **inverse sine function**, is defined by

$$y = \arcsin x \quad \text{if and only if} \quad x = \sin y \quad\quad (1)$$

where $-1 \le x \le 1$ and $-\pi/2 \le y \le \pi/2$.

In words:

The arcsine of the number x is that number y (or radian-measured angle) satisfying $-\pi/2 \leq y \leq \pi/2$ whose sine is x.

Note of Caution: ▶

$$(\sin x)^{-1} = \frac{1}{\sin x} \neq \sin^{-1} x$$

When using the notation $\sin^{-1} x$ it is important to realize that "−1" is not an exponent; rather, it denotes an inverse function. The notation arcsin x has an advantage over the notation $\sin^{-1} x$ in that there is no "−1" and hence no potential for misinterpretation; moreover, the prefix "arc" refers to an angle—*the* angle whose sine is x. But since $y = \arcsin x$ and $y = \sin^{-1} x$ are used interchangeably in calculus and in applications, we will continue to alternate their use so that you become comfortable with both notations.

■ EXAMPLE 1   Evaluating the Inverse Sine Function

Find **(a)** $\arcsin \frac{1}{2}$, **(b)** $\sin^{-1}\left(-\frac{1}{2}\right)$, and **(c)** $\sin^{-1}(-1)$.

Solution (a) If we let $y = \arcsin \frac{1}{2}$, then by (1) we must find the number y (or radian-measured angle) that satisfies $\sin y = \frac{1}{2}$ *and* $-\pi/2 \leq y \leq \pi/2$. Since $\sin(\pi/6) = \frac{1}{2}$ and $\pi/6$ satisfies the inequality $-\pi/2 \leq y \leq \pi/2$ it follows that $y = \pi/6$.

(b) If we let $y = \sin^{-1}\left(-\frac{1}{2}\right)$, then $\sin y = -\frac{1}{2}$. Since we must choose y such that $-\pi/2 \leq y \leq \pi/2$, we find that $y = -\pi/6$.

(c) Letting $y = \sin^{-1}(-1)$, we have that $\sin y = -1$ and $-\pi/2 \leq y \leq \pi/2$.

Hence $y = -\pi/2$. ■

Read this paragraph several times. ▶

In parts (b) and (c) of Example 1 we were careful to choose y so that $-\pi/2 \leq y \leq \pi/2$. For example, it is a common error to think that because $\sin(3\pi/2) = -1$, then necessarily $\sin^{-1}(-1)$ can be taken to be $3\pi/2$. Remember: If $y = \sin^{-1} x$, then y is subject to the restriction $-\pi/2 \leq y \leq \pi/2$ and $3\pi/2$ does not satisfy this inequality.

■ EXAMPLE 2   Evaluating a Composition

Without using a calculator, find $\tan\left(\sin^{-1}\frac{1}{4}\right)$.

Solution We must find the tangent of the angle of t radians with sine equal to $\frac{1}{4}$, that is, $\tan t$, where $t = \sin^{-1}\frac{1}{4}$. The angle t is shown in FIGURE 4.7.3. Since

$$\tan t = \frac{\sin t}{\cos t} = \frac{\frac{1}{4}}{\cos t},$$

we want to determine the value of $\cos t$. From Figure 4.7.3 and the Pythagorean identity $\sin^2 t + \cos^2 t = 1$, we see that

$$\left(\frac{1}{4}\right)^2 + \cos^2 t = 1 \qquad \text{or} \qquad \cos t = \frac{\sqrt{15}}{4}.$$

FIGURE 4.7.3 The angle $t = \sin^{-1}\frac{1}{4}$ in Example 2

Hence we have

$$\tan t = \frac{1/4}{\sqrt{15}/4} = \frac{1}{\sqrt{15}} = \frac{\sqrt{15}}{15},$$

and so

$$\tan\left(\sin^{-1}\frac{1}{4}\right) = \tan t = \frac{\sqrt{15}}{15}. \qquad ■$$

□ **Arccosine Function** If we restrict the domain of the cosine function to the closed interval $[0, \pi]$, the resulting function is one-to-one and thus has an inverse. We denote this inverse by

$$\arccos x \quad \text{or} \quad \cos^{-1}x.$$

By interchanging the symbols x and y in $y = \cos x$, $0 \le x \le \pi$, the inverse function $y = \arccos x$ is defined implicitly by

$$x = \cos y, \ 0 \le y \le \pi.$$

DEFINITION 4.7.2 Arccosine Function

The **arccosine function**, or **inverse cosine function**, is defined by

$$y = \arccos x \quad \text{if and only if} \quad x = \cos y \quad\quad (2)$$

where $-1 \le x \le 1$ and $0 \le y \le \pi$.

The graphs shown in FIGURE 4.7.4 illustrate how the function $y = \cos x$ restricted to the interval $[0, \pi]$ becomes a one-to-one function. The inverse of the function shown in Figure 4.7.4(b) is $y = \arccos x$.

(a) Not a one-to-one function

(b) A one-to-one function

FIGURE 4.7.4 Restricting the domain of $y = \cos x$ to produce a one-to-one function

By reflecting the graph of the one-to-one function in Figure 4.7.4(b) in the line $y = x$ we obtain the graph of $y = \arccos x$ shown in FIGURE 4.7.5. Note that the figure clearly shows that the domain and range of $y = \arccos x$ are $[-1, 1]$ and $[0, \pi]$, respectively.

FIGURE 4.7.5 Graph of $y = \arccos x$

▌**EXAMPLE 3**   **Evaluating the Inverse Cosine Function**

Find **(a)** $\arccos\left(\sqrt{2}/2\right)$ **(b)** $\cos^{-1}(-\sqrt{3}/2)$.

Solution (a) If we let $y = \arccos(\sqrt{2}/2)$, then $\cos y = \sqrt{2}/2$ and $0 \le y \le \pi$. Thus $y = \pi/4$.
(b) Letting $y = \cos^{-1}(-\sqrt{3}/2)$, we have that $\cos y = -\sqrt{3}/2$, and we must find y such that $0 \le y \le \pi$. Therefore, $y = 5\pi/6$ since $\cos(5\pi/6) = -\sqrt{3}/2$.   ▬

▌**EXAMPLE 4**   **Evaluating the Compositions of Functions**

Write $\sin(\cos^{-1}x)$ as an algebraic expression in x.

FIGURE 4.7.6 The angle $t = \cos^{-1}x$ in Example 4

Solution In FIGURE 4.7.6 we have constructed an angle of t radians with cosine equal to x. Then $t = \cos^{-1}x$, or $x = \cos t$, where $0 \le t \le \pi$. Now to find $\sin(\cos^{-1}x) = \sin t$, we use the identity $\sin^2 t + \cos^2 t = 1$. Thus

$$\sin^2 t + x^2 = 1$$
$$\sin^2 t = 1 - x^2$$
$$\sin t = \sqrt{1 - x^2}$$
$$\sin(\cos^{-1}x) = \sqrt{1 - x^2}.$$

We use the positive square root of $1 - x^2$, since the range of $\cos^{-1}x$ is $[0, \pi]$, and the sine of an angle t in the first or second quadrant is positive. ■

☐ **Arctangent Function** If we restrict the domain of $\tan x$ to the open interval $(-\pi/2, \pi/2)$, then the resulting function is one-to-one and thus has an inverse. This inverse is denoted by

$$\arctan x \qquad \text{or} \qquad \tan^{-1}x.$$

DEFINITION 4.7.3 Arctangent Function

The **arctangent**, or **inverse tangent**, function is defined by

$$y = \arctan x \qquad \text{if and only if} \qquad x = \tan y \qquad (3)$$

where $-\infty < x < \infty$ and $-\pi/2 < y < \pi/2$.

The graphs shown in FIGURE 4.7.7 illustrate how the function $y = \tan x$ restricted to the open interval $(-\pi/2, \pi/2)$ becomes a one-to-one function.

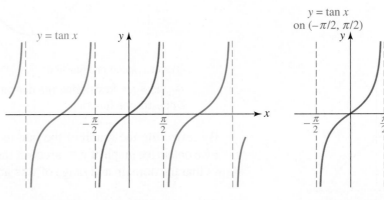

(a) Not a one-to-one function

(b) A one-to-one function

FIGURE 4.7.7 Restricting the domain of $y = \tan x$ to produce a one-to-one function

By reflecting the graph of the one-to-one function in Figure 4.7.7(b) in the line $y = x$ we obtain the graph of $y = \arctan x$ shown in FIGURE 4.7.8. We see in the figure that the domain and range of $y = \arctan x$ are, in turn, the intervals $(-\infty, \infty)$ and $(-\pi/2, \pi/2)$.

FIGURE 4.7.8 Graph of $y = \arctan x$

EXAMPLE 5   **Evaluating the Inverse Tangent Function**

Find $\tan^{-1}(-1)$.

Solution If $\tan^{-1}(-1) = y$, then $\tan y = -1$, where $-\pi/2 < y < \pi/2$. It follows that $\tan^{-1}(-1) = y = -\pi/4$. ■

CHAPTER 4 TRIGONOMETRIC FUNCTIONS

EXAMPLE 6   Evaluating the Compositions of Functions

Without using a calculator, find $\sin\left(\arctan\left(-\frac{5}{3}\right)\right)$.

Solution If we let $t = \arctan\left(-\frac{5}{3}\right)$, then $\tan t = -\frac{5}{3}$. The Pythagorean identity $1 + \tan^2 t = \sec^2 t$ can be used to find $\sec t$:

$$1 + \left(-\frac{5}{3}\right)^2 = \sec^2 t$$

$$\sec t = \sqrt{1 + \frac{25}{9}} = \sqrt{\frac{34}{9}} = \frac{\sqrt{34}}{3}.$$

In the preceding line we take the positive square root because $t = \arctan\left(-\frac{5}{3}\right)$ is in the interval $(-\pi/2, \pi/2)$ (the range of the arctangent function) and the secant of an angle t in the first or fourth quadrant is positive. Also, from $\sec t = \sqrt{34}/3$ we find the value of $\cos t$ from the reciprocal identity:

$$\cos t = \frac{1}{\sec t} = \frac{1}{\sqrt{34}/3} = \frac{3}{\sqrt{34}}.$$

Finally, we can use the identity $\tan t = \sin t/\cos t$ in the form $\sin t = \tan t \cos t$ to compute $\sin\left(\arctan\left(-\frac{5}{3}\right)\right)$. It follows that

$$\sin t = \tan t \cos t = \left(-\frac{5}{3}\right)\left(\frac{3}{\sqrt{34}}\right) = -\frac{5}{\sqrt{34}}. \qquad \blacksquare$$

☐ **Properties of the Inverses** Recall from Section 2.8 that $f^{-1}(f(x)) = x$ and $f(f^{-1}(x)) = x$ hold for any function f and its inverse under suitable restrictions on x. Thus for the inverse trigonometric functions, we have the following properties.

THEOREM 4.7.1   Properties of Inverse Trigonometric Functions

(*i*) $\arcsin(\sin x) = \sin^{-1}(\sin x) = x$ if $-\pi/2 \le x \le \pi/2$
(*ii*) $\sin(\arcsin x) = \sin(\sin^{-1} x) = x$ if $-1 \le x \le 1$
(*iii*) $\arccos(\cos x) = \cos^{-1}(\cos x) = x$ if $0 \le x \le \pi$
(*iv*) $\cos(\arccos x) = \cos(\cos^{-1} x) = x$ if $-1 \le x \le 1$
(*v*) $\arctan(\tan x) = \tan^{-1}(\tan x) = x$ if $-\pi/2 < x < \pi/2$
(*vi*) $\tan(\arctan x) = \tan(\tan^{-1} x) = x$ if $-\infty < x < \infty$

EXAMPLE 7   Using the Inverse Properties

Without using a calculator, evaluate:

(a) $\sin^{-1}\left(\sin\frac{\pi}{12}\right)$ **(b)** $\cos\left(\cos^{-1}\frac{1}{3}\right)$ **(c)** $\tan^{-1}\left(\tan\frac{3\pi}{4}\right)$.

Solution In each case we use the properties of the inverse trigonometric functions given in Theorem 4.7.1.

(a) Because $\pi/12$ satisfies $-\pi/2 \le x \le \pi/2$ it follows from property (*i*) that

$$\sin^{-1}\left(\sin\frac{\pi}{12}\right) = \frac{\pi}{12}.$$

(b) By property (*iv*), $\cos\left(\cos^{-1}\frac{1}{3}\right) = \frac{1}{3}$.

(c) In this case we *cannot* apply property (v), because $3\pi/4$ is not in the interval $(-\pi/2, \pi/2)$. If we first evaluate $\tan(3\pi/4) = -1$, then we have

see Example 5
$$\downarrow$$
$$\tan^{-1}\left(\tan\frac{3\pi}{4}\right) = \tan^{-1}(-1) = -\frac{\pi}{4}.$$

In the next section we illustrate how inverse trigonometric functions can be used to solve trigonometric equations.

☐ **Postscript—The Other Inverse Trig Functions** The functions $\cot x$, $\sec x$, and $\csc x$ also have inverses when their domains are suitably restricted. See Problems 49–51 in Exercises 4.7. Because these functions are not used as often as arctan, arccos, and arcsin, most scientific calculators do not have keys for them. However, any calculator that computes arcsin, arccos, and arctan can be used to obtain values for **arccsc, arcsec,** and **arccot.** Unlike the fact that $\sec x = 1/\cos x$, we note that $\sec^{-1}x \neq 1/\cos^{-1}x$; rather, $\sec^{-1}x = \cos^{-1}(1/x)$ for $|x| \geq 1$. Similar relationships hold for $\csc^{-1}x$ and $\cot^{-1} x$. See Problems 56–58 in Exercises 4.7.

Exercises 4.7   Answers to selected odd-numbered problems begin on page ANS–17.

In Problems 1–14, find the exact value of the given trigonometric expression. Do not use a calculator.

1. $\sin^{-1}0$
2. $\tan^{-1}\sqrt{3}$
3. $\arccos(-1)$
4. $\arcsin\dfrac{\sqrt{3}}{2}$
5. $\arccos\frac{1}{2}$
6. $\arctan\left(-\sqrt{3}\right)$
7. $\sin^{-1}\left(-\dfrac{\sqrt{3}}{2}\right)$
8. $\cos^{-1}\dfrac{\sqrt{3}}{2}$
9. $\tan^{-1}1$
10. $\sin^{-1}\dfrac{\sqrt{2}}{2}$
11. $\arctan\left(-\dfrac{\sqrt{3}}{3}\right)$
12. $\arccos(-\frac{1}{2})$
13. $\sin^{-1}\left(-\dfrac{\sqrt{2}}{2}\right)$
14. $\arctan 0$

In Problems 15–32, find the exact value of the given trigonometric expression. Do not use a calculator.

15. $\sin\left(\cos^{-1}\frac{3}{5}\right)$
16. $\cos\left(\sin^{-1}\frac{1}{3}\right)$
17. $\tan\left(\arccos\left(-\frac{2}{3}\right)\right)$
18. $\sin\left(\arctan\frac{1}{4}\right)$
19. $\cos(\arctan(-2))$
20. $\tan\left(\sin^{-1}\left(-\frac{1}{6}\right)\right)$
21. $\csc\left(\sin^{-1}\frac{3}{5}\right)$
22. $\sec(\tan^{-1}4)$
23. $\sin\left(\sin^{-1}\frac{1}{5}\right)$
24. $\cos\left(\cos^{-1}\left(-\frac{4}{5}\right)\right)$
25. $\tan(\tan^{-1}1.2)$
26. $\sin(\arcsin 0.75)$
27. $\arcsin\left(\sin\dfrac{\pi}{16}\right)$
28. $\arccos\left(\cos\dfrac{2\pi}{3}\right)$

29. $\tan^{-1}(\tan\pi)$

30. $\sin^{-1}\left(\sin\dfrac{5\pi}{6}\right)$

31. $\cos^{-1}\left(\cos\left(-\dfrac{\pi}{4}\right)\right)$

32. $\arctan\left(\tan\dfrac{\pi}{7}\right)$

In Problems 33–40, write the given expression as an algebraic expression in x.

33. $\sin(\tan^{-1}x)$
34. $\cos(\tan^{-1}x)$
35. $\tan(\arcsin x)$
36. $\sec(\arccos x)$
37. $\cot(\sin^{-1}x)$
38. $\cos(\sin^{-1}x)$
39. $\csc(\arctan x)$
40. $\tan(\arccos x)$

In Problems 41–48, sketch the graph of the given function.

41. $y = \arctan|x|$

42. $y = \dfrac{\pi}{2} - \arctan x$

43. $y = |\arcsin x|$
44. $y = \sin^{-1}(x + 1)$
45. $y = 2\cos^{-1}x$
46. $y = \cos^{-1}2x$
47. $y = \arccos(x - 1)$
48. $y = \cos(\arcsin x)$

49. The **arccotangent** function can be defined by $y = \text{arccot}\,x$ (or $y = \cot^{-1}x$) if and only if $x = \cot y$, where $0 < y < \pi$. Graph $y = \text{arccot}\,x$, and give the domain and the range of this function.

50. The **arccosecant** function can be defined by $y = \text{arccsc}\,x$ (or $y = \csc^{-1}x$) if and only if $x = \csc y$, where $-\pi/2 \le y \le \pi/2$ and $y \ne 0$. Graph $y = \text{arccsc}\,x$, and give the domain and the range of this function.

51. One definition of the **arcsecant** function is $y = \text{arcsec}\,x$ (or $y = \sec^{-1}x$) if and only if $x = \sec y$, where $0 \le y \le \pi$ and $y \ne \pi/2$. (See Problem 52 for an alternative definition.) Graph $y = \text{arcsec}\,x$, and give the domain and the range of this function.

52. An alternative definition of the arcsecant function can be made by restricting the domain of the secant function to $[0, \pi/2) \cup [\pi, 3\pi/2)$. Under this restriction, define the arcsecant function. Graph $y = \text{arcsec}\,x$, and give the domain and the range of this function.

53. Using the definition of the arccotangent function from Problem 49, for what values of x is it true that **(a)** $\cot(\text{arccot}\,x) = x$ and **(b)** $\text{arccot}(\cot x) = x$?

54. Using the definition of the arccosecant function from Problem 50, for what values of x is it true that **(a)** $\csc(\text{arccsc}\,x) = x$ and **(b)** $\text{arccsc}(\csc x) = x$?

55. Using the definition of the arcsecant function from Problem 51, for what values of x is it true that **(a)** $\sec(\text{arcsec}\,x) = x$ and **(b)** $\text{arcsec}(\sec x) = x$?

56. Verify that $\text{arccot}\,x = \dfrac{\pi}{2} - \arctan x$, for all real numbers x.

57. Verify that $\text{arccsc}\,x = \arcsin(1/x)$ for $|x| \ge 1$.

58. Verify that $\text{arcsec}\,x = \arccos(1/x)$ for $|x| \ge 1$.

In Problems 59–64, use the results of Problems 56–58 and a calculator to find the indicated value.

59. $\cot^{-1}0.75$
60. $\csc^{-1}(-1.3)$
61. $\text{arccsc}(-1.5)$
62. $\text{arccot}(-0.3)$
63. $\text{arcsec}(-1.2)$
64. $\sec^{-1}2.5$

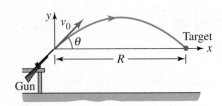

FIGURE 4.7.9 Angle of elevation in Problem 65

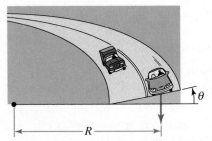

FIGURE 4.7.10 Banked curve in Problem 67

FIGURE 4.7.11 Ladder in Problem 69

Applications

65. Projectile Motion The angle of elevation θ for a gun firing a bullet to hit a target at a horizontal distance R (assuming that the target and the muzzle of the gun are at the same height) satisfies

$$R = \frac{v_0^2 \sin 2\theta}{g},$$

where v_0 is the muzzle velocity and g is the acceleration due to gravity. See FIGURE 4.7.9. Find the angle of elevation θ if the target is 800 ft from the gun and the muzzle velocity is 200 ft/s. Use $g = 32$ ft/s². [*Hint*: There are two solutions.]

66. Olympic Sports For the Olympic event, the hammer throw, it can be shown that the maximum distance is achieved for the release angle θ (measured from the horizontal) that satisfies

$$\cos 2\theta = \frac{gh}{v_0^2 + gh},$$

where h is the height of the hammer above the ground at release, v_0 is the initial velocity, and g is the acceleration due to gravity. For $v_0 = 13.7$ m/s and $h = 2.25$ m, find the optimal release angle. Use $g = 9.81$ m/s².

67. Highway Design In the design of highways and railroads, curves are banked to provide centripetal force for safety. The optimal banking angle θ is given by $\tan \theta = v^2/Rg$, where v is the speed of the vehicle, R is the radius of the curve, and g is the acceleration due to gravity. See FIGURE 4.7.10. As the formula indicates, for a given radius there is no one correct angle for all speeds. Consequently, curves are banked for the average speed of the traffic over them. Find the correct banking angle for a curve of radius 600 ft on a country road where speeds average 30 mi/h. Use $g = 32$ ft/s². [*Hint*: Use consistent units.]

68. Highway Design—Continued If μ is the coefficient of friction between the car and the road, then the maximum velocity v_m that a car can travel around a curve without slipping is given by $v_m^2 = gR\tan(\theta + \tan^{-1}\mu)$, where θ is the banking angle of the curve. Find v_m for the country road in Problem 67 if $\mu = 0.26$.

69. Ladder About to Slip Consider a ladder of length L leaning against a house with a load at point P a distance x measured from the bottom of the ladder. See FIGURE 4.7.11. The angle θ at which the ladder is at the verge of slipping can be shown to be related to x and the coefficient of friction c between the ladder and the ground by

$$\frac{x}{L} = \frac{c}{1 + c^2}(c + \tan \theta).$$

(a) Find θ when $c = 1$ and the load is at the top of the ladder.
(b) Find θ when $c = 0.5$ and the load is $\frac{3}{4}$ of the way up the ladder.

For Discussion

70. Using a calculator set in radian mode, evaluate $\arctan(\tan 1.8)$, $\arccos(\cos 1.8)$, and $\arcsin(\sin 1.8)$. Explain the results.

71. Using a calculator set in radian mode, evaluate $\tan^{-1}(\tan(-1))$, $\cos^{-1}(\cos(-1))$, and $\sin^{-1}(\sin(-1))$. Explain the results.

72. In Section 4.3 we saw that the graphs of $y = \sin x$ and $y = \cos x$ are related by shifting and reflecting. Justify the identity

$$\arcsin x + \arccos x = \frac{\pi}{2},$$

for all x in $[-1, 1]$, by finding a similar relationship between the graphs of $y = \arcsin x$ and $y = \arccos x$.

73. With a calculator set in radian mode determine which of the following inverse trigonometric evaluations result in an error message: **(a)** $\sin^{-1}(-2)$, **(b)** $\cos^{-1}(-2)$, **(c)** $\tan^{-1}(-2)$. Explain.

74. Discuss: Can any periodic function be one-to-one?

75. Show that $\arcsin\frac{3}{5} + \arcsin\frac{5}{13} = \arcsin\frac{56}{65}$. [*Hint*: See (4) of Section 4.5.]

76. The **angle of inclination of a line** is defined to be the angle θ measured counterclockwise between the line and the x-axis $0 \leq \theta < \pi$ (or $0° \leq \theta < 180°$). See FIGURE 4.7.12. The angle of inclination of a horizontal line is 0 and the angle of inclination of a vertical line is $\pi/2$. If L is any line with slope m (that is, not vertical), use FIGURE 4.7.13 to show that $m = \tan\theta$.

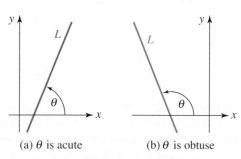

(a) θ is acute   (b) θ is obtuse

FIGURE 4.7.12 Angles of inclination in Problem 76

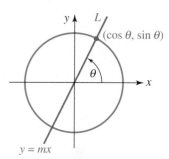

$y = mx$

FIGURE 4.7.13 Line passing through the origin and a unit circle in Problem 76

77. Use Problem 76 to find the angle of inclination of the given line.
    (a) $3x - 2y = 6$ **(b)** $y = -2x + 5$

78. Suppose that θ is the acute angle between the two nonperpendicular intersecting lines L_1 and L_2 in FIGURE 4.7.14. If L_1 and L_2 have slopes m_1 and m_2, respectively, then show that

$$\tan\theta = \left|\frac{m_1 - m_2}{1 + m_1 m_2}\right|.$$

FIGURE 4.7.14 Two lines in Problem 78

79. Use Problem 78 to find the acute angle θ between the given pair of lines.
    (a) $-x + 2y = 10, x + 2y = -5$ **(b)** $x + 2y = 6, 4x - 3y = 1$

80. **(a)** Show that

$$\tan(\tan^{-1}1 + \tan^{-1}2 + \tan^{-1}3) = 0.$$

(b) Discuss: How does the result in part (a) prove that

$$\tan^{-1}1 + \tan^{-1}2 + \tan^{-1}3 = \pi?$$

4.8 Trigonometric Equations

INTRODUCTION In Sections 4.5 and 4.6 we considered **trigonometric identities**, which are equations involving trigonometric functions that are satisfied by all values of the variable for which both sides of the equality are defined. In this section we examine **conditional trigonometric equations**, that is, equations that are true for only certain values of the variable. We discuss techniques for finding those values of the variable (if any) that satisfy the equation.

We begin by considering the problem of finding all real numbers x that satisfy $\sin x = \sqrt{2}/2$. Interpreted as the x-coordinates of the points of intersection of the graphs

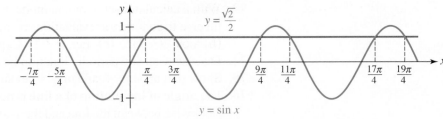

FIGURE 4.8.1 Graphs of $y = \sin x$ and $y = \frac{\sqrt{2}}{2}$

of $y = \sin x$ and $y = \sqrt{2}/2$, FIGURE 4.8.1 shows that there exists infinitely many solutions of the equation $\sin x = \sqrt{2}/2$:

$$\ldots, -\frac{7\pi}{4}, \frac{\pi}{4}, \frac{9\pi}{4}, \frac{17\pi}{4}, \ldots \tag{1}$$

$$\ldots, -\frac{5\pi}{4}, \frac{3\pi}{4}, \frac{11\pi}{4}, \frac{19\pi}{4}, \ldots. \tag{2}$$

Note that in each of the lists (1) and (2), two successive solutions differ by $2\pi = 8\pi/4$. This is a consequence of the periodicity of the sine function. It is common for trigonometric equations to have an infinite number of solutions because of the periodicity of the trigonometric functions. In general, to obtain solutions of an equation such as $\sin x = \sqrt{2}/2$, it is more convenient to use a unit circle and reference angles rather than a graph of the trigonometric function. We illustrate this approach in the following example.

EXAMPLE 1   Using the Unit Circle

Find all real numbers x satisfying $\sin x = \sqrt{2}/2$.

Solution If $\sin x = \sqrt{2}/2$, the reference angle for x is $\pi/4$ radian. Since the value of $\sin x$ is positive, the terminal side of the angle x lies in either the first or second quadrant. Thus, as shown in FIGURE 4.8.2, the only solutions between 0 and 2π are

$$x = \frac{\pi}{4} \qquad \text{and} \qquad x = \frac{3\pi}{4}.$$

Since the sine function is periodic with period 2π, all of the remaining solutions can be obtained by adding integer multiples of 2π to these solutions. The two solutions are

$$x = \frac{\pi}{4} + 2n\pi \qquad \text{and} \qquad x = \frac{3\pi}{4} + 2n\pi, \tag{3}$$

where n is an integer. The numbers that you see in (1) and (2) correspond, respectively, to letting $n = -1$, $n = 0$, $n = 1$, and $n = 2$ in the first and second formulas in (3). ▬

FIGURE 4.8.2 Unit circle in Example 1

When we are faced with a more complicated equation, such as

$$4\sin^2 x - 8\sin x + 3 = 0,$$

the basic approach is to solve for a single trigonometric function (in this case, it would be $\sin x$) by using methods similar to those for solving algebraic equations.

EXAMPLE 2   Solving a Trigonometric Equation by Factoring

Find all solutions of $4\sin^2 x - 8\sin x + 3 = 0$.

Solution We first observe that this is a quadratic equation in $\sin x$, and that it factors as

$$(2\sin x - 3)(2\sin x - 1) = 0.$$

FIGURE 4.8.3 Unit circle in Example 2

This implies that either

$$\sin x = \frac{3}{2} \quad \text{or} \quad \sin x = \frac{1}{2}.$$

The first equation has no solution since $|\sin x| \le 1$. As we see in FIGURE 4.8.3 the two angles between 0 and 2π for which $\sin x$ equals $\frac{1}{2}$ are

$$x = \frac{\pi}{6} \quad \text{and} \quad x = \frac{5\pi}{6}.$$

Therefore, by the periodicity of the sine function, the solutions are

$$x = \frac{\pi}{6} + 2n\pi \quad \text{and} \quad x = \frac{5\pi}{6} + 2n\pi,$$

where n is an integer.

EXAMPLE 3   Checking for Lost Solutions

Find all solutions of

$$\sin x = \cos x. \tag{4}$$

Solution In order to work with a single trigonometric function, we divide both sides of the equation by $\cos x$ to obtain

$$\tan x = 1. \tag{5}$$

Equation (5) is equivalent to (4) *provided* that $\cos x \ne 0$. We observe that if $\cos x = 0$, then as we have seen in (2) of Section 4.3, $x = (2n + 1)\pi/2 = \pi/2 + n\pi$, for n an integer. By the sum formula for the sine,

$$\sin\left(\frac{\pi}{2} + n\pi\right) = \overset{\underset{\mathrm{see\ (4)\ of\ Section\ 4.5}}{\downarrow}}{\sin\frac{\pi}{2}}\overset{\underset{(-1)^n}{\downarrow}}{\cos n\pi} + \cos\frac{\pi}{2}\overset{\underset{0}{\downarrow}}{\sin n\pi} = (-1)^n \ne 0,$$

we see that these values of x do not satisfy the original equation. Thus we will find *all* the solutions to (4) by solving equation (5).

Now $\tan x = 1$ implies that the reference angle for x is $\pi/4$ radian. Since $\tan x = 1 > 0$, the terminal side of the angle of x radians can lie either in the first or in the third quadrant, as shown in FIGURE 4.8.4. Thus the solutions are

$$x = \frac{\pi}{4} + 2n\pi \quad \text{and} \quad x = \frac{5\pi}{4} + 2n\pi,$$

where n is an integer. We can see from Figure 4.8.4 that these two sets of numbers can be written more compactly as

$$x = \frac{\pi}{4} + n\pi,$$

where n is an integer.

▶ $\cos 0 = 1, \quad \cos \pi = -1, \quad \cos 2\pi = 1,$ $\cos 3\pi = -1,$ and so on. In general, $\cos n\pi = (-1)^n$, where n is an integer.

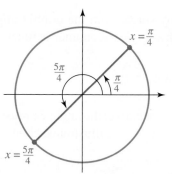

FIGURE 4.8.4 Unit circle in Example 3

This also follows from the fact that $\tan x$ is ▶ π-periodic.

▢ **Losing Solutions** When solving an equation, if you divide by an expression containing a variable, you may lose some solutions of the original equation. For example, in algebra a common mistake in solving equations such as $x^2 = x$ is to divide by x to obtain $x = 1$. But by writing $x^2 = x$ as $x^2 - x = 0$ or $x(x - 1) = 0$ we see that in fact $x = 0$ or $x = 1$. To prevent the loss of a solution you must determine the values that make the expression zero and check to see whether they are solutions of the original equation. Note that in Example 3, when we divided by $\cos x$, we took care to check that no solutions were lost.

4.8 Trigonometric Equations

Whenever possible, it is preferable to avoid dividing by a variable expression. As illustrated with the algebraic equation $x^2 = x$, this can frequently be accomplished by collecting all nonzero terms on one side of the equation and then factoring (something we could not do in Example 3). Example 4 illustrates this technique.

EXAMPLE 4   Solving a Trigonometric Equation by Factoring

Solve
$$2\sin x\cos^2 x = -\frac{\sqrt{3}}{2}\cos x. \tag{6}$$

Solution To avoid dividing by $\cos x$, we write the equation as

$$2\sin x\cos^2 x + \frac{\sqrt{3}}{2}\cos x = 0$$

and factor:
$$\cos x\left(2\sin x\cos x + \frac{\sqrt{3}}{2}\right) = 0.$$

Thus either

$$\cos x = 0 \qquad \text{or} \qquad 2\sin x\cos x + \frac{\sqrt{3}}{2} = 0.$$

Since the cosine is zero for all odd multiples of $\pi/2$, the solutions of $\cos x = 0$ are

$$x = (2n + 1)\frac{\pi}{2} = \frac{\pi}{2} + n\pi,$$

where n is an integer.

See (12) in Section 4.5. ▶   In the second equation we replace $2\sin x\cos x$ by $\sin 2x$ from the double-angle formula for the sine function to obtain an equation with a single trigonometric function:

$$\sin 2x + \frac{\sqrt{3}}{2} = 0 \qquad \text{or} \qquad \sin 2x = -\frac{\sqrt{3}}{2}.$$

Thus the reference angle for $2x$ is $\pi/3$. Since the sine is negative, the angle $2x$ must be in either the third quadrant or the fourth quadrant. As FIGURE 4.8.5 illustrates, either

$$2x = \frac{4\pi}{3} + 2n\pi \qquad \text{or} \qquad 2x = \frac{5\pi}{3} + 2n\pi.$$

$2x = \frac{4\pi}{3}$ $2x = \frac{5\pi}{3}$

FIGURE 4.8.5 Unit circle in Example 4

Dividing by 2 gives

$$x = \frac{2\pi}{3} + n\pi \qquad \text{or} \qquad x = \frac{5\pi}{6} + n\pi.$$

Therefore, all solutions of (6) are

$$x = \frac{\pi}{2} + n\pi, \quad x = \frac{2\pi}{3} + n\pi, \quad \text{and} \quad x = \frac{5\pi}{6} + n\pi,$$

where n is an integer.

In Example 4 had we simplified the equation by dividing by $\cos x$ and not checked to see whether the values of x for which $\cos x = 0$ satisfied equation (6), we would have lost the solutions $x = \pi/2 + n\pi$, where n is an integer.

EXAMPLE 5   Using a Trigonometric Identity

Solve $3\cos^2 x - \cos 2x = 1$.

Solution We observe that the given equation involves both the cosine of x and the cosine of $2x$. Consequently, we use the double-angle formula for the cosine in the form

$$\cos 2x = 2\cos^2 x - 1 \qquad \leftarrow \text{See (14) of Section 4.5}$$

to replace the equation by an equivalent equation that involves $\cos x$ only. We find that

$$3\cos^2 x - (2\cos^2 x - 1) = 1 \qquad \text{becomes} \qquad \cos^2 x = 0.$$

Therefore, $\cos x = 0$, and the solutions are

$$x = (2n + 1)\frac{\pi}{2} = \frac{\pi}{2} + n\pi,$$

where n is an integer.   ■

We are often interested in finding roots of an equation only in a specified interval.

EXAMPLE 6   Using a Trigonometric Identity

Find all solutions of the equation $\cos t - \cos 5t = 0$ in the interval $[0, 2\pi)$.

Solution In this case it is helpful to use a sum-to-product formula. In Example 3 of Section 4.6 we saw that by identifying $x_1 = t$ and $x_2 = 5t$, (9) of Theorem 4.6.2 gives

$$\cos t - \cos 5t = -2\sin\frac{t + 5t}{2}\sin\frac{t - 5t}{2} = -2\sin 3t \sin(\overset{\underset{\displaystyle \sin(-2t)\, =\, -\sin 2t}{\downarrow}}{-2t}) = 2\sin 3t \sin 2t$$

Replacing the sum $\cos t - \cos 5t$ in the given equation by the product $2\sin 3t \sin 2t$ gives the equivalent equation $2\sin 3t \sin 2t = 0$, or

$$\sin 3t \sin 2t = 0.$$

The last equation is satisfied if either $\sin 3t = 0$ or $\sin 2t = 0$. Then from (1) in Section 4.3 we see that $\sin 3t = 0$ implies

$$3t = n\pi, \, n = 0, 1, 2, 3, \ldots \qquad (7)$$

whereas $\sin 2t = 0$ implies

$$2t = n\pi, \, n = 0, 1, 2, 3, \ldots. \qquad (8)$$

If you think in terms of angles measured in radians and the unit circle, then the only angles satisfying the condition that t be in the interval $[0, 2\pi)$ correspond to $n = 0, 1, 2, 3, 4, 5$ in (7) and $n = 1, 2, 3$ in (8):

$$t = 0, \pi/3, 2\pi/3, \pi, 4\pi/3, 5\pi/3, \qquad (9)$$

or

$$t = \pi/2, \pi, 3\pi/2. \qquad (10)$$

The solution set of the original equation is then the union of the two sets defined by the numbers in (9) and (10), that is

$$\{0, \pi/3, \pi/2, 2\pi/3, \pi, 4\pi/3, 3\pi/2, 5\pi/3\}. \qquad ■$$

So far in this section we have viewed the variable in the trigonometric equation as representing either a real number or an angle measured in radians. If the variable represents an angle measured in degrees, the technique for solving is the same.

FIGURE 4.8.6 Unit circle in Example 7

The figure labels: $2\theta = 120°$, $120°$, $240°$, $2\theta = 240°$

EXAMPLE 7   **Equation When the Angle Is in Degrees**

Solve $\cos 2\theta = -\frac{1}{2}$, where θ is an angle measured in degrees.

Solution Since $\cos 2\theta = -\frac{1}{2}$, the reference angle for 2θ is $60°$ and the angle 2θ must be in either the second or the third quadrant. **FIGURE 4.8.6** illustrates that either $2\theta = 120°$ or $2\theta = 240°$. Any angle that is coterminal with one of these angles will also satisfy $\cos 2\theta = -\frac{1}{2}$. These angles are obtained by adding any integer multiple of $360°$ to $120°$ or to $240°$:

$$2\theta = 120° + 360°n \qquad \text{or} \qquad 2\theta = 240° + 360°n,$$

where n is an integer. Dividing by 2 the last line yields the two solutions

$$\theta = 60° + 180°n \qquad \text{and} \qquad \theta = 120° + 180°n. \qquad \blacksquare$$

☐ **Extraneous Solutions** When an equation is squared or multiplied by a variable expression, the resulting equation may not be equivalent to the original. In other words, the new equation may possess solutions that are not solutions of the given equation. Such numbers are called **extraneous solutions**. For example, the simple equation $x = 1$ has only one solution but by squaring both sides the resulting equation $x^2 = 1$ is now satisfied by $x = 1$ and $x = -1$. Thus number $x = -1$ is an extraneous solution of the original equation. The next example illustrates the same idea for trigonometric equations.

EXAMPLE 8   **Extraneous Roots**

Find all solutions of $1 + \tan \alpha = \sec \alpha$, where α is an angle measured in degrees.

Solution The equation does not factor, but we see that if we square both sides, we can use a fundamental identity to obtain an equation involving a single trigonometric function:

$$(1 + \tan \alpha)^2 = (\sec \alpha)^2$$
$$1 + 2\tan \alpha + \tan^2\alpha = \sec^2\alpha \qquad \leftarrow \text{now use (9) of Section 4.4}$$
$$1 + 2\tan \alpha + \tan^2\alpha = 1 + \tan^2\alpha$$
$$2\tan \alpha = 0$$
$$\tan \alpha = 0.$$

The values of α for $0° \leq \alpha < 360°$ at which $\tan \alpha = 0$ are

$$\alpha = 0° \qquad \text{and} \qquad \alpha = 180°.$$

Since we squared each side of the original equation, we may have introduced extraneous solutions. Therefore, it is important that we check all solutions in the original equation. Substituting $\alpha = 0°$ into $1 + \tan \alpha = \sec \alpha$, we obtain the *true* statement $1 + 0 = 1$. But after substituting $\alpha = 180°$, we obtain the *false* statement $1 + 0 = -1$. Therefore, $180°$ is an extraneous solution and $\alpha = 0°$ is the only solution satisfying $0° \leq \alpha < 360°$. Thus, all the solutions of the equation are given by

$$\alpha = 0° + 360°n = 360°n,$$

where n is an integer. For $n \neq 0$, these are the angles that are coterminal with $0°$. $\blacksquare$

Recall from Section 2.1 that to find the x-intercepts of the graph of a function $y = f(x)$ we find the zeros of f, that is, we must solve the equation $f(x) = 0$. The following example makes use of this fact.

CHAPTER 4 TRIGONOMETRIC FUNCTIONS

EXAMPLE 9 **Intercepts of a Graph**

Find the first three x-intercepts of the graph of $f(x) = \sin 2x \cos x$ on the positive x-axis.

Solution We must solve $f(x) = 0$, that is, $\sin 2x \cos x = 0$. It follows that either $\sin 2x = 0$ or $\cos x = 0$.

From $\sin 2x = 0$, we obtain $2x = n\pi$, where n is an integer, or $x = n\pi/2$, where n is an integer. From $\cos x = 0$, we find $x = \pi/2 + n\pi$, where n is an integer. Then for $n = 2$, $x = n\pi/2$ gives $x = \pi$, whereas for $n = 0$ and $n = 1$, $x = \pi/2 + n\pi$ gives $x = \pi/2$ and $x = 3\pi/2$, respectively. Thus the first three x-intercepts on the positive x-axis are $(\pi/2, 0)$, $(\pi, 0)$, and $(3\pi/2, 0)$. ■

☐ **Using Inverse Functions** So far all of the trigonometric equations have had solutions that were related by reference angles to the special angles 0, $\pi/6$, $\pi/4$, $\pi/3$, or $\pi/2$. If this is not the case, we will see in the next example how to use inverse trigonometric functions and a calculator to find solutions.

EXAMPLE 10 **Solving Equations Using Inverse Functions**

Find the solutions of $4\cos^2 x - 3\cos x - 2 = 0$ in the interval $[0, \pi]$.

Solution We recognize that this is a quadratic equation in $\cos x$. Since the left-hand side of the equation does not readily factor, we apply the quadratic formula to obtain

$$\cos x = \frac{3 \pm \sqrt{41}}{8}.$$

At this point we can discard the value $(3 + \sqrt{41})/8 \approx 1.18$, because $\cos x$ cannot be greater than 1. We then use the inverse cosine function and a calculator to solve the remaining equation:

$$\cos x = \frac{3 - \sqrt{41}}{8} \quad \text{which implies} \quad x = \cos^{-1}\left(\frac{3 - \sqrt{41}}{8}\right) \approx 2.01. \quad ■$$

Of course in Example 10, had we attempted to compute $\cos^{-1}[(3 + \sqrt{41})/8]$ with a calculator, we would have received an error message.

Exercises 4.8 Answers to selected odd-numbered problems begin on page ANS–17.

In Problems 1–6, find all solutions of the given trigonometric equation if x represents an angle measured in radians.

1. $\sin x = \sqrt{3}/2$ **2.** $\cos x = -\sqrt{2}/2$
3. $\sec x = \sqrt{2}$ **4.** $\tan x = -1$
5. $\cot x = -\sqrt{3}$ **6.** $\csc x = 2$

In Problems 7–12, find all solutions of the given trigonometric equation if x represents a real number.

7. $\cos x = -1$ **8.** $2\sin x = -1$
9. $\tan x = 0$ **10.** $\sqrt{3}\sec x = 2$
11. $-\csc x = 1$ **12.** $\sqrt{3}\cot x = 1$

In Problems 13–18, find all solutions of the given trigonometric equation if θ represents an angle measured in degrees.

13. $\csc \theta = 2\sqrt{3}/3$

14. $2\sin \theta = \sqrt{2}$

15. $1 + \cot \theta = 0$

16. $\sqrt{3} \sin \theta = \cos \theta$

17. $\sec \theta = -2$

18. $2\cos \theta + \sqrt{2} = 0$

In Problems 19–46, find all solutions of the given trigonometric equation if x is a real number and θ is an angle measured in degrees.

19. $\cos^2 x - 1 = 0$

20. $2\sin^2 x - 3\sin x + 1 = 0$

21. $3\sec^2 x = \sec x$

22. $\tan^2 x + (\sqrt{3} - 1)\tan x - \sqrt{3} = 0$

23. $2\cos^2 \theta - 3\cos \theta - 2 = 0$

24. $2\sin^2 \theta - \sin \theta - 1 = 0$

25. $\cot^2 \theta + \cot \theta = 0$

26. $2\sin^2 \theta + (2 - \sqrt{3})\sin \theta - \sqrt{3} = 0$

27. $\cos 2x = -1$

28. $\sec 2x = 2$

29. $2\sin 3\theta = 1$

30. $\tan 4\theta = -1$

31. $\cot (x/2) = 1$

32. $\csc (\theta/3) = -1$

33. $\sin 2x + \sin x = 0$

34. $\cos 2x + \sin^2 x = 1$

35. $\cos 2\theta = \sin \theta$

36. $\sin 2\theta + 2\sin \theta - 2\cos \theta = 2$

37. $\sin^4 x - 2\sin^2 x + 1 = 0$

38. $\tan^4 \theta - 2\sec^2 \theta + 3 = 0$

39. $\sec x \sin^2 x = \tan x$

40. $\dfrac{1 + \cos \theta}{\cos \theta} = 2$

41. $\sin \theta + \cos \theta = 1$

42. $\sin x + \cos x = 0$

43. $\sqrt{\dfrac{1 + 2\sin x}{2}} = 1$

44. $\sin x + \sqrt{\sin x} = 0$

45. $\cos \theta - \sqrt{\cos \theta} = 0$

46. $\cos \theta \sqrt{1 + \tan^2 \theta} = 1$

In Problems 47–52, use a sum-to-product formula (as in Example 6) to solve the given equation on the indicated interval.

47. $\sin 6t - \sin 4t = 0$, $[0, 2\pi)$

48. $\cos 2t + \cos 3t = 0$, $[0, 2\pi)$

49. $\cos \theta - \cos 4\theta = 0$, $[-\pi, \pi)$

50. $\sin 5\alpha + \sin 3\alpha = 0$, $[0, 2\pi)$

51. $\sin 7x - \sin x - 2\sin 3x = 0$, $(-\pi, \pi)$

52. $\sin x + \cos 2x - \sin 3x = 0$, $[0, 3\pi)$

In Problems 53–60, find the first three x-intercepts of the graph of the given function on the positive x-axis.

53. $f(x) = -5\sin (3x + \pi)$

54. $f(x) = 2\cos \left(x + \dfrac{\pi}{4}\right)$

55. $f(x) = 2 - \sec \dfrac{\pi}{2} x$

56. $f(x) = 1 + \cos \pi x$

57. $f(x) = \sin x + \tan x$

58. $f(x) = 1 - 2\cos \left(x + \dfrac{\pi}{3}\right)$

59. $f(x) = \sin x - \sin 2x$

60. $f(x) = \cos x + \cos 3x$
[*Hint*: Write $3x = x + 2x$.]

In Problems 61–64, by graphing determine whether the given equation has any solutions.

61. $\tan x = x$ [*Hint*: Graph $y = \tan x$ and $y = x$ on the same coordinate axes.]

62. $\sin x = x$

63. $\cot x - x = 0$

64. $\cos x + x + 1 = 0$

In Problems 65–70, using an inverse trigonometric function find the solutions of the given equation in the indicated interval. Round your answers to two decimal places.

65. $20\cos^2 x + \cos x - 1 = 0$, $[0, \pi]$

66. $3\sin^2 x - 8\sin x + 4 = 0$, $[-\pi/2, \pi/2]$

67. $\tan^2 x + \tan x - 1 = 0, \quad (-\pi/2, \pi/2)$
68. $3 \sin 2x + \cos x = 0, \quad [-\pi/2, \pi/2]$
69. $5 \cos^3 x - 3 \cos^2 x - \cos x = 0, \quad [0, \pi]$
70. $\tan^4 x - 3 \tan^2 x + 1 = 0, \quad (-\pi/2, \pi/2)$

Applications

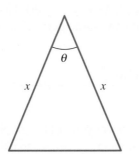

FIGURE 4.8.7 Isosceles triangle in Problem 71

71. **Isosceles Triangle** By methods that will be discussed in Section 4.10 it can be shown that the area of the isosceles triangle shown in FIGURE 4.8.7 is given by $A = \frac{1}{2}x^2 \sin \theta$. If the length of side x is 4, what value of θ will give a triangle with area 4?

72. **Circular Motion** An object travels in a circular path centered at the origin with constant angular speed. The y-coordinate of the object at any time t seconds is given by $y = 8 \cos(\pi t - \pi/12)$. At what time(s) does the object cross the x-axis?

73. **Mach Number** Use Problem 73 in Exercises 4.5 to find the vertex angle of the cone of sound waves made by an airplane flying at Mach 2.

74. **Alternating Current** An electric generator produces a 60-cycle alternating current given by $I(t) = 30 \sin 120\pi \left(t - \frac{7}{36}\right)$, where $I(t)$ is the current in amperes at t seconds. Find the smallest positive value of t for which the current is 15 amperes.

75. **Electrical Circuits** If the voltage given by $V = V_0 \sin(\omega t + \alpha)$ is impressed on a series circuit, an alternating current is produced. If $V_0 = 110$ volts, $\omega = 120\pi$ radians per second, and $\alpha = -\pi/6$, when is the voltage equal to zero?

FIGURE 4.8.8 Light rays in Problem 76

76. **Refraction of Light** Consider a ray of light passing from one medium (such as air) into another medium (such as a crystal). Let ϕ be the angle of incidence and θ the angle of refraction. As shown in FIGURE 4.8.8, these angles are measured from a vertical line. According to **Snell's law**, there is a constant c that depends on the two mediums, such that $\dfrac{\sin \phi}{\sin \theta} = c$. Assume that for light passing from air into a crystal, $c = 1.437$. Find ϕ and θ such that the angle of incidence is twice the angle of refraction.

77. **Snow Cover** On the basis of data collected from 1966 to 1980, the extent of snow cover S in the northern hemisphere, measured in millions of square kilometers, can be modeled by the function

$$S(w) = 25 + 21 \cos \frac{\pi}{26}(w - 5),$$

where w is the number of weeks past January 1.
   (a) How much snow cover does this formula predict for April Fool's Day? (Round w to the nearest integer.)
   (b) In which week does the formula predict the least amount of snow cover?
   (c) What month does this fall in?

4.9 Simple Harmonic Motion

INTRODUCTION Many physical objects vibrate or oscillate in a regular manner, repeatedly moving back and forth over a definite time interval. Some examples are clock pendulums, a mass on a spring, sound waves, strings on a guitar when plucked, the human heart, tides, and alternating current. In this section we will focus on mathematical models of the undamped oscillatory motion of a mass on a spring.

Before proceeding with the main discussion we need to discuss the graph of the sum of constant multiples of $\cos Bx$ and $\sin Bx$, that is, $y = c_1 \cos Bx + c_2 \sin Bx$, where c_1 and c_2 are constants.

☐ **Addition of Two Sinusoidal Functions** In Section 4.3 we examined the graphs of horizontally shifted sine and cosine graphs. It turns out that any linear combination of a sine function and a cosine function of the form

$$y = c_1\cos Bx + c_2\sin Bx, \tag{1}$$

where c_1 and c_2 are constants, can be expressed either as a shifted sine function $y = A\sin(Bx + \phi), B > 0$, or as a shifted cosine function $y = A\cos(Bx + \phi)$. Note that in (1) the sine and cosine functions $\sin Bx$ and $\cos Bx$ have the same period $2\pi/B$.

EXAMPLE 1　　　Addition of a Sine and a Cosine

Graph the function $y = \cos 2x - \sqrt{3}\sin 2x$.

Solution Using a graphing utility we have shown in FIGURE 4.9.1 four cycles of the graphs of $y = \cos 2x$ (in red) and $y = -\sqrt{3}\sin 2x$ (in green). It is apparent in FIGURE 4.9.2 that the period of the sum of these two functions is π, the common period of $\cos 2x$ and $\sin 2x$. Also apparent is that the blue graph is a horizontally shifted sine (or cosine) function. Although Figure 4.9.2 suggests that the amplitude of the function $y = \cos 2x - \sqrt{3}\sin 2x$ is 2, the exact phase shift of the graph is certainly *not* apparent.

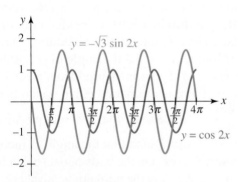

FIGURE 4.9.1 Superimposed graphs of $y = \cos 2x$ and $y = -\sqrt{3}\sin 2x$ in Example 1

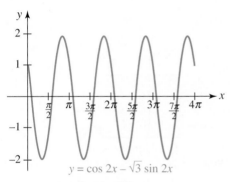

FIGURE 4.9.2 Graph of the sum $y = \cos 2x - \sqrt{3}\sin 2x$ in Example 1

▶ The sine form $y = A\sin(Bx + \phi)$ is slightly easier to use than the cosine form $y = A\cos(Bx + \phi)$.

☐ **Reduction to a Sine Function** We examine only the reduction of (1) to the form $y = A\sin(Bx + \phi), B > 0$.

THEOREM 4.9.1　Reduction of (1) to (2)

For real numbers c_1, c_2, B, and x,

$$c_1\cos Bx + c_2\sin Bx = A\sin(Bx + \phi) \tag{2}$$

where A and ϕ are defined by

$$A = \sqrt{c_1^2 + c_2^2} \tag{3}$$

and

$$\left.\begin{array}{l} \sin\phi = \dfrac{c_1}{A} \\[2mm] \cos\phi = \dfrac{c_2}{A} \end{array}\right\} \tan\phi = \dfrac{c_1}{c_2} \tag{4}$$

PROOF: To prove (2), we use the sum formula (4) of Section 4.5:

$$A\sin(Bx + \phi) = A\sin Bx\cos\phi + A\cos Bx\sin\phi$$
$$= (A\sin\phi)\cos Bx + (A\cos\phi)\sin Bx$$
$$= c_1\cos Bx + c_2\sin Bx$$

and identify $A\sin\phi = c_1$, $A\cos\phi = c_2$. Thus, $\sin\phi = c_1/A = c_1/\sqrt{c_1^2 + c_2^2}$ and $\cos\phi = c_2/A = c_2/\sqrt{c_1^2 + c_2^2}$. ■

■ EXAMPLE 2   Example 1 Revisited

Express $y = \cos 2x - \sqrt{3}\sin 2x$ as a single sine function.

Solution With the identifications $c_1 = 1$, $c_2 = -\sqrt{3}$, and $B = 2$, we have from (3) and (4),

$$A = \sqrt{c_1^2 + c_2^2} = \sqrt{1^2 + (-\sqrt{3})^2} = \sqrt{4} = 2,$$

$$\left.\begin{array}{l} \sin\phi = \dfrac{1}{2} \\[2mm] \cos\phi = -\dfrac{\sqrt{3}}{2} \end{array}\right\} \quad \tan\phi = -\dfrac{1}{\sqrt{3}}.$$

Although $\tan\phi = -1/\sqrt{3}$ we cannot blindly assume that $\phi = \tan^{-1}(-1/\sqrt{3})$. The value we take for ϕ must be consistent with the algebraic signs of $\sin\phi$ and $\cos\phi$. Thinking in terms of radian-measured angles, because $\sin\phi > 0$ and $\cos\phi < 0$ the terminal side of the angle ϕ lies in the second quadrant. But since the range of the inverse tangent function is the interval $(-\pi/2, \pi/2)$, $\tan^{-1}(-1/\sqrt{3}) = -\pi/6$ is a fourth-quadrant angle. The correct angle is found by using the reference angle $\pi/6$ for $\tan^{-1}(-1/\sqrt{3})$ to find the second-quadrant angle

$$\phi = \pi - \frac{\pi}{6} = \frac{5\pi}{6} \text{ radians.}$$

Therefore $y = \cos 2x - \sqrt{3}\sin 2x$ can be rewritten as

$$y = 2\sin\left(2x + \frac{5\pi}{6}\right) \quad \text{or} \quad y = 2\sin 2\left(x + \frac{5\pi}{12}\right).$$

Hence the graph of $y = \cos 2x - \sqrt{3}\sin 2x$ is the graph of $y = 2\sin 2x$, which has amplitude 2, period $2\pi/2 = \pi$, and is shifted $5\pi/12$ units to the left. ■

■ EXAMPLE 3   Example 2 Revisited

Find the first two x-intercepts on the positive x-axis of the graph of the function in Example 2.

Solution The first alternative form $y = 2\sin(2x + 5\pi/6)$ of the function in Example 2 can be used to find the x-intercepts of its graph. Recall, $\sin x = 0$ when $x = n\pi$, $n = 0, \pm 1, \pm 2, \ldots$. So by replacing the symbol x by $2x + 5\pi/6$, we see

$$\sin\left(2x + \frac{5\pi}{6}\right) = 0 \quad \text{implies} \quad 2x + \frac{5\pi}{6} = n\pi.$$

Solving

$$2x + \frac{5\pi}{6} = \pi \quad \text{and} \quad 2x + \frac{5\pi}{6} = 2\pi$$

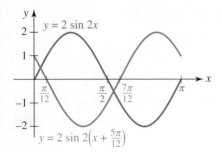

FIGURE 4.9.3 Graph of function in Example 2

FIGURE 4.9.4 An undamped spring/mass system exhibits simple harmonic motion

yields $x = \pi/12$ and $7\pi/12$. Thus the first two intercepts on the positive x-axis are $(\pi/12, 0)$ and $(7\pi/12, 0)$. The blue graph in FIGURE 4.9.3 is the portion of the graph in Figure 4.9.2 on the interval $[0, \pi]$.

We note that the x-coordinates of the x-intercepts of $y = 2\sin(2x + 5\pi/6)$ can also be obtained by subtracting the phase shift $5\pi/12$ from the x-coordinates of the x-intercepts of the red graph of $y = 2\sin 2x$ in Figure 4.9.3. ∎

□ **Simple Harmonic Motion** Consider the motion of a mass on a spring as shown in FIGURE 4.9.4. In the absence of frictional or damping forces, a mathematical model for the displacement (or directed distance) of the mass measured from a position called the **equilibrium position** is given by the function

$$y(t) = y_0\cos \omega t + \frac{v_0}{\omega}\sin \omega t, \tag{5}$$

where t is time, y_0 and v_0 are, respectively, the initial ($t = 0$) displacement and velocity of the mass.

Oscillatory motion modeled by the function (5) is said to be **simple harmonic motion**.

More precisely, we have the following definition.

DEFINITION 4.9.1 Simple Harmonic Motion

A point moving on a coordinate line whose position at time t is given by

$$y(t) = A\sin(\omega t + \phi) \qquad \text{or} \qquad y(t) = A\cos(\omega t + \phi) \tag{6}$$

where A, $\omega > 0$, and ϕ are constants, is said to exhibit **simple harmonic motion**.

Special cases of the trigonometric functions in (6) are $y(t) = A\sin \omega t$, $y(t) = A\cos \omega t$, and $y(t) = c_1\cos \omega t + c_2\sin \omega t$.

□ **Terminology** The function (5) is said to be the **equation of motion** of the mass. Also, in (5), $\omega = \sqrt{k/m}$, where k is the **spring constant** (an indicator of the stiffness of the spring), m is the **mass** attached to the spring (measured in slugs or kilograms), y_0 is the **initial displacement** of the mass (measured above or below the equilibrium position), v_0 is the **initial velocity** of the mass, t is **time** measured in seconds, and the **period** p of motion is $p = 2\pi/\omega$ seconds. The number $f = 1/p = 1/(2\pi/\omega) = \omega/2\pi$ is called the **frequency** of motion. The frequency indicates the number of cycles completed by the graph per unit time. For example, if the period of (5) is, say, $p = 2$ seconds, then we know that one cycle of the function is complete in 2 seconds. The frequency $f = 1/p = \frac{1}{2}$ means one-half of a cycle is complete in 1 second.

In the study of simple harmonic motion it is convenient to recast the equation of motion (5) as a single expression involving only the sine function:

$$y(t) = A\sin(\omega t + \phi). \tag{7}$$

The reduction of (5) to the sine function (7) can be done in exactly the same manner as illustrated in Example 2. In this situation we make the following identifications between (2) and (5):

$$c_1 = y_0, \quad c_2 = v_0/\omega, \quad A = \sqrt{c_1^2 + c_2^2}, \quad \text{and} \quad B = \omega.$$

CHAPTER 4 TRIGONOMETRIC FUNCTIONS

EXAMPLE 4   Equation of Motion

(a) Find the equation of simple harmonic motion (5) for a spring mass system if $m = \frac{1}{16}$ slug, $y_0 = -\frac{2}{3}$ ft, $k = 4$ lb/ft, and $v_0 = \frac{4}{3}$ ft/s.
(b) Find the period and frequency of motion.

Solution (a) We begin with the simple harmonic motion equation (5). Since $k/m = 4/(\frac{1}{16}) = 64$, $\omega = \sqrt{k/m} = 8$, and $v_0/\omega = (\frac{4}{3})/8 = \frac{1}{6}$, therefore (5) becomes

$$y(t) = -\tfrac{2}{3}\cos 8t + \tfrac{1}{6}\sin 8t. \tag{8}$$

(b) The period of motion is $2\pi/8 = \pi/4$ second; the frequency is $4/\pi \approx 1.27$ cycles per second.   ■

EXAMPLE 5   Example 4 Continued

Express the equation of motion (8) as a single sine function (7).

Solution With $c_1 = -\frac{2}{3}$, $c_2 = \frac{1}{6}$, we find the amplitude of motion is

$$A = \sqrt{(-\tfrac{2}{3})^2 + (\tfrac{1}{6})^2} = \tfrac{1}{6}\sqrt{17} \text{ ft.}$$

Then from

$$\left.\begin{array}{l} \sin\phi = -\frac{2}{3}/\frac{\sqrt{17}}{6} < 0 \\[2mm] \cos\phi = \frac{1}{6}/\frac{\sqrt{17}}{6} > 0 \end{array}\right\} \tan\phi = -4$$

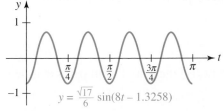

$$y = \frac{\sqrt{17}}{6}\sin(8t - 1.3258)$$

FIGURE 4.9.5 Graph of the equation of motion in Example 5

we can see from algebraic signs $\sin\phi < 0$ and $\cos\phi > 0$ that the terminal side of the angle ϕ lies in the fourth quadrant. Hence the correct value of ϕ is $\tan^{-1}(-4) \approx -1.3258$. The equation of motion is then $y(t) = \frac{1}{6}\sqrt{17}\sin(8t - 1.3258)$. As shown in FIGURE 4.9.5, the amplitude of motion is $A = \sqrt{17}/6 \approx 0.6872$. Since we are assuming that that is no resistance to the motion, once the spring/mass system is set in motion the model indicates it stays in motion bouncing back and forth between its maximum displacement $\sqrt{17}/6$ feet above the equilibrium position and a minimum of $-\sqrt{17}/6$ feet below the equilibrium position.   ■

Only in the two cases, $c_1 > 0$, $c_2 > 0$ or $c_1 < 0$, $c_2 > 0$, can we use $\tan\phi$ in (4) to write $\phi = \tan^{-1}(c_1/c_2)$. (Why?) Correspondingly, ϕ is a first or a fourth-quadrant angle.

Exercises 4.9   Answers to selected odd-numbered problems begin on page ANS–18.

In Problems 1–6, proceed as in Example 2 and write the given trigonometric function in the form $y = A\sin(Bx + \phi)$. Sketch the graph and give the amplitude, the period, and the phase shift.

1. $y = \cos \pi x - \sin \pi x$ **2.** $y = \sin\frac{\pi}{2}x - \sqrt{3}\cos\frac{\pi}{2}x$

3. $y = \sqrt{3}\sin 2x + \cos 2x$ **4.** $y = \sqrt{3}\cos 4x - \sin 4x$

5. $y = \frac{\sqrt{2}}{2}(-\sin x - \cos x)$ **6.** $y = \sin x + \cos x$

In Problems 7 and 8, proceed as in Example 3 and find the first two x-intercepts on the positive x-axis of the graph of the function.

7. $y = -\cos 2\pi x + \sin 2\pi x$ **8.** $y = \frac{1}{\sqrt{3}}\cos \pi x - \sin \pi x$

In Problems 9 and 10, use (2), (3), and (4) to write the left-hand side of the given equation in the form $A \sin(Bx + \phi)$. Then find the solutions of the equation in the indicated interval.

9. $-\cos 2x + \sin 2x = 1$; $[0, \pi]$ **10.** $\cos \dfrac{x}{2} + \sqrt{3} \sin \dfrac{x}{2} = 2$; $[0, 4]$

In Problems 11–14, proceed as in Examples 4 and 5 and use the given information to express the equation of simple harmonic motion (5) for a spring/mass system in the trigonometric form (7). Give the amplitude, period, and frequency of motion.

11. $m = \frac{1}{4}$ slug, $y_0 = \frac{1}{2}$ ft, $k = 1$ lb/ft, and $v_0 = \frac{3}{2}$ ft/s
12. $m = 1.6$ slug, $y_0 = -\frac{1}{3}$ ft, $k = 40$ lb/ft, and $v_0 = -\frac{5}{4}$ ft/s
13. $m = 1$ slug, $y_0 = -1$ ft, $k = 16$ lb/ft, and $v_0 = -2$ ft/s
14. $m = 2$ slug, $y_0 = -\frac{2}{3}$ ft, $k = 200$ lb/ft, and $v_0 = 5$ ft/s

15. The equation of simple harmonic motion of a spring/mass system is $y(t) = \frac{5}{2}\sin(2t - \pi/3)$. Determine the initial displacement y_0 and initial velocity v_0 of the mass. [*Hint*: Use (5).]
16. Use the equation of simple harmonic motion of the spring/mass system given in Problem 15 to find the times for which the mass passes through the equilibrium position $y = 0$.

Applications

17. Current Under certain conditions, the current $I(t)$ measured in amperes at time t in an electrical circuit is given by

$$I(t) = I_0[\sin(\omega t + \theta)\cos\phi + \cos(\omega t + \theta)\sin\phi].$$

Express $I(t)$ as a single sine function of the form given in (7). [*Hint*: Review the sum formula in (4) of Theorem 4.5.2.]
18. More Current In a certain kind of electrical circuit, the current $I(t)$ measured in amperes at time t seconds is given by

$$I(t) = 10\cos\left(120\pi t + \frac{\pi}{3}\right).$$

(a) Change the form of the function $I(t)$ to the sine function form given in (7).
(b) Give the period of the sine function in part (a) and the phase shift. Use the sine function to sketch two cycles of the graph of $I(t)$.

19. Pendulum Motion An object that swings back and forth is called a physical pendulum. A **simple pendulum** is a special case of the physical pendulum and consists of a rod of length L with a mass m attached at one end. See FIGURE 4.9.6. If the motion of a simple pendulum takes place in a vertical plane and it is assumed that the mass of the rod is negligible and no damping forces act on the system, then it can be shown that the displacement angle θ of the pendulum as a function of time t, measured from the vertical, is given by

$$\theta(t) = \theta_0\cos\omega t + \frac{v_0}{\omega}\sin\omega t.$$

Here θ_0 and v_0 are the initial displacement and velocity and $\omega = \sqrt{g/L}$, and $g = 32$ ft/s² is the acceleration due to gravity.
(a) Find $\theta(t)$ if $\theta_0 > 0$ and $v_0 = 0$ at $t = 0$.
(b) Show that the period (in seconds) of motion of a simple pendulum is

$$T = 2\pi\sqrt{\frac{L}{g}}$$

Clock pendulum

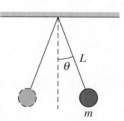

FIGURE 4.9.6 Pendulum in Problem 19

20. Pendulum Motion on the Moon Use Problem 19 to describe the motion of a simple pendulum:
 (a) On the Moon where the acceleration of due to gravity is $\frac{1}{6}$ that of the Earth.
 (b) If its length is increased to $4L$.
 (c) If its length is decreased to $\frac{1}{4}L$.

Calculator/Computer Problems

In Problems 21 and 22, use a graphing utility to obtain the graph of the given function f on the interval $[0, 2\pi]$. Use (2), (3), and (4) to write f in the form $f(x) = A\sin(Bx + \phi)$. Then find approximate solutions of indicated equation in the interval.

21. $f(x) = 3\cos 2x + 4\sin 2x$; $f(x) = 5$
22. $f(x) = 5\cos 3x - 12\sin 3x$; $f(x) = -13$

4.10 Right Triangle Trigonometry

INTRODUCTION The word *trigonometry* (from the Greek *trigonon* meaning "triangle" and *metria* meaning "measurement") refers to the measurement of triangles. In Section 4.2 we defined the trigonometric functions using coordinates of points on the unit circle and by using radian measure we were able to define the trigonometric functions of any angle. In this section we will show that the trigonometric functions of an acute angle in a right triangle have an equivalent definition in terms of the lengths of the sides of the triangle.

☐ **Terminology** In FIGURE 4.10.1(a) we have drawn a right triangle with sides labeled a, b, and c (indicating their respective lengths) and one of the acute angles denoted by θ. From the Pythagorean theorem we know that $a^2 + b^2 = c^2$. The side opposite the right angle is called the **hypotenuse**; the remaining sides are referred to as the **legs** of the triangle. The legs labeled a and b are, in turn, said to be the side **adjacent** to the angle θ and the side **opposite** the angle θ. We will also use the abbreviations **hyp**, **adj**, and **opp** to denote the lengths of these sides.

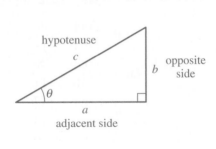

(a) A right triangle with an acute angle θ

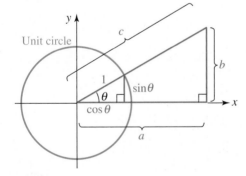

(b) The unit circle with angle θ placed in standard position

FIGURE 4.10.1 In (a) and (b) the right triangles are the same

If we place θ in standard position and draw a unit circle centered at the origin, we see from Figure 4.10.1(b) that there are two similar right triangles containing the same angle θ. Since corresponding sides of similar triangles are proportional, it follows that

$$\frac{\sin\theta}{1} = \frac{b}{c} = \frac{\text{opp}}{\text{hyp}} \qquad \text{and} \qquad \frac{\cos\theta}{1} = \frac{a}{c} = \frac{\text{adj}}{\text{hyp}}.$$

FIGURE 4.10.2 Defining the trigonometric functions of θ

Also, we have

$$\frac{\tan\theta}{1} = \frac{\sin\theta}{\cos\theta} = \frac{b/c}{a/c} = \frac{b}{a} = \frac{\text{opp}}{\text{adj}}.$$

Then, applying the reciprocal identities (12) in Section 4.4, each trigonometric function of θ can be written as the ratio of the lengths of the sides of a right triangle as follows. See FIGURE 4.10.2.

DEFINITION 4.10.1 Trigonometric Functions of θ in a Right Triangle

For an acute angle θ in a right triangle as shown in Figure 4.10.2,

$$\sin\theta = \frac{\text{opp}}{\text{hyp}} \qquad \cos\theta = \frac{\text{adj}}{\text{hyp}}$$

$$\tan\theta = \frac{\text{opp}}{\text{adj}} \qquad \cot\theta = \frac{\text{adj}}{\text{opp}} \tag{1}$$

$$\sec\theta = \frac{\text{hyp}}{\text{adj}} \qquad \csc\theta = \frac{\text{hyp}}{\text{opp}}$$

EXAMPLE 1   Values of the Six Trigonometric Functions

Find the exact values of the six trigonometric functions of the angle θ in the right triangle shown in FIGURE 4.10.3.

Solution From Figure 4.10.3 we see that the side opposite θ has length 8 and the side adjacent has length 15. From the Pythagorean theorem the hypotenuse c is

$$c^2 = 8^2 + 15^2 = 289 \qquad \text{and so} \qquad c = \sqrt{289} = 17.$$

FIGURE 4.10.3 Right triangle in Example 1

Thus from (1) the values of the six trigonometric functions are

$$\sin\theta = \frac{\text{opp}}{\text{hyp}} = \frac{8}{17}, \qquad \cos\theta = \frac{\text{adj}}{\text{hyp}} = \frac{15}{17},$$

$$\tan\theta = \frac{\text{opp}}{\text{adj}} = \frac{8}{15}, \qquad \cot\theta = \frac{\text{adj}}{\text{opp}} = \frac{15}{8},$$

$$\sec\theta = \frac{\text{hyp}}{\text{adj}} = \frac{17}{15}, \qquad \csc\theta = \frac{\text{hyp}}{\text{opp}} = \frac{17}{8}.$$

EXAMPLE 2   Using a Right Triangle Sketch

If θ is an acute angle and $\sin\theta = \frac{2}{7}$, find the values of the other trigonometric functions of θ.

Solution We sketch a right triangle with an acute angle θ satisfying $\sin\theta = \frac{2}{7}$, by making opp $= 2$ and hyp $= 7$ as shown in FIGURE 4.10.4. From the Pythagorean theorem we have

$$2^2 + (\text{adj})^2 = 7^2 \qquad \text{so that} \qquad (\text{adj})^2 = 7^2 - 2^2 = 45.$$

FIGURE 4.10.4 Right triangle in Example 2

Thus,

$$\text{adj} = \sqrt{45} = 3\sqrt{5}.$$

The values of the remaining five trigonometric functions are obtained from the definitions in (1):

$$\cos\theta = \frac{\text{adj}}{\text{hyp}} = \frac{3\sqrt{5}}{7}, \qquad\qquad \sec\theta = \frac{\text{hyp}}{\text{adj}} = \frac{7}{3\sqrt{5}} = \frac{7\sqrt{5}}{15},$$

$$\tan\theta = \frac{\text{opp}}{\text{adj}} = \frac{2}{3\sqrt{5}} = \frac{2\sqrt{5}}{15}, \qquad \cot\theta = \frac{\text{adj}}{\text{opp}} = \frac{3\sqrt{5}}{2},$$

$$\csc\theta = \frac{\text{hyp}}{\text{opp}} = \frac{7}{2}.$$

FIGURE 4.10.5 Standard labeling for a right triangle

☐ **Solving Right Triangles** Applications of right triangle trigonometry in fields such as surveying and navigation involve **solving right triangles**. The expression "to solve a triangle" means that we wish to find the length of each side and the measure of each angle in the triangle. We can solve any right triangle if we know either two sides or one acute angle and one side. As the following examples will show, sketching and labeling the triangle is an essential part of the solution process. It will be our general practice to label a right triangle as shown in FIGURE 4.10.5. The three vertices will be denoted by A, B, and C, with C at the vertex of the right angle. We denote the angles at A and B by α and β and the lengths of the sides opposite these angles by a and b, respectively. The length of the side opposite the right angle at C is denoted by c.

| EXAMPLE 3 | Solving a Right Triangle |

Solve the right triangle having a hypotenuse of length $4\sqrt{3}$ and one $60°$ angle.

Solution First we make a sketch of the triangle and label it as shown in FIGURE 4.10.6. We wish to find a, b, and β. Since α and β are complementary angles, $\alpha + \beta = 90°$ yields

$$\beta = 90° - \alpha = 90° - 60° = 30°.$$

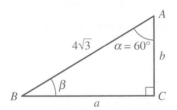

FIGURE 4.10.6 Right triangle in Example 3

We are given the length of the hypotenuse, namely, hyp $= 4\sqrt{3}$. To find a, the length of the side opposite the angle $\alpha = 60°$, we select the sine function. From $\sin\alpha = \text{opp}/\text{hyp}$, we obtain

$$\sin 60° = \frac{a}{4\sqrt{3}} \qquad \text{or} \qquad a = 4\sqrt{3}\sin 60°.$$

Since $\sin 60° = \sqrt{3}/2$, we have

$$a = 4\sqrt{3}\sin 60° = 4\sqrt{3}\left(\frac{\sqrt{3}}{2}\right) = 6.$$

Finally, to find the length b of the side adjacent to the $60°$ angle, we select the cosine function. From $\cos\alpha = \text{adj}/\text{hyp}$, we obtain

$$\cos 60° = \frac{b}{4\sqrt{3}} \qquad \text{or} \qquad b = 4\sqrt{3}\cos 60°.$$

Because $\cos 60° = \frac{1}{2}$, we find

$$b = 4\sqrt{3}\cos 60° = 4\sqrt{3}\left(\tfrac{1}{2}\right) = 2\sqrt{3}.$$

In Example 3 once we determined a, we could have found b by using either the Pythagorean theorem or the tangent function. In general, there are usually several ways to solve a triangle.

Use of a Calculator If angles other than 30°, 45°, or 60° are involved in a problem, we can obtain approximations of the desired trigonometric function values with a calculator. For the remainder of this chapter, whenever an approximation is used, we will round the final results to the nearest hundredth unless the problem specifies otherwise. To take full advantage of the calculator's accuracy, store the computed values of the trigonometric functions in the calculator for subsequent calculations. If, instead, a rounded version of a displayed value is written down and then later keyed back into the calculator, the accuracy of the final result may be diminished.

▮EXAMPLE 4   Solving a Right Triangle

Solve the right triangle with legs of length 4 and 5.

Solution After sketching and labeling the triangle as shown in FIGURE 4.10.7, we see that we need to find c, α, and β. From the Pythagorean theorem, the hypotenuse c is given by

$$c = \sqrt{5^2 + 4^2} = \sqrt{41} \approx 6.40.$$

To find β, we use $\tan\beta = $ opp/adj. (By choosing to work with the given quantities, we avoid error due to previous approximations.) Thus we have

$$\tan\beta = \tfrac{4}{5} = 0.8.$$

From a calculator set in degree mode, we find $\beta \approx 38.66°$. Since $\alpha = 90° - \beta$, we obtain $\alpha \approx 51.34°$. ▮

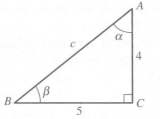

FIGURE 4.10.7 Right triangle in Example 4

Exercises 4.10   Answers to selected odd-numbered problems begin on page ANS–18.

In Problems 1–10, find the values of the six trigonometric functions of the angle θ in the given triangle.

1.

FIGURE 4.10.8 Triangle for Problem 1

2.

FIGURE 4.10.9 Triangle for Problem 2

3.

FIGURE 4.10.10 Triangle for Problem 3

4.

FIGURE 4.10.11 Triangle for Problem 4

5.

FIGURE 4.10.12 Triangle for Problem 5

6.

FIGURE 4.10.13 Triangle for Problem 6

7.

FIGURE 4.10.14 Triangle for Problem 7

8.

FIGURE 4.10.15 Triangle for Problem 8

9.

FIGURE 4.10.16 Triangle for Problem 9

10.

FIGURE 4.10.17 Triangle for Problem 10

FIGURE 4.10.18 Triangle for Problems 11–22

In Problems 11–22, find the indicated unknowns. Each problem refers to the triangle shown in **FIGURE 4.10.18**.

11. $a = 4, \beta = 27°; b, c$
12. $c = 10, \beta = 49°; a, b$
13. $b = 8, \beta = 34.33°; a, c$
14. $c = 25, \alpha = 50°; a, b$
15. $b = 1.5, c = 3; \alpha, \beta, a$
16. $a = 5, b = 2; \alpha, \beta, c$
17. $a = 4, b = 10; \alpha, \beta, c$
18. $b = 4, \alpha = 58°; a, c$
19. $a = 9, c = 12; \alpha, \beta, b$
20. $b = 3, c = 6; \alpha, \beta, a$
21. $b = 20, \alpha = 23°; a, c$
22. $a = 11, \alpha = 33.5°; b, c$

In Problems 23 and 24, solve for x in the given triangle.

23.

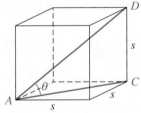

FIGURE 4.10.19 Triangle for Problem 23

24.

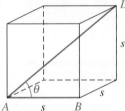

FIGURE 4.10.20 Triangle for Problem 24

In Problems 25 and 26, the cube given in the figure has a side of length s. Find the angle θ between the diagonal AC of its base and the diagonal of the cube AD (Problem 25) and the angle θ between the edge AB and the diagonal of the cube AD (Problem 26).

25.

FIGURE 4.10.21 Cube in Problem 25

26.

FIGURE 4.10.22 Cube in Problem 26

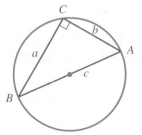

FIGURE 4.10.23 Inscribed right triangle in Problem 27

27. Inscribed Right Triangle If a right triangle is inscribed in a circle, then its hypotenuse is a diameter of the circle. In **FIGURE 4.10.23** the blue dot is the center of the circle. Find the area of the red right triangle with sides of length a, b, and c inscribed in a unit circle when the angle at vertex A is 54°.

28. Isosceles Triangle An isosceles triangle is a triangle with two sides of the same length s. As a consequence of this definition, the angles opposite the equal sides have the same measure θ. See FIGURE 4.10.24. Express the area A of an isosceles triangle in terms of s and θ.

29. Equilateral Triangle An equilateral triangle is a triangle with all three sides of the same length s. As a consequence of this definition, the angles opposite the sides have the same measure $60°$. See FIGURE 4.10.25. Express the area A of an equilateral triangle as a function of s.

FIGURE 4.10.24 Isosceles triangle in Problem 28

FIGURE 4.10.25 Equilateral triangle in Problem 29

30. Stacked Circles Suppose 10 circles of radius r are stacked within the red triangle shown in FIGURE 4.10.26. Each circle is externally tangent to its neighboring circles and each of the 9 outer circles in the stack is tangent to the red line(s).

(a) Show that the triangle is equilateral.

(b) Express the area A of the equilateral triangle as a function of the radius r.

FIGURE 4.10.26 Stacked circles in Problem 30

4.11 Applications of Right Triangles

INTRODUCTION Right triangle trigonometry can be used to solve many practical problems, particularly those involving lengths, heights, and distances.

▮EXAMPLE 1   Finding the Height of a Tree

A kite is caught in the top branches of a tree. If the 90-ft kite string makes an angle of $22°$ with the ground, estimate the height of the tree by finding the distance from the kite to the ground.

Solution Let h denote the height of the kite. From FIGURE 4.11.1 we see that

$$\frac{h}{90} = \sin 22° \quad \text{or} \quad h = 90\sin 22°.$$

A calculator set in degree mode gives $h \approx 33.71$ ft. ▮

FIGURE 4.11.1 Tree in Example 1

▮EXAMPLE 2   Length of a Saw Cut

A carpenter cuts the end of a 4-in.-wide board on a $25°$ bevel from the vertical, starting at a point $1\frac{1}{2}$ in. from the end of the board. Find the lengths of the diagonal cut and the remaining side. See FIGURE 4.11.2.

Solution Let x, y, and z be the (unknown) dimensions, as labeled in Figure 4.11.2. It follows from the definition of the tangent function that

$$\frac{x}{4} = \tan 25° \quad \text{so therefore} \quad x = 4\tan 25° \approx 1.87 \text{ in.}$$

FIGURE 4.11.2 Saw cut in Example 2

To find y we observe that

$$\frac{4}{y} = \cos 25° \qquad \text{so} \qquad y = \frac{4}{\cos 25°} \approx 4.41 \text{ in.}$$

Since $z = \frac{3}{2} + x$ and $x \approx 1.87$ in., we see that $z \approx 1.5 + 1.87 \approx 3.37$ in.

☐ **Angles of Elevation and Depression** The angle between an observer's line of sight to an object and the horizontal is given a special name. As FIGURE 4.11.3 illustrates, if the line of sight is to an object above the horizontal, the angle is called an **angle of elevation**, whereas if the line of sight is to an object below the horizontal, the angle is called an **angle of depression**.

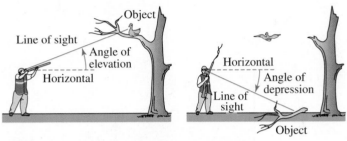

FIGURE 4.11.3 Angles of elevation and depression

EXAMPLE 3　Using Angles of Elevation

A surveyor uses an instrument called a theodolite to measure the angle of elevation between ground level and the top of a mountain. At one point the angle of elevation is measured to be 41°. A half kilometer farther from the base of the mountain, the angle of elevation is measured to be 37°. How high is the mountain?

Solution Let h represent the height of the mountain. FIGURE 4.11.4 shows that there are two right triangles sharing the common side h, so we obtain two equations in two unknowns z and h:

$$\frac{h}{z + 0.5} = \tan 37° \qquad \text{and} \qquad \frac{h}{z} = \tan 41°.$$

We can solve each of these for h, obtaining, respectively,

$$h = (z + 0.5)\tan 37° \qquad \text{and} \qquad h = z\tan 41°.$$

Equating the last two results gives an equation from which we can determine the distance z:

$$(z + 0.5)\tan 37° = z\tan 41°.$$

Solving for z gives us

$$z = \frac{-0.5\tan 37°}{\tan 37° - \tan 41°}.$$

Using $h = z\tan 41°$ we find the height h of the mountain to be

$$h = \frac{-0.5\tan 37° \tan 41°}{\tan 37° - \tan 41°} \approx 2.83 \text{ km.}$$

FIGURE 4.11.4 Mountain in Example 3

4.11 Applications of Right Triangles

EXAMPLE 4　　　　**Glide Path**

Most airplanes approach San Francisco International Airport (SFO) on a straight 3° glide path starting at a point 5.5 mi from the field. A few years ago, the FAA experimented with a computerized two-segment approach where a plane approaches the field on a 6° glide path starting at a point 5.5 mi out and then switches to a 3° glide path 1.5 mi from the point of touchdown. The point of this experimental approach was to reduce the noise of the planes over the outlying residential areas. Compare the height of a plane P' using the standard 3° approach with the height of a plane P using the experimental approach when both planes are 5.5 mi from the airport.

Solution For purposes of illustration, the angles and distances shown in FIGURE 4.11.5 are exaggerated.

FIGURE 4.11.5 Glide paths in Example 4

First, suppose y is the height of plane P' on the standard approach when it is 5.5 mi out from the airport. As we see in Figure 4.11.5(a),

$$\frac{y}{5.5} = \tan 3° \qquad \text{or} \qquad y = 5.5 \tan 3°.$$

Because distances from the airport are measured in miles, we convert y to feet

$$y = 5.5(5280)\tan 3° \,\text{ft} \approx 1522 \,\text{ft}.$$

Now, suppose z is the height of plane P on the experimental approach when it is 5.5 mi out from the airport. As shown in Figure 4.11.5(b), $z = x + w$, so we use two right triangles to obtain

$$\frac{x}{1.5} = \tan 3° \qquad \text{or} \qquad x = 1.5 \tan 3°$$

and

$$\frac{w}{4} = \tan 6° \qquad \text{or} \qquad w = 4 \tan 6°.$$

Hence the approximate height of plane P at a point 5.5 mi out from the airport is

$$z = x + w$$
$$= 1.5 \tan 3° + 4 \tan 6° \qquad \leftarrow 1 \text{ mile} = 5280 \text{ feet}$$
$$= 1.5(5280)\tan 3° + 4(5280)\tan 6° \approx 2635 \,\text{ft}.$$

In other words, plane P is approximately 1113 ft higher than plane P'. ∎

☐ **Building a Function** Section 2.9 was devoted to setting up or constructing functions that were described or expressed in words. As emphasized in that section, this is a task that you will surely face in a course in calculus. Our final example illustrates a recommended procedure of sketching a figure and labeling quantities of interest with appropriate variables.

EXAMPLE 5   Functions That Involve Trigonometry

A plane flying horizontally at an altitude of 2 miles approaches a radar station as shown in FIGURE 4.11.6.

(a) Express the distance d between the plane and the radar station as a function of the angle of elevation θ.

(b) Express the angle of elevation θ of the plane as a function of the horizontal separation x between the plane and the radar station.

Solution As shown in Figure 4.11.6, θ is an acute angle in a right triangle.

(a) We can relate the distance d and the angle θ by $\sin\theta = 2/d$. Solving for d gives

$$d(\theta) = \frac{2}{\sin\theta} \quad \text{or} \quad d(\theta) = 2\csc\theta,$$

where $0 < \theta \le 90°$.

(b) The horizontal separation x and θ are related by $\tan\theta = 2/x$. We make use of the inverse tangent function to solve for θ:

$$\theta(x) = \tan^{-1}\frac{2}{x},$$

where $0 < x < \infty$.

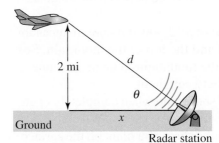

FIGURE 4.11.6 Plane in Example 5

Exercises 4.11   Answers to selected odd-numbered problems begin on page ANS–18.

FIGURE 4.11.7 Trees and river in Problem 2

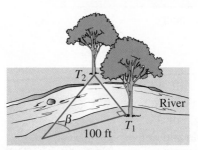

FIGURE 4.11.8 Mountain in Problem 4

FIGURE 4.11.9 Mountain in Problem 6

1. A building casts a shadow 20 m long. If the angle from the tip of the shadow to a point on top of the building is 69°, how high is the building?
2. Two trees are on opposite sides of a river, as shown in FIGURE 4.11.7. A baseline of 100 ft is measured from tree T_1, and from that position the angle β to T_2 is measured to be 29.7°. If the baseline is perpendicular to the line segment between T_1 and T_2, find the distance between the two trees.
3. A 50-ft tower is located on the edge of a river. The angle of elevation between the opposite bank and the top of the tower is 37°. How wide is the river?
4. A surveyor uses a geodometer to measure the straight-line distance from a point on the ground to a point on top of a mountain. Use the information given in FIGURE 4.11.8 to find the height of the mountain.
5. An observer on the roof of building A measures a 27° angle of depression between the horizontal and the base of building B. The angle of elevation from the same point to the roof of the second building is 41.42°. What is the height of building B if the height of building A is 150 ft? Assume buildings A and B are on the same horizontal plane.
6. Find the height h of a mountain using the information given in FIGURE 4.11.9.
7. The top of a 20-ft ladder is leaning against the edge of the roof of a house. If the angle of inclination of the ladder from the horizontal is 51°, what is the approximate height of the house and how far is the bottom of the ladder from the base of the house?
8. An airplane flying horizontally at an altitude of 25,000 ft approaches a radar station located on a 2000-ft-high hill. At one instant in time, the angle between the radar dish pointed at the plane and the horizontal is 57°. What is the

FIGURE 4.11.10 Box in Problem 10

FIGURE 4.11.11 Mountain in Problem 11

straight-line distance in miles between the airplane and the radar station at that particular instant?

9. A 5-mi straight segment of a road climbs a 4000-ft hill. Determine the angle that the road makes with the horizontal.

10. A box has dimensions as shown in FIGURE 4.11.10. Find the length of the diagonal between the corners P and Q. What is the angle θ formed between the diagonal and the bottom edge of the box?

11. Observers in two towns A and B on either side of a 12,000-ft mountain measure the angles of elevation between the ground and the top of the mountain. See FIGURE 4.11.11. Assuming that the towns and the mountaintop lie in the same vertical plane, find the horizontal distance between them.

12. A drawbridge* measures 7.5 m from shore to shore, and when completely open it makes an angle of 43° with the horizontal. See FIGURE 4.11.12(a). When the bridge is closed, the angle of depression from the shore to a point on the surface of the water below the opposite end is 27°. See Figure 4.11.12(b). When the bridge is fully open, what is the distance d between the highest point of the bridge and the water below?

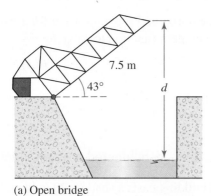

(a) Open bridge   (b) Closed bridge

FIGURE 4.11.12 Drawbridge in Problem 12

13. A flagpole is located at the edge of a sheer 50-ft cliff at the bank of a river of width 40 ft. See FIGURE 4.11.13. An observer on the opposite side of the river measures an angle of 9° between her line of sight to the top of the flagpole and her line of sight to the top of the cliff. Find the height of the flagpole.

14. From an observation site 1000 ft from the base of Mt. Rushmore the angle of elevation to the top of the sculpted head of George Washington is measured to be 80.05°, whereas the angle of elevation to the bottom of his head is 79.946°. Determine the height of George Washington's head.

Bust of George Washington on Mt. Rushmore

FIGURE 4.11.13 Flagpole in Problem 13

15. The length of a Boeing 747 airplane is 231 ft. What is the plane's altitude if it subtends an angle of 2° when it is directly above an observer on the ground? See FIGURE 4.11.14.

16. The height of a gnomon (pin) of a sundial is 4 in. If it casts a 6-in. shadow, what is the angle of elevation of the Sun?

Sundial

*The drawbridge shown in Figure 4.11.12, where the span is continuously balanced by a counterweight, is called a *bascule* bridge.

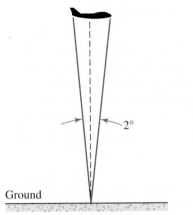

FIGURE 4.11.14 Airplane in Problem 15

FIGURE 4.11.15 Searchlight in Problem 18

17. Weather radar is capable of measuring both the angle of elevation to the top of a thunderstorm and its range (the horizontal distance to the storm). If the range of a storm is 90 km and the angle of elevation is 4°, can a passenger plane that is able to climb to 10 km fly over the storm?

18. Cloud ceiling is the lowest altitude at which solid cloud is present. The cloud ceiling at airports must be sufficiently high for safe takeoffs and landings. At night the cloud ceiling can be determined by illuminating the base of the clouds with a searchlight pointed vertically upward. If an observer is 1 km from the searchlight and the angle of elevation to the base of the illuminated cloud is 8°, find the cloud ceiling. See FIGURE 4.11.15. (During the day cloud ceilings are generally estimated by sight. However, if an accurate reading is required, a balloon is inflated so that it will rise at a known constant rate. Then it is released and timed until it disappears into the cloud. The cloud ceiling is determined by multiplying the rate by the time of the ascent; trigonometry is not required for this calculation.)

19. On a rescue flight, a U.S. Coast Guard helicopter approaches a container ship at an altitude of 1800 ft as shown in FIGURE 4.11.16. Measured from the front the helicopter, the angle of depression of the ship's stern is 35.3°and the angle of depression of its bow is 26.6°. Approximately how long is the container ship? [*Hint*: Ignore the height of the bow and stern above sea level.]

20. From a room on the top floor of a nearby hotel, the angle of elevation to the top of the Eiffel Tower is 17.5°; from street level in front of the hotel the angle of elevation to the top of the tower is 21.4°. See FIGURE 4.11.17. What is the distance from the hotel to the midpoint M, shown in the figure, of the base of the tower? Approximately how high is the hotel building?

FIGURE 4.11.16 Container ship in Problem 19

FIGURE 4.11.17 Eiffel Tower in Problem 20

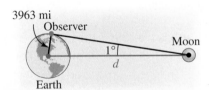

FIGURE 4.11.18 Angle in Problem 21

FIGURE 4.11.19 Lava flow in Problem 22

21. The distance between the Earth and the Moon varies as the Moon revolves around the Earth. At a particular time the **geocentric parallax** angle shown in FIGURE 4.11.18 is measured to be 1°. Calculate to the nearest hundred miles the distance between the center of the Earth and the center of the Moon at this instant. Assume that the radius of the Earth is 3963 miles.

22. The final length of a volcanic lava flow seems to decrease as the elevation of the lava vent from which it originates increases. An empirical study of Mt. Etna gives the final lava flow length L in terms of elevation h by the formula

$$L = 23 - 0.0053h,$$

where L is measured in kilometers and h is measured in meters. Suppose that a Sicilian village at elevation 750 m is on a 10° slope directly below a lava vent at 2500 m. See FIGURE 4.11.19. According to the formula, how close will the lava flow get to the village?

Mt. Etna

FIGURE 4.11.20 Weather balloon in Problem 23

23. As shown in FIGURE 4.11.20, two tracking stations S_1 and S_2 sight a weather balloon between them at elevation angles α and β, respectively. Express the height h of the balloon in terms of α and β, and the distance c between the tracking stations.

24. An entry in a soapbox derby rolls down a hill. Using the information given in FIGURE 4.11.21, find the total distance $d_1 + d_2$ that the soapbox travels.

25. Find the height and area of the isosceles trapezoid shown in FIGURE 4.11.22.

26. An escalator between the first and second floors of a department store is 58 ft long and makes an angle of $20°$ with the first floor. See FIGURE 4.11.23. Find the vertical distance between the floors.

FIGURE 4.11.21 Soapbox in Problem 24

FIGURE 4.11.22 Trapezoid in Problem 25

FIGURE 4.11.23 Escalator in Problem 26

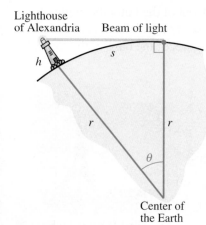

FIGURE 4.11.24 Lighthouse in Problem 28

27. Recent History According to the online encyclopedia *Wikipedia*, a French helicopter flown by Jean Boulet attained the world's record height of 12,442 m in 1972. What would the angle of elevation to the helicopter have been from a point P on the ground 2000 m from the point directly beneath the helicopter?

28. Ancient History In an article from the online encyclopedia *Wikipedia*, the height h of the Lighthouse of Alexandria, one of the Seven Wonders of the Ancient World built between 280 and 247 B.C.E., is estimated to have been between 393 ft and 450 ft. The article goes on to say that there are ancient claims that the light could be seen on the ocean up to 29 miles away. Use the right triangle in FIGURE 4.11.24 along with the two given heights h to determine the accuracy of the 29 mile claim. Assume that the radius of the Earth is $r = 3963$ mi and s is distance measured in miles on the ocean. [*Hint:* Use 1 ft = 1/5280 mi and (7) of Section 4.1.]

Artist's rendering of the Lighthouse of Alexandria

29. Really Ancient History The Great Pyramid of Giza is believed to be the tomb of the Pharaoh Khufu (AKA Cheops) and is the oldest of the Seven Wonders of the Ancient World. When completed around 2560 B.C.E. the pyramid had a square base with length 756 ft on a side and estimated height of 481 ft.

(a) Find the angle of inclination θ shown in FIGURE 4.11.25(a) that a side face of the pyramid makes with level ground.

The Great Pyramid of Giza

FIGURE 4.11.25 Pyramid in Problem 29

(b) In one of several theories on how the pyramid was constructed, workers pulled massive stone blocks up an inclined ramp made out of mud bricks and stone rumple. See Figure 4.11.25(b). The height and width of the ramp had to be changed as the height of the pyramid increased. Suppose at the point in time when the unfinished pyramid was 300 ft high, the base of the ramp measured 984 ft. What was the angle of inclination ϕ of the inclined ramp indicated in Figure 4.11.25(b)?

(c) What was the length of the ramp at the 300 ft stage? [*Hint*: Look at Figure 4.11.25(b) in cross section and use part (a).]

30. **Medieval History** The construction of the campanile, or bell tower, for the Cathedral of Pisa, Italy began in 1173 C.E. After the construction of the second floor one side of its base started sinking into the soft marshy ground. Because it continued to sink long after the completion of its construction in 1372 C.E. the campanile came to be dubbed the Leaning Tower of Pisa. Over the centuries the tower has defied many ingenious attempts to correct its tilt, but after the most recent restoration in 2001 the tower leans at angle 3.99° (corrected from 5.5°) measured from the vertical. This angle is shown in FIGURE 4.11.26 along with the heights, measured from the ground, of the high and low sides of tower belfry. Find the horizontal displacement d of the center P of the roof of the belfry from the vertical. [*Hint*: Keep in mind that the height of the point P measured from the ground is neither of the two heights given in the figure.]

Leaning Tower of Pisa

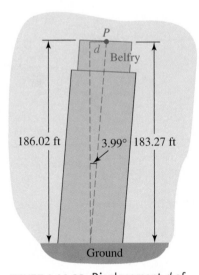

FIGURE 4.11.26 Displacement d of the point P on the belfry roof in Problem 30

FIGURE 4.11.27 Searchlight in Problem 32

In Problems 31–34, proceed as in Example 5 and translate the words into an appropriate function.

31. A tracking telescope, located 1.25 km from the point of a rocket launch, follows a vertically ascending rocket. Express the height h of the rocket as a function of the angle of elevation θ.

32. A searchlight one-half mile offshore illuminates a point P on the shore. Express the distance d from the searchlight to the point of illumination P as a function of the angle θ shown in FIGURE 4.11.27.

33. A statue is placed on a pedestal as shown in FIGURE 4.11.28. Express the viewing angle θ as a function of the distance x from the pedestal.

34. A woman on an island wishes to reach a point R on a straight shore on the mainland from a point P on the island. The point P is 9 mi from the shore and 15 mi from point R. See FIGURE 4.11.29. If the woman rows a boat at a rate of 3 mi/h to a point Q on the mainland, then walks the rest of the way at a rate of 5 mi/h, express the total time it takes the woman to reach point R as a function of the indicated angle θ. [*Hint*: Distance = rate × time.]

FIGURE 4.11.28 Viewing angle in Problem 33

FIGURE 4.11.29 Woman rowing to shore in Problem 34

For Discussion

35. Consider the blue rectangle circumscribed around the red rectangle in FIGURE 4.11.30. With the aid of calculus it can be shown that the area of the blue rectangle is greatest when $\theta = \pi/4$. Find this area in terms of a and b.

36. Home heating oil is often stored in a right circular cylindrical tank of diameter D that rests horizontally. As FIGURE 4.11.31 shows, the depth of the oil can be measured by inserting a dipstick down a vertical diameter. If the dipstick indicates that the depth of the oil is d inches, then show that the volume V of the oil is given by

$$V = \frac{V_0}{\pi}\left[\cos^{-1}\left(1 - \frac{2d}{D}\right) - 2\left(1 - \frac{2d}{D}\right)\sqrt{\left(1 - \frac{d}{D}\right)\frac{d}{D}}\right],$$

where V_0 is the volume of the tank. Although not necessary, assume for simplicity that the tank is less than half full as shown in the figure.

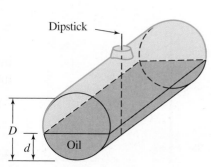

FIGURE 4.11.30 Rectangles in Problem 35

FIGURE 4.11.31 Oil tank in Problem 36

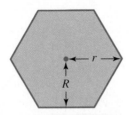

FIGURE 4.11.32 Hexagon in Problem 37

FIGURE 4.11.33 Octagon in Problem 38

37. Regular Polygon Recall, in geometry a polygon is a closed plane figure whose sides are straight-line segments. If all sides of a polygon have equal lengths and if all angles have the same measure, then the polygon is said to be a regular polygon. A regular polygon with three sides is an equilateral triangle; a four-sided regular polygon is a square. The six-sided regular polygon shown FIGURE 4.11.32 is a **hexagon**. In Figure 4.11.32 the perpendicular distance R from the center to a side of the polygon is called the *inradius* or the *apothem*. The distance r from the center to a vertex in Figure 4.11.32 is called the *circumradius* of the polygon.

(a) Use trigonometry to express the area A of a hexagon as a function of the inradius R.

(b) Express the area A of a hexagon as a function of the circumradius r.

38. Express the area A of a eight-sided regular polygon, or **octagon**, shown in FIGURE 4.11.33 as a function of the length S of one side.

39. Using trigonometric functions, generalize the three area formulas in Problems 37 and 38 to any regular polygon having n sides. [*Hint*: You might want to keep track of the number 6 in Problem 37 and the number 8 in Problem 38.]

40. Intersection of Circles–Again The centers of the red and blue circles in FIGURE 4.11.34 are B and C and have coordinates (x_0, y_0) and (x_1, y_1), respectively. Let d denote the distance between the centers.

(a) Suppose each circle has radius r. Use a triangle in Figure 4.11.34(b) to express the area A of the intersection of the circles, the yellow region in the Figure 4.11.34(a), in terms of r and θ.

(b) Show that the answer to Problem 95 in Exercises 4.1 is special case of the area formula in part (a).

(c) Find area of the intersection of the circles

$$(x - 4)^2 + (y - 4)^2 = 9 \quad \text{and} \quad (x - 6)^2 + (y - 8)^2 = 9.$$

(a)

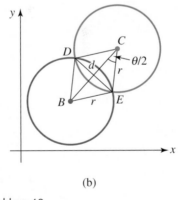

(b)

FIGURE 4.11.34 Intersecting circles in Problem 40

4.12 Law of Sines

INTRODUCTION In Section 4.10 we saw how to solve *right* triangles. In this and the next section we consider two techniques for solving an *oblique* triangle, that is, a triangle that has no right angle. An oblique triangle has either three acute angles or two acute angles and an obtuse angle.

☐ **Area of an Oblique Triangle** Before introducing the principal topic of this section, let us first examine a familiar area problem. From geometry we know that the area A of any triangle is $A = \frac{1}{2}(\text{base} \times \text{height})$. In the case of an oblique triangle we must generally use trigonometry to obtain the height h. As shown in FIGURE 4.12.1(a) we begin

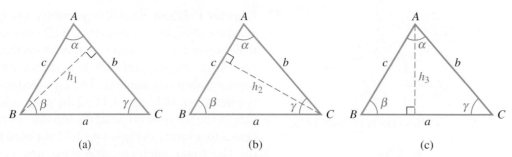

FIGURE 4.12.1 Oblique triangle with three acute angles

by constructing an altitude with length h_1 in the blue triangle from vertex B to side AC. From the right triangle whose hypotenuse has length c we see that

$$\frac{h_1}{c} = \sin\alpha \qquad \text{and so} \qquad h_1 = c\sin\alpha. \qquad (1)$$

Next, in Figure 4.12.1(b) let h_2 denote the length of the altitude from vertex C to side AB. Then from the right triangle whose hypotenuse has length a we get

$$\frac{h_2}{a} = \sin\beta \qquad \text{and so} \qquad h_2 = a\sin\beta. \qquad (2)$$

Finally, by constructing an altitude from vertex A to side BC we see from Figure 4.12.1(c) that its length h_3 can be expressed in terms γ:

$$\frac{h_3}{b} = \sin\gamma \qquad \text{and so} \qquad h_3 = b\sin\gamma. \qquad (3)$$

Using $A = \frac{1}{2}(\text{base} \times \text{height})$ for the three heights h_1, h_2, and h_3 in (1), (2), and (3) along with the respective bases of lengths b, c, and a the area A of the oblique triangle can be expressed in terms of the lengths of two sides and the included angle:

$$A = \frac{1}{2}bc\sin\alpha, \quad A = \frac{1}{2}ac\sin\beta, \quad A = \frac{1}{2}ab\sin\gamma. \qquad (4)$$

Although we used an oblique triangle in which all angles were acute, the results in (4) are equally valid for a triangle with an obtuse angle.

The formulas in (4) are certainly useful in their own right in finding area, but coincidentally we have proved a more important result.

☐ **Law of Sines** Consider the oblique triangle ABC, shown in FIGURE 4.12.2, with angles α, β, and γ, sides BC, AC, and AB with corresponding lengths a, b, and c. If we know the length of one side and two other parts of the triangle, that is, either

- one side and two angles (SAA or ASA), or
- two sides and an angle opposite one of the sides (SSA),

then the remaining three parts can be found using the **Law of Sines**.

FIGURE 4.12.2 Oblique triangle

THEOREM 4.12.1 The Law of Sines

Suppose angles α, β, and γ, opposite sides of length a, b, and c are as shown in Figure 4.12.2. Then

$$\frac{\sin\alpha}{a} = \frac{\sin\beta}{b} = \frac{\sin\gamma}{c} \qquad (5)$$

CHAPTER 4 TRIGONOMETRIC FUNCTIONS

Because the three formulas in (4) give the same area we see immediately that

$$\frac{1}{2}bc \sin \alpha = \frac{1}{2}ac \sin \beta = \frac{1}{2}ab \sin \gamma.$$

By dividing each term in this double equality by $\frac{1}{2}abc$ we obtain (5).

■ EXAMPLE 1   Solving an Oblique Triangle (SAA)

Find the remaining parts of the triangle shown in FIGURE 4.12.3.

Solution Let $\beta = 20°$, $\alpha = 130°$, and $b = 6$. Because the sum of the angles in a triangle is $180°$ we have $\gamma + 20° + 130 = 180°$ and so $\gamma = 180° - 20° - 130° = 30°$. From (5) we then see that

$$\frac{\sin 130°}{a} = \frac{\sin 20°}{6} = \frac{\sin 30°}{c}. \tag{6}$$

We use the first equality in (6) to solve for a:

$$a = 6\frac{\sin 130°}{\sin 20°} \approx 13.44.$$

The second equality in (6) gives c:

$$c = 6\frac{\sin 30°}{\sin 20°} \approx 8.77.$$ ■

A

c   6

B 20°   130°   γ   C

a

FIGURE 4.12.3 Triangle in Example 1

■ EXAMPLE 2   Height of a Building (ASA)

A building is situated on the side of a hill that slopes downward at an angle of $15°$. The Sun is uphill from the building at an angle of elevation of $42°$. Find the building's height if it casts a shadow 36 ft long.

Solution Denote the height of the building on the downward slope by h and construct a right triangle QPS as shown in FIGURE 4.12.4. Now $\alpha + 15° = 42°$ so that $\alpha = 27°$. Since ΔQPS is a right triangle, $\gamma + 42° = 90°$ gives $\gamma = 90° - 42° = 48°$. From the Law of Sines (5),

$$\frac{\sin 27°}{h} = \frac{\sin 48°}{36} \qquad \text{so} \qquad h = 36\frac{\sin 27°}{\sin 48°} \approx 21.99 \text{ ft.}$$ ■

In Examples 1 and 2, where we were given *two angles and a side opposite one of these angles*, each triangle had a unique solution. However, this may not always be true for triangles where we know *two sides and an angle opposite one of these sides*. The next example illustrates the latter situation.

Light beam

42°   Q

Building   γ h

42°   α

15°

P   S

36 ft shadow

FIGURE 4.12.4 Triangle QPS in Example 2

■ EXAMPLE 3   Two Triangles (SSA)

Find the remaining parts of the triangle with $\beta = 50°$, $b = 5$, and $c = 6$.

Solution From the Law of Sines, we have

$$\frac{\sin 50°}{5} = \frac{\sin \gamma}{6} \qquad \text{or} \qquad \sin \gamma = \frac{6}{5}\sin 50° \approx 0.9193.$$

From a calculator set in degree mode, we obtain $\gamma \approx 66.82°$. At this point it is essential to recall that the sine function is also positive for second quadrant angles. In other

FIGURE 4.12.5 Triangles in Example 3

FIGURE 4.12.6 Horizontal base, the angle β, and side AB

words, there is another angle satisfying $0° \leq \gamma \leq 180°$ for which $\sin\gamma \approx 0.9193$. Using $66.82°$ as a reference angle we find the second quadrant angle to be $180° - 66.82° = 113.18°$. Therefore, the two possibilities for γ are $\gamma_1 \approx 66.82°$ and $\gamma_2 \approx 113.18°$. Thus, as shown in FIGURE 4.12.5, there are two possible triangles ABC_1 and ABC_2 satisfying the given three conditions.

To complete the solution of triangle ABC_1 (Figure 4.12.5(a)), we first find $\alpha_1 = 180° - \gamma_1 - \beta$ or $\alpha_1 \approx 63.18°$. To find the side opposite this angle we use

$$\frac{\sin 63.18°}{a_1} = \frac{\sin 50°}{5} \qquad \text{which gives} \qquad a_1 = 5\left(\frac{\sin 63.18°}{\sin 50°}\right)$$

or $a_1 \approx 5.82$.

To complete the solution of triangle ABC_2 (Figure 4.12.5(b)), we find $\alpha_2 = 180° - \gamma_2 - \beta$ or $\alpha_2 \approx 16.82°$. Then from

$$\frac{\sin 16.82°}{a_2} = \frac{\sin 50°}{5} \qquad \text{we find} \qquad a_2 = 5\left(\frac{\sin 16.82°}{\sin 50°}\right)$$

or $a_2 \approx 1.89$. ▬

☐ **Ambiguous Case** When solving triangles, the situation where two sides and an angle opposite one of these sides are given (SSA) is called the **ambiguous case**. We have just seen in Example 3 that the given information may determine two different triangles. In the ambiguous case other complications can arise. For instance, suppose that the length of sides AB and AC (that is, c and b, respectively) and the angle β in triangle ABC are specified. As shown in FIGURE 4.12.6, we draw the angle β and mark off side AB with length c to locate the vertices A and B. The third vertex C is located on the base by drawing an arc of a circle of radius b (the length of AC) with center A. As shown in FIGURE 4.12.7, there are four possible outcomes of this construction:

- The arc does not intersect the base and no triangle is formed.
- The arc intersects the base in two distinct points C_1 and C_2 and two triangles are formed (as in Example 3).
- The arc intersects the base in one point and one triangle is formed.
- The arc is tangent to the base and a single right triangle is formed.

  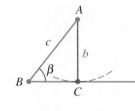

(a) No triangle   (b) Two triangles   (c) Single triangle   (d) Right triangle

FIGURE 4.12.7 Solution possibilities for the ambiguous case in the Law of Sines

EXAMPLE 4   **Determining the Parts of a Triangle (SSA)**

Find the remaining parts of the triangle with $\beta = 40°$, $b = 5$, and $c = 9$.

Solution From the Law of Sines (1), we have

$$\frac{\sin 40°}{5} = \frac{\sin\gamma}{9} \qquad \text{and so} \qquad \sin\gamma = \frac{9}{5}\sin 40° \approx 1.1570.$$

Since the sine of any angle must be between -1 and 1, $\sin\gamma \approx 1.1570$ is impossible. This means the triangle has no solution; the side with length b is not long enough to reach the base. This is the case illustrated in Figure 4.12.7(a). ▬

Exercises 4.12   Answers to selected odd-numbered problems begin on page ANS-18.

In Problems 1–4, find the area A of the given triangle.

1.

FIGURE 4.12.8 Triangle for Problem 1

2.

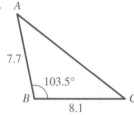

FIGURE 4.12.9 Triangle for Problem 2

3.

FIGURE 4.12.10 Triangle for Problem 3

4.

FIGURE 4.12.11 Triangle for Problem 4

In Problems 5–20, refer to Figure 4.12.2. ▶
In Problems 5–20, use the Law of Sines to solve the triangle.

5. $\alpha = 80°, \beta = 20°, b = 7$

6. $\alpha = 60°, \beta = 15°, c = 30$

7. $\beta = 37°, \gamma = 51°, a = 5$

8. $\alpha = 30°, \gamma = 75°, a = 6$

9. $\beta = 72°, b = 12, c = 6$

10. $\alpha = 120°, a = 9, c = 4$

11. $\gamma = 62°, b = 7, c = 4$

12. $\beta = 110°, \gamma = 25°, a = 14$

13. $\gamma = 15°, a = 8, c = 5$

14. $\alpha = 55°, a = 20, c = 18$

15. $\gamma = 150°, b = 7, c = 5$

16. $\alpha = 35°, a = 9, b = 12$

17. $\beta = 30°, a = 10, b = 7$

18. $\alpha = 140°, \gamma = 20°, c = 12$

19. $\alpha = 20°, a = 8, c = 27$

20. $\alpha = 75°, \gamma = 45°, b = 8$

In Problems 21 and 22, solve for x in the given triangle.

21.

FIGURE 4.12.12 Triangle for Problem 21

22.

FIGURE 4.12.13 Triangle for Problem 22

Applications

23. Length of a Pool A 10-ft rope that is available to measure the length between two points A and B at opposite ends of a kidney-shaped swimming pool is not long enough. A third point C is found such that the distance from A to C is 10 ft. It is determined that angle ACB is 115° and angle ABC is 35°. Find the distance from A to B. See FIGURE 4.12.14.

FIGURE 4.12.14 Pool in Problem 23

FIGURE 4.12.15 Telephone pole in Problem 25

24. Width of a River Two points A and B lie on opposite sides of a river. Another point C is located on the same side of the river as B at a distance of 230 ft from B. If angle ABC is $105°$ and angle ACB is $20°$, find the distance across the river from A to B.

25. Length of a Telephone Pole A telephone pole makes an angle of $82°$ with the level ground. As shown in FIGURE 4.12.15, the angle of elevation of the Sun is $76°$. Find the length of the telephone pole if its shadow is 3.5 m. (Assume that the tilt of the pole is away from the Sun and in the same plane as the pole and the Sun.)

26. Not on the Level A man 5 ft 9 in. tall stands on a sidewalk that slopes down at a constant angle. A vertical street lamp directly behind him causes his shadow to be 25 ft long. The angle of depression from the top of the man to the tip of his shadow is $31°$. Find the angle α, as shown in FIGURE 4.12.16, that the sidewalk makes with the horizontal.

27. How High? If the man in Problem 26 is 20 ft down the sidewalk from the street lamp, find the height of the light above the sidewalk.

28. Plane with an Altitude Angles of elevation to an airplane are measured from the top and the base of a building that is 20 m tall. The angle from the top of the building is $38°$, and the angle from the base of the building is $40°$. Find the altitude of the airplane.

29. Angle of Drive The distance from the tee to the green on a particular golf hole is 370 yd. A golfer hits his drive and paces its distance off at 210 yd. From the point where the ball lies, he measures an angle of $160°$ between the tee and the green. Find the angle of his drive off the tee measured from the dashed line from the tee to the green shown in FIGURE 4.12.17.

FIGURE 4.12.16 Sloping sidewalk in Problem 26

FIGURE 4.12.17 Angle of drive in Problem 29

30. In Problem 29, what is the distance from the ball to the green?

31. Help! One Coast Guard vessel is located 4 nautical miles due south of a second Coast Guard vessel when they receive a distress signal from a sailboat. To offer assistance, the first vessel sails on a bearing of S50°E at 5 knots and the second vessel sails S10°E at 10 knots. Which one of the Coast Guard vessels reaches the sailboat first? [*Hint*: The concept of bearing is reviewed on page 306.]

4.13 Law of Cosines

FIGURE 4.13.1 Right triangle

INTRODUCTION In the right triangle shown in FIGURE 4.13.1, the length c of the hypotenuse is related to the lengths a and b of the other two sides by the Pythagorean theorem

$$c^2 = a^2 + b^2. \tag{1}$$

In this section we will see that (1) is just a special case of a general formula that relates the lengths of the sides of *any* triangle.

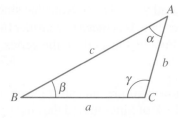

FIGURE 4.13.2 Oblique triangle

An oblique triangle, such as that in FIGURE 4.13.2, for which we know either

- three sides (SSS), or
- two sides and the included angle (that is, the angle formed by the given sides) (SAS),

cannot be solved directly using the Law of Sines. The **Law of Cosines** that we consider next can be used to solve triangles in these two cases. Like the Law of Sines, (5) of Section 4.12, the Law of Cosines is valid for any oblique triangle, but for convenience we prove the last two equations in (2) using a triangle in which the angles α, β, and γ, are acute.

THEOREM 4.13.1 The Law of Cosines

Suppose angles α, β, and γ, and opposite sides of length a, b, and c are as shown in Figure 4.13.2. Then

$$a^2 = b^2 + c^2 - 2bc\cos\alpha$$
$$b^2 = a^2 + c^2 - 2ac\cos\beta \qquad (2)$$
$$c^2 = a^2 + b^2 - 2ab\cos\gamma$$

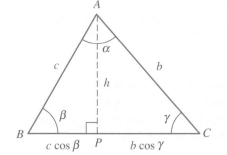

FIGURE 4.13.3 Acute triangle

PROOF: Let P denote the point where the altitude from the vertex A intersects side BC. Then, since both $\triangle BPA$ and $\triangle CPA$ in FIGURE 4.13.3 are right triangles we have from (1),

$$c^2 = h^2 + (c\cos\beta)^2 \qquad (3)$$

and

$$b^2 = h^2 + (b\cos\gamma)^2. \qquad (4)$$

Now the length of BC is $a = c\cos\beta + b\cos\gamma$ so that

$$c\cos\beta = a - b\cos\gamma. \qquad (5)$$

Moreover, from (4),

$$h^2 = b^2 - (b\cos\gamma)^2. \qquad (6)$$

Substituting (5) and (6) into (3) and simplifying yields the third equation in (2):

$$c^2 = b^2 - (b\cos\gamma)^2 + (a - b\cos\gamma)^2$$
$$= b^2 - b^2\cos^2\gamma + a^2 - 2ab\cos\gamma + b^2\cos^2\gamma$$

or

$$c^2 = a^2 + b^2 - 2ab\cos\gamma. \qquad (7)$$

Note that equation (7) reduces to the Pythagorean theorem (1) when $\gamma = 90°$.

Similarly, if we use $b\cos\gamma = a - c\cos\beta$ and $h^2 = c^2 - (c\cos\beta)^2$ to eliminate $b\cos\gamma$ and h^2 in (4), we obtain the second equation in (2). ∎

EXAMPLE 1 **Solving an Oblique Triangle (SAS)**

Find the remaining parts of the triangle shown in FIGURE 4.13.4.

Solution First, if we call the unknown side b and identify $a = 12, c = 10,$ and $\beta = 26°$, then from the second equation in (2) we can write

$$b^2 = (12)^2 + (10)^2 - 2(12)(10)\cos 26°.$$

Therefore, $b^2 \approx 28.2894$ and so $b \approx 5.32$.

Next, we use the Law of Cosines to determine the remaining angles in the triangle in Figure 4.13.4. If γ is the angle at the vertex C, then the third equation in (2) gives

$$10^2 = 12^2 + (5.32)^2 - 2(12)(5.32)\cos\gamma \qquad \text{or} \qquad \cos\gamma \approx 0.5663.$$

FIGURE 4.13.4 Triangle in Example 1

With the aid of a calculator and the inverse cosine we find $\gamma \approx 55.51°$. Note that since the cosine of an angle between $90°$ and $180°$ is negative, there is no need to consider two possibilities as we did in Example 3 in Section 4.12. Finally, the angle at the vertex A is $\alpha = 180° - \beta - \gamma$ or $\alpha \approx 98.49°$.　■

In Example 1, observe that after b is found, we know two sides and an angle opposite one of these sides. Hence we could have used the Law of Sines to find the angle γ.

In the next example we consider the case in which the lengths of the three sides of a triangle are given.

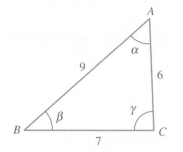

FIGURE 4.13.5 Triangle in Example 2

EXAMPLE 2　　　Determining the Angles in a Triangle (SSS)

Find the angles α, β, and γ in the triangle shown in FIGURE 4.13.5.

Solution We use the Law of Cosines to find the angle opposite the longest side:

$$9^2 = 6^2 + 7^2 - 2(6)(7)\cos\gamma \qquad \text{or} \qquad \cos\gamma = \tfrac{1}{21}.$$

A calculator then gives $\gamma \approx 87.27°$. Although we could use the Law of Cosines, we choose to find β by the Law of Sines:

$$\frac{\sin\beta}{6} = \frac{\sin 87.27°}{9} \qquad \text{or} \qquad \sin\beta = \frac{6}{9}\sin 87.27° \approx 0.6659.$$

Since γ is the angle opposite the longest side it is the largest angle in the triangle, so β must be an acute angle. Thus, $\sin\beta \approx 0.6659$ yields $\beta \approx 41.75°$. Finally, from $\alpha = 180° - \beta - \gamma$ we find $\alpha \approx 50.98°$.　■

□ **Bearing** In navigation directions are given using bearings. A **bearing** designates the acute angle that a line makes with the north–south line. For example, FIGURE 4.13.6(a) illustrates a bearing of S40°W, meaning 40 degrees west of south. The bearings in Figures 4.13.6(b) and 4.13.6(c) are N65°E and S80°E, respectively.

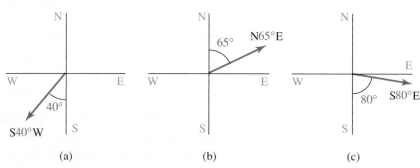

FIGURE 4.13.6 Three examples of bearings

EXAMPLE 3　　　Bearings of Two Ships (SAS)

Two ships leave a port at 7:00 AM, one traveling at 12 knots (nautical miles per hour) and the other at 10 knots. If the faster ship maintains a bearing of N47°W and the other ship maintains a bearing of S20°W, what is their separation (to the nearest nautical mile) at 11:00 AM that day?

Solution Since the elapsed time is 4 hours, the faster ship has traveled $4 \cdot 12 = 48$ nautical miles from port and the slower ship $4 \cdot 10 = 40$ nautical miles. Using these distances and the given bearings, we can sketch the triangle (valid at 11:00 AM) shown in FIGURE 4.13.7. In the triangle, c denotes the distance separating the ships and γ is the

FIGURE 4.13.7 Ships in Example 3

angle opposite that side. Since $47° + \gamma + 20° = 180°$ we find $\gamma = 113°$. Finally, the Law of Cosines

$$c^2 = 48^2 + 40^2 - 2(48)(40)\cos 113°$$

gives $c^2 \approx 5404.41$ or $c \approx 73.51$. Thus the distance between the ships (to the nearest nautical mile) is 74 nautical miles.

NOTES FROM THE CLASSROOM

(*i*) An important first step in solving a triangle is determining which of the three approaches we have discussed to use: right triangle trigonometry, the Law of Sines, or the Law of Cosines. The following table describes the various types of problems and gives the most appropriate approach for each.

Type of Triangle	Information Given	Technique
Right	Two sides or an angle and a side	Basic definitions of sine, cosine, and tangent; the Pythagorean theorem
Oblique	Two angles and a side (SAA or ASA)	Law of Sines
Oblique	Two sides and an angle opposite one of the sides (SSA)	Law of Sines (if the given angle is acute, it is an ambiguous case)
Oblique	Three sides (SSS)	Law of Cosines
Oblique	Two sides and the included angle (SAS)	Law of Cosines

(*ii*) Here are some additional bits of advice for solving triangles.

- Students will frequently use the Law of Sines when a right triangle trigonometric function could have been used. A right triangle approach is the simplest and most efficient.
- In applying the Law of Sines, if you obtain a value greater than 1 for the sine of an angle, there is no solution.
- In the ambiguous case of the Law of Sines, when solving for the first unknown angle, you must consider *both the acute angle found from your calculator and its supplement as possible solutions.* The supplement will be a solution if the sum of the supplement and the angle given in the triangle is less than 180°.
- When three sides are given, check first to see whether the length of the longest side is greater than or equal to the sum of the lengths of the other two sides. If it is, there can be no solution (even though the given information indicates a Law of Cosines approach). This is because the shortest distance between two points is the length of the line segment joining them.

Exercises 4.13   Answers to selected odd-numbered problems begin on page ANS-18.

In Problems 1–16, refer to Figure 4.13.2. ▶

In Problems 1–16, use the Law of Cosines to solve the triangle.

1. $\gamma = 65°$, $a = 5$, $b = 8$
2. $\beta = 48°$, $a = 7$, $c = 6$
3. $a = 8$, $b = 10$, $c = 7$
4. $\gamma = 31.5°$, $a = 4$, $b = 8$
5. $\gamma = 97.33°$, $a = 3$, $b = 6$
6. $a = 7$, $b = 9$, $c = 4$
7. $a = 11$, $b = 9.5$, $c = 8.2$
8. $\alpha = 162°$, $b = 11$, $c = 8$
9. $a = 5$, $b = 7$, $c = 10$
10. $a = 6$, $b = 5$, $c = 7$
11. $a = 3$, $b = 4$, $c = 5$
12. $a = 5$, $b = 12$, $c = 13$
13. $a = 6$, $b = 8$, $c = 12$
14. $\beta = 130°$, $a = 4$, $c = 7$
15. $\alpha = 22°$, $b = 3$, $c = 9$
16. $\beta = 100°$, $a = 22.3$, $b = 16.1$

Applications

17. **How Far?** A ship sails due west from a harbor for 22 nautical miles. It then sails S62°W for another 15 nautical miles. How far is the ship from the harbor?

18. **How Far Apart?** Two hikers leave their camp simultaneously, taking bearings of N42°W and S20°E, respectively. If they each average a rate of 5 km/h, how far apart are they after 1 h?

19. **Bearings** On a hiker's map point A is 2.5 in. due west of point B and point C is 3.5 in. from B and 4.2 in. from A, respectively. See FIGURE 4.13.8. Find **(a)** the bearing of A from C, and **(b)** the bearing of B from C.

20. **How Long Will It Take?** Two ships leave port simultaneously, one traveling at 15 knots and the other at 12 knots. They maintain bearings of S42°W and S10°E, respectively. After 3 h the first ship runs aground and the second ship immediately goes to its aid.
    (a) How long will it take the second ship to reach the first ship if it travels at 14 knots?
    (b) What bearing should it take?

21. **A Robotic Arm** A two-dimensional robot arm "knows" where it is by keeping track of a "shoulder" angle α and an "elbow" angle β. As shown in FIGURE 4.13.9, this arm has a fixed point of rotation at the origin. The shoulder angle is measured counterclockwise from the x-axis, and the elbow angle is measured counterclockwise from the upper to the lower arm. Suppose that the upper and lower arms are both of length 2 and that the elbow angle β is prevented from "hyperextending" beyond 180°. Find the angles α and β that will position the robot's hand at the point (1, 2).

22. **Which Way?** Two lookout towers are situated on mountain tops A and B, 4 mi from each other. A helicopter firefighting team is located in a valley at point C, 3 mi from A and 2 mi from B. Using the line between A and B as a reference, a lookout spots a fire at an angle of 40° from tower A and 82° from tower B. See FIGURE 4.13.10. At what angle, measured from CB, should the helicopter fly in order to head directly for the fire?

23. **Making a Kite** For the kite shown in FIGURE 4.13.11, use the Law of Cosines to find the lengths of the two dowels required for the diagonal supports.

24. **Bermuda Triangle** The Bermuda Triangle, also known as the Devil's Triangle, is a triangular patch of the Atlantic Ocean where it is claimed by some to be a place of paranormal activities. This claim is bolstered by the fact that, over the years, many planes and ships have disappeared without a trace within this region. The three vertices of the triangle are generally taken to be at Miami (Florida), San Juan (Puerto Rico), and the islands of Bermuda.

FIGURE 4.13.8 Triangle in Problem 19

FIGURE 4.13.9 Robotic arm in Problem 21

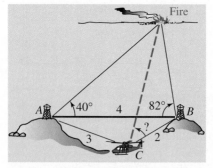

FIGURE 4.13.10 Fire in Problem 22

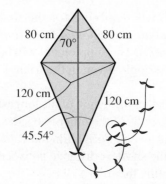

FIGURE 4.13.11 Kite in Problem 23

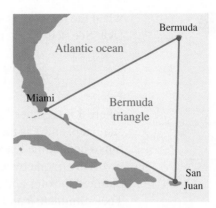

FIGURE 4.13.12 Bermuda Triangle in Problem 24

See FIGURE 4.13.12. The distance from Miami to Bermuda is 1035 mi, the distance from Bermuda to San Juan is 962 mi, and the distance from San Juan to Miami is 1033 mi.

(a) Find the three acute angles in the triangle.

(b) Determine the approximate area of the triangle.

25. **Playing Hardball** (a) A professional baseball diamond is a square 90 feet on a side with a base at each vertex. The rubber (or plate) on the pitcher's mound is 60.5 ft from home base on the line between home base and second base. See FIGURE 4.13.13. What is the distance s between the pitching rubber and first base?

(b) In softball (where the ball is *not* soft), the dimensions of the diamond vary according to age and gender of the players and whether it is fast pitch or slow pitch. What is the distance s for fast-pitch women's college softball where the diamond is 60 ft on a side and the distance from home base to the pitching rubber is 43 ft?

26. **On the Clock** Suppose the lengths of the minute and hour hands of an analog clock are 6 inches and 4.5 inches, respectively. Find the distance d between the tips of the minute and hour hands at 4 PM.

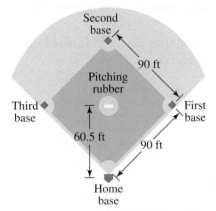

FIGURE 4.13.13 Baseball diamond in Problem 25

For Discussion

27. **Heron's Formula** Use the Law of Cosines to derive the formula

$$A = \sqrt{s(s - a)(s - b)(s - c)}$$

for the area of a triangle with sides a, b, c where $s = \frac{1}{2}(a + b + c)$. This formula is named after the Greek mathematician and inventor **Heron of Alexandria** (c. 20–62 C.E.) but should actually be credited to Archimedes.

28. **Garden Plot** Use Heron's formula in Problem 27 to find the area of a triangular garden plot if the lengths of the three sides are 25, 32, and 41 m, respectively.

29. **Corner Lot** Find the area of the irregular corner lot shown in FIGURE 4.13.14. [*Hint*: Divide the lot into two triangular lots as shown and then find the area of each triangle. Use Heron's formula in Problem 27 for the area of the acute triangle.]

30. Use Heron's formula in Problem 27 to find the area of a triangle with vertices located at $(3, 2)$, $(-3, -6)$, and $(0, 6)$ in a rectangular coordinate system.

31. **Blue Man** The effort in climbing a flight of stairs depends largely on the flexing angle of the leading knee. A simplified blue stick-figure model of a person walking up a staircase indicates that the maximum flexing of the knee occurs

FIGURE 4.13.14 Corner lot in Problem 29

FIGURE 4.13.15 Blue man in
Problem 31

when the back leg is straight and the hips are directly over the heel of the front foot. See FIGURE 4.13.15. Show that

$$\cos\theta = \left(\frac{R}{a}\right)\sqrt{4 - \left(\frac{T}{a}\right)^2} + \frac{(T/a)^2 - (R/a)^2}{2} - 1,$$

where θ is the knee joint angle, $2a$ is the length of the leg, R is the rise of a single stair step, and T is the width of a step. [*Hint*: Let h be the vertical distance from hip to heel of the leading leg, as shown in the figure. Set up two equations involving h: one by applying the Pythagorean theorem to the right triangle outlined in color and the other by using the Law of Cosines on the angle θ. Then eliminate h and solve for $\cos\theta$.]

32. For a triangle with sides of lengths a, b, and c and γ is the angle opposite c, we have seen on page 305 that when γ is a right angle the Law of Cosines reduces to the Pythagorean theorem $c^2 = a^2 + b^2$. How is c^2 related to $a^2 + b^2$ when
   (a) γ is an acute angle   (b) γ is an obtuse angle?

33. **Intersection of Circles–Finale** The centers of the red and blue circles in FIGURE 4.13.16 are B and C and have coordinates (x_0, y_0) and (x_1, y_1), respectively. Let d denote the distance between the centers. In Problem 95 in Exercises 4.1 and in Problem 40 in Exercises 4.11 you were asked to find the area of the intersection of two circles when each circle had the same radius r. Now suppose that the radii of the circles are r_0 and r_1, $r_0 \neq r_1$.
   (a) Use a triangle in Figure 4.13.16 to find a general formula for the area of the intersection of the circles in terms of r_0, r_1, θ, and ϕ.
   (b) Find area of the intersection of the circles

$$(x - 4)^2 + (y - 5)^2 = 4 \quad \text{and} \quad (x - 6)^2 + (y - 8)^2 = 9.$$

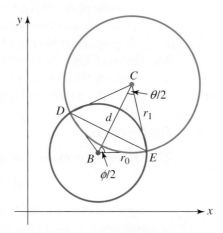

FIGURE 4.13.16 Intersecting circles in Problem 33

34. **Rhombus** A rhombus is a parallelogram in which the four sides have the same length s. See FIGURE 4.13.17. Since there is no agreement on whether a square is a rhombus, for purposes of this problem we will exclude it. In a rhombus opposite angles are equal—acute angles at vertices B and D and obtuse angles at A and B. The acute and obtuse angles are supplementary.
   (a) Use geometry to show that the area of a rhombus is the product of the base times the height, that is, $A = sh$.
   (b) Use trigonometry to show that the area of a rhombus is $A = s^2\sin\theta$, where θ is any of the four interior angles. Explain why this area formula is valid for any vertex angle in the rhombus.

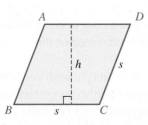

FIGURE 4.13.17 Rhombus in
Problem 34

CHAPTER 4 TRIGONOMETRIC FUNCTIONS

FIGURE 4.13.18 Quadrilateral in Problem 35

(c) Show that the area of a rhombus is $A = \frac{1}{2}d_1d_2$, where d_1 and d_2 are the lengths of the line segments, or diagonals, AC and BD.

(d) Use geometry to show the diagonals connecting opposite vertices bisect the vertex angles.

(e) Use trigonometry to show that the diagonal lines are perpendicular.
   [*Hint*: See Problem 76 in Exercises 4.7.]

35. **Quadrilateral** A quadrilateral (also called a quadrangle) is any polygon with four sides. The rhombus discussed in Problem 34 is a quadrilateral. Discuss how to find the area of the quadrilateral given in FIGURE 4.13.18. Carry out your ideas.

4.14 The Limit Concept Revisited

INTRODUCTION As we saw in Section 2.10, the fundamental motivating problem of differential calculus, *find a tangent line to the graph of the function*, is answered by the concept of a *limit*. In that section we purposely kept the discussion about limits at an intuitive level; our emphasis was on reviewing the appropriate algebra, such as factoring and rationalization, necessary to be able to compute a limit analytically. In the study of the calculus of the trigonometric functions you will, of course, be expected to compute limits involving trigonometric functions. As the examples in this section will illustrate, computation of trigonometric limits entail both algebraic manipulations and knowledge of basic trigonometric identities.

We begin with a fundamental limit result for the sine function.

☐ **An Important Trigonometric Limit** To do the calculus of the trigonometric functions, $\sin x$, $\cos x$, $\tan x$, and so on, it is important to realize that the variable x is a real number or an angle x measured in radians. With that in mind, consider the numerical values of $(\sin x)/x$ as x approaches 0 from the right ($x \to 0^+$) given in the table that follows.

$x \to 0^+$	0.1	0.01	0.001	0.0001
$\dfrac{\sin x}{x}$	0.99833416	0.99998333	0.99999983	0.99999999

It is easy to see that the same results given in the table hold as $x \to 0^-$. Because $\sin x$ is an odd function, for $x > 0$ and $-x < 0$ we have $\sin(-x) = -\sin x$ and as a consequence $\dfrac{\sin(-x)}{-x} = \dfrac{\sin x}{x}$. In other words, when the value of x is small in absolute value

$$\frac{\sin x}{x} \approx 1.$$

While numerical calculations such as this do not constitute a proof, they do suggest that $\dfrac{\sin x}{x} \to 1$ as $x \to 0$. Using the limit symbol, we have motivated the following result

$$\lim_{x \to 0} \frac{\sin x}{x} = 1. \tag{1}$$

Problem 40 in Exercises 4.14 gives a guided tour through the basic steps of a proof of (1) that is usually presented in calculus.

Important ▶   In this discussion we make the same assumption that we did in Sections 1.5 and 2.10, namely, that all limits under consideration actually exist. Everything that we do, algebraic manipulations, taking limits of products and quotients in the examples in this section is predicated on this assumption.

Other limits of importance are

$$\lim_{x \to a} \sin x = \sin a, \tag{2}$$

$$\lim_{x \to a} \cos x = \cos a. \tag{3}$$

The results (2) and (3) are immediate consequences of the fact that $f(x) = \sin x$ and $g(x) = \cos x$ are continuous functions for all x. As we have seen in Section 4.3 the graphs of $\sin x$ and $\cos x$ are smooth and unbroken. For example, from (2),

$$\lim_{x \to \pi/6} \sin x = \sin \frac{\pi}{6} = \frac{1}{2}$$

and

$$\lim_{x \to 0} \sin x = \sin 0 = 0. \tag{4}$$

Also, from (3),

$$\lim_{x \to 0} \cos x = \cos 0 = 1. \tag{5}$$

The results in (1), (2), and (3) are used often to compute other limits. As in Section 1.5 many of the limits considered in this section are limits of fractional expressions where *both* the numerator and the denominator are approaching 0. Recall, these kinds of limits are said to have the **indeterminate form** 0/0. Note that the limit (1) is of this indeterminate form.

▌EXAMPLE 1   Using (1)

Find $\displaystyle\lim_{x \to 0} \frac{10x - 3\sin x}{x}$.

Solution We rewrite the fractional expression as two fractions with the same denominator x:

$$\lim_{x \to 0} \frac{10x - 3\sin x}{x} = \lim_{x \to 0} \left[\frac{10x}{x} - \frac{3\sin x}{x} \right]$$

$$= \lim_{x \to 0} \frac{10x}{x} - 3\lim_{x \to 0} \frac{\sin x}{x} \qquad \leftarrow \text{cancel the } x \text{ in the first expression}$$

$$= \lim_{x \to 0} 10 - 3\lim_{x \to 0} \frac{\sin x}{x} \qquad \leftarrow \text{now use (1)}$$

$$= 10 - 3 \cdot 1$$

$$= 7. \qquad\qquad\blacksquare$$

▌EXAMPLE 2   Using the Double-Angle Formula

Find $\displaystyle\lim_{x \to 0} \frac{\sin 2x}{x}$.

Solution To evaluate the given limit, we make use of the double-angle formula $\sin 2x = 2\sin x \cos x$ of Section 4.5 and the results in (1) and (5):

$$\lim_{x \to 0} \frac{\sin 2x}{x} = \lim_{x \to 0} \frac{2\cos x \sin x}{x} = 2 \lim_{x \to 0} \cos x \cdot \overset{\displaystyle\substack{\text{from (5)} \quad \text{from (1)}\\ \downarrow \qquad \downarrow}}{\frac{\sin x}{x}} = 2 \cdot 1 \cdot 1 = 2.$$

Thus,

$$\lim_{x \to 0} \frac{\sin 2x}{x} = 2. \tag{6} \blacksquare$$

☐ **Using a Substitution** We are often interested in limits similar to that considered in Example 2. But if we wish to find, say, $\lim\limits_{x\to 0} \dfrac{\sin 5x}{x}$ the procedure employed in Example 2 breaks down at a practical level since we have not developed a trigonometric identity for sin 5x. There is an alternative procedure that allows us to quickly find $\lim\limits_{x\to 0} \dfrac{\sin kx}{x}$, where $k \neq 0$ is any real constant, by simply changing the variable by means of a **substitution**. If we let $t = kx$, then $x = t/k$. Notice that as $x \to 0$ then necessarily $t \to 0$. Thus we can write

$$\lim_{x\to 0} \frac{\sin kx}{x} = \lim_{t\to 0} \frac{\sin t}{t/k} = \lim_{t\to 0} \frac{\sin t}{1}\cdot\frac{k}{t} = k\lim_{t\to 0} \overset{\overset{\text{this limit is 1 from (1)}}{\downarrow}}{\frac{\sin t}{t}} = k.$$

Thus we have proved the general result

$$\lim_{x\to 0} \frac{\sin kx}{x} = k. \tag{7}$$

Hence $\lim\limits_{x\to 0} \dfrac{\sin 5x}{x} = 5$. See Problem 25 in Exercises 4.14.

EXAMPLE 3   **Trigonometric Limit**

Find $\lim\limits_{x\to 0} \dfrac{\tan x}{x}$.

Solution Using the definition $\tan x = \sin x/\cos x$ we can write

$$\lim_{x\to 0} \frac{\tan x}{x} = \lim_{x\to 0} \frac{\dfrac{\sin x}{\cos x}}{x} = \lim_{x\to 0} \frac{1}{\cos x}\cdot\frac{\sin x}{x} = \lim_{x\to 0} \frac{1}{\cos x}\cdot\frac{\sin x}{x}.$$

From (5) and (1) we know that $\cos x \to 1$ and $(\sin x)/x \to 1$ as $x \to 0$, and so the preceding line becomes

$$\lim_{x\to 0} \frac{\tan x}{x} = \frac{1}{1}\cdot 1 = 1. \qquad \blacksquare$$

EXAMPLE 4   **Using a Pythagorean Identity**

Find $\lim\limits_{x\to 0} \dfrac{1 - \cos x}{x}$.

Solution To compute this limit we start with a bit of algebraic cleverness by multiplying the numerator and denominator by the conjugate factor of the numerator. Next we use the fundamental Pythagorean identity $\sin^2 x + \cos^2 x = 1$ in the form $1 - \cos^2 x = \sin^2 x$:

$$\lim_{x\to 0} \frac{1 - \cos x}{x} = \lim_{x\to 0} \frac{1 - \cos x}{x}\cdot\frac{1 + \cos x}{1 + \cos x}$$

$$= \lim_{x\to 0} \frac{1 - \cos^2 x}{x(1 + \cos x)}$$

$$= \lim_{x\to 0} \frac{\sin^2 x}{x(1 + \cos x)}.$$

For the next step we resort back to algebra to rewrite the fractional expression as a product, then use the results in (1), (4), and (5):

$$\lim_{x\to 0} \frac{1 - \cos x}{x} = \lim_{x\to 0} \frac{\sin^2 x}{x(1 + \cos x)}$$

$$= \lim_{x\to 0} \frac{\sin x}{x} \cdot \frac{\sin x}{1 + \cos x}$$

$$= 1 \cdot \frac{0}{2} \tag{8}$$

$$= 0.$$

That is,
$$\lim_{x\to 0} \frac{1 - \cos x}{x} = 0. \qquad\blacksquare$$

From (8) we obtain a limit result that is used in calculus to find the derivatives of the sine and cosine functions. Since the limit in (8) is equal to 0, we can write

$$\lim_{x\to 0} \frac{1 - \cos x}{x} = \lim_{x\to 0} \frac{-(\cos x - 1)}{x} = (-1)\lim_{x\to 0} \frac{\cos x - 1}{x} = 0.$$

Dividing by -1 then gives

$$\lim_{x\to 0} \frac{\cos x - 1}{x} = 0. \tag{9}$$

☐ **The Calculus Connection** In Section 2.10 we saw that the derivative of a function $y = f(x)$ is the function $f'(x)$ defined by a limit of a difference quotient:

$$f'(x) = \lim_{h\to 0} \frac{f(x + h) - f(x)}{h}. \tag{10}$$

In computing this limit we shrink h to zero, but x is held fixed. Recall too, if a number $x = a$ is in the domains of f and f', then $f(a)$ is the y-coordinate of the point of tangency $(a, f(a))$ and $f'(a)$ is the slope of the tangent line at that point.

☐ **Derivatives of $f(x) = \sin x$ and $f(x) = \cos x$** To find the derivative of $f(x) = \sin x$ we use the four-step process illustrated in Example 3 of Section 2.10. In the first step we use from Section 4.5 the sum formula for the sine function:

$$\sin(x_1 + x_2) = \sin x_1 \cos x_2 + \cos x_1 \sin x_2. \tag{11}$$

(*i*) With x and h playing the parts of x_1 and x_2, we have from (11):

$$f(x + h) = \sin(x + h) = \sin x \cos h + \cos x \sin h.$$

(*ii*) $f(x + h) - f(x) = \sin x \cos h + \cos x \sin h - \sin x$
$$= \sin x(\cos h - 1) + \cos x \sin h.$$

As we see in the next line, we cannot cancel the h's in the difference quotient, but we can rewrite the expression to make use of the limit results in (1) and (9).

(*iii*) $\dfrac{f(x + h) - f(x)}{h} = \dfrac{\sin x(\cos h - 1) + \cos x \sin h}{h}$

$$= \sin x \frac{\cos h - 1}{h} + \cos x \frac{\sin h}{h}.$$

(iv) In this line, the symbol h plays the part of the symbol x in (1) and (9):

$$f'(x) = \lim_{h \to 0} \frac{f(x + h) - f(x)}{h} = \sin x \lim_{h \to 0} \frac{\cos h - 1}{h} + \cos x \lim_{h \to 0} \frac{\sin h}{h}.$$

From the limit results in (1) and (9), the last line is the same as

$$f'(x) = \lim_{h \to 0} \frac{f(x + h) - f(x)}{h} = \sin x \cdot 0 + \cos x \cdot 1 = \cos x.$$

In summary:

- The derivative of $f(x) = \sin x$ is $f'(x) = \cos x$.   (12)

It is left to you the student to show that:

- The derivative of $f(x) = \cos x$ is $f'(x) = -\sin x$.   (13)

See Problems 23 and 24 in Exercises 4.14.

EXAMPLE 5   Equation of a Tangent Line

Find an equation of the tangent line to the graph of $f(x) = \sin x$ at $x = 4\pi/3$.

Solution We start by finding the point of tangency. From

$$f\left(\frac{4\pi}{3}\right) = \sin\frac{4\pi}{3} = -\frac{\sqrt{3}}{2}$$

we see that the point of tangency is $\left(4\pi/3, -\sqrt{3}/2\right)$. The slope of the tangent line at that point is the derivative of $f(x) = \sin x$ evaluated at the x-coordinate. From (12) we know that $f'(x) = \cos x$ and so the slope at $\left(4\pi/3, -\sqrt{3}/2\right)$ is

$$f'\left(\frac{4\pi}{3}\right) = \cos\frac{4\pi}{3} = -\frac{1}{2}.$$

From the point-slope form of a line, an equation of the tangent line is

$$y + \frac{\sqrt{3}}{2} = -\frac{1}{2}\left(x - \frac{4\pi}{3}\right) \quad \text{or} \quad y = -\frac{1}{2}x + \frac{2\pi}{3} - \frac{\sqrt{3}}{2}.$$

See FIGURE 4.14.1.

FIGURE 4.14.1 Tangent line in Example 5

Exercises 4.14   Answers to selected odd-numbered problems begin on page ANS–18.

In Problems 1–18, use the results in (1), (2), (3), (7), and (9) to find the indicated limit.

1. $\lim_{x \to 0} \dfrac{\sin\frac{1}{2}x}{x}$

2. $\lim_{x \to 0} \dfrac{\sin \pi x}{x}$

3. $\lim_{\theta \to 0} \dfrac{\sin(-\theta)}{\theta}$

4. $\lim_{t \to 0} \dfrac{\sin 3t}{4t}$

5. $\lim_{x \to 5\pi/6} \cos x$

6. $\lim_{x \to \pi/4} \sin x$

7. $\lim_{x \to \pi/2} (\cos x + 5 \sin x)$

8. $\lim_{x \to \pi/6} \cos x \sin x$

9. $\lim_{x \to 0} \dfrac{\cos x - 1}{10x}$

10. $\lim_{\theta \to 0} \dfrac{8(1 - \cos\theta)}{\theta}$

11. $\displaystyle\lim_{x \to 0} \frac{4x^2 - 2\sin x}{x}$

12. $\displaystyle\lim_{x \to 0} \frac{2\sin 4x + 1 - \cos x}{x}$

13. $\displaystyle\lim_{x \to 0} \frac{\sin^2 x}{x}$

14. $\displaystyle\lim_{x \to 0} \frac{\sin^2 x}{x^2}$

15. $\displaystyle\lim_{x \to \pi/2} \frac{\cos x}{\cot x}$

16. $\displaystyle\lim_{x \to 0} \frac{\cos x \tan x}{x}$

17. $\displaystyle\lim_{x \to 0} x \cot x$

18. $\displaystyle\lim_{x \to \pi/4} \frac{\cos 2x}{\cos x - \sin x}$

In Problems 19–22, proceed as in Example 5 to find an equation of the tangent line to the graph of $f(x) = \sin x$ at the indicated value of x.

19. $x = 0$

20. $x = \pi/2$

21. $x = \pi/6$

22. $x = 2\pi/3$

23. Proceed as on pages 314–315 and find the derivative of $f(x) = \cos x$.

24. Use the result of Problem 23 to find an equation of the tangent line to the graph of $f(x) = \cos x$ at $x = \pi/3$.

25. Use the facts that

$$\lim_{x \to 0} \frac{\cos 5x - 1}{x} = 0 \qquad \text{and} \qquad \lim_{x \to 0} \frac{\sin 5x}{x} = 5$$

to find the derivative of $f(x) = \sin 5x$.

26. Use the result of Problem 25 to find an equation of the tangent line to the graph of $f(x) = \sin 5x$ at $x = \pi$.

Calculator/Computer Problems

In Problems 27 and 28, use a calculator or computer to estimate the given limit by completing each table. Round the entries in each table to eight decimal places.

27. $\displaystyle\lim_{x \to 0} \frac{1 - \cos x}{x^2}$

$x \to 0^+$	0.1	0.01	0.001	0.0001	0.00001
$\dfrac{1 - \cos x}{x^2}$					

Explain why we do not have to consider $x \to 0^-$.

28. $\displaystyle\lim_{x \to 2} \frac{x^2 - 4}{\sin(x - 2)}$

$x \to 2^+$	2.1	2.01	2.001	2.0001	2.00001
$\dfrac{x^2 - 4}{\sin(x - 2)}$					

$x \to 2^-$	1.9	1.99	1.999	1.9999	1.99999
$\dfrac{x^2 - 4}{\sin(x - 2)}$					

For Discussion

In Problems 29–36, discuss how to use the result in (1) along with some clever algebra, trigonometry, or a substitution to find the given limit.

29. $\displaystyle\lim_{x\to 0}\frac{x}{\sin 3x}$

30. $\displaystyle\lim_{x\to 0}\frac{\sin 4x}{\sin 5x}$

31. $\displaystyle\lim_{x\to 0}\frac{\sin x^2}{x^2}$

32. $\displaystyle\lim_{x\to \pi}\frac{\sin x}{\pi - x}$

33. $\displaystyle\lim_{x\to 0}\frac{x^2}{1 - \cos x}$

34. $\displaystyle\lim_{x\to 0}\frac{\cos\left(x + \frac{1}{2}\pi\right)}{x}$

35. $\displaystyle\lim_{x\to 0^+}\frac{\sin x}{\sqrt{x}}$

36. $\displaystyle\lim_{x\to 1}\frac{\sin(x - 1)}{x^2 + 2x - 3}$

37. Using what you have learned in Problems 29 and 36, find the limit

$$\lim_{x\to 2}\frac{x^2 - 4}{\sin(x - 2)}$$

without the aid of the numerical table in Problem 28.

38. **(a)** Use a calculator to complete the following table.

$x \to 0^+$	0.1	0.01	0.001	0.0001	0.00001
$\dfrac{1 - \cos x^2}{x^4}$					

(b) Find the limit $\displaystyle\lim_{x\to 0}\frac{1 - \cos x^2}{x^4}$ using the method given in Example 4.

(c) Discuss any differences that you observe between parts **(a)** and **(b)**.

39. **(a)** A regular n-gon is an n-sided polygon inscribed in a circle; the polygon is formed by n equally spaced points on the circle. Suppose the polygon shown in FIGURE 4.14.2 represents a regular n-gon inscribed in a circle of radius r. Use trigonometry to show that the area $A(n)$ of the n-gon is given by

$$A(n) = \frac{n}{2}r^2 \sin\left(\frac{2\pi}{n}\right).$$

See Problems 37–39 in Exercises 4.11.

(b) It stands to reason that the area $A(n)$ approaches the area of the circle as the number of sides of the n-gon increases. Compute A_{100} and A_{1000}.

(c) Let $x = 2\pi/n$ in $A(n)$ and note that as $n \to \infty$ then $x \to 0$. Use (1) of this section to show that $\displaystyle\lim_{n\to\infty}A(n) = \pi r^2$.

40. Consider a circle centered at the origin O with radius 1. As shown in FIGURE 4.14.3(a), let the shaded region OPR be a sector of the circle with central angle t such that $0 < t < \pi/2$. We see from Figures 4.14.3(a)–4.14.3(d) that

$$\text{area of } \triangle OPR < \text{area of sector } OPR < \text{area of } \triangle OQR. \tag{12}$$

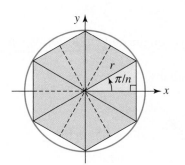

FIGURE 4.14.2 Inscribed n-gon in Problem 39

| (a) Unit circle | (b) Triangle OPR | (c) Sector OPR | (d) Right triangle OQR |

FIGURE 4.14.3 Unit circle in Problem 40

(a) Show that the area of $\triangle OPR$ is $\frac{1}{2}\sin t$ and that the area of $\triangle OQR$ is $\frac{1}{2}\tan t$.

(b) Since the area of a sector of a circle is $\frac{1}{2}r^2\theta$, where r is its radius and θ is measured in radians, it follows that the area of sector OPR is $\frac{1}{2}t$. Use this result, along with the areas in part (a), to show that the inequality in (12) yields

$$\cos t < \frac{\sin t}{t} < 1.$$

(c) Discuss how the preceding inequality proves (1) when we let $t \to 0^+$.

Chapter 4 Review Exercises   Answers to selected odd-numbered problems begin on page ANS-19.

A. Fill in the Blanks

In Problems 1–20, fill in the blanks.

1. $\pi/5$ radians = _____ degrees.
2. 10 degrees = _____ radians.
3. The exact values of the coordinates of the point $P(t)$ on the unit circle corresponding to $t = 5\pi/6$ are _____.
4. The reference angle for $4\pi/3$ radians is _____ radians.
5. $\tan\dfrac{\pi}{3} = $ _____.
6. In standard position, the terminal side of the angle $\dfrac{8\pi}{5}$ radians lies in the _____ quadrant.
7. If $\sin\theta = -\frac{1}{3}$ and θ is in quadrant IV, then $\sec\theta = $ _____.
8. If $\tan t = 2$ and t is in quadrant III, then $\cos t = $ _____.
9. The y-intercept for the graph of the function $y = 2\sec(x + \pi)$ is _____.
10. The values of t in the interval $[0, 2\pi]$ that satisfy $\sin 2t = \frac{1}{2}$ are_____.
11. If $\sin u = \frac{3}{5}, 0 < u < \pi/2$, and $\cos v = 1/\sqrt{5}, 3\pi/2 < v < 2\pi$, then $\cos(u + v) = $ _____.
12. If $\cos t = -\frac{2}{3}, \pi < t < 3\pi/2$, then $\cos\frac{1}{2}t = $ _____.
13. A sine function with period 4 and amplitude 6 is given by _____.
14. The first vertical asymptote for the graph of $y = \tan\left(x - \dfrac{\pi}{4}\right)$ to the right of the y-axis is _____.

15. $\sin t + \cos t = \underline{\quad} \sin\left(t + \dfrac{\pi}{4}\right)$.

16. If $\sin t = \frac{1}{6}$, then $\cos\left(t - \dfrac{\pi}{2}\right) = \underline{\quad}$.

17. The amplitude of $y = -10\cos\dfrac{\pi}{3}x$ is $\underline{\quad}$.

18. $\cos\left(\dfrac{\pi}{6} - \dfrac{5\pi}{4}\right) = \underline{\quad}$.

19. The exact value of $\arccos\left(\cos\dfrac{9\pi}{5}\right) = \underline{\quad}$.

20. The period of the function $y = 2\sin\dfrac{\pi}{3}t$ is $\underline{\quad}$.

B. True/False

In Problems 1–20, answer true or false.

1. If $\tan t = \frac{3}{4}$, then $\sin t = 3$ and $\cos t = 4$. $\underline{\quad}$
2. In a right triangle, If $\sin\theta = \frac{11}{61}$, then $\cot\theta = \frac{60}{11}$. $\underline{\quad}$
3. $\sec(-\pi) = \csc\dfrac{3\pi}{2}$ $\underline{\quad}$
4. There is no angle t such that $\sec t = \frac{1}{2}$. $\underline{\quad}$
5. $\sin(2\pi - t) = -\sin t$ $\underline{\quad}$
6. $1 + \sec^2\theta = \tan^2\theta$ $\underline{\quad}$
7. $(5, 0)$ is an x-intercept of the graph of $y = 3\sin\pi x$. $\underline{\quad}$
8. $\left(2\pi/3, -1/\sqrt{3}\right)$ is a point on the graph of $y = \cot x$. $\underline{\quad}$
9. The range of the function $y = \csc x$ is $(-\infty, -1] \cup [1, \infty)$. $\underline{\quad}$
10. The graph of $y = \csc x$ does not intersect the y-axis. $\underline{\quad}$
11. The line $x = \pi/2$ is a vertical asymptote for the graph of $y = \tan x$. $\underline{\quad}$
12. If $\tan(x + \pi) = 0.3$, then $\tan x = 0.3$. $\underline{\quad}$
13. For the sine function $y = -2\sin x$ we have $-2 \le y \le 2$. $\underline{\quad}$
14. $\sin 6x = 2\sin 3x\cos 3x$ $\underline{\quad}$
15. The graph of $y = \sin\left(2x - \dfrac{\pi}{3}\right)$ is the graph of $y = \sin 2x$ shifted $\pi/3$ units to the right. $\underline{\quad}$
16. Since $\tan(5\pi/4) = 1$, then $\arctan(1) = 5\pi/4$. $\underline{\quad}$
17. $\arcsin\left(\frac{1}{2}\right) = 30°$ $\underline{\quad}$
18. $f(x) = \arcsin x$ is not periodic. $\underline{\quad}$
19. $f(x) = x\sin x$ is 2π periodic. $\underline{\quad}$
20. $f(x) = \sin(\cos x)$ is an even function. $\underline{\quad}$

C. Review Exercises

In Problems 1–6, find all t in the interval $[0, 2\pi]$ that satisfy the given equation.

1. $\cos t\sin t - \cos t + \sin t - 1 = 0$ **2.** $\cos t - \sin t = 0$
3. $4\sin^2 t - 1 = 0$ **4.** $\sin t = 2\tan t$
5. $\sin t + \cos t = 1$ **6.** $\tan t - 3\cot t = 2$

In Problems 7–10, solve the triangle satisfying the given conditions.

7. $\alpha = 30°, \beta = 70°, b = 10$
8. $\gamma = 145°, a = 25, c = 20$
9. $\alpha = 51°, b = 20, c = 10$
10. $a = 4, b = 6, c = 3$

In Problems 11–18, find the indicated value without using a calculator.

11. $\cos^{-1}\left(-\frac{1}{2}\right)$
12. $\arcsin(-1)$
13. $\cot\left(\cos^{-1}\frac{3}{4}\right)$
14. $\cos\left(\arcsin\frac{2}{5}\right)$
15. $\sin^{-1}(\sin\pi)$
16. $\cos(\arccos 0.42)$
17. $\sin\left(\arccos\left(\frac{5}{13}\right)\right)$
18. $\arctan(\cos\pi)$

In Problems 19 and 20, write the given expression as an algebraic expression in x.

19. $\sin(\arccos x)$
20. $\sec\left(\tan^{-1}x\right)$

In Problems 21–24, the given graph can be interpreted as a rigid/nonrigid transformation of the graph of $y = \sin x$ and of the graph of $y = \cos x$. Find an equation of the graph using the sine function. Then find an equation of the same graph using the cosine function.

21.

FIGURE 4.R.1 Graph for Problem 21

22.

FIGURE 4.R.2 Graph for Problem 22

23.

FIGURE 4.R.3 Graph for Problem 23

24.

FIGURE 4.R.4 Graph for Problem 24

FIGURE 4.R.5 Cliff in Problem 25

FIGURE 4.R.6 Water-skier in Problem 27

25. A surveyor 100 m from the base of an overhanging cliff measures a 28° angle of elevation from that point to the top of the cliff. See FIGURE 4.R.5. If the cliff makes an angle of 65° with the horizontal ground, determine its height h.

26. A rocket is launched from ground level at an angle of elevation of 43°. If the rocket hits a drone target plane flying at 20,000 ft, find the horizontal distance between the rocket launch site and the point directly beneath the plane. What is the straight-line distance between the rocket launch site and the target plane?

27. A competition water-skier leaves a ramp at point R and lands at S. See FIGURE 4.R.6. A judge at point J measures an $\angle RJS$ as 47°. If the distance from the ramp to the judge is 110 ft, find the length of the jump. Assume that $\angle SRJ$ is 90°.

28. The angle between two sides of a parallelogram is 40°. If the lengths of the sides are 5 and 10 cm, find the lengths of the two diagonals.

29. A weather satellite orbiting the equator of the Earth at a height of $H = 36,000$ km spots a thunderstorm to the north at P at an angle of $\theta = 6.5°$ from its vertical. See FIGURE 4.R.7.

CHAPTER 4 TRIGONOMETRIC FUNCTIONS

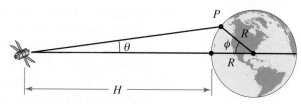

FIGURE 4.R.7 Satellite in Problem 29

(a) Given that the Earth's radius is approximately $R = 6370$ km, find the latitude ϕ of the thunderstorm.

(b) Show that angles θ and ϕ are related by

$$\tan\theta = \frac{R\sin\phi}{H + R(1 - \cos\phi)}.$$

FIGURE 4.R.8 Basketball in Problem 30

30. It can be shown that a basketball of diameter d approaching the basket from an angle θ to the horizontal will pass through a hoop of diameter D if $D\sin\theta > d$, where $0° \leq \theta \leq 90°$. See FIGURE 4.R.8. If the basketball has diameter 24.6 cm and the hoop has diameter 45 cm, what range of approach angles θ will result in a basket?

FIGURE 4.R.9 GPS satellite in Problem 31

31. Each of the 24 NAVSTAR Global Positioning System (GPS) satellites orbits the Earth at an altitude of $h = 20{,}200$ km. Using this network of satellites, an inexpensive hand-held GPS receiver can determine its position on the surface of the Earth to within 10 m. Find the greatest distance s (in km) on the surface of the Earth that can be observed from a single GPS satellite. See FIGURE 4.R.9. Take the radius of the Earth to be 6370 km. [*Hint*: Find the central angle θ subtended by s.]

32. An airplane flying horizontally at a speed of 400 miles per hour is climbing at an angle of 6° from the horizontal. When it passes directly over a car traveling 60 miles per hour, it is 2 miles above the car. Assuming that the airplane and the car remain in the same vertical plane, find the angle of elevation from the car to the airplane after 30 minutes.

33. A house measures 45 ft from front to back. The roof measures 32 ft from the front of the house to the peak and 18 ft from the peak to the back of the house. See FIGURE 4.R.10. Find the angles of elevation of the front and back parts of the roof.

34. A regular five-sided polygon is called a regular pentagon. See Figure 4.14.2. Using radian measure, determine the sum of the vertex angles in a regular pentagon.

35. In FIGURE 4.R.11 the blue, green, and red circles are of radii 3, 4, and 6, respectively. The dots represent the centers of the circles. Let d denote the distance between the centers of the blue and red circles. Determine the angle θ shown in the figure that corresponds to $d = 14$.

FIGURE 4.R.10 House in Problem 33

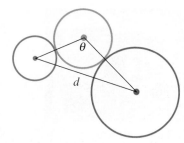

FIGURE 4.R.11 Circles in Problem 35

36. **Navigator's Error** An airplane is supposed to fly 500 mi due west to a refueling rendezvous point. If a 5°error is made in the heading, how far is the plane from the rendezvous point after flying 400 mi? Through what angle must the airplane turn in order to correct its course at that point?

37. **National Historic Landmark** Completed in 1902 on a triangular city block, the Flatiron Building in New York City was declared a National Historic Landmark in 1989. See FIGURE 4.R.12. The original 21 story stone clad steel-frame building is considered to be one of the first skyscrapers built in the city. The sides of the building measure 173 ft along Fifth Avenue, 87 ft along East 22nd Street, and 190 ft along Broadway.

(a) Show that the base of the building is approximately a right triangle.

(b) Assuming that the base of the building is a right triangle, find the two acute angles in it.

Flatiron Building in
New York, NY

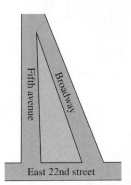

FIGURE 4.R.12 Triangular
block in Problem 37

38. **Volcanic Cones** Viewed from the side, a volcanic cinder cone usually looks like an isosceles trapezoid. See FIGURE 4.R.13. Studies of cinder cones that are less than 50,000 years old indicate that cone height H_{co} and crater width W_{cr} are related to the cone width W_{co} by the equations $H_{co} = 0.18\ W_{co}$ and $W_{cr} = 0.40\ W_{co}$. If $W_{co} = 1.00$, use these equations to determine the base angle ϕ of the trapezoid in Figure 4.R.13.

Volcanic cinder cones in
Haleakala Crater, Maui, Hawaii

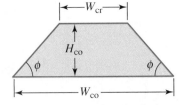

FIGURE 4.R.13 Volcanic cinder
cone in Problem 38

39. **Angels Flight** Claimed to be the world's shortest railway, Angels Flight is a funicular railway consisting of two cars (named Olivet and Sinai) that transports people up and down the steep hill between Hill Street and California Plaza in downtown Los Angeles, CA. The original railway dates back to 1901 and, in its present form, is only 298 ft long. If the angle of elevation of the tracks at its base on Hill Street is 33°, then how high is the hill?

40. **Distance Across a Canyon** From the floor of a canyon it takes 62 ft of rope to reach the top of one canyon wall and 86 ft to reach the top of the opposite wall.

Angels Flight in downtown
Los Angeles

FIGURE 4.R.14 Canyon in Problem 40

See FIGURE 4.R.14. If the two ropes make an angle of 123°, what is the distance d from the top of one canyon wall to the other?

41. How Fast? An observer at a horizontal distance of 1 mile (5280 ft) watches an F-15E Strike Eagle fighter jet go into a vertical climb. See FIGURE 4.R.15. If the angle of elevation of the jet at the observer changes from the initial measurement of 43.44° to 75.21° in 30 seconds, then how fast (in feet per minute) is its rate of climb?

US Air Force F-15E Strike Eagle

FIGURE 4.R.15 Climbing F-15E in Problem 41

42. Building Height Two buildings were constructed on a inclined lot as shown in FIGURE 4.R.16. The angle of elevation from the right side of the roof of the brown building to the left side of the roof of the gray building is 23°. From the same spot on the roof of the brown building, the angle of depression to the base of the gray building is 48°. Use the additional information in the figure to determine the heights of the facing sides of the buildings relative to the inclined lot.

FIGURE 4.R.16 Buildings in Problem 42

One way of finding the height of a tree: climb it

43. Estimating Tree Height There are many sophisticated instruments, such as range finders and inclinometers (or clinometer), that are invaluable in determining an accurate measurement of the height of an object. A nontechnical method for determining the height of, say, a tree is to climb the tree and then drop the weighted end of a measuring-tape line to the ground. Assuming that you have no measuring devices and that climbing a tree is not practical (or even legal), explain why the following method gives an approximation to the height of a tree:

Find a patch of level ground containing the tree. Guess the distance from the ground to your eye level and the length of your walking stride. Find two straight sticks, back away from the tree (counting your strides) holding the sticks at eye level like this ∠ *(one parallel to the ground and the other adjusted by following the top of the tree). Stop when you think the angle between the sticks is* 45°. *The height of the tree is approximately:*

(*number of strides*) × (*stride length*) + *eye level height.*

44. **Gate of Europe** The *Puerta de Europa* towers are twin office buildings in Madrid, Spain. To accommodate a required setback from the wide *Paseo de la Castellana*, the towers were built in 1996 at an angle of 15° from the vertical. In the photo, note the vertical line on the side of each building. The sides of the buildings shown in FIGURE 4.R.17 are congruent parallelograms. Use the information in the figure to find the lengths s_1 and s_2 of the sides of the parallelogram and the distance d between the roofs of the towers.

Puerta de Europa towers and the *Paseo de la Castellana* in Madrid, Spain

FIGURE 4.R.17 Towers in Problem 44

45. **Stairs in Homes** In home construction, stairs are usually constructed using two stringers, which are boards that have been notched to accommodate the treads (steps) and the risers. Suppose the stairs consists of 9 risers, the riser height is 7.75 inches, and the tread depth is 10 inches. See FIGURE 4.R.18. Use two different methods to find the approximate length L of a stringer.

Stairs showing a stringer

FIGURE 4.R.18 Stringer in Problem 45

46. **El Castillo** The most prominent feature in the archaeological site of Chichén Itzá is a step pyramid on whose flat top rests a temple to the Mayan feathered-serpent god Kukulkán. The Kukulkán pyramid, more commonly known by the Spanish name *El Castillo*, was built around 900 C.E. and is located in the Mexican state of Yucátan. On each of the four faces of the pyramid there is a protruding stone-block stairway rising to the 79 ft high temple level, although only two of the staircases

have been completely restored. The angle of inclination a stairway relative to the ground is 45° whereas the angle of inclination of a face is 53.3°. Prior to 2006, tourists were allowed to climb one of the steep stairways aided by a rope positioned in the middle of the stairway and anchored at the base of the pyramid and at the temple level. Find the approximate length *L* of the rope. See **FIGURE 4.R.19**.

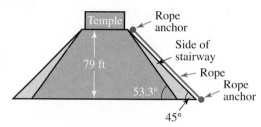

Tourists are no longer
allowed to climb *El Castillo*

FIGURE 4.R.19 Pyramid in Problem 46

In Problems 47–60, translate the words into an appropriate function.

47. A 20-ft-long water trough has ends in the form of isosceles triangles with sides that are 4 ft long. See Figure 2.9.27 in Exercises 2.9. As shown in **FIGURE 4.R.20**, let θ denote the angle between the vertical and one of the sides of a triangular end. Express the volume of the trough as a function of 2θ.

48. A person driving a car approaches a freeway sign as shown in **FIGURE 4.R.21**. Let θ be her viewing angle of the sign and let x represent her horizontal distance (measured in feet) to that sign. Express θ as a function of x.

49. As shown in **FIGURE 4.R.22**, a plank is supported by a sawhorse so that one end rests on the ground and the other end rests against a building. Express the length of the plank as a function of the indicated angle θ.

FIGURE 4.R.20 End of water trough in Problem 47

FIGURE 4.R.21 Freeway sign in Problem 48

FIGURE 4.R.22 Plank in Problem 49

50. A farmer wishes to enclose a pasture in the form of a right triangle using 2000 ft of fencing on hand. See **FIGURE 4.R.23**. Show that the area of the pasture as a function of the indicated angle θ is

$$A(\theta) = \frac{1}{2}\cot\theta \cdot \left(\frac{2000}{1 + \cot\theta + \csc\theta}\right)^2.$$

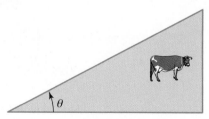

FIGURE 4.R.23 Pasture in Problem 50

FIGURE 4.R.24 Box in Problem 51

51. Express the volume of the box shown in FIGURE 4.R.24 as a function of the indicated angle θ.

52. A corner of an 8.5 in. × 11 in. piece of paper is folded over to the other edge of the paper as shown in FIGURE 4.R.25. Express the length L of the crease as a function of the angle θ shown in the figure.

53. A gutter is to be made from a sheet of metal 30 cm wide by turning up the edges of width 10 cm along each side so that the sides make equal angles ϕ with the vertical. See FIGURE 4.R.26. Express the cross-sectional area of the gutter as a function of the angle ϕ.

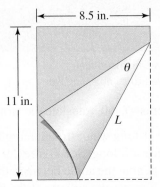

FIGURE 4.R.25 Folded paper in Problem 52

FIGURE 4.R.26 Gutter in Problem 53

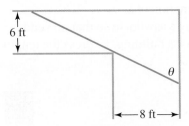

FIGURE 4.R.27 Pipe in Problem 54

54. A metal pipe is to be carried horizontally around a right-angled corner from a hallway 8 feet wide into a hallway that is 6 feet wide. See FIGURE 4.R.27. Express the length L of the pipe as a function of the angle θ shown in the figure.

55. Circle of Latitude A circle of latitude is 2 circle that connects all locations on the Earth that have the same latitude ϕ. A circle of latitude is also referred to as a line of latitude, or a parallel, because if the Earth is represented as a circle in a two-dimensional coordinate system, then the circles of latitude appear as (parallel) horizontal lines. See FIGURE 4.R.28.

(a) If the radius R of the Earth is taken to be 3959 miles, express the radius r of a circle of latitude as a function of ϕ.

(b) Find the radius of the Artic Circle if its latitude is 66°33′44″N.

(c) A line of longitude, or meridian, is one half of a great circle whose center is the center of the Earth. Longitudes are measured east/west from the prime meridian that runs through the Royal Observatory at Greenwich, England. See Figure 4.2.14 in Exercises 4.2. The longitudes of Boston, Massachusetts and Detroit, Michigan are, respectively, 71°3′37″W and 83°2′44″W but the latitude of both cities is approximately the same. What is the distance between Boston and Detroit measured on the circle of latitude at 42°20′N?

(d) Measured on a meridian, what is the distance between two points that differ by one degree of latitude?

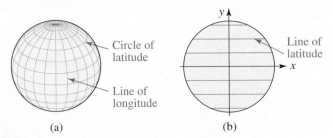

FIGURE 4.R.28 Planet Earth in Problem 55

CHAPTER 4 TRIGONOMETRIC FUNCTIONS

200 ft

Ground

15 ft

15 ft

θ

s

FIGURE 4.R.29 Diver in Problem 57

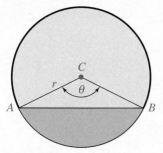

FIGURE 4.R.30 Circular segment in Problem 58

56. More Latitude **(a)** Use Problem 55 to show that the circumference of a circle of latitude as a function of the latitude angle ϕ is given by $C_\phi = C_e \cos \phi$, where C_e is the circumference of the Earth at the equator. Find C_e.
(b) Use part (a) to find the circumference of the Arctic Circle.
(c) Use part (a) to find the distance "around the world" at the latitude $52°45'$ N.

57. High Dive A diver jumps from a high platform with an initial downward velocity of 1 ft/s toward the center of a large circular tank of water. See FIGURE 4.R.29. From physics, the height of the diver above ground level is given by $s(t) = -16t^2 - t + 200$, where s is measured in feet and $t \geq 0$ is time in seconds. See (5) in Section 2.4.
(a) Express the angle θ shown in the figure as a function of s.
(b) For the function in part (a), what value does θ approach as $s \rightarrow 15$?

58. Circular Segment A circular segment is the region formed between a chord AB of a circle and its associated arc $\overset{\frown}{AB}$. This is the light red region in FIGURE 4.R.30. For a circle of fixed radius r, express the area of a circular segment as a function of the central angle θ, where $0 < \theta < \pi$.

59. Equilateral Arch An equilateral arch is obtained by constructing circular arcs on two sides of an equilateral triangle. Let ABC be an equilateral triangle with sides of length s as shown in red in FIGURE 4.R.31. A circular arc $\overset{\frown}{CB}$ is drawn from vertex C to vertex B using a circle of radius s centered at A. In like manner, $\overset{\frown}{CA}$ is an arc of the circle of radius s centered at B. Equilateral arches were used extensively in Gothic architecture in the design of church windows and doorways. Use the concept of a circular segment discussed in Problem 58 to express the area of an equilateral arch as a function of the length s.

FIGURE 4.R.31 Equilateral arch in Problem 59

The equilateral arch is found throughout the *Doumo di Milano*, the largest Gothic cathedral in the world.

60. Express the perimeter of the equilateral arch in Problem 59 as a function of s.

© Gunnar Kullenberg/RGB Ventures/SuperStock/Alamy Images

5 Exponential and Logarithmic Functions

Chapter Contents

5.1 Exponential Functions

INTRODUCTION In the preceding chapters we considered functions such as $f(x) = x^2$, that is, a function with a variable base x and constant power or exponent 2. We now examine functions having a constant base b and a variable exponent x.

DEFINITION 5.1.1 Exponential Function

If $b > 0$ and $b \neq 1$, then an **exponential function** $y = f(x)$ is a function of the form

$$f(x) = b^x \tag{1}$$

The number b is called the **base** and x is called the **exponent**.

The **domain** of an exponential function f defined in (1) of Definition 5.1.1 is the set of all real numbers $(-\infty, \infty)$.

In (1) the base b is restricted to positive numbers in order to guarantee that b^x is always a real number. For example, with this restriction we avoid complex numbers such as $(-4)^{1/2}$. Also, the base $b = 1$ is of little interest to us since it can be shown that f is the constant function $f(x) = 1^x = 1$. Moreover, for $b > 0$, we have $f(0) = b^0 = 1$.

☐ **Exponents** As just mentioned, the domain of an exponential function (1) is the set of all real numbers. This means that the exponent x can be either a rational or an irrational number. For example, if the base $b = 3$ and the exponent x is a *rational number*, say, $x = \frac{1}{5}$ and $x = 1.4$, then

$$3^{1/5} = \sqrt[5]{3} \qquad \text{and} \qquad 3^{1.4} = 3^{14/10} = 3^{7/5} = \sqrt[5]{3^7}.$$

For an exponent x that is an *irrational number*, b^x is defined, but its precise definition is beyond the scope of this text. We can, however, suggest a procedure for defining a number such as $3^{\sqrt{2}}$. From the decimal representation $\sqrt{2} = 1.414213562 \ldots$ we see that the rational numbers

$$1, 1.4, 1.41, 1.414, 1.4142, 1.41421, \ldots$$

are successively better approximations to $\sqrt{2}$. By using these rational numbers as exponents, we would expect that the numbers

$$3^1, 3^{1.4}, 3^{1.41}, 3^{1.414}, 3^{1.4142}, 3^{1.41421}, \ldots$$

are then successively better approximations to $3^{\sqrt{2}}$. In fact, this can be shown to be true with a precise definition of b^x for an irrational value of x. But on a practical level, we can use the $\boxed{y^x}$ key on a calculator to obtain the approximation 4.728804388 to $3^{\sqrt{2}}$.

☐ **Laws of Exponents** In most algebra texts the laws of exponents are stated first for integer exponents and then for rational exponents. Since b^x can be defined for all real numbers x when $b > 0$, it can be proved that these same **laws of exponents** hold for all real number exponents.

THEOREM 5.1.1   Laws of Exponents

If $a > 0$, $b > 0$ and x, x_1, and x_2 denote real numbers, then

(i) $b^{x_1} \cdot b^{x_2} = b^{x_1 + x_2}$

(ii) $\dfrac{b^{x_1}}{b^{x_2}} = b^{x_1 - x_2}$

(iii) $\dfrac{1}{b^x} = b^{-x}$

(iv) $(b^{x_1})^{x_2} = b^{x_1 x_2}$

(v) $(ab)^x = a^x b^x$

(vi) $\left(\dfrac{a}{b}\right)^x = \dfrac{a^x}{b^x}$

EXAMPLE 1   Rewriting a Function

At times, we will use the laws of exponents to rewrite a function in a different form. For example, neither $f(x) = 2^{3x}$ nor $g(x) = 4^{-2x}$ has the precise form of the exponential function defined in (1). However, by the laws of exponents given in Theorem 5.1.1, f can be rewritten as $f(x) = 8^x$ ($b = 8$ in (1)), and g can be recast as $g(x) = \left(\frac{1}{16}\right)^x$ ($b = \frac{1}{16}$ in (1)). The details are shown below:

$$f(x) = 2^{3x} \overset{\text{by }(iv)}{=} (2^3)^x = \overset{\text{form is now }b^x}{8^x}$$

$$g(x) = 4^{-2x} \overset{\text{by }(iv)}{=} (4^{-2})^x \overset{\text{by }(iii)}{=} \left(\frac{1}{4^2}\right)^x = \overset{\text{form is now }b^x}{\left(\frac{1}{16}\right)^x}.$$

☐ **Graphs** We distinguish two types of graphs for (1) depending on whether the base b satisfies $b > 1$ or $0 < b < 1$. The next two examples illustrate, in turn, the graphs of $f(x) = 3^x$ and $f(x) = \left(\frac{1}{3}\right)^x$. Before graphing, we can make some intuitive observations about both functions. Since the bases $b = 3$ and $b = \frac{1}{3}$ are positive, the values of 3^x and $\left(\frac{1}{3}\right)^x$ are *positive* for every real number x. As a consequence, there are no real numbers x_1 and x_2 for which 3^{x_1} and $\left(\frac{1}{3}\right)^{x_2}$ are zero. Graphically, this means that the graphs of $f(x) = 3^x$ and $f(x) = \left(\frac{1}{3}\right)^x$ have no x-intercepts. Also, $3^0 = 1$ and $\left(\frac{1}{3}\right)^0 = 1$, and so $f(0) = 1$ in each case. This means that the graphs of $f(x) = 3^x$ and $f(x) = \left(\frac{1}{3}\right)^x$ have the same y-intercept $(0, 1)$.

EXAMPLE 2   Graph for $b > 1$

Graph the function $f(x) = 3^x$.

Solution We first construct a table of some function values corresponding to preselected values of x. As shown in FIGURE 5.1.1, we plot the corresponding points obtained from the table and connect them with a continuous curve. The graph shows that f is an increasing function on the interval $(-\infty, \infty)$.

x	-3	-2	-1	0	1	2
$f(x)$	$\frac{1}{27}$	$\frac{1}{9}$	$\frac{1}{3}$	1	3	9

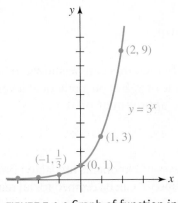

FIGURE 5.1.1 Graph of function in Example 2

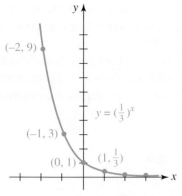

FIGURE 5.1.2 Graph of function in Example 3

Review Theorem 2.2.3 in Section 2.2 ▶ for reflections in the *x*- and *y*-axes.

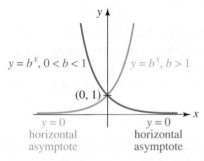

FIGURE 5.1.3 f increasing for $b > 1$; f decreasing for $0 < b < 1$

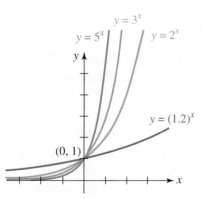

FIGURE 5.1.4 Graphs of $y = b^x$ for $b = 1.2, 2, 3, 5$

EXAMPLE 3　　　Graph for $0 < b < 1$

Graph the function $f(x) = \left(\frac{1}{3}\right)^x$.

Solution Proceeding as in Example 2, we construct a table of some function values corresponding to preselected values of *x*. Note, for example, by the laws of exponents

$$f(-2) = \left(\tfrac{1}{3}\right)^{-2} = (3^{-1})^{-2} = 3^2 = 9.$$

As shown in FIGURE 5.1.2, we plot the corresponding points obtained from the table and connect them with a continuous curve. In this case the graph shows that f is a decreasing function on the interval $(-\infty, \infty)$.

x	-3	-2	-1	0	1	2
$f(x)$	27	9	3	1	$\frac{1}{3}$	$\frac{1}{9}$

☐ **Reflections** Exponential functions with bases satisfying $0 < b < 1$, such as $b = \frac{1}{3}$, are frequently written in an alternative manner. We note that $y = \left(\frac{1}{3}\right)^x$ is the same as $y = 3^{-x}$. From this last result we see that the graph of $y = 3^{-x}$ is simply the graph of $y = 3^x$ reflected in the *y*-axis.

☐ **Horizontal Asymptote** FIGURE 5.1.3 illustrates the two general shapes that the graph of an exponential function $f(x) = b^x$ can have; but there is one more important aspect of all such graphs. Observe in Figure 5.1.3 that for $b > 1$,

$$f(x) = b^x \to 0 \quad \text{as} \quad x \to -\infty, \quad ← \text{blue graph}$$

whereas for $0 < b < 1$,

$$f(x) = b^x \to 0 \quad \text{as} \quad x \to \infty. \quad ← \text{red graph}$$

In other words, the line $y = 0$ (the *x*-axis) is a **horizontal asymptote** for both types of exponential graphs.

☐ **Properties** The following list summarizes some of the important properties of the exponential function $f(x) = b^x$. Reexamine the graphs in Figures 5.1.1–5.1.3 as you read this list.

Properties of the Exponential Function

- The domain of f is the set of real numbers, that is, $(-\infty, \infty)$.
- The range of f is the set of positive real numbers, that is, $(0, \infty)$.
- The *y*-intercept of f is $(0, 1)$. The graph of f has no *x*-intercepts.
- The function f is increasing for $b > 1$ and decreasing for $0 < b < 1$.
- The *x*-axis, that is, $y = 0$, is a horizontal asymptote for the graph of f.
- The function f is continuous on $(-\infty, \infty)$.
- The function f is one-to-one.

Although the graphs $y = b^x$ in the case, say, when $b > 1$, all share the same basic shape and all pass through the same point $(0, 1)$, there are subtle differences. The larger the base *b* the more steeply the graph rises as *x* increases. In FIGURE 5.1.4 we compare the graphs of $y = 5^x$, $y = 3^x$, $y = 2^x$, and $y = (1.2)^x$ in green, blue, gold, and red, respectively, on the same coordinate axes. We see from its graph that the values of $y = (1.2)^x$ increase slowly as *x* increases. For example, for $y = (1.2)^x$, $f(3) = (1.2)^3 = 1.728$, whereas, for $y = 5^x$, $f(3) = 5^3 = 125$.

The fact that (1) is a one-to-one function follows from the horizontal line test discussed in Section 2.8. Note in Figures 5.1.1–5.1.4 that a horizontal line can cross or intersect an exponential graph in at most one point.

Of course, we can obtain other kinds of graphs by rigid and nonrigid transformations, or when an exponential function is combined with other functions by either an arithmetic operation or by function composition. In the next several examples we examine variations of the exponential graph.

EXAMPLE 4   Horizontally Shifted Graph

Graph the function $f(x) = 3^{x+2}$.

Solution From the discussion in Section 2.2 you should recognize that the graph of $f(x) = 3^{x+2}$ is the graph of $y = 3^x$ shifted 2 units to the left. Recall, since the shift is a rigid transformation to the left, the points on the graph of $f(x) = 3^{x+2}$ are the points on the graph of $y = 3^x$ moved horizontally 2 units to the left. This means that the y-coordinates of points (x, y) on the graph of $y = 3^x$ remain unchanged but 2 is subtracted from all the x-coordinates of the points. Thus we see from FIGURE 5.1.5 that the points $(0, 1)$ and $(2, 9)$ on the blue graph of $y = 3^x$ are moved, in turn, to the points $(-2, 1)$ and $(0, 9)$ on the red graph of $f(x) = 3^{x+2}$.

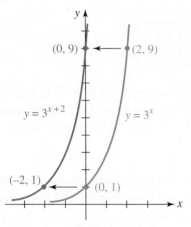

FIGURE 5.1.5 Shifted graph in Example 4

The function $f(x) = 3^{x+2}$ in Example 4 can be rewritten, if desired, as $f(x) = 9 \cdot 3^x$. By (*i*) of the laws of exponents, $3^{x+2} = 3^2 3^x = 9 \cdot 3^x$. In this manner we can reinterpret the graph of $f(x) = 3^{x+2}$ as a vertical stretch of the graph of $y = 3^x$ by a factor of 9. For example, $(1, 3)$ is on the graph of $y = 3^x$, whereas $(1, 9 \cdot 3) = (1, 27)$ is on the graph of $f(x) = 3^{x+2}$.

☐ **The Number *e*** Most every student of mathematics has heard of, and has likely worked with, the famous irrational number $\pi = 3.141592654\ldots$. Recall, that an irrational number is a nonrepeating and nonterminating decimal. In calculus and applied mathematics the irrational number

$$e = 2.718281828459\ldots$$

arguably plays a role more important than the number π. The usual definition of the number e is the number that the function $f(x) = (1 + 1/x)^x$ approaches as we let x become large without bound in the positive direction, that is, $f(x) \to e$ as $x \to \infty$. Using the limit notation introduced in Sections 1.5 and 2.10, we write

$$e = \lim_{x \to \infty} \left(1 + \frac{1}{x}\right)^x. \tag{3}$$

See Problems 53 and 55 in Exercises 5.1. You will often see an alternative definition of the number e. If we let $h = 1/x$ in (3), then as $x \to \infty$ we have simultaneously $h \to 0$. Hence an equivalent form of (3) is

$$e = \lim_{h \to 0} (1 + h)^{1/h}. \tag{4}$$

See Problems 54 and 56 in Exercises 5.1. Of course, advancing (3) and (4) as *definitions* of the number e raises the obvious question: Where do these strange limits come from? An unsatisfying partial answer is: Definitions (3) and (4) come from calculus. While we cannot prove in this course that the limits in (3) and (4) exist, we will, however, discuss the origins of e in Section 5.5.

☐ **The Natural Exponential Function** When the base in (1) is chosen to be $b = e$, the function

$$f(x) = e^x \qquad (5)$$

is called the **natural exponential function**. Since $b = e > 1$ and $b = 1/e < 1$, the graphs of $y = e^x$ and $y = e^{-x}$ (or $y = (1/e)^x = 1/e^x$) are given in FIGURE 5.1.6. Alternatively, because $f(-x) = e^{-x}$ the graph of $y = e^{-x}$ is the graph of $f(x) = e^x$ reflected in the y-axis.

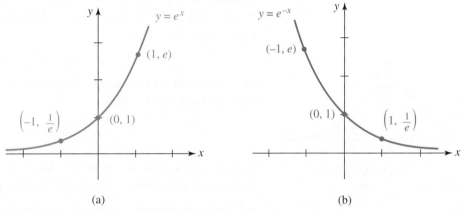

(a)   (b)

FIGURE 5.1.6 Graphs of the natural exponential function (in (a)) and its reciprocal (in (b))

On the face of it, the natural exponential function (5) possesses no noticeable graphical characteristic that distinguishes it from, say, the function $f(x) = 3^x$, and has no special properties other than the ones given in (2), above. Questions as to why (5) is a "natural" and frankly, the most important exponential function, can only be answered fully in courses in calculus and beyond. We will explore some of the importance of the number e in Sections 5.4 and 5.5.

EXAMPLE 5   **Reflection and Vertical Shift**

Graph the function $f(x) = 2 - e^{-x}$. State the range.

Solution We first draw the graph of $y = e^{-x}$ as shown in FIGURE 5.1.7(a). Then we reflect the first graph in the x-axis to obtain the graph of $y = -e^{-x}$ in Figure 5.1.7(b). Finally, the graph in Figure 5.1.7(c) is obtained by shifting the graph in part (b) upward 2 units.

The y-intercept $(0, -1)$ of $y = -e^{-x}$ when shifted upward 2 units returns us to the original y-intercept in Figure 5.1.7(a). Finally, because of the vertical shift the horizontal asymptote, which was $y = 0$ in parts (a) and (b) of the figure, becomes $y = 2$ in Figure 5.1.7(c).

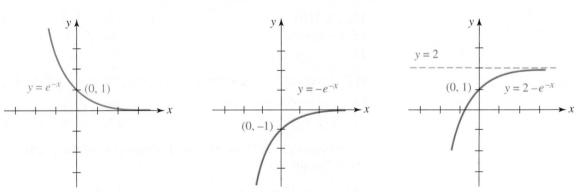

(a) Start with graph of $y = e^{-x}$ (b) Graph in (a) reflected in x-axis   (c) Graph in (b) shifted upward 2 units

FIGURE 5.1.7 Graph of function in Example 5

From the last graph we can conclude that the range of the function $f(x) = 2 - e^{-x}$ is the set of real numbers defined by $y < 2$, that is, the interval $(-\infty, 2)$ on the y-axis.

In the next example we graph the function composition of the natural exponential function $y = e^x$ with the simple quadratic polynomial function $y = -x^2$.

EXAMPLE 6   A Function Composition

Graph the function $f(x) = e^{-x^2}$.

Solution Because $f(0) = e^{-0^2} = e^0 = 1$, the y-intercept of the graph is $(0, 1)$. Also, $f(x) \neq 0$ since $e^{-x^2} \neq 0$ for every real number x. This means that the graph of f has no x-intercepts. Then from

$$f(-x) = e^{-(-x)^2} = e^{-x^2} = f(x)$$

we conclude that f is an even function and so its graph is symmetric with respect to the y-axis. Lastly, observe that

$$f(x) = \frac{1}{e^{x^2}} \to 0 \quad \text{as} \quad x \to \infty.$$

By symmetry we can also conclude that $f(x) \to 0$ as $x \to -\infty$. This shows that $y = 0$ is a horizontal asymptote for the graph of f. The graph of f is given in FIGURE 5.1.8.

Bell-shaped graphs such as that given in Figure 5.1.8 are very important in the study of probability and statistics.

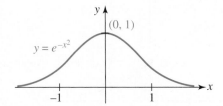

FIGURE 5.1.8 Graph of function in Example 6

Exercises 5.1   Answers to selected odd-numbered problems begin on page ANS–19.

In Problems 1–6, graph the given functions on the same rectangular coordinate system.

1. $y = 3^x$, $y = 3^{-x}$

2. $y = -2^x$, $y = -2^{-x}$

3. $y = \left(\frac{3}{4}\right)^x$, $y = \left(\frac{4}{3}\right)^x$

4. $y = -\left(\frac{1}{3}\right)^x$, $y = \left(\frac{1}{3}\right)^{-x}$

5. $y = 3^{x-1}$, $y = 3^{-x+1}$

6. $y = 2^{x-2}$, $y = -2^{x+2}$

In Problems 7–12, sketch the graph of the given function f. Find the y-intercept and the horizontal asymptote of the graph. State whether the function is increasing or decreasing.

7. $f(x) = -5 + 3^x$

8. $f(x) = 2 + 3^{-x}$

9. $f(x) = 3 - \left(\frac{1}{5}\right)^x$

10. $f(x) = 9 - e^x$

11. $f(x) = -1 + e^{x-3}$

12. $f(x) = -3 - e^{x+5}$

In Problems 13–18, find an exponential function $f(x) = b^x$ such that the graph of f passes through the given point.

13. $(3, 216)$

14. $(-1, 5)$

15. $(-1, e^2)$

16. $(2, e)$

17. $(-2, 9)$

18. $\left(\frac{1}{2}, 6\right)$

In Problems 19–22, determine the range of the given function.

19. $f(x) = 5 + e^{-x}$

20. $f(x) = 4 - 2^{-x}$

21. $f(x) = 3^x - 2$

22. $f(x) = -e^x - 3$

In Problems 23–28, find the x- and y-intercepts of the graph of the given function. Do not graph.

23. $f(x) = 2^x - 4$

24. $f(x) = -3^{2x} + 9$

25. $f(x) = xe^x + 10e^x$

26. $f(x) = x^2 2^x - 2^x$

27. $f(x) = x^3 8^x + 5x^2 8^x + 6x 8^x$

28. $f(x) = 4^x x^4 - 4^{x+1}$

In Problems 29–32, use a graph to solve the given inequality.

29. $2^x > 16$ **30.** $e^x \le 1$

31. $e^{x-2} < 1$ **32.** $\left(\frac{1}{2}\right)^x \ge 8$

In Problems 33–36, use the graph in Figure 5.1.8 to sketch the graph of the function f.

33. $f(x) = e^{-(x-3)^2}$ **34.** $f(x) = -e^{-(x+2)^2}$

35. $f(x) = 3 - e^{-(x+1)^2}$ **36.** $f(x) = -1 + e^{-(x-2)^2}$

In Problems 37 and 38, use $f(-x) = f(x)$ to demonstrate that the given function is even. Sketch the graph of f.

37. $f(x) = e^{x^2}$ **38.** $f(x) = e^{-|x|}$

In Problems 39–42, use the graphs obtained in Problems 37 and 38 as an aid in sketching the graph of the given function f.

39. $f(x) = 1 - e^{x^2}$ **40.** $f(x) = 2 + 3e^{|x|}$

41. $f(x) = -e^{|x-3|}$ **42.** $f(x) = e^{(x+2)^2}$

43. Show that $f(x) = \frac{1}{2}(3^x + 3^{-x})$ is an even function. Sketch the graph of f.

44. Show that $g(x) = \frac{1}{2}(3^x - 3^{-x})$ is an odd function. Sketch the graph of g.

45. For the functions f and g in Problems 43 and 44, show that $[f(x)]^2 - [g(x)]^2 = 1$.

46. For $f(x) = b^x$, show that:

$$\textbf{(a)} \quad \frac{f(x+h) - f(x)}{h} = b^x\left(\frac{b^h - 1}{h}\right) \qquad \textbf{(b)} \quad f(x_1 + x_2) = f(x_1)f(x_2).$$

47. Show that the function $f(x) = e^{\sin \pi x}$ is periodic. Find the amplitude A of the graph of f.

48. The graphs of the exponential functions $y = e^{-x/7}$ and $y = -e^{-x/7}$ are called **envelope curves** for the graph of $y = e^{-x/7} \sin x$. As shown in FIGURE 5.1.9 the red envelope curves are tangent to the blue curve at certain points. Find the coordinates of the first four points of tangency for $x > 0$.

FIGURE 5.1.9 Graph for Problem 48

In Problems 49 and 50, sketch the graph of the given piecewise-defined function f.

49. $f(x) = \begin{cases} -e^x, & x < 0 \\ -e^{-x}, & x \ge 0 \end{cases}$ **50.** $f(x) = \begin{cases} e^{-x}, & x \le 0 \\ -e^x, & x > 0 \end{cases}$

51. Find an equation of the red line in FIGURE 5.1.10.

52. Find the total area of the shaded region in FIGURE 5.1.11.

FIGURE 5.1.10 Graph for
Problem 51

FIGURE 5.1.11 Graph for
Problem 52

5.1 Exponential Functions

Calculator Problems

In Problems 53 and 54, use a calculator to fill out the given table.

53.

x	10	100	1000	10,000	100,000	1,000,000
$(1 + 1/x)^x$						

54.

h	0.1	0.01	0.001	0.0001	0.00001	0.000001
$(1 + h)^{1/h}$						

55. (a) Use a graphing utility to graph the functions $f(x) = (1 + 1/x)^x$ and $g(x) = e$ on the same set of coordinate axes. Use the intervals $(0, 10]$, $(0, 100]$, $(0, 1000]$. Describe the behavior of f for large values of x. In graphical terms, what is $g(x) = e$?

  (b) Graph the function f in part (a) on the interval $[-10, 0)$. Superimpose that graph with the graph of f on $(0, 10]$ obtained in part (a). Is f a continuous function?

56. Use a graphing utility to graph the function $f(x) = (1 + x)^{1/x}$ on the intervals $[0.1, 1]$, $[0.01, 1]$, and $[0.001, 1]$. Describe the behavior of f near $x = 0$.

In Problems 57 and 58, use a graphing utility as an aid in approximating the x-coordinates of the points of intersection of the graphs of the functions f and g.

57. $f(x) = x^2$, $g(x) = 2^x$ **58.** $f(x) = x^3$, $g(x) = 3^x$

For Discussion

In Problems 59–64, assume that $2^t = a$ and $6^t = b$. Use the laws of exponents given in this section to express the value of the given expression in terms of a and b.

59. 12^t **60.** 3^t

61. 6^{-t} **62.** 6^{3t}

63. $2^{-3t}2^{7t}$ **64.** 18^t

In Problems 65 and 66, the given fractional expression can be decomposed into partial fractions. See Section 3.6. Discuss how this can be done and carry out your ideas.

65. $\dfrac{e^t}{(e^t + 1)^2(e^t - 2)}$ **66.** $\dfrac{e^{2t}}{(e^t + 1)^3}$

5.2 Logarithmic Functions

INTRODUCTION Since an exponential function $y = b^x$ is one-to-one, we know that it has an inverse function. To find this inverse, we interchange the variables x and y to obtain $x = b^y$. This last formula defines y implicitly as a function of x:

y is that exponent of the base b that produces x.

By replacing the word *exponent* with the word *logarithm*, we can rephrase the preceding line as:

y is that logarithm of the base b that produces x.

This last line is abbreviated by the notation $y = \log_b x$ and is called the logarithmic function.

DEFINITION 5.2.1 Logarithmic Function

The **logarithmic function** with base $b > 0$, $b \neq 1$, is defined by

$$y = \log_b x \qquad \text{if and only if} \qquad x = b^y \qquad (1)$$

For $b > 0$ there is no real number y for which b^y can be either 0 or negative. It then follows from $x = b^y$ that $x > 0$. In other words, the **domain** of a logarithmic function $y = \log_b x$ is the set of positive real numbers $(0, \infty)$.

For emphasis, all that is being said in the preceding sentences is:

The logarithmic expression $y = \log_b x$ and the exponential expression $x = b^y$ are equivalent.

That is, both symbols mean the same thing. As a consequence, within a specific context such as solving a problem, we can use whichever form happens to be more convenient. The following table lists several examples of equivalent logarithmic and exponential statements.

Logarithmic Form	Exponential Form
$\log_3 9 = 2$	$9 = 3^2$
$\log_8 2 = \frac{1}{3}$	$2 = 8^{1/3}$
$\log_{10} 0.001 = -3$	$0.001 = 10^{-3}$
$\log_b 5 = -1$	$5 = b^{-1}$

☐ **Graphs** Recall from Section 2.8 that the graph of an inverse function can be obtained by reflecting the graph of the original function in the line $y = x$. This technique was used to obtain the red graphs from the blue graphs in FIGURE 5.2.1. As you inspect the two graphs in Figure 5.2.1(a) and in Figure 5.2.1(b), remember that the domain $(-\infty, \infty)$ and range $(0, \infty)$ of $y = b^x$ become, in turn, the range $(-\infty, \infty)$ and domain $(0, \infty)$ of $y = \log_b x$. Also note that the y-intercept $(0, 1)$ for the exponential function (blue graphs) becomes the x-intercept $(1, 0)$ for the logarithmic function (red graphs).

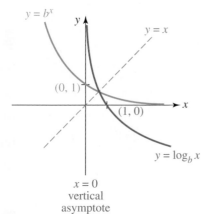

(a) Base $b > 1$ 　　　　　　　　 (b) Base $0 < b < 1$

FIGURE 5.2.1 Graphs of logarithmic functions

☐ **Vertical Asymptote** When the exponential function is reflected in the line $y = x$, the horizontal asymptote $y = 0$ for the graph of $y = b^x$ becomes a vertical asymptote for the graph of $y = \log_b x$. In Figure 5.2.1 we see that for $b > 1$,

$$\log_b x \to -\infty \qquad \text{as} \qquad x \to 0^+, \quad \leftarrow \text{red graph in (a)}$$

whereas for $0 < b < 1$,

$$\log_b x \to \infty \qquad \text{as} \qquad x \to 0^+. \quad \leftarrow \text{red graph in (b)}$$

From (7) of Section 3.5 we conclude that $x = 0$, which is the equation of the y-axis, is a **vertical asymptote** for the graph of $y = \log_b x$.

☐ **Properties** The following list summarizes some of the important properties of the logarithmic function $f(x) = \log_b x$.

Properties of the Logarithmic Function

- The domain of f is the set of positive real numbers, that is, $(0, \infty)$.
- The range of f is the set of real numbers, that is, $(-\infty, \infty)$.
- The x-intercept of f is $(1, 0)$. The graph of f has no y-intercept.
- The function f is increasing for $b > 1$ and decreasing for $0 < b < 1$.   (2)
- The y-axis, that is, $x = 0$, is a vertical asymptote for the graph of f.
- The function f is continuous on $(0, \infty)$.
- The function f is one-to-one.

We would like to call attention to the third entry in the foregoing list (2) for special emphasis:

$$\log_b 1 = 0 \qquad \text{since} \qquad b^0 = 1. \tag{3}$$

Also,
$$\log_b b = 1 \qquad \text{since} \qquad b^1 = b. \tag{4}$$

Thus, in addition to $(1, 0)$ the graph of any logarithmic function (1) with base b also contains the point $(b, 1)$. The equivalence of $y = \log_b x$ and $x = b^y$ also yields two sometimes-useful identities. By substituting $y = \log_b x$ into $x = b^y$, and then $x = b^y$ into $y = \log_b x$ gives

$$x = b^{\log_b x} \qquad \text{and} \qquad y = \log_b b^y. \tag{5}$$

■ EXAMPLE 1   Using Properties

Simplify **(a)** $\log_6 1$ **(b)** $\log_{\sqrt{3}} \sqrt{3}$ **(c)** $8^{\log_8 7}$ **(d)** $\log_{10} 10^5$.

Solution (a) With $b = 6$, property (3) gives $\log_6 1 = 0$.

(b) With $b = \sqrt{3}$, property (4) gives $\log_{\sqrt{3}} \sqrt{3} = 1$.

(c) With $b = 8$ and $x = 7$, the first property in (5) gives $8^{\log_8 7} = 7$.

(d) With $b = 10$ and $y = 5$, the second property in (5) gives $\log_{10} 10^5 = 5$.   ■

■ EXAMPLE 2   Logarithmic Graph for $b > 1$

Graph $f(x) = \log_{10}(x + 10)$.

Solution This is the graph of $y = \log_{10} x$, which has the shape shown in Figure 5.2.1(a), shifted 10 units to the left. To reinforce the fact that the domain of a logarithmic function $y = \log_{10} x$ is the set of positive real numbers, that is, $x > 0$, we can obtain the domain of $f(x) = \log_{10}(x + 10)$ by replacing x by $x + 10$ and requiring that $x + 10 > 0$ or $x > -10$. In interval notation, the domain of f is $(-10, \infty)$. In the

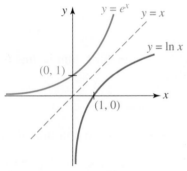

FIGURE 5.2.2 Graph of function in Example 1

FIGURE 5.2.3 Graph of the natural logarithm is shown in red

short accompanying table, we have chosen convenient values of x in order to plot a few points.

x	-9	0	90
$f(x)$	0	1	2

Notice,

$$f(-9) = \log_{10}1 = 0 \quad \leftarrow \text{by (3)}$$
$$f(0) = \log_{10}10 = 1. \quad \leftarrow \text{by (4)}$$

The vertical asymptote $x = 0$ for the graph of $y = \log_{10}x$ becomes $x = -10$ for the shifted graph. This asymptote is the red dashed vertical line in FIGURE 5.2.2.

☐ **Natural Logarithm** Logarithms with base $b = 10$ are called **common logarithms** and logarithms with base $b = e$ are called **natural logarithms**. Furthermore, it is customary to write the natural logarithm

$$\log_{e}x \quad \text{as} \quad \ln x.$$

The symbol "$\ln x$" is usually read phonetically as "ell-en of x." Since $b = e > 1$, the graph of $y = \ln x$ has the characteristic logarithmic shape shown in Figure 5.2.1(a). See FIGURE 5.2.3. For base $b = e$, (1) of Definition 5.2.1 becomes

$$y = \ln x \quad \text{if and only if} \quad x = e^{y}. \tag{6}$$

The analogs of properties (3) and (4) for the natural logarithm are

$$\ln 1 = 0 \quad \text{since} \quad e^{0} = 1, \tag{7}$$
$$\ln e = 1 \quad \text{since} \quad e^{1} = e. \tag{8}$$

The identities in (5) become

$$x = e^{\ln x} \quad \text{and} \quad y = \ln e^{y}. \tag{9}$$

For example, from (9), $e^{\ln 13} = 13$.

Common and natural logarithms can be found on all calculators. Often the symbol for the common logarithm is written without a subscript, that is, $\log_{10}x$ is simply written $\log x$. But in this text we will continue to use $\log_{10}x$. See (v) in *Notes from the Classroom* at the end of this section.

☐ **Laws of Logarithms** The laws of exponents given in Theorem 5.1.1 can be restated in an equivalent manner as the laws of logarithms. To see this, suppose we write $M = b^{x_1}$ and $N = b^{x_2}$. Then by (1), $x_1 = \log_{b} M$ and $x_2 = \log_{b} N$.

Product: By (i) of Theorem 5.1.1, $MN = b^{x_1 + x_2}$. Expressed as a logarithm this is $x_1 + x_2 = \log_{b} MN$. Substituting for x_1 and x_2 gives

$$\log_{b}M + \log_{b}N = \log_{b}MN.$$

Quotient: By (ii) of Theorem 5.1.1, $M/N = b^{x_1 - x_2}$. Expressed as a logarithm this is $x_1 - x_2 = \log_{b}(M/N)$. Substituting for x_1 and x_2 gives

$$\log_{b}M - \log_{b}N = \log_{b}(M/N).$$

Power: By (iv) of Theorem 5.1.1, $M^{c} = b^{cx_1}$. Expressed as a logarithm this is $cx_1 = \log_{b}M^{c}$. Substituting for x_1 gives

$$c\log_{b}M = \log_{b}M^{c}.$$

For convenience and future reference, we summarize these product, quotient, and power laws of logarithms next.

5.2 Logarithmic Functions

> **THEOREM 5.2.1** Laws of Logarithms
>
> For any base $b > 0$, $b \neq 1$, and positive numbers M and N:
>
>   (*i*) $\log_b MN = \log_b M + \log_b N$
>
>   (*ii*) $\log_b\left(\dfrac{M}{N}\right) = \log_b M - \log_b N$
>
>   (*iii*) $\log_b M^c = c\log_b M$, for c any real number

EXAMPLE 3   Using the Laws of Logarithms

Simplify and write as a single logarithm

$$\tfrac{1}{2}\ln 36 + 2\ln 4 - \ln 4.$$

Solution There are several ways to approach this problem. Note, for example, that the second and third terms can be combined arithmetically as

$$2\ln 4 - \ln 4 = \ln 4. \quad \leftarrow \text{analogous to } 2x - x = x$$

Alternatively, we can use (*iii*) followed by (*ii*) of Theorem 5.2.1 to combine these terms:

$$
\begin{aligned}
2\ln 4 - \ln 4 &= \ln 4^2 - \ln 4 \\
&= \ln 16 - \ln 4 \\
&= \ln \tfrac{16}{4} \\
&= \ln 4.
\end{aligned}
$$

Hence,
$$
\begin{aligned}
\tfrac{1}{2}\ln 36 + 2\ln 4 - \ln 4 &= \ln(36)^{1/2} + \ln 4 \quad \leftarrow \text{by (\textit{iii}) of Theorem 5.2.1} \\
&= \ln 6 + \ln 4 \\
&= \ln 24. \quad\quad\quad \leftarrow \text{by (\textit{i}) of Theorem 5.2.1} \blacksquare
\end{aligned}
$$

EXAMPLE 4   Rewriting Logarithmic Expressions

Use the laws of logarithms to rewrite each expression and evaluate.

(a) $\ln\sqrt{e}$ **(b)** $\ln 5e$ **(c)** $\ln\dfrac{1}{e}$

Solution (a) Since $\sqrt{e} = e^{1/2}$ we have from (*iii*) of Theorem 5.2.1:

$$\ln\sqrt{e} = \ln e^{1/2} = \tfrac{1}{2}\ln e = \tfrac{1}{2}. \quad \leftarrow \text{from (8), } \ln e = 1$$

(b) From (*i*) of Theorem 5.2.1 and a calculator:

$$\ln 5e = \ln 5 + \ln e = \ln 5 + 1 \approx 2.6094.$$

(c) From (*ii*) of Theorem 5.2.1:

$$\ln\frac{1}{e} = \ln 1 - \ln e = 0 - 1 = -1. \quad \leftarrow \text{from (7) and (8)}$$

Note that (*iii*) of Theorem 5.2.1 can also be used here:

$$\ln\frac{1}{e} = \ln e^{-1} = (-1)\ln e = -1. \quad \leftarrow \text{from (8), } \ln e = 1 \blacksquare$$

EXAMPLE 5   **Value of a Logarithm**

If $\log_b 2 = 0.4307$ and $\log_b 3 = 0.6826$, then find $\log_b \sqrt[3]{18}$.

Solution We begin by rewriting $\sqrt[3]{18}$ as $(18)^{1/3}$. Then by the laws of logarithms

$$
\begin{aligned}
\log_b(18)^{1/3} &= \tfrac{1}{3}\log_b 18 &&\leftarrow \text{by } (iii) \text{ of Theorem 5.2.1}\\
&= \tfrac{1}{3}\log_b(2 \cdot 3^2)\\
&= \tfrac{1}{3}[\log_b 2 + \log_b 3^2] &&\leftarrow \text{by } (i) \text{ of Theorem 5.2.1}\\
&= \tfrac{1}{3}[\log_b 2 + 2\log_b 3] &&\leftarrow \text{by } (iii) \text{ of Theorem 5.2.1}\\
&= \tfrac{1}{3}[0.4307 + 2(0.6826)]\\
&= 0.5986.
\end{aligned}
$$

NOTES FROM THE CLASSROOM

(*i*) Students often struggle with the concept of a *logarithm*. It may help if you repeat to yourself a few dozen times, "A logarithm is an exponent." It may also help if you begin reading a statement such as $3 = \log_{10}1000$ as "3 is the exponent of 10 that. . . ."

(*ii*) Be *very* careful applying the laws of logarithms. The logarithm does *not* distribute over addition,

$$\log_b(M + N) \neq \log_b M + \log_b N.$$

In other words, the exponent of a sum is not the sum of the exponents. Also,

$$\frac{\log_b M}{\log_b N} \neq \log_b M - \log_b N.$$

In general, there is no property of real numbers that enables us to rewrite either

$$\log_b(M + N) \qquad \text{or} \qquad \frac{\log_b M}{\log_b N}.$$

(*iii*) Also be careful when rewriting logarithmic functions using the laws of logarithms. For example, the domain of the function $y = \log_b x^2$ is the set of real numbers satisfying $x \neq 0$ but the domain of $y = 2\log_b x$ is the set of real numbers satisfying $x > 0$. See Problem 83 in Exercises 5.2.

(*iv*) In calculus, the first step in a procedure known as *logarithmic differentiation* requires the student to take the natural logarithm of both sides of a complicated function such as $y = \dfrac{x^{10}\sqrt{x^2 + 5}}{\sqrt[3]{8x^3 + 2}}$. The idea is to use the laws of logarithms to transform powers into constant multiples, products into sums, and quotients into differences. See Problems 69–72 in Exercises 5.2.

(*v*) You may see different notations for the natural exponential function and for the natural logarithm. For example, on some calculators you may see $y = \exp x$ instead of $y = e^x$. In the computer algebra system *Mathematica* the natural exponential function is written Exp[x] and the natural logarithm is written Log[x].

Exercises 5.2   Answers to selected odd-numbered problems begin on page ANS–19.

In Problems 1–6, rewrite the given exponential expression as an equivalent logarithmic expression.

1. $4^{-1/2} = \left(\frac{1}{2}\right)$

2. $9^0 = 1$

3. $10^4 = 10,000$

4. $10^{0.3010} = 2$

5. $t^{-s} = v$

6. $(a + b)^2 = a^2 + 2ab + b^2$

In Problems 7–12, rewrite the given logarithmic expression as an equivalent exponential expression.

7. $\log_2 128 = 7$

8. $\log_5 \frac{1}{25} = -2$

9. $\log_{\sqrt{3}} 81 = 8$

10. $\log_{16} 2 = \frac{1}{4}$

11. $\log_b u = v$

12. $\log_b b^2 = 2$

In Problems 13–18, find the exact value of the given logarithm.

13. $\log_{10}(0.0000001)$

14. $\log_4 64$

15. $\log_2(2^2 + 2^2)$

16. $\log_9 \frac{1}{3}$

17. $\ln e^e$

18. $\ln(e^4 e^9)$

In Problems 19–22, find the exact value of the given expression.

19. $10^{\log_{10} 6^2}$

20. $25^{\log_5 8}$

21. $e^{-\ln 7}$

22. $e^{\frac{1}{2}\ln \pi}$

In Problems 23 and 24, find a logarithmic function $f(x) = \log_b x$ such that the graph of f passes through the given point.

23. $(49, 2)$

24. $\left(4, \frac{1}{3}\right)$

In Problems 25–32, find the domain of the given function f. Find the x-intercept and the vertical asymptote of the graph. Use transformations to graph the given function f.

25. $f(x) = -\log_2 x$

26. $f(x) = -\log_2(x + 1)$

27. $f(x) = \log_2(-x)$

28. $f(x) = \log_2(3 - x)$

29. $f(x) = 3 - \log_2(x + 3)$

30. $f(x) = 1 - 2\log_4(x - 4)$

31. $f(x) = -1 + \ln x$

32. $f(x) = 1 + \ln(x - 2)$

In Problems 33 and 34, use a graph to solve the given inequality.

33. $\ln(x + 1) < 0$

34. $\log_{10}(x + 3) > 1$

35. Show that $f(x) = \ln|x|$ is an even function. Rewrite f as a piecewise-defined function and sketch its graph. Find the x-intercepts and the vertical asymptote of the graph.

36. Use the graph obtained in Problem 35 to sketch the graph of $y = \ln|x - 2|$. Find the x-intercept and the vertical asymptote of the graph.

In Problems 37 and 38, sketch the graph of the given function f.

37. $f(x) = |\ln x|$

38. $f(x) = |\ln(x + 1)|$

CHAPTER 5 EXPONENTIAL AND LOGARITHMIC FUNCTIONS

In Problems 39–44, find the domain of the given function f.

39. $f(x) = \ln(2x - 3)$

40. $f(x) = \ln(3 - x)$

41. $f(x) = \ln(9 - x^2)$

42. $f(x) = \ln(x^2 - 2x)$

43. $f(x) = \sqrt{\ln x}$

44. $f(x) = \dfrac{1}{\ln x}$

In Problems 45 and 46, graph the given equations on the same rectangular coordinate system.

45. $y = 3^x,\ x = 3^y$

46. $y = 3^{-x},\ x = 3^{-y}$

In Problems 47–50, the given function f is one-to-one. Find f^{-1} and give its domain and range.

47. $f(x) = 2 + 4^x$

48. $f(x) = 10^{x+3} - 10$

49. $f(x) = 1 + \ln(x - 2)$

50. $f(x) = 5 + \log_2 \dfrac{1}{x}$

In Problems 51–56, use the laws of logarithms in Theorem 5.2.1 to rewrite the given expression as one logarithm.

51. $\log_{10} 2 + 2\log_{10} 5$

52. $\frac{1}{2}\log_5 49 - \frac{1}{3}\log_5 8 + 13\log_5 1$

53. $\ln(x^4 - 4) - \ln(x^2 + 2)$

54. $\ln\left(\dfrac{x}{y}\right) - 2\ln x^3 - 4\ln y$

55. $\ln 5 + \ln 5^2 + \ln 5^3 - \ln 5^6$

56. $5\ln 2 + 2\ln 3 - 3\ln 4$

In Problems 57–68, use $\log_b 4 = 0.6021$ and $\log_b 5 = 0.6990$ to evaluate the given logarithm. Round your answer to four decimal places.

57. $\log_b 2$

58. $\log_b 20$

59. $\log_b 64$

60. $\log_b 625$

61. $\log_b \sqrt{5}$

62. $\log_b \frac{5}{4}$

63. $\log_b \sqrt[3]{4}$

64. $\log_b 80$

65. $\log_b 0.8$

66. $\log_b 3.2$

67. $\log_4 b$

68. $\log_5 5b$

In Problems 69–72, use the laws of logarithms in Theorem 5.2.1 so that $\ln y$ contains no products, quotients, or powers.

69. $y = \dfrac{x^{10}\sqrt{x^2 + 5}}{\sqrt[3]{8x^3 + 2}}$

70. $y = \sqrt{\dfrac{(2x + 1)(3x + 2)}{4x + 3}}$

71. $y = \dfrac{(x^3 - 3)^5(x^4 + 3x^2 + 1)^8}{\sqrt{x}(7x + 5)^9}$

72. $y = 64x^6\sqrt{x + 1}\sqrt[3]{x^2 + 2}$

In Problems 73–76, verify the given identity.

73. $\ln|\sec x| = -\ln|\cos x|$

74. $\ln|\cot x| = -\ln|\tan x|$

75. $\ln|\sec x - \tan x| = -\ln|\sec x + \tan x|$

76. $\ln|1 + \cos x| + \ln|1 - \cos x| = 2\ln|\sin x|$

77. What is the domain of the function $f(x) = \ln|\sin x|$?

78. For $f(x) = \log_b x$, show that $f(x_1 x_2) = f(x_1) + f(x_2)$.

FIGURE 5.2.4 Graph for Problem 79

For Discussion

79. In science it is sometimes useful to display data using logarithmic coordinates. Which of the following equations determines the graph shown in FIGURE 5.2.4?
    (i) $y = 2x + 1$ 　　　　　(ii) $y = e + x^2$
    (iii) $y = ex^2$ 　　　　　(iv) $x^2y = e$

80. (a) Use a graphing utility to obtain the graph of the function $f(x) = \ln(x + \sqrt{x^2 + 1})$.
    (b) Show that f is an odd function, that is, $f(-x) = -f(x)$.

81. If $a > 0$ and $b > 0$, $a \neq b$, then $\log_a x$ is a constant multiple of $\log_b x$. That is, $\log_a x = k \log_b x$. Find k.

82. Show that $(\log_{10} e)(\log_e 10) = 1$. Can you generalize this result?

83. The following question appeared on an examination:

$$\text{Find the domain of the function } f(x) = \ln\left(\frac{x - 3}{x}\right).$$

One student reasoned that using the laws of logarithms the function f could be rewritten as

$$f(x) = \ln(x - 3) - \ln x.$$

Because the domain of $\ln(x - 3)$ is the interval $(3, \infty)$ and the domain of $\ln x$ is the interval $(0, \infty)$, the domain of f is the intersection $(0, \infty) \cap (3, \infty) = (3, \infty)$. Discuss: Is the student's reasoning valid?

84. Find the vertical asymptotes for the graph of $f(x) = \ln\left(\frac{x - 3}{x}\right)$. Sketch the graph of f. Do not use a graphing utility.

In Problems 85–88, discuss how the graph of the given function can be obtained from the graph of $f(x) = \ln x$ by means of a rigid transformation (a shift or a reflection)?

85. $y = \ln 5x$ 　　　　　86. $y = \ln\dfrac{x}{4}$

87. $y = \ln x^{-1}$ 　　　　　88. $y = \ln(-x)$

In Problems 89 and 90, discuss how the graph of the given function can be obtained from the graph of $f(x) = \ln x$ by means of a rigid or nonrigid transformation.

89. $y = \log_{0.5} x$ 　　　　　90. $y = \log_{3.5} x$

5.3 Exponential and Logarithmic Equations

INTRODUCTION Since exponential and logarithmic functions appear in the context of many different applications, we are often called upon to solve equations that involve these functions. While we postpone applications until Section 5.4, we examine in the present section some of the ways that can be used to solve a variety of exponential and logarithmic equations.

☐ **Solving Equations** Here is a brief list of equation-solving strategies.

Solving Exponential and Logarithmic Equations

(*i*) Rewrite an exponential expression as a logarithmic expression.
(*ii*) Rewrite a logarithmic expression as an exponential expression.
(*iii*) Use the one-to-one properties of b^x and $\log_b x$.
(*iv*) For equations for the form $a^{x_1} = b^{x_2}$, where $a \neq b$, take the natural logarithm of both sides of the equality and simplify using (*iii*) of the laws of logarithms given in Theorem 5.2.1 of Section 5.2.

Of course, this list is not comprehensive and does not reflect the fact that in solving equations involving exponential and logarithmic functions we may also have to employ standard algebraic procedures such as *factoring* and using the *quadratic formula*.

In the first two examples we use the equivalence

$$y = \log_b x \quad \text{if and only if} \quad x = b^y \tag{1}$$

to toggle between logarithmic and exponential expressions.

■ EXAMPLE 1   Rewriting an Exponential Expression

Solve $e^{10k} = 7$ for k.

Solution We use (1), with $b = e$, to rewrite the given exponential expression as a logarithmic expression:

$$e^{10k} = 7 \quad \text{means} \quad 10k = \ln 7.$$

Therefore, with the aid of a calculator $k = \frac{1}{10}\ln 7 \approx 0.1946$.   ■

■ EXAMPLE 2   Rewriting a Logarithmic Expression

Solve $\log_2 x = 5$ for x.

Solution We use (1), with $b = 2$, to rewrite the logarithmic expression in its equivalent exponential form:

$$x = 2^5 = 32.$$ ■

☐ **One-to-One Properties** Recall from (1) of Section 2.8 that a one-to-one function f possesses the property that if $f(x_1) = f(x_2)$, then necessarily $x_1 = x_2$. We have seen in Sections 5.1 and 5.2 that both the exponential function $y = b^x$, $b > 0$, $b \neq 1$, and the logarithmic function $y = \log_b x$ are one-to-one. As a consequence we have:

$$\text{If } b^{x_1} = b^{x_2}, \text{ then } x_1 = x_2. \tag{2}$$
$$\text{If } \log_b x_1 = \log_b x_2, \text{ then } x_1 = x_2. \tag{3}$$

■ EXAMPLE 3   Using the One-to-One Property (2)

Solve $2^{x-3} = 8^{x+1}$ for x.

Solution Observe on the right-hand side of the given equality that 8 can be written as a power of 2, that is, $8 = 2^3$. Furthermore, by (*iv*) of the laws of exponents given in Theorem 5.1.1,

$$\overset{\text{multiply exponents}}{\overset{\downarrow \quad \downarrow}{8^{x+1} = (2^3)^{x+1} = 2^{3x+3}}}.$$

Thus, the equation is the same as

$$2^{x-3} = 2^{3x+3}.$$

From the one-to-one property (2) it follows that the exponents are equal, that is, $x - 3 = 3x + 3$. Solving for x then gives $2x = -6$ or $x = -3$. You are encouraged to check this answer by substituting -3 for x in the original equation. ▬

EXAMPLE 4　　　Using the One-to-One Property (2)

Solve $7^{2(x+1)} = 343$ for x.

Solution By noting that $343 = 7^3$, we have the same base on both sides of the equality:

$$7^{2(x+1)} = 7^3.$$

Thus by (2) we can equate exponents and solve for x:

$$2(x + 1) = 3$$
$$2x + 2 = 3$$
$$2x = 1$$
$$x = \frac{1}{2}.$$ ▬

EXAMPLE 5　　　Using the One-to-One Property (3)

Solve $\ln 2 + \ln(4x - 1) = \ln(2x + 5)$ for x.

Solution By (i) of the laws of logarithms in Theorem 5.2.1, the left-hand side of the equation can be written

$$\ln 2 + \ln(4x - 1) = \ln 2(4x - 1) = \ln(8x - 2).$$

The original equation is then

$$\ln(8x - 2) = \ln(2x + 5).$$

Since two logarithms with the same base are equal, it follows immediately from the one-to-one property (3) that $8x - 2 = 2x + 5$ or $6x = 7$ or $x = \frac{7}{6}$. ▬

☐ **Extraneous Solutions** For logarithmic equations, especially of the kind in Example 5, you should get accustomed to checking your answer by substituting it back into the original equation. It is possible for a logarithmic equation to have an **extraneous solution**.

EXAMPLE 6　　　An Extraneous Solution

Solve $\log_2 x + \log_2(x - 2) = 3$.

Solution We start using again that the sum of logarithms on the left-hand side of the equation is the logarithm of a product:

$$\log_2 x(x - 2) = 3.$$

With $b = 2$ we use (1) to rewrite the last equation in the equivalent exponential form

$$x(x - 2) = 2^3.$$

By ordinary algebra we then have

$$x^2 - 2x = 8$$
$$x^2 - 2x - 8 = 0$$
$$(x - 4)(x + 2) = 0.$$

From the last equation we conclude that either $x = 4$ or $x = -2$. However, we must rule out $x = -2$ as a solution. In other words, the number $x = -2$ is an extraneous solution because, when substituted into the original equation, the very first term, $\log_2(-2)$, is not defined. Thus the only solution of the given equation is $x = 4$.

Check:
$$\log_2 4 + \log_2 2 = \log_2 2^2 + \log_2 2$$
$$= \log_2 2^3 = 3\log_2 2 = 3 \cdot 1 = 3. \qquad \blacksquare$$

When we use the phrase "take the logarithm of both sides of an equality" we are actually using the property that if M and N are two positive numbers such that $M = N$, then $\log_b M = \log_b N$.

■ **EXAMPLE 7**   **Taking the Natural Logarithm of Both Sides**

Solve $e^{2x} = 3^{x-4}$.

Solution Since the bases of the exponential expression on each side of the equality are different, one way to proceed is to take the natural logarithm (the common logarithm could also be used) of both sides. From the equality

$$\ln e^{2x} = \ln 3^{x-4}$$

and (*iii*) of the laws of logarithms in Theorem 5.2.1, we get

$$2x \ln e = (x - 4)\ln 3.$$

Now using $\ln e = 1$ and the distributive law, the last equation becomes

$$2x = x \ln 3 - 4\ln 3.$$

Gathering the terms involving the symbol x to one side of the equality then gives

$$\overbrace{2x - x\ln 3}^{\substack{\text{factor } x \text{ out} \\ \text{of these terms}}} = -4\ln 3 \quad \text{or} \quad (2 - \ln 3)x = -4\ln 3 \quad \text{or} \quad x = \frac{-4\ln 3}{2 - \ln 3}.$$

You are encouraged to verify the calculation that $x \approx -4.8752$. $\qquad \blacksquare$

■ **EXAMPLE 8**   **Using the Quadratic Formula**

Solve $5^x - 5^{-x} = 2$.

Solution Because $5^{-x} = 1/5^x$, the equation is

$$5^x - \frac{1}{5^x} = 2.$$

Multiplying both sides of the foregoing equation by 5^x then gives

$$(5^x)^2 - 1 = 2(5^x) \quad \text{or} \quad (5^x)^2 - 2(5^x) - 1 = 0.$$

If we let $X = 5^x$, then the last equation can be interpreted as a quadratic equation $X^2 - 2X - 1 = 0$. Using the quadratic formula to solve for X yields

$$X = \frac{2 \pm \sqrt{4 + 4}}{2} = 1 \pm \sqrt{2} \quad \text{or} \quad 5^x = 1 \pm \sqrt{2}.$$

Because $1 - \sqrt{2}$ is a negative number and 5^x is positive for every real number x there are no real solutions of $5^x = 1 - \sqrt{2}$ and so

$$5^x = 1 + \sqrt{2}. \qquad (4)$$

Now by taking the natural logarithm of both sides of the equality we obtain

$$\ln 5^x = \ln\left(1 + \sqrt{2}\right)$$
$$x \ln 5 = \ln\left(1 + \sqrt{2}\right)$$
$$x = \frac{\ln\left(1 + \sqrt{2}\right)}{\ln 5}.$$

(5)

Using the $\boxed{\ln}$ key of a calculator, the division yields $x \approx 0.5476$. ▬

☐ **Change of Base** In (4) of Example 8 it follows from (1) that a perfectly valid solution of the equation $5^x - 5^{-x} = 2$ is $x = \log_5\left(1 + \sqrt{2}\right)$. But from a computational viewpoint (that is, expressing x as a number), the last answer is not desirable since no calculator has a logarithmic function with base 5. But by equating $x = \log_5\left(1 + \sqrt{2}\right)$ with the result in (5) we have discovered that logarithm with base 5 can be expressed in terms of the natural logarithm:

$$\log_5\left(1 + \sqrt{2}\right) = \frac{\ln\left(1 + \sqrt{2}\right)}{\ln 5}.$$

(6)

The result given in (6) is just a special case of a more general result known as the **change-of-base formula**.

THEOREM 5.3.1   Change-of-Base Formula

If $a \neq 1$, $b \neq 1$, and M are positive numbers, then

$$\log_a M = \frac{\log_b M}{\log_b a}$$

(7)

PROOF: If we let $y = \log_a M$, then from (1), $a^y = M$. Then

$$\log_b a^y = \log_b M$$
$$y \log_b a = \log_b M \qquad \leftarrow \text{by (iii) of Theorem 5.2.1}$$
$$y = \frac{\log_b M}{\log_b a} \qquad \leftarrow \text{by assumption } y = \log_a M$$
$$\log_a M = \frac{\log_b M}{\log_b a}. \qquad ▬$$

In order to obtain the numerical value of a logarithm using a calculator, we usually choose $b = 10$ or $b = e$ in (7):

$$\log_a M = \frac{\log_{10} M}{\log_{10} a} \qquad \text{or} \qquad \log_a M = \frac{\ln M}{\ln a}.$$

(8)

▬ **EXAMPLE 9**   **Changing the Base**

Find the numerical value of $\log_2 50$.

Solution We can use either formula in (8). If we choose the first formula in (8) with $M = 50$ and $a = 2$, we have

$$\log_2 50 = \frac{\log_{10} 50}{\log_{10} 2}.$$

Using the $\boxed{\log}$ key to calculate the two common logarithms and then dividing yields the approximation

$$\log_2 50 \approx 5.6439.$$

Alternatively, the second formula in (8) gives the same result:

$$\log_2 50 = \frac{\ln 50}{\ln 2} \approx 5.6439.$$

We can check the answer in Example 9 on a calculator by using the $\boxed{y^x}$ key. You are urged to verify that $2^{5.6439} \approx 50$.

■ EXAMPLE 10   Changing the Base

Find the number x in the domain of $f(x) = 6^x$ for which $f(x) = 73$.

Solution We must find a solution of the equation $6^x = 73$. One way of proceeding is to rewrite the exponential expression as an equivalent logarithmic expression:

$$x = \log_6 73.$$

Then with the identification $a = 6$ it follows from the second equation in (8) and the aid of a calculator that

$$x = \log_6 73 = \frac{\ln 73}{\ln 6} \approx 2.3946.$$

You should verify that $f(2.3946) = 6^{2.3946} \approx 73$.

Exercises 5.3   Answers to selected odd-numbered problems begin on page ANS-20.

In Problems 1–20, solve the given exponential equation.

1. $5^{x-2} = 1$

2. $3^x = 27^{x^2}$

3. $10^{-2x} = \dfrac{1}{10,000}$

4. $27^x = \dfrac{9^{2x-1}}{3^x}$

5. $e^{5x-2} = 30$

6. $\left(\dfrac{1}{e}\right)^x = e^3$

7. $2^x \cdot 3^x = 36$

8. $\dfrac{4^x}{3^x} = \dfrac{9}{16}$

9. $2^{x^2} = 8^{2x-3}$

10. $\frac{1}{4}(10^{-2x}) = 25(10^x)$

11. $5 - 10^{2x} = 0$

12. $7^{-x} = 9$

13. $3^{2(x-1)} = 7^2$

14. $\left(\frac{1}{2}\right)^{-x+2} = 8(2^{x-1})^3$

15. $\dfrac{1}{3} = (2^{|x|-2} - 1)^{-1}$

16. $\left(\frac{1}{3}\right)^x = 9^{1-2x}$

17. $5^{|x|-1} = 25$

18. $(e^2)^{x^2} - \dfrac{1}{e^{5x+3}} = 0$

19. $4^x = 5^{2x+1}$

20. $3^{x+4} = 2^{x-16}$

In Problems 21–40, solve the given logarithmic equation.

21. $\log_3 5x = \log_3 160$

22. $\ln(10 + x) = \ln(3 + 4x)$

23. $\ln x = \ln 5 + \ln 9$

24. $3 \log_8 x = \log_8 36 + \log_8 12 - \log_8 2$

25. $\log_{10} \dfrac{1}{x^2} = 2$

26. $\log_3 \sqrt{x^2 + 17} = 2$

27. $\log_2(\log_3 x) = 2$

28. $\log_5|1 - x| = 1$

29. $\log_3 81^x - \log_3 3^{2x} = 3$

30. $\dfrac{\log_2 8^x}{\log_2 \frac{1}{4}} = \dfrac{1}{2}$

31. $\log_{10} x = 1 + \log_{10} \sqrt{x}$

32. $\log_2(x - 3) - \log_2(2x + 1) = -\log_2 4$

33. $\log_2 x + \log_2(10 - x) = 4$

34. $\log_8 x + \log_8 x^2 = 1$

35. $\log_6 2x - \log_6(x + 1) = 0$

36. $\log_{10} 54 - \log_{10} 2 = 2\log_{10} x - \log_{10} \sqrt{x}$

37. $\log_9 \sqrt{10x + 5} - \dfrac{1}{2} = \log_9 \sqrt{x + 1}$

38. $\log_{10} x^2 + \log_{10} x^3 + \log_{10} x^4 - \log_{10} x^5 = \log_{10} 16$

39. $\ln 3 + \ln(2x - 1) = \ln 4 + \ln(x + 1)$

40. $\ln(x + 3) + \ln(x - 4) - \ln x = \ln 3$

In Problems 41–50, either use factoring or the quadratic formula to solve the given equation.

41. $(5^x)^2 - 26(5^x) + 25 = 0$

42. $64^x - 10(8^x) + 16 = 0$

43. $\log_4 x^2 = (\log_4 x)^2$

44. $(\log_{10} x)^2 + \log_{10} x = 2$

45. $(5^x)^2 - 2(5^x) - 1 = 0$

46. $2^{2x} - 12(2^x) + 35 = 0$

47. $(\ln x)^2 + \ln x = 2$

48. $(\log_{10} 2x)^2 = \log_{10}(2x)^2$

49. $2^x + 2^{-x} = 2$

50. $10^{2x} - 103(10^x) + 300 = 0$

In Problems 51–56, find the x-intercepts of the graph of the given function.

51. $f(x) = e^{x+4} - e$

52. $f(x) = 1 - \frac{1}{5}(0.1)^x$

53. $f(x) = 4^{x-1} - 3$

54. $f(x) = -3^{2x} + 5$

55. $f(x) = \dfrac{10}{2 + e^{-2x}} - 1$

56. $f(x) = \dfrac{2^x - 6 + 2^{3-x}}{x + 2}$

In Problems 57 and 58, find the x- and y-intercepts of the given graphs.

57.

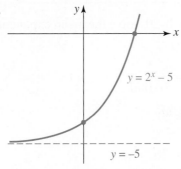

FIGURE 5.3.1 Graph for Problem 57

58.

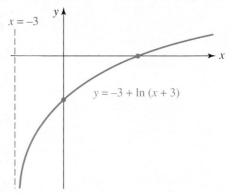

FIGURE 5.3.2 Graph for Problem 58

In Problems 59 and 60, find the zeros of the given function.

59. $f(x) = 5 - \log_2|x + 4|$

60. $f(x) = -1 + e^{|3 - \frac{1}{2}x| - 2}$

In Problems 61–66, graph the given functions. Determine the approximate x-coordinates of the points of intersection of their graphs.

61. $f(x) = 4e^x, \quad g(x) = 3^{-x}$

62. $f(x) = 2^x, \quad g(x) = 3 - 2^x$

63. $f(x) = 3^{x^2}, \quad g(x) = 2(3^x)$

64. $f(x) = \dfrac{1}{3} \cdot 2^{x^2}, \quad g(x) = 2^{x^2} - 1$

65. $f(x) = \log_{10} \dfrac{10}{x}, \quad g(x) = \log_{10} x$

66. $f(x) = \log_{10} \dfrac{x}{2}, \quad g(x) = \log_2 x$

In Problems 67–70, solve the given equation.

67. $x^{\ln x} = e^9$

68. $x^{\log_{10} x} = \dfrac{1000}{x^2}$

69. $\log_x 81 = 2$

70. $\log_5 125^x = -2$

In Problems 71–76, find the points on the graph of the given function that have the indicated y-coordinate.

71. $f(x) = 6^x$; 51

72. $f(x) = (\tfrac{1}{2})^x$; 7

73. $f(x) = \log_3(x + 2)$; 2

74. $f(x) = 5 - 2\ln x$; 4

75. $f(x) = 1 - e^{-x^2}$; $\tfrac{1}{2}$

76. $f(x) = 25^x - 5^{x+1}$; -6

In Problems 77 and 78, find the numerical value of the given logarithm.

77. $\log_\pi 4$

78. $\log_8 e$

For Discussion

In Problems 79 and 80, discuss how to solve the given equation. Carry out your ideas.

79. $\log_2 x + \log_4 x = 6$

80. $\log_3 x - \log_6 x - 2 = 0$

81. Use a graphing utility to obtain the graph of the function $f(x) = \log_{x+2}(3 - x)$. Give the domain of the function f.

82. Discuss: Are the given two equations equivalent, that is, do they have the same solution set?
  (a) $\log_5(x - 2)^2 = 2$; $2\log_5(x - 2) = 2$
  (b) $\log_5(x - 2)^3 = 3$; $3\log_5(x - 2) = 3$

5.4 Exponential and Logarithmic Models

INTRODUCTION In this section we consider some **mathematical models** utilizing exponential or logarithmic functions. Roughly speaking, a mathematical model is a mathematical description of something that we will call a *system*. To construct a mathematical model we start with a set of reasonable assumptions about the system that we are trying to describe. These assumptions include any empirical laws that are applicable to the system. The end result could be a description as simple as a single function.

☐ **Exponential Models** In the physical sciences, the exponential expression Ce^{kt}, where C and k are constants, frequently appears in mathematical models of systems that change with time t. As a consequence, mathematical models are often used to predict a future state of a system. For example, extremely complicated mathematical models are used to predict the weather over various regions of the country for, say, the next week.

☐ **Population Growth** In one model of a growing population, it is assumed that the *rate* of growth of the population is proportional to the *number present* at time t. If $P(t)$ denotes the population or number present at time t, then with the aid of calculus it can be shown that this assumption gives rise to

$$P(t) = P_0 e^{kt}, \quad k > 0, \tag{1}$$

where t is time, and P_0 and k are constants. The function (1) is used to describe the growth of populations of bacteria, small animals, and, in some rare circumstances, humans. Setting $t = 0$ gives $P(0) = P_0$, and so P_0 is called the **initial population**. The constant

E. Coli bacteria

When working problems such as this, be ▶
sure to store the value of k in the memory
of your calculator.

$k > 0$ is called the **growth constant** or **growth rate**. Since e^{kt}, $k > 0$, is an increasing function on the interval $[0, \infty)$, the model in (1) describes uninhibited growth.

EXAMPLE 1   Bacterial Growth

It is known that the doubling time* of *E. Coli* bacteria, which reside in the large intestine (colon) of healthy people, is just 20 minutes. Use the exponential growth model (1) to find the number of *E. Coli* bacteria in a culture after 6 hours.

Solution Let us use hours as our unit of time, so that 20 min $= \frac{1}{3}$ h. Because the initial number of *E. Coli* in the culture is not specified, we will simply denote the initial size of the culture as P_0. Now using (1), a function interpretation of the first sentence in this example is $P\left(\frac{1}{3}\right) = 2P_0$. This means $P_0 e^{k/3} = 2P_0$ or $e^{k/3} = 2$. Solving this last equation for k gives the growth constant

$$\frac{k}{3} = \ln 2 \quad \text{or} \quad k = 3\ln 2 \approx 2.0794.$$

A model for the size of the culture after t hours is then $P(t) = P_0 e^{2.0794t}$. Setting $t = 6$ gives $P(6) = P_0 e^{2.0794(6)} \approx 262{,}144 P_0$. Put another way, if the culture consists of only *one* bacterium at $t = 0$, then (with $P_0 = 1$) the model predicts that there will be $262{,}144$ cells 6 hours later. ▪

In the early nineteenth century the English clergyman and economist Thomas R. Malthus used the growth model (1) to predict the world population. For specific values of P_0 and k, the function values $P(t)$ were actually reasonable approximations to the world population for a period of time during the nineteenth century. Since $P(t)$ is an increasing function, Malthus predicted that the future population growth would surpass the world's ability to produce food. As a consequence he also predicted wars and worldwide famine. More a doomsayer than a seer, Malthus failed to foresee that the food supply would keep pace with the increased population through simultaneous advances in science and technology.

In 1840, a more realistic model for predicting human populations in small countries was advanced by the Belgian mathematician/biologist **P. F. Verhulst** (1804–1849). The so-called **logistic function**

$$P(t) = \frac{K}{1 + ce^{rt}}, \quad r < 0, \tag{2}$$

Thomas R. Malthus
(1776–1834)

where K, c, and r are constants, has over the years proved to be an accurate growth model for populations of protozoa, bacteria, fruit flies, water fleas, fish, and animals confined to limited spaces. In contrast to uninhibited growth of the Malthusian model (1), (2) exhibits bounded growth. More specifically, the population predicted by (2) will not increase beyond the number K, called the **carrying capacity** of the ecosystem. For $r < 0$, $e^{rt} \to 0$ and $P(t) \to K$ as $t \to \infty$. You are asked to graph a special case of (2) in Problem 7 in Exercises 5.4.

☐ **Radioactive Decay** Element 88, better known as **radium**, was discovered by Pierre and Marie Curie in 1898. Radium is a radioactive element, which means that a radium atom spontaneously **decays**, or disintegrates, by emitting radiation in the form of alpha particles, beta particles, and gamma rays. When an atom disintegrates in this manner, its nucleus is transmuted into a nucleus of another element. For example, the nucleus of an atom of the most stable isotope of radium, Ra-226, is transmuted into the

Pierre and Marie Curie

---

*In biology the doubling time is sometimes referred to as the **generation time**.

nucleus of a radon atom Rn-222. Radon is a heavy, odorless, colorless, and highly dangerous radioactive gas that usually originates in the ground. Because it can penetrate a sealed concrete floor, radon frequently accumulates in the basements of some new and highly insulated homes. Some medical organizations have claimed that after cigarette smoking, exposure to radon gas is the second leading cause of lung cancer.

If it is assumed that the rate of decay of a radioactive substance is proportional to the amount remaining or present at time t, then we arrive at basically the same model as in (1). The important difference is that $k < 0$. If $A(t)$ represents the amount of the decaying substance that remains at time t, then

$$A(t) = A_0 e^{kt}, \quad k < 0, \tag{3}$$

where A_0 is the initial amount of the substance present, that is, $A(0) = A_0$. The constant $k < 0$ in (3) is called the **decay constant** or **decay rate**.

▌EXAMPLE 2   Decay of Radium

Suppose there are 20 grams of radium on hand initially. After t years the amount remaining is modeled by the function $A(t) = 20e^{-0.000418t}$. Find the amount of radium remaining after 100 years. What percent of the original 20 grams has decayed after 100 years?

Solution Using a calculator, we find that after 100 years there remains

$$A(100) = 20e^{-0.000418(100)} \approx 19.18 \text{ g}.$$

Thus, only

$$\frac{20 - 19.18}{20} \times 100\% \approx 4.1\%$$

of the initial 20 grams has decayed. ▬

FIGURE 5.4.1 Time T is the half-life

☐ **Half-Life** The **half-life** of a radioactive substance is the time T it takes for one-half of a given amount of that element to disintegrate and change into a new element. See FIGURE 5.4.1. Half-life is a measure of the stability of an element; that is, the shorter the half-life, the more unstable the element. For example, the half-life of the highly radioactive strontium-90, Sr-90, produced in nuclear explosions, is 29 days, whereas the half-life of the uranium isotope U-238 is 4,560,000 years. The half-life of californium, Cf-244, first discovered in 1950, is only 45 minutes. Polonium, Po-213, has a half-life of 0.000004 seconds.

▌EXAMPLE 3   Half-Life of Radium

Use the exponential model in Example 2 to determine the half-life of radium.

Solution If $A(t) = 20e^{-0.000418t}$, then we must find the time T for which

$$\overset{\text{one-half the initial amount}}{\underset{\downarrow}{}}$$

$$A(T) = \tfrac{1}{2}(20) = 10.$$

From $20e^{-0.000418T} = 10$ we get $e^{-0.000418T} = \tfrac{1}{2}$. By rewriting the last expression in the logarithmic form $-0.000418T = \ln\tfrac{1}{2}$ we can solve for T:

$$T = \frac{\ln\tfrac{1}{2}}{-0.000418} \approx 1660 \text{ years.} \quad ▬$$

A careful reading of Example 3 reveals that the initial amount present plays no part in the actual calculation of the half-life. Since the solution of $A(T) = A_0 e^{-0.000418T} = \tfrac{1}{2}A_0$

Ibuprofen is an NSAID

Willard Libby
(1908–1980)

The Psalms scroll

leads to $e^{-0.000418T} = \frac{1}{2}$, we see that T is independent of A_0. In other words, the half-life of 1 gram, 20 grams, or 10,000 grams of radium is the same. It takes about 1660 years for one-half of *any* given quantity of radium to transmute into radon.

Medications also have half-lives. In this case, the half-life of a drug is the time T that it takes for the body to eliminate, by metabolism or excretion, one-half of the amount of the drug taken. For example, the most popular NSAIDs (nonsteroidal anti-inflammatory drugs such as aspirin and ibuprofen) taken for the relief of continuing pain, have relatively short half-lives of a few hours and as a consequence must be taken several times a day. The NSAID naproxen has a longer half-life and is usually taken once every 12 hours. See Problem 33 in Exercises 5.4.

☐ **Carbon Dating** The approximate age of fossils of once-living matter can sometimes be determined by a method known as **carbon dating**. The radioactive isotope of carbon, carbon-14 or C-14, is formed presumably at a constant rate in the atmosphere by the interaction of cosmic rays on nitrogen-14. The carbon-dating method, invented by the chemist Willard Libby around 1950, is based on the fact that a plant or an animal absorbs C-14 through the process of breathing and eating, and ceases to absorb C-14 when it dies. As the next example shows, the carbon-dating procedure is based on the knowledge that the half-life of C-14 is about 5730 years. Carbon-14 decays back to the original nitrogen-14.

Libby won the 1960 Nobel Prize in chemistry for his work. Libby's method has been used to date wooden furniture found in Egyptian tombs, the Dead Sea scrolls written on papyrus and animal skin, the famous linen Shroud of Turin, and a recently discovered copy of the Gnostic Gospel of Judas written on papyrus.

EXAMPLE 4   Carbon Dating a Fossil

A fossilized bone is found to contain $\frac{1}{1000}$ of the initial amount of C-14 that the organism contained while it was alive. Determine the approximate age of the fossil.

Solution If A_0 denotes an initial amount A_0, measured in grams, of C-14 in the organism, then t years after its death there are $A(t) = A_0 e^{kt}$ grams remaining. When $t = 5730$, $A(5730) = \frac{1}{2}A_0$, and so $\frac{1}{2}A_0 = A_0 e^{5730k}$. Solving this last equation for the decay constant k gives

$$e^{5730k} = \frac{1}{2} \quad \text{and so} \quad k = \frac{\ln\frac{1}{2}}{5730} \approx -0.00012097.$$

Hence a model for the amount of C-14 remaining is $A(t) = A_0 e^{-0.00012097t}$. Using this model, we now solve $A(t) = \frac{1}{1000}A_0$ for t:

$$A_0 e^{-0.00012097t} = \frac{1}{1000}A_0 \quad \text{implies} \quad t = \frac{\ln\frac{1}{1000}}{-0.00012097} \approx 57{,}100 \text{ years.} \quad ▬$$

The age determined in the last example is actually beyond the border of accuracy for the carbon-14-dating method. After 9 half-lives of the isotope, or about 52,000 years, about 99.7% of carbon-14 has decayed making its measurement in a fossil nearly impossible.

☐ **Newton's Law of Cooling/Warming** Suppose an object or body is placed in a medium (air, water, etc.) that is held at constant temperature T_m, called the **ambient temperature**. If the initial temperature T_0 of the body or object at the moment it is

We take this moment to correspond to the time $t = 0$. ▶

placed into the medium is greater than the ambient temperature T_m, then the body will cool. On the other hand, if T_0 is less than T_m, then it will warm up. For example, in an office kept at, say, 70°F, a steaming cup of coffee will cool off, whereas a glass of ice water will warm up. The usual cooling/warming assumption is that the rate at which an object cools/warms is proportional to the difference $T(t) - T_m$, where $T(t)$ represents the temperature of the object at time t. In either case, cooling or warming, this assumption leads to $T(t) - T_m = (T_0 - T_m)e^{kt}$, where k is a negative constant. Observe that since $e^{kt} \to 0$ for $k < 0$, the last expression is consistent with one's intuitive expectation that $T(t) - T_m \to 0$, or equivalently $T(t) \to T_m$, as $t \to \infty$ (the coffee cools to room temperature; the ice water warms to room temperature). Solving for $T(t)$ we obtain a function for the temperature of the object,

$$T(t) = T_m + (T_0 - T_m)e^{kt}, \quad k < 0. \tag{4}$$

The mathematical model in (4), named after its discoverer, is called **Newton's law of cooling/warming**. Note that $T(0) = T_0$.

▮ EXAMPLE 5   Cooling of a Cake

A cake is removed from an oven where the temperature was 350°F into a kitchen where the temperature is 75°F. One minute later the temperature of the cake is measured to be 300°F. Assume that the temperature of the cake in the kitchen is given by (4).

(a) What is the temperature of the cake after 6 minutes?
(b) At what time is the temperature of the cake 80°F?
(c) Graph $T(t)$.

Solution (a) When the cake is removed from the oven its temperature is also 350°F, that is, $T_0 = 350$. The ambient temperature is the temperature of the kitchen $T_m = 75$. Thus (4) becomes $T(t) = 75 + 275e^{kt}$. The measurement that $T(1) = 300$ is the condition that determines the value of k. From $T(1) = 75 + 275e^k = 300$ we find

$$e^k = \frac{225}{275} = \frac{9}{11} \quad \text{or} \quad k = \ln\frac{9}{11} \approx -0.2007.$$

The mathematical model $T(t) = 75 + 275e^{-0.2007t}$ then predicts that the temperature of the cake 6 minutes after it is removed from the oven will be

$$T(6) = 75 + 275e^{-0.2007(6)} \approx 157.5°. \tag{5}$$

(b) To determine when the temperature of the cake will be 80°F, we solve the equation $T(t) = 80$ for t. Rewriting $T(t) = 75 + 275e^{-0.2007t} = 80$ as

$$e^{-0.2007t} = \frac{5}{275} = \frac{1}{55} \quad \text{we find} \quad t = \frac{\ln\frac{1}{55}}{-0.2007} \approx 20 \text{ min.}$$

(c) With the aid of a graphing utility we obtain the graph of $T(t)$ shown in blue in **FIGURE 5.4.2**. Since $T(t) = 75 + 275e^{-0.2007t} \to 75$ as $t \to \infty$, $T = 75$, shown in red in Figure 5.4.2, is a horizontal asymptote for the graph of $T(t) = 75 + 275e^{-0.2007t}$. ▮

Cake will cool off to room temperature

FIGURE 5.4.2 Graph of $T(t)$ in Example 5

☐ **Compound Interest** Investments such as savings accounts pay an annual rate of interest that can be compounded annually, quarterly, monthly, weekly, daily, and so on. In general, if a principal of P dollars is invested at an annual rate r of interest that is compounded n times a year, then the amount S accrued at the end of t years is given by

$$S = P\left(1 + \frac{r}{n}\right)^{nt}. \tag{6}$$

S is called the **future value** of the principal P. If the number n is increased without bound, then interest is said to be **compounded continuously**. To find the future value of P in this case, we let $m = n/r$. Then $n = mr$ and

$$\left(1 + \frac{r}{n}\right)^{nt} = \left(1 + \frac{1}{m}\right)^{mrt} = \left[\left(1 + \frac{1}{m}\right)^{m}\right]^{rt}.$$

Since $n \to \infty$ implies that $m \to \infty$, we see from page 332 of Section 5.1 that $(1 + 1/m)^m \to e$. The right-hand side of (6) becomes

$$P\left[\left(1 + \frac{1}{m}\right)^{m}\right]^{rt} \to P[e]^{rt} \quad \text{as} \quad m \to \infty.$$

Thus, if an annual rate r of interest is compounded continuously, the future value S of a principal P in t years is

$$S = Pe^{rt}. \tag{7}$$

EXAMPLE 6   Comparison of Future Values

Suppose that $1000 is deposited in a savings account whose annual rate of interest is 3%. Compare the future value of this principal in 10 years **(a)** if interest is compounded monthly and **(b)** if interest is compounded continuously.

Solution (a) Since there are 12 months in a year, we identify $n = 12$. Furthermore, with $P = 1000$, $r = 0.03$, and $t = 10$, (6) becomes

$$S = 1000\left(1 + \frac{0.03}{12}\right)^{12(10)} = 1000(1.0025)^{120} \approx \$1{,}349.35.$$

(b) From (7),

$$S = 1000e^{(0.03)(10)} = 1000e^{0.3} \approx \$1{,}349.86.$$

Thus over 10 years we have gained only $0.51 by compounding continuously rather than monthly. ▬

☐ **Logarithmic Models** Probably the most famous application of the base 10 logarithm, or common logarithm, is the **Richter scale**. In 1935, the American seismologist Charles F. Richter devised a logarithmic scale for comparing the energies of different earthquakes. The magnitude M of an earthquake is defined by

$$M = \log_{10}\frac{A}{A_0}, \tag{8}$$

where A is the amplitude of the largest seismic wave of the earthquake and A_0 is a reference amplitude that corresponds to the magnitude $M = 0$. The number M is calculated to one decimal place. Earthquakes of magnitude 6 or greater are considered potentially destructive.

Since 1979 the USGS has used the **moment magnitude scale** to assign a magnitude to earthquakes that range from strong to massive. This scale devised by the seismologists Thomas C. Hanks and Hiroo Kanamori corrects certain deficiencies in the Richter scale at that level. The moment magnitude M_w is defined by

$$M_w = \tfrac{2}{3}\log_{10}M_0 - 10.7,$$

where M_0 denotes the magnitude of the seismic moment. For moderate earthquakes the magnitudes given by moment magnitude scale and Richter scale are about the same. Using the moment magnitude scale the USGS assigned a magnitude of $M_w = 9.0$ to the Japan earthquake on March 11, 2011.

Charles F. Richter
(1900–1985)

EXAMPLE 7   Comparing Intensities

The earthquake on December 26, 2004, off the west coast of Northern Sumatra, which spawned a tsunami causing over 200,000 deaths, was initially classified as a 9.3 on the Richter scale. On March 28, 2005, an aftershock in the same area was classified as an 8.7 on the Richter scale. How many times more intense was the 2004 earthquake?

Solution From (8) we have

$$9.3 = \log_{10}\left(\frac{A}{A_0}\right)_{2004} \quad \text{and} \quad 8.7 = \log_{10}\left(\frac{A}{A_0}\right)_{2005}.$$

This means, in turn, that

$$\left(\frac{A}{A_0}\right)_{2004} = 10^{9.3} \quad \text{and} \quad \left(\frac{A}{A_0}\right)_{2005} = 10^{8.7}.$$

Now, since $9.3 = 0.6 + 8.7$, it follows from the laws of exponents that

$$\left(\frac{A}{A_0}\right)_{2004} = 10^{9.3} = 10^{0.6}10^{8.7} = 10^{0.6}\left(\frac{A}{A_0}\right)_{2005} \approx 3.98\left(\frac{A}{A_0}\right)_{2005}.$$

Thus the original earthquake in 2004 was approximately 4 times as intense as the aftershock in 2005. ■

You can see from Example 7 that if, say, one earthquake is a 6.0 and another is a 4.0 on the Richter scale, then the 6.0 earthquake is $10^2 = 100$ times more intense than the 4.0 earthquake.

☐ **pH of a Solution** In chemistry, the hydrogen potential, or **pH**, of a solution is defined as

$$pH = -\log_{10}[H^+], \tag{9}$$

where the symbol $[H^+]$ denotes the concentration of hydrogen ions in a solution measured in moles per liter. The pH scale was invented in 1909 by the Danish biochemist Søren Sørensen. Solutions are classified according to their pH value as *acidic, base,* or *neutral.* A solution with a pH in the range $0 < pH < 7$ is said to be acidic; when $pH > 7$, the solution is base (or alkaline). In the case when $pH = 7$, the solution is neutral. Water, if uncontaminated by other solutions or by acid rain, is an example of a neutral solution, whereas undiluted lemon juice is highly acidic and has a pH in the range $pH \leq 3$. A solution with $pH = 6$ is ten times more acidic than a neutral solution. See Problems 49–52 in Exercises 5.4.

As the next example illustrates, pH values are usually calculated to one decimal place.

Søren Sørensen
(1868–1939)

EXAMPLE 8   pH of Human Blood

The concentration of hydrogen ions in the blood of a healthy person is found to be $[H^+] = 3.98 \times 10^{-8}$ moles/liter. Find the pH of blood.

Solution From (9) and the laws of logarithms (Theorem 5.2.1),

$$
\begin{aligned}
pH &= -\log_{10}[3.98 \times 10^{-8}] \\
&= -[\log_{10}3.98 + \log_{10}10^{-8}] \\
&= -[\log_{10}3.98 - 8\log_{10}10] \quad \leftarrow \log_{10}10 = 1 \\
&= -[\log_{10}3.98 - 8].
\end{aligned}
$$

With the help of the base 10 log key on a calculator, we find that

$$pH \approx -[0.5999 - 8] \approx 7.4. \quad\blacksquare$$

Human blood in usually a base solution. The pH values of blood usually fall within the rather narrow range $7.2 < \text{pH} < 7.6$. A person with a blood pH outside these limits can suffer illness and even death.

Exercises 5.4 Answers to selected odd-numbered problems begin on page ANS–20.

Population Growth

1. After 2 hours the number of bacteria in a culture is observed to have doubled.
   (a) Find an exponential model (1) for the number of bacteria in the culture at time t.
   (b) Find the number of bacteria present in the culture after 5 hours.
   (c) Find the time that it takes the culture to grow to 20 times its initial size.
2. A model for the number of bacteria in a culture after t hours is given by (1).
   (a) Find the growth constant k if it is known that after 1 hour the colony has expanded to 1.5 times its initial population.
   (b) Find the time that it takes for the culture to quadruple in size.
3. A model for the population in a small community is given by $P(t) = 1500e^{kt}$. If the initial population increases by 25% in 10 years, what will the population be in 20 years?
4. A model for the population in a small community after t years is given by (1).
   (a) If the initial population has doubled in 5 years, how long will it take to triple? To quadruple?
   (b) If the population of the community in part (a) is 10,000 after 3 years, what was the initial population?
5. A model for the number of bacteria in a culture after t hours is given by $P(t) = P_0 e^{kt}$. After 3 hours it is observed that 400 bacteria are present. After 10 hours 2000 bacteria are present. What was the initial number of bacteria?
6. In genetic research a small colony of drosophila (small two-winged fruit flies) is grown in a laboratory environment. After 2 days it is observed that the population of flies in the colony has increased to 200. After 5 days the colony has 400 flies.
   (a) Find a model $P(t) = P_0 e^{kt}$ for the population of the fruit-fly colony after t days.
   (b) What will be the population of the colony in 10 days?
   (c) When will the population of the colony be 5000 fruit flies?
7. A student sick with a flu virus returns to an isolated college campus of 2000 students. A model for the number of students infected with the flu t days after the student's return is given by the logistic function

$$P(t) = \frac{2000}{1 + 1999e^{-0.8905t}}.$$

   (a) According to this model, how many students will be infected with the flu after 5 days?
   (b) How long will it take for one-half of the student population to become infected?
   (c) How many students does the model predict will become infected after a very long period of time?
   (d) Sketch a graph of $P(t)$.
8. In 1920, Raymond Pearl and Lowell Reed proposed a logistic model for the population of the United States based on the years 1790, 1850, and 1910. The logistic function they proposed was

$$P(t) = \frac{2930.3009}{0.014854 + e^{-0.0313395t}},$$

   where P is measured in thousands and t represents the number of years past 1780.

(a) The model agrees quite well with the census figures between 1790 and 1910. Determine the population figures for 1790, 1850, and 1910.

(b) What does this model predict for the population of the United States after a very long time? How does this prediction compare with the 2000 census population of 281 million?

Radioactive Decay and Half-Life

9. Initially 200 milligrams of a radioactive substance was present. After 6 hours the mass had decreased by 3%. Construct an exponential model $A(t) = A_0 e^{kt}$ for the amount remaining of the decaying substance after t hours. Find the amount remaining after 24 hours.

10. Determine the half-life of the substance in Problem 9.

11. Do this problem without using the exponential model (3). Initially there are 400 grams of a radioactive substance on hand. If the half-life of the substance is 8 hours, give an educated guess of how much remains (approximately) after 17 hours. After 23 hours. After 33 hours.

12. Construct an exponential model $A(t) = A_0 e^{kt}$ for the amount remaining of the decaying substance in Problem 11. Compare the predicted values $A(17)$, $A(23)$, and $A(33)$ with your guesses.

13. Iodine-131, used in nuclear medicine procedures, is radioactive and has a half-life of 8 days. Find the decay constant k for iodine-131. If the amount remaining of an initial sample after t days is given by the exponential model $A(t) = A_0 e^{kt}$, how long will it take for 95% of the sample to decay?

14. The amount remaining of a radioactive substance after t hours is given by $A(t) = 100 e^{kt}$. After 12 hours, the initial amount has decreased by 7%. How much remains after 48 hours? What is the half-life of the substance?

15. The half-life of polonium-210, Po-210, is 140 days. If $A(t) = A_0 e^{kt}$ represents the amount of Po-210 remaining after t days, what is the amount remaining after 80 days? After 300 days?

16. Strontium-90 is a dangerous radioactive substance found in acid rain. As such it can make its way into the food chain by polluting the grass in a pasture on which milk cows graze. The half-life of strontium-90 is 29 years.

(a) Find an exponential model (3) for the amount remaining after t years.

(b) Suppose a pasture is found to contain Str-90 that is 3 times a safe level A_0. How long will it be before the pasture can be used again for grazing cows?

Carbon Dating

17. Charcoal drawings were discovered on walls and ceilings in a cave in Lascaux, France. Determine the approximate age of the drawings, if it was found that 86% of C-14 in a piece of charcoal found in the cave had decayed through radioactivity.

18. Analysis on an animal bone fossil at an archeological site reveals that the bone has lost between 90% and 95% of C-14. Give an interval for the possible ages of the bone.

19. The Shroud of Turin shows the negative image of the body of a man who appears to have been crucified. It is believed by many to be the burial shroud of Jesus of Nazareth. In 1988 the Vatican granted permission to have the shroud carbon dated. Several independent scientific laboratories analyzed the cloth and the consensus opinion was that the shroud is approximately 660 years old, an age consistent with its historical appearance. This age has been disputed by many scholars. Using this age, determine what percentage of the original amount of C-14 remained in the cloth as of 1988.

Charcoal drawing in Problem 17

Shroud image in Problem 19

The iceman in Problem 20

20. In 1991 hikers found a preserved body of a man partially frozen in a glacier in the Austrian Alps. Through carbon-dating techniques it was found that the body of Ötzi—the iceman, as he came to be called—contained 53% as much C-14 as found in a living person. What is the approximate date of his death?

Newton's Law of Cooling/Warming

21. Suppose a pizza is removed from an oven at 400°F into a kitchen whose temperature is a constant 80°F. Three minutes later the temperature of the pizza is found to be 275°F.
   (a) What is the temperature $T(t)$ of the pizza after 5 minutes?
   (b) Determine the time when the temperature of the pizza is 150°F.
   (c) After a very long period of time, what is the approximate temperature of the pizza?

22. A glass of cold water is removed from a refrigerator whose interior temperature is 39°F into a room maintained at 72°F. One minute later the temperature of the water is 43°F. What is the temperature of the water after 10 minutes? After 25 minutes?

23. A thermometer is brought from the outside, where the air temperature is $-20°F$, into a room where the air temperature is a constant 70°F. After 1 minute inside the room the thermometer reads 0°F. How long will it take for the thermometer to read 60°F?

Thermometer in Problem 24

24. A thermometer is taken from inside a house to the outside, where the air temperature is 5°F. After 1 minute outside the thermometer reads 59°F, and after 5 minutes it reads 32°F. What is the temperature inside the house?

25. A dead body was found within a closed room of a house where the temperature was a constant 70°F. At the time of discovery, the core temperature of the body was determined to be 85°F. One hour later a second measurement showed that the core temperature of the body was 80°F. Assume that the time of death corresponds to $t = 0$ and that the core temperature at that time was 98.6°F. Determine how many hours elapsed before the body was found.

26. Repeat Problem 25 if evidence indicated that the dead person was running a fever of 102°F at the time of death.

Compound Interest

27. Suppose that 1¢ is deposited in a savings account paying 1% annual interest compounded continuously. How much money will have accrued in the account after 2000 years? What is the future value of 1¢ in 2000 years if the account pays 2% annual interest compounded continuously?

28. Suppose that $100,000 is invested at an annual interest rate of 5%. Use (6) and (7) to compare the future values of that amount in 1 year by completing the following table.

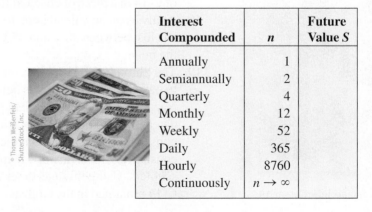

Interest Compounded	n	Future Value S
Annually	1	
Semiannually	2	
Quarterly	4	
Monthly	12	
Weekly	52	
Daily	365	
Hourly	8760	
Continuously	$n \to \infty$	

29. Suppose that $5000 is deposited in a savings account paying 6% annual interest compounded continuously. How much interest will be earned in 8 years?

30. If (7) is solved for P, that is, $P = Se^{-rt}$, we obtain the amount that should be invested now at an annual rate r of interest in order to be worth S dollars after t years. We say that P is the **present value** of the amount S. What is the present value of $100,000 at an annual rate of 3% compounded continuously for 30 years?

Additional Exponential Models

31. **Potassium-40 Decay** Potassium is one of the most abundant metals found throughout the Earth's crust and oceans. Although potassium occurs naturally in the form of three isotopes, only the isotope potassium-40 (K-40) is radioactive. This isotope is unusual in that it decays by two different nuclear reactions. By emitting a beta particle a great percentage of an initial amount of K-40 decays over time into the stable isotope calcium-40 (Ca-40), whereas by electron capture a smaller percentage of K-40 decays into the stable isotope argon-40 (Ar-40). If it is assumed that rates at which the amounts $C(t)$ of Ca-40 and $A(t)$ of Ar-40 *increase* are proportional to the amount $P(t)$ of potassium present at time t, and that the rate at which $P(t)$ *decays* is also proportional to $P(t)$, then it can be shown that

$$C(t) = \frac{k_C}{k_A + k_C}P_0\left[1 - e^{-(k_A + k_C)t}\right],$$

$$A(t) = \frac{k_A}{k_A + k_C}P_0\left[1 - e^{-(k_A + k_C)t}\right],$$

$$P(t) = P_0 e^{-(k_A + k_C)t},$$

where $P(0) = P_0$, $k_C = 4.962 \times 10^{-10}$, and $k_A = 0.581 \times 10^{-10}$.

(a) Find the half-life of K-40.

(b) Determine the percentage of an initial amount P_0 of K-40 that decays into Ca-40 and the percentage that decays into Ar-40 over a very long period of time, that is, as $t \to \infty$.

32. **Potassium-Argon Dating** The potassium-argon dating method can be used to find the ages of igneous rocks, that is, rocks formed in the cooling of magma or lava.

(a) Use the functions $A(t)$ and $P(t)$ in Problem 31 to show that

$$\frac{A(t)}{P(t)} = \frac{k_A}{k_A + k_C}\left[e^{(k_A + k_C)t} - 1\right].$$

(b) Solve the equation in part (a) for t in terms $A(t)$, $P(t)$, k_A, and k_C. Explain the significance of the value of t found from this equation when $A(0) = 0$ and $P(0) = P_0$.

(c) Suppose it is found that each gram of a rock sample contains 8.6×10^{-7} grams of Ar-40 and 5.3×10^{-6} grams of K-40. Use the equation obtained in part (b) to determine the approximate age T of the rock.

(d) Use the answer in part (c) and the function $P(t)$ in Problem 31 to determine the initial amount P_0 of potassium in the rock sample.

33. **Effective Half-life** Radioactive substances are removed from living organisms by two processes: natural physical decay and biological metabolism. Each process contributes to an effective half-life E that is defined by

$$1/E = 1/P + 1/B,$$

where P is the physical half-life of the radioactive substance and B is the biological half-life.

Lava flow on the island of Hawaii

(a) Radioactive iodine, I-131, is used to treat hyperthyroidism (overactive thyroid). It is known that for human thyroids, $P = 8$ days and $B = 24$ days. Find the effective half-life of I-131 in the thyroid.

(b) Suppose the amount of I-131 in the human thyroid after t days is modeled by $A(t) = A_0 e^{kt}$, $k < 0$. Use the effective half-life found in part (a) to determine the percentage of radioactive iodine remaining in the human thyroid gland two weeks after its ingestion.

34. Newton's Law of Cooling Revisited The rate at which a body cools also depends on its exposed surface area S. If S is a constant, then a modification of (4) is

$$T(t) = T_m + (T_0 - T_m)e^{kSt}, \quad k < 0.$$

Suppose two cups A and B are filled with coffee at the same time. Initially the temperature of the coffee is 150°F. The exposed surface area of the coffee in cup B is twice the surface area of the coffee in cup A. After 30 min, the temperature of the coffee in cup A is 100°F. If $T_m = 70°F$, what is the temperature of the coffee in cup B after 30 min?

35. Series Circuit In a simple series circuit consisting of a constant voltage E, an inductance of L henries, and a resistance of R ohms, it can be shown that the current $I(t)$ is given by

$$I(t) = \frac{E}{R}(1 - e^{-(R/L)t}).$$

Solve for t in terms of the other symbols.

36. Drug Concentration Under some conditions the concentration of a drug at time t after injection is given by

$$C(t) = \frac{a}{b} + \left(C_0 - \frac{a}{b}\right)e^{-bt}.$$

Here a and b are positive constants and C_0 is the concentration of the drug at $t = 0$. Determine the steady-state concentration of a drug, that is, the limiting value of $C(t)$ as $t \to \infty$. Determine the time t at which $C(t)$ is one-half the steady-state concentration.

Richter Scale

37. Two of the most devastating earthquakes in the San Francisco Bay area occurred in 1906 along the San Andreas fault and in 1989 in the Santa Cruz Mountains near Loma Prieta peak. The 1906 and 1989 earthquakes measured 8.5 and 7.1 on the Richter scale, respectively. How much greater was the intensity of the 1906 earthquake compared to the 1989 earthquake?

38. How much greater was the intensity of the 2004 Northern Sumatra earthquake (Example 7) compared to the 1964 Alaskan earthquake of magnitude 8.9?

39. If an earthquake has a magnitude 4.2 on the Richter scale, what is the magnitude on the Richter scale of an earthquake that has an intensity 20 times greater? [*Hint*: First solve the equation $10^x = 20$.]

40. Show that the Richter scale defined in (8) of this section can be written

$$M = \frac{\ln A - \ln A_0}{\ln 10}.$$

pH of a Solution

In Problems 41–44, determine the pH of a solution with the given hydrogen-ion concentration [H^+].

41. 10^{-6} **42.** 4×10^{-7} **43.** 2.8×10^{-8} **44.** 5.1×10^{-5}

Marina district in San Francisco, 1989

In Problems 45–48, determine the hydrogen-ion concentration $[H^+]$ of a solution with the given pH.

45. 3.3   **46.** 7.3   **47.** 6.6   **48.** 8.1

In Problems 49–52, determine how many more times acidic the first substance is compared to the second substance.

49. lemon juice, pH = 2.3;   vinegar, pH = 3.3
50. battery acid, pH = 1;   lye, pH = 13
51. acidic rain, pH = 3.8;   clean rain, pH = 5.6
52. HCL, $[H^+] = 10^{-1.5}$;   NaOH, $[H^+] = 10^{-14}$

Additional Logarithmic Models

53. Richter Scale and Energy **(a)** Charles Richter working with Beno Gutenberg developed the model

$$M = \tfrac{2}{3}[\log_{10} E - 11.8]$$

that relates the Richter magnitude M of an earthquake and its seismic energy E (measured in ergs). Calculate the seismic energy E of the 2004 Northern Sumatra earthquake where $M = 9.3$.
 (b) Show that the proportional energy difference $f_{\Delta E} = E_1/E_2$ between two different earthquakes of Richter magnitudes M_1 and M_2 is given by

$$f_{\Delta E} = E_1/E_2 = 10^{\frac{3}{2}[M_1 - M_2]}.$$

 (c) Use part (b) to show that if M_1 is one unit more than M_2 then the seismic energy E_1 of an earthquake of magnitude M_1 is approximately 32 times the seismic energy E_2 of the earthquake of magnitude M_2. Repeat the calculation if M_1 is two units more than M_2.

54. Intensity Level The **intensity level** b of a sound measured in decibels (dB) is defined by

$$b = 10 \log_{10} \frac{I}{I_0}, \qquad (10)$$

where I is the **intensity of the sound** measured in watts/cm^2 and $I_0 = 10^{-16}$ watts/cm^2 is the intensity of the faintest sound that can be heard (0 dB). Use (10) and complete the following table.

Sound	Intensity I (watts/cm^2)	Intensity Level b (dB)
Whisper	10^{-14}	
Conversation	10^{-11}	
TV commercials	10^{-10}	
Smoke alarm	10^{-9}	
Jet plane taking off	10^{-7}	
Rock band	10^{-4}	

55. Threshold of Pain The threshold of pain is generally taken to be around 140 dB. Find the intensity of sound I corresponding to 140 dB.

56. Intensity Levels The intensity of sound I is inversely proportional to the square of the distance d from its source, that is,

$$I = \frac{k}{d^2}, \tag{11}$$

where k is the constant of proportionality. Suppose d_1 and d_2 are distances from a source of sound, and that the corresponding intensity levels of the sounds are b_1 and b_2. Use (11) in (10) to show that b_1 and b_2 are related by

$$b_2 = b_1 + 20 \log_{10} \frac{d_1}{d_2}. \tag{12}$$

57. Intensity Level As part of a 4.76 billion dollar upgrade of Los Angeles International Airport (LAX), the Los Angeles City Council recently gave its final approval to lengthen and move the northernmost runway (24R) 260 to 350 ft closer to the bordering communities of Westchester and Playa del Rey. Spokespersons for these communities say that such a move will, amongst other things, increase the noise level there. Suppose that the sound intensity level b_1 of a jet airplane taking off from runway 24R is 100 dB when measured at a distance of 1000 ft. Use (12) in Problem 56 to find the intensity level b_2 of the same plane at a point that is 350 ft closer to the runway.

Los Angeles International Airport (LAX)

The shortest runway (24R) handles most of the A380 landings at LAX

58. Use (12) in Problem 56 to compare the intensity levels b_1 and b_2 of a sound if $d_2 = 10 d_1$.

59. Pupil of the Eye An empirical model published in 1952 by S. G. De Groot and J. W. Gebhard in the *Journal of the Optical Society of America* relates the diameter d of the pupil of the eye (measured in millimeters, mm) to the luminance B of light source (measured in millilambert's, mL):

$$\log_{10} d = 0.8558 - 0.000401 \, (8.1 + \log_{10} B)^3. \tag{13}$$

See **FIGURE 5.4.3**.

(a) The average luminance of clear sky is approximately $B = 255$ mL. Use (13) to find the corresponding pupil diameter.

(b) The luminance of the Sun varies from approximately $B = 190{,}000$ mL at sunrise to $B = 51{,}000{,}000$ mL at noon. Find the corresponding pupil diameters.

(c) Find the luminance B corresponding to a pupil diameter of 7 mm.

60. Body Surface Area A mathematical model for estimating body surface area S (in square meters) is given by

$$\log_{10} S = -0.69364 + (0.425) \log_{10} w + (0.725) \log_{10} h, \tag{14}$$

where w and h are a person's weight (in kilograms) and height (in meters), respectively. This empirical formula, due to D. Dubois and E. F. Dubois, first published in the *Archives of Internal Medicine* in 1916, is still used today by medical researchers.

(a) Use (14) to estimate the body surface area of a person whose weight is $w = 70$ kg and who is $h = 1.75$ m tall.

(b) Determine your weight and height and estimate your own body surface area.

FIGURE 5.4.3 Pupil diameter in Problem 59

61. Eliminate the logarithms in the surface area formula (14) and express S as a function of two variables w and h.

62. **Intensity of Light** According to the Beer-Lambert law, the intensity I (measured in lumens) of a vertical beam of light passing through a transparent substance decreases according to the exponential function $I(x) = I_0 e^{kx}$, $k < 0$, where I_0 is the intensity of the incident beam and x is the depth measured in meters. Suppose the intensity of light 1 meter below the surface of water is 30% of I_0.

   (a) What is the intensity 3 meters below the surface?
   (b) At what depth is the intensity 50% of that incident on the surface?

5.5 The Hyperbolic Functions

Leonhard Euler

INTRODUCTION Logarithms were invented in the late sixteenth century by the Scottish lord—and nonmathematician—**John Napier** (1550–1617). It was he who coined the word "logarithm" from the two Greek words *logos*, meaning ratio, and *arithmos*, meaning number or power. But it took almost two centuries and the genius of the Swiss mathematician **Leonhard Euler** (1707–1783) before the mathematical community became fully aware of the irrational number e and its importance. It is his work that we emulate below in showing why the number e is the natural choice of base for the exponential and logarithmic functions.

☐ **Difference Quotient Revisited** We return to the difference quotient concept first introduced in Section 2.10. Recall that we compute

$$\frac{f(x + h) - f(x)}{h} \tag{1}$$

in three steps. For the exponential function $f(x) = b^x$, we have

$(i)\ f(x + h) = b^{x+h} = b^x b^h$ ← laws of exponents

$(ii)\ f(x + h) - f(x) = b^{x+h} - b^x$ ← $\begin{cases} \text{law of exponents} \\ \text{and factoring} \end{cases}$
$\qquad = b^x b^h - b^x = b^x(b^h - 1)$

$(iii)\ \dfrac{f(x + h) - f(x)}{h} = \dfrac{b^x(b^h - 1)}{h} = b^x \dfrac{b^h - 1}{h}.$

In the fourth step, the calculus step, we let $h \to 0$ but, unlike all the problems given in Exercises 2.10, there is no apparent way of canceling the h in (iii). Nonetheless, the derivative of $f(x) = b^x$ is

$$f'(x) = \lim_{h \to 0} b^x \cdot \frac{b^h - 1}{h}. \tag{2}$$

Since b^x does not depend on the variable h, we can rewrite (2) as

$$f'(x) = b^x \cdot \lim_{h \to 0} \frac{b^h - 1}{h}. \tag{3}$$

Now here are the amazing results. The limit in (3),

$$\lim_{h \to 0} \frac{b^h - 1}{h}, \tag{4}$$

can be shown to exist for every positive base b. However, as one might expect, we will get a different answer for each base b. So let's denote the expression in (4) by the symbol $m(b)$. The derivative of $f(x) = b^x$ in (3) is then

$$f'(x) = b^x m(b). \tag{5}$$

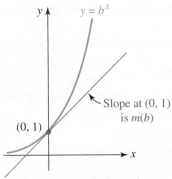

FIGURE 5.5.1 Find a base b so that the slope $m(b)$ of tangent line at $(0, 1)$ is 1

You are asked to approximate the value of $m(b)$ in the four cases $b = 1.5, 2, 3,$ and 5 in Problems 39–42 of Exercises 5.5. For example, it can be shown that $m(10) \approx 2.302585\ldots$, and as a consequence the derivative of $f(x) = 10^x$ is

$$f'(x) = (2.302585\ldots)10^x. \tag{6}$$

We can get a better understanding of what $m(b)$ is by evaluating (5) at $x = 0$. Since $b^0 = 1$, we have $f'(0) = m(b)$. In other words, $m(b)$ is the slope of the tangent line to the graph of $f(x) = b^x$ at $x = 0$, that is, at the y-intercept $(0, 1)$. See FIGURE 5.5.1. Given that we have to calculate a different $m(b)$ for each base b, and that $m(b)$ is likely to be an "ugly" number as in (6), over time the following question arose naturally:

Is there a base b for which $m(b) = 1$? (7)

☐ **The Answer** To answer the question posed in (7), we must return to the definitions of e given in Section 5.1. Specifically, (4) of Section 5.1,

$$e = \lim_{h \to 0} (1 + h)^{1/h}, \tag{8}$$

provides the means for answering the question posed in (7). If you have studied Sections 1.5, 2.10, and 4.14 you should have an intuitive understanding that the equality in (8) means that as h gets closer and closer to 0 then $(1 + h)^{1/h}$ can be made arbitrarily close to the number e. Thus for values of h near 0, we have the approximation $(1 + h)^{1/h} \approx e$, and so it follows that $1 + h \approx e^h$. By rewriting the last expression in the form

$$\frac{e^h - 1}{h} \approx 1 \tag{9}$$

we can conclude that

$$1 = \lim_{h \to 0} \frac{e^h - 1}{h}. \tag{10}$$

Since the right-hand side of (10) is $m(e)$, we have the answer to the question in (7):

The base b for which $m(b) = 1$ is $b = e$. (11)

In addition, from (3) we have discovered a wonderfully simple result: The derivative of $f(x) = e^x$ is

$$f'(x) = e^x. \tag{12}$$

The result in (12) is the same as

$$f'(x) = f(x).$$

Moreover, the only other nonzero function f in calculus whose derivative is equal to itself is $f(x) = ce^x$, where $c \neq 0$ is a constant.

☐ **What's Next?** Since the functions $y = \log_b x$ and $y = b^x$ are inverses of each other, one would expect that since the simplest derivative of $y = b^x$ is obtained when $b = e$ that the simplest derivative of $y = \log_b x$ also occurs for that base. That is indeed the case. You are encouraged to reexamine (3) of Section 5.1 and then work Problems 1–4 in Exercises 5.5.

☐ **Hyperbolic Functions** We have already seen in Section 5.4 the usefulness of the exponential function e^x in various mathematical models. As a further application, consider a long rope or a flexible wire, such as a telephone wire hanging only under its own

Telephone wires

CHAPTER 5 EXPONENTIAL AND LOGARITHMIC FUNCTIONS

weight between two fixed supports. It can be shown that under certain conditions the hanging wire assumes the shape of the graph of the function

$$f(x) = c\frac{e^{x/c} + e^{-x/c}}{2}. \tag{13}$$

The symbol c stands for a positive constant that depends on the physical characteristics of the wire. Functions such as (13), consisting of certain combinations of e^x and e^{-x}, appear in so many applications that mathematicians have given them names.

☐ **Hyperbolic Functions** In particular, when $c = 1$ in (13), the resulting function $f(x) = \dfrac{e^x + e^{-x}}{2}$ is one of six **hyperbolic functions**.

DEFINITION 5.5.1 Hyperbolic Functions

For any real number x, the **hyperbolic sine** of x, denoted $\sinh x$ is

$$\sinh x = \frac{e^x - e^{-x}}{2} \tag{14}$$

and the **hyperbolic cosine** of x, denoted $\cosh x$, is

$$\cosh x = \frac{e^x + e^{-x}}{2} \tag{15}$$

Analogous to the trigonometric functions $\tan x$, $\cot x$, $\sec x$, and $\csc x$ that are defined in terms of $\sin x$ and $\cos x$, there are four additional hyperbolic functions $\tanh x$, $\coth x$, $\text{sech}\, x$, and $\text{csch}\, x$ that are defined in terms of $\sinh x$ and $\cosh x$:

$$\tanh x = \frac{\sinh x}{\cosh x} = \frac{e^x - e^{-x}}{e^x + e^{-x}} \qquad \coth x = \frac{1}{\tanh x} = \frac{e^x + e^{-x}}{e^x - e^{-x}} \tag{16}$$

$$\text{sech}\, x = \frac{1}{\cosh x} = \frac{2}{e^x + e^{-x}} \qquad \text{csch}\, x = \frac{1}{\sinh x} = \frac{2}{e^x - e^{-x}}. \tag{17}$$

☐ **Graphs** The graph of the hyperbolic cosine, shown in FIGURE 5.5.2, is called a **catenary**. The word *catenary* derives from the Latin word for a chain, *catena*. The shape of the famous Gateway Arch in St. Louis, Missouri, is an inverted catenary. Compare the shape in Figure 5.5.2 with that in the accompanying photo. The graph of $y = \sinh x$ is given in FIGURE 5.5.3. See Problem 35 in Exercises 5.5.

Gateway Arch in St. Louis, MO

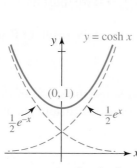

FIGURE 5.5.2 Graph of $y = \cosh x$

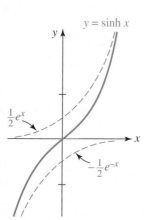

FIGURE 5.5.3 Graph of $y = \sinh x$

5.5 The Hyperbolic Functions

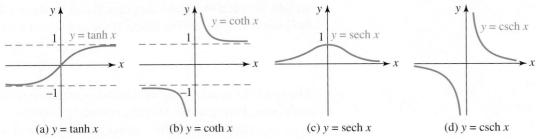

(a) $y = \tanh x$ (b) $y = \coth x$ (c) $y = \operatorname{sech} x$ (d) $y = \operatorname{csch} x$

FIGURE 5.5.4 Graphs of the hyperbolic tangent (a), cotangent (b), secant (c), and cosecant (d)

The graphs of the hyperbolic tangent, cotangent, secant, and cosecant are given in FIGURE 5.5.4. Observe that $y = 1$ and $y = -1$ are horizontal asymptotes for the graphs of $y = \tanh x$ and $y = \coth x$ and that $x = 0$ is a vertical asymptote for the graphs of $y = \coth x$ and $y = \operatorname{csch} x$.

☐ **Identities** Although the hyperbolic functions are not periodic, they possess identities that are very similar to trigonometric identities. Analogous to the basic Pythagorean identity of trigonometry $\cos^2 x + \sin^2 x = 1$, for the hyperbolic sine and cosine we have

$$\cosh^2 x - \sinh^2 x = 1. \tag{18}$$

See Problems 9–14 in Exercises 5.5.

Exercises 5.5   Answers to selected odd-numbered problems begin on page ANS–21.

1. Use the laws of logarithms to show that for $f(x) = \log_b x$,

$$\frac{f(x + h) - f(x)}{h} = \frac{1}{h}\log_b\left(1 + \frac{h}{x}\right) = \frac{1}{x}\log_b\left(1 + \frac{h}{x}\right)^{x/h}.$$

2. From Problem 1, the derivative of $f(x) = \log_b x$ is

$$f'(x) = \lim_{h \to 0}\frac{f(x + h) - f(x)}{h} = \frac{1}{x}\lim_{h \to 0}\log_b\left(1 + \frac{h}{x}\right)^{x/h}.$$

Let us assume that the limiting process can be taken inside the logarithm:

$$f'(x) = \frac{1}{x}\log_b\left[\lim_{h \to 0}\left(1 + \frac{h}{x}\right)^{x/h}\right].$$

Rewrite the foregoing result using the substitution $n = x/h$. Notice that since x is held fixed, as $h \to 0$ we must have $n \to \infty$. Give the precise value of $f'(x)$.

In Problems 3 and 4, use the result of Problem 2 to find $f'(x)$ for the given function.

3. $f(x) = \log_{10} x$

4. $f(x) = \ln x$

In Problems 5 and 6, use the result of Problems 1 and 2 to find $f'(x)$ for the given function. Before using the difference quotient, use the laws of logarithms to rewrite the function.

5. $f(x) = \ln\dfrac{x}{9}$

6. $f(x) = \log_{10} 6x$

In Problems 7 and 8, compute $\dfrac{f(x+h) - f(x)}{h}$ for the given function.

7. $f(x) = e^{5x}$

8. $f(x) = e^{-x+4}$

In Problems 9–14, use the definitions of $\sinh x$ and $\cosh x$ in (14) and (15) to verify the given identity.

9. $\cosh^2 x - \sinh^2 x = 1$

10. $1 - \tanh^2 x = \operatorname{sech}^2 x$.

11. $\cosh(-x) = \cosh x$

12. $\sinh(-x) = -\sinh x$

13. $\sinh 2x = 2 \sinh x \cosh x$

14. $\cosh 2x = \cosh^2 x + \sinh^2 x$

15. (a) If $\sinh x = -\frac{3}{2}$, use the identity given in Problem 9 to find the value of $\cosh x$.
   (b) Use the result of part (a) to find the numerical values of $\tanh x$, $\coth x$, $\operatorname{sech} x$, and $\operatorname{csch} x$.

16. (a) If $\tanh x = \frac{1}{2}$, use the identity given in Problem 10 to find the value of $\operatorname{sech} x$.
   (b) Use the result of part (a) to find the numerical values of $\cosh x$, $\sinh x$, $\coth x$, and $\operatorname{csch} x$.

In Problems 17–20, find the exact numerical value of the given quantity.

17. $\cosh(\ln 4)$

18. $\sinh(\ln 0.5)$

19. $\sinh(\ln 4 - \ln 3)$

20. $\cosh(-\ln 3)$

In Problems 21–24, express the given composition of functions as a rational function of x, where $x > 0$.

21. $\sinh(\ln x)$

22. $\tanh(3 \ln x)$

23. $\coth(\ln 2x)$

24. $\operatorname{sech}(\ln x)$

25. As can be seen in Figure 5.5.3, the hyperbolic sine function $y = \sinh x$ is one-to-one. By interchanging x and y in the definition of the hyperbolic sine in (14) the equation $e^y - 2x - e^{-y} = 0$ defines implicitly the **inverse hyperbolic sine** $\sinh^{-1} x$. Show that $\sinh^{-1} x$ can be expressed in terms of the natural logarithm:

$$\sinh^{-1} x = \ln\left(x + \sqrt{x^2 + 1}\right). \qquad (19)$$

26. (a) Use the graph of $y = \sinh x$ in Figure 5.5.3 to sketch the graph of the inverse hyperbolic sine $y = \sinh^{-1} x$ defined in Problem 25.
   (b) Give the domain and range of $y = \sinh^{-1} x$.

27. The function $y = \cosh x$ is one-to-one on the restricted domain $[0, \infty)$. Proceed as in Problem 25 to show that the **inverse hyperbolic cosine** $\cosh^{-1} x$ can be expressed in terms of the natural logarithm:

$$\cosh^{-1} x = \ln\left(x + \sqrt{x^2 - 1}\right). \qquad (20)$$

28. (a) Use the graph of $y = \cosh x$ in Figure 5.5.2 to sketch the graph of the inverse hyperbolic cosine $y = \cosh^{-1} x$ defined in Problem 27.
   (b) Give the domain and range of $y = \cosh^{-1} x$.

In Problems 29 and 30, use (19) to approximate the given quantity.

29. $\sinh^{-1}(-2)$

30. $\sinh^{-1} 4$

In Problems 31 and 32, use (20) to approximate the given quantity.

31. $\cosh^{-1}(1.5)$

32. $\cosh^{-1} 10$

33. As can be seen from its graph in Figure 5.5.4(a) the function $y = \tanh x$ is one-to-one. Proceed as in Problem 25 to show that the **inverse hyperbolic tangent** $\tanh^{-1} x$ can be expressed in terms of the natural logarithm:

$$\tanh^{-1} x = \frac{1}{2}\ln\left(\frac{1 + x}{1 - x}\right). \tag{21}$$

34. (a) Use the graph of $y = \tanh x$ in Figure 5.5.4(a) to sketch the graph of the inverse hyperbolic tangent $y = \tanh^{-1} x$ defined in (21).
(b) Give the domain and range of $y = \tanh^{-1} x$.

Applications

35. Gateway Arch The stainless steel Gateway Arch built on the west bank of the Mississippi River is a monument to the fact the area around St. Louis, Missouri, where the Mississippi and Missouri Rivers merge, was the staging point for much of the westward population migration in the years following the Louisiana Purchase in 1803. The arch, designed by the Finish-American architect **Eero Saarinen** (1910–1961), was completed in 1965. A mathematical model for the shape of the Gateway Arch is given by

$$f(x) = 694 - 69 \cosh\frac{x}{100}.$$

The graph of f, the red curve shown in FIGURE 5.5.5, is called the **centroid curve** of the arch because it passes through the centroids of the equilateral triangular cross-sections of the arch.
(a) Use the model to find the approximate height of the arch.
(b) Use the model and the logarithmic identity for the inverse hyperbolic cosine given in (20) to find the x-intercept $(b, 0)$ of the red graph on the positive x-axis. What is the approximate width of the base of the arch?

A triangular cross-section
of The Gateway Arch

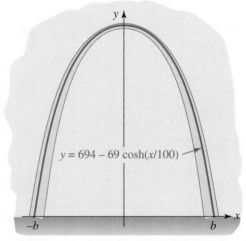

$y = 694 - 69\cosh(x/100)$

FIGURE 5.5.5 Gateway Arch in Problem 35

36. Terminal Velocity When air resistance is ignored, the velocity v of a body of mass m dropped from a specified height is given by the simple function $v(t) = gt$, where $g = 32$ ft/s^2 is the acceleration due to gravity and t is measured in seconds. The model $v(t) = gt$ predicts over time that the velocity of the body increases until it eventually hits the ground. But if it is assumed that air resistance is proportional to the square of the velocity v, then it can be shown that

$$v(t) = \sqrt{\frac{mg}{k}} \tanh\sqrt{\frac{kg}{m}}t, \tag{22}$$

where k is a positive constant of proportionality. The model (22) predicts that the velocity of a body falling from a great height will now attain a terminal, or limiting, velocity v_{term}. Determine v_{term} using the end behavior of the graph of the hyperbolic tangent given in Figure 5.5.4(a) for $t \to \infty$.

Calculator Problems

37. Use a calculator to investigate $\lim\limits_{h\to 0}\dfrac{f(x + h) - f(x)}{h}$ for the function in Problem 7. Determine $f'(x)$.

38. Use a calculator to investigate $\lim\limits_{h\to 0}\dfrac{f(x + h) - f(x)}{h}$ for the function in Problem 8. Determine $f'(x)$.

In Problems 39–42, use a calculator to approximate the value $m(b) = \lim\limits_{h\to 0}\dfrac{b^h - 1}{h}$ for $b = 1.5, b = 2, b = 3$, and $b = 5$ by filling out the given table.

39.

$h \to 0$	0.1	0.01	0.001	0.0001	0.00001	0.000001
$\dfrac{(1.5)^h - 1}{h}$						

40.

$h \to 0$	0.1	0.01	0.001	0.0001	0.00001	0.000001
$\dfrac{2^h - 1}{h}$						

41.

$h \to 0$	0.1	0.01	0.001	0.0001	0.00001	0.000001
$\dfrac{3^h - 1}{h}$						

42.

$h \to 0$	0.1	0.01	0.001	0.0001	0.00001	0.000001
$\dfrac{5^h - 1}{h}$						

43. Fill out a table of the kind in Problems 39–42, but this time use $\dfrac{e^h - 1}{h}$.

44. Historical Curiosity The logarithm developed by John Napier (see page 365) was actually

$$10^7 \log_{1/e}\left(\frac{x}{10^7}\right).$$

Use (8) of Section 5.3 to express this logarithm in terms of the natural logarithm.

For Discussion

45. At the beginning of this section we saw that the derivative of $f(x) = e^x$ is $f'(x) = e^x$. Use this information to find all tangent lines to the graph of $f(x) = e^x$ that pass through the origin.

46. Show that the x-intercept of the tangent line to the graph of $f(x) = e^x$ at $x = x_0$ is one unit to the left of $(x_0, 0)$.

Chapter 5 Review Exercises   Answers to selected odd-numbered Problems begin on page ANS-21.

A. Fill in the Blanks

In Problems 1–25, fill in the blanks.

1. The graph of $y = 6 - e^{-x}$ has the y-intercept _____ and horizontal asymptote $y = $ _____.

2. The x-intercept of the graph of $y = -10 + 10^{5x}$ is _____.

3. The graph of $y = \ln(x + 4)$ has the x-intercept _____ and vertical asymptote $x = $ _____.

4. The y-intercept of the graph of $y = \log_8(x + 2)$ is _____.

5. $\log_5 2 - \log_5 10 = $ _____

6. $6 \ln e + 3 \ln \dfrac{1}{e} = $ _____

7. $e^{3 \ln 10} = $ _____

8. $10^{\log_{10} 4.89} = $ _____

9. $\log_4(4 \cdot 4^2 \cdot 4^3) = $ _____

10. $\dfrac{\log_5 625}{\log_5 125} = $ _____

11. If $\log_3 N = -2$, then $N = $ _____.

12. If $\log_b 6 = \frac{1}{2}$, then $b = $ _____.

13. If $\ln e^3 = y$, then $y = $ _____.

14. If $\ln 3 + \ln(x - 1) = \ln 2 + \ln x$, then $x = $ _____.

15. If $-1 + \ln(x - 3) = 0$, then $x = $ _____.

16. If $\ln(\ln x) = 1$, then $x = $ _____.

17. If $100 - 20e^{-0.15t} = 35$, then to four rounded decimals $t = $ _____.

18. If $3^x = 5$, then $3^{-2x} = $ _____.

19. $f(x) = 4^{3x} = ($ _____ $)^x$

20. $f(x) = (e^2)^{x/6} = ($ _____ $)^x$

21. If the graph of $y = e^{x-2} + C$ passes through $(2, 9)$, then $C = $ _____.

22. By rigid transformations, the point $(0, 1)$ on the graph of $y = e^x$ is moved to the point _____ on the graph of $y = 4 + e^{x-3}$.

23. $\log_2(\log_3(\log_4 64)) = $ _____.

24. If $f(x) = 8 + \log_2(x + 5)$, then $f(11) = $ _____.

25. The function $f(x) = 2 - \ln x$ is one-to-one, so its inverse is $f^{-1}(x) = $ _____.

B. True/False

In Problems 1–25, answer true or false.

1. $y = \ln x$ and $y = e^x$ are inverse functions. _____

2. The point $(b, 1)$ is on the graph of $f(x) = \log_b x$. _____

3. $y = 10^{-x}$ and $y = (0.1)^x$ are the same function. _____

4. If $f(x) = e^{x^2} - 1$, then $f(x) = 1$ when $x = \pm \ln \sqrt{2}$. _____

5. $4^{x/2} = 2^x$ _____

6. $\dfrac{2^{x^2}}{2^x} = 2^x$ _____

7. $2^x + 2^{-x} = (2 + 2^{-1})^x$ _____

8. $2^{3+3x} = 8^{1+x}$ _____

9. $-\ln 2 = \ln\left(\frac{1}{2}\right)$ _____

10. $\ln\dfrac{e^a}{e^b} = a - b$ _____

11. $\ln(\ln e) = 1$ _____

12. $\ln\sqrt{43} = \dfrac{\ln 43}{2}$ _____

13. $\ln(e + e) = 1 + \ln 2$ _____

14. $\log_6(36)^{-1} = -2$ _____

15. $\ln\dfrac{1}{e} = -1$ _____

16. $\dfrac{\ln 10}{\ln 2} = \ln 5$ _____

17. The range of the function $f(x) = 4^{-x} - 5$ is the interval $(-5, \infty)$ on the y-axis. _____

18. The point $(-5, 40)$ is on the graph of $f(x) = \left(\frac{1}{2}\right)^x + 8$. _____

19. If $f(x) = b^x$, then $f(nx) = f(x)^n$. _____

20. $f(x) = \log_b x$, then $f(x^n) = nf(x)$. _____

21. $\ln y = 2\ln x + \ln 5$, then $y = 2x + 5$. _____

22. $\ln e \cdot \ln e^2 \cdots \ln e^n = n!$ _____

23. If $a > 0$ and $b > 0$, then $\log_{b^2} a^2 = \log_b a$. _____

24. The domain of the function $f(x) = \ln(\ln x)$ is $(0, \infty)$. _____

25. The function $f(x) = 3 + x + 5e^{x-1}$ is one-to-one, so $f^{-1}(9) = 1$. _____

C. Review Exercises

In Problems 1 and 2, rewrite the given exponential expression as an equivalent logarithmic expression.

1. $5^{-1} = 0.2$

2. $\sqrt[3]{512} = 8$

In Problems 3 and 4, rewrite the given logarithmic expression as an equivalent exponential expression.

3. $\log_9 27 = 1.5$

4. $\log_6(36)^{-2} = -4$

In Problems 5–12, solve for x.

5. $2^{1-x} = 8$

6. $3^{2x} = 81$

7. $e^{1-2x} = e^2$

8. $e^{x^2} - e^5 e^{x-1} = 0$

9. $2^{1-x} = 7$

10. $3^x = 7^{x-1}$

11. $e^{x+2} = 6$

12. $3e^x = 4e^{-3x}$

In Problems 13 and 14, solve for the indicated variable.

13. $P = Se^{-rm}$; for m

14. $P = \dfrac{K}{1 + ce^{rt}}$; for t

In Problems 15 and 16, graph the given functions on the same coordinate axes.

15. $y = 4^x$, $y = \log_4 x$

16. $y = \left(\frac{1}{2}\right)^x$, $y = \log_{1/2} x$

17. Match the letter of the graph in FIGURE 5.R.1 with the appropriate function.

(*i*) $f(x) = b^x, b > 2$ (*ii*) $f(x) = b^x, 1 < b < 2$

(*iii*) $f(x) = b^x, \frac{1}{2} < b < 1$ (*iv*) $f(x) = b^x, 0 < b < \frac{1}{2}$

18. In FIGURE 5.R.2, fill in the blanks for the coordinates of the points on each graph.

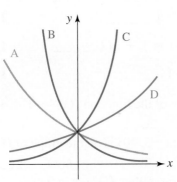

FIGURE 5.R.1 Graph for Problem 17

FIGURE 5.R.2 Graph for Problem 18

In Problems 19 and 20, find the slope of the red line L given in each figure.

19. $f(x) = 3^{-(x+1)}$

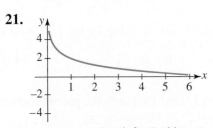

FIGURE 5.R.3 Graph for Problem 19

20.

FIGURE 5.R.4 Graph for Problem 20

In Problems 21–26, match each of the following functions with one of the given graphs.

(*i*) $y = \ln(x - 2)$ (*ii*) $y = 2 - \ln x$

(*iii*) $y = 2 + \ln(x + 2)$ (*iv*) $y = -2 - \ln(x + 2)$

(*v*) $y = -\ln(2x)$ (*vi*) $y = 2 + \ln(-x + 2)$

21.

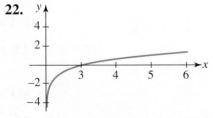

FIGURE 5.R.5 Graph for Problem 21

22.

FIGURE 5.R.6 Graph for Problem 22

23.

FIGURE 5.R.7 Graph for Problem 23

24.

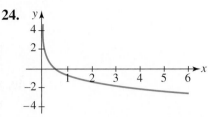

FIGURE 5.R.8 Graph for Problem 24

25.

FIGURE 5.R.9 Graph for Problem 25

26.

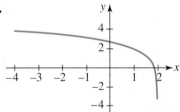

FIGURE 5.R.10 Graph for Problem 26

In Problems 27–36, eliminate the logarithms in the given expression. Express your answer in the form $y = f(x)$.

27. $\ln y = \ln(x - 3)$

28. $\ln y = 2\ln(x + 5)$

29. $\ln y = \ln x + 3\ln 2$

30. $\ln y = 4\ln x - \ln 5$

31. $\ln(y - 6) = \ln(x - 1) + \ln x + 2\ln 2$

32. $\ln y - \ln(x - 1) = \ln(x + 4) - \ln 2$

33. $\ln(y - 3) = \ln y - \ln(x^2 + 1) + 2\ln x$

34. $\ln(y/4) = \frac{1}{2}\ln(x^2 + 1) + \ln x$

35. $\ln y = -3x + \ln 5$

36. $\ln(y - x) = x^2 + 2\ln x - \ln\frac{1}{2}$

In Problems 37 and 38, find f^{-1} for the given one-to-one function f.

37. $f(x) = 5 + e^{-x}$

38. $f(x) = -3 + \ln(x + 10)$

In Problems 39 and 40, rewrite the given function f as a piecewise-defined function.

39. $f(x) = \ln|2 - x|$

40. $f(x) = \ln|x^2 - 1|$

In Problems 41 and 42, in words describe the graph of the function f in terms of a transformation of the graph of $y = \ln x$.

41. $f(x) = \ln ex$

42. $f(x) = \ln x^3$

43. Find a function $f(x) = Ae^{kx}$ if $(0, 5)$ and $(6, 1)$ are points on the graph of f.

44. Find a function $f(x) = A10^{kx}$ if $f(3) = 8$ and $f(0) = \frac{1}{2}$.

45. Find a function $f(x) = a + b^x, 0 < b < 1$, if $f(1) = 5.5$ and the graph of f has a horizontal asymptote $y = 5$.

46. Find a function $f(x) = a + \log_3(x - c)$ if $f(11) = 10$ and the graph of f has a vertical asymptote $x = 2$.

47. Doubling Time If the initial number of bacteria present in a culture doubles after 9 hours, how long will it take for the number of bacteria in the culture to double again?

48. **Got Bait?** A commercial fishing lake is stocked with 10,000 fingerlings. Find a model $P(t) = P_0 e^{kt}$ for the fish population of the lake at time t if the owner of the lake estimates that there will be 5,000 fish left after six months. After how many months does the model predict that there will be 1000 fish left?

49. **Radioactive Decay** Tritium, an isotope of hydrogen, has a half-life of approximately 12.5 years. How much of an initial quantity of this element remains after 50 years?

50. **Old Bones** It is found that 97% of C-14 has been lost in a human skeleton found at an archeological site. What is the approximate age of the skeleton?

51. **Wishful Thinking** A person facing retirement invests $650,000 in a savings account. She wants the account to be worth $1,000,000 in 10 years. What annual rate r of interest compounded continuously will achieve this dream?

52. **Photic Zone** The photic zone (also called the euphotic zone) is the depth of water in a lake or an ocean which is exposed to sufficient sunlight to allow photosynthesis to take place. The photic zone extends from surface of the water to a depth where the light intensity is 1% of the intensity of light I_0 incident on the surface. According to the Beer-Lambert Law the intensity of light at a depth of x meters is $I(x) = I_0 e^{kx}$, where $k < 0$. See FIGURE 5.R.11 and Problem 62 in Exercises 5.4.

   (a) Find the depth of the photic zone in the Atlantic ocean near Cape Cod, Massachusetts where $k = -0.28782$.

   (b) Using calculus and the depth found in part (a) it can be shown that the average intensity over the entire photic zone is 21.5% of I_0. At what depth does this average intensity occur?

FIGURE 5.R.11 Photic zone in Problem 52

53. The **Gompertz function** $y = ae^{-be^{-ct}}$, where a, b, and c are positive constants, is named after the self-educated mathematician **Benjamin Gompertz** (1779–1865) and was used initially in the study of population demographics. Today Gompertz's function is used as a mathematical model in diverse areas such as economics, statistics, and oncological studies of the growth of tumors. Solve for t in terms of the other symbols.

54. The graph of a Gompertz function is naturally called a **Gompertz curve**. Sketch the Gompertz curve in the following cases. In each case, superimpose the three graphs on the same rectangular coordinate system.

   (a) $y = ae^{-e^{-t}}, a = \frac{1}{2}, a = 1, a = 2$

   (b) $y = e^{-be^{-t}}, b = \frac{1}{2}, b = 1, b = 2$

   (c) $y = e^{-e^{-ct}}, c = \frac{1}{2}, c = 1, c = 2$ [*Hint*: The graph has two horizontal asymptotes.]

55. Seriously Saving An annuity is a savings plan where the same amount of money P is deposited into an account at n equally spaced periods (say, years) of time. If the annual rate r of interest is compounded continuously, then the amount S accrued in the account immediately after the nth deposit is

$$S = P + Pe^r + Pe^{2r} + \cdots + Pe^{(n-1)r}.$$

What is the value of such an annuity in 15 years if $P = \$3000$ and the annual rate of interest is 2%.

56. Curve Fitting Logarithms can be used to find a function that approximately fits a set of numerical data. For example, a relationship between length L and weight W of yellowfin tuna (or ahi, as it is called in most restaurants) in a mid-Pacific fishery is given in the following table.

TABLE 5.R.1

Length L (cm)	70	80	90	100	110	120	130	140	160	180
Weight W (lb)	14.3	24.6	29.3	42.5	56.8	74.1	94.7	103.6	179	256

(a) If $x = \ln L$ and $y = \ln W$, then use Table 5.R.1 to fill out the remaining entries in Table 5.R.2.

TABLE 5.R.2

x	4.25									
y	2.66									

(b) Use the first and last points (x, y) defined by Table 5.R.2 to show that the slope of a line through these points is approximately $m = 3.074$. Find an equation of the line through the first and last points.

(c) Use a graphing utility to plot the points defined in Table 5.R.2 and the line in part (b) on the same rectangular coordinate system.

(d) Use the equation of the line obtained in part (b) to write W as an explicit function of L. Use this function and the lengths L in Table 5.R.1 to compute the corresponding weights. Compare these weights with those given in Table 5.R.1.

(e) Use a graphing utility to plot the points defined by Table 5.R.1 and the function W in part (d) on the same rectangular coordinate system.

Yellowfin tuna in the Pacific Ocean near Hawaii

Courtesy of NASA/JPL-Caltech/SSI

6 Conic Sections

Chapter Contents

6.1 The Parabola

Hypatia

INTRODUCTION **Hypatia** is the first woman in the history of mathematics about whom we have considerable knowledge. Born in 370 C.E. in Alexandria, she was renowned as a mathematician and philosopher. Among her writings is *On the Conics of Apollonius*, which popularized **Apollonius'** (200 B.C.E.) work on curves that can be obtained by intersecting a double-napped cone with a plane: the circle, parabola, ellipse, and hyperbola. Note in FIGURE 6.1.1(a) that the plane does not pass through the vertex of the cone. When the plane passes through the vertex, the resulting figures: a single point, a single line, or two intersecting lines are commonly called **degenerate conics** (see Figure 6.1.1(b)). With the close of the Greek period, interest in conic sections waned; after Hypatia the study of these curves was neglected for over 1000 years.

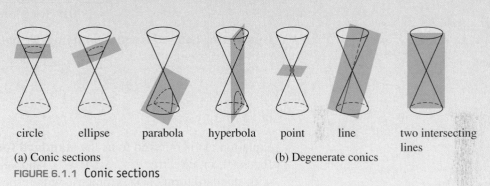

circle   ellipse   parabola   hyperbola   point   line   two intersecting lines

(a) Conic sections   (b) Degenerate conics

FIGURE 6.1.1 Conic sections

Solar System

In the seventeenth century, the Italian physicist and mathematician **Galileo Galilei** (1564–1642) showed that in the absence of air resistance the path of a projectile follows a parabolic arc. At about the same time, the German mathematician, astronomer, and astrologist **Johannes Kepler** (1571–1630) hypothesized that the orbits of planets around the Sun are ellipses with the Sun at one focus. This was later verified by the English mathematician **Sir Isaac Newton** (1642–1726), using the methods of the newly developed calculus. Kepler also experimented with the reflecting properties of parabolic mirrors; these investigations sped the development of the reflecting telescope. The Greeks had known little of these practical applications. They had studied the conics for their beauty and fascinating properties. In the first three sections of this chapter, we will examine both the ancient properties and the modern applications of these curves. Rather than using a cone, we shall see how the parabola, ellipse, and hyperbola are defined by means of distance. Using a rectangular coordinate system and the distance formula, we obtain equations for the conics. Each of these equations will be in the form of a quadratic equation in variables x and y:

$$Ax^2 + Bxy + Cy^2 + Dx + Ey + F = 0,$$

where A, B, C, D, E, and F are constants. We have already studied the special case $y = ax^2 + bx + c$ of the foregoing equation in Section 2.4.

DEFINITION 6.1.1   Parabola

A **parabola** is the set of points $P(x, y)$ in the plane that are equidistant from a fixed line L, called the **directrix**, and a fixed point F, called the **focus**.

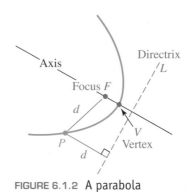

FIGURE 6.1.2 A parabola

A parabola is shown in FIGURE 6.1.2. The line through the focus perpendicular to the directrix is called the **axis** of the parabola. The point of intersection of the parabola and the axis is called the **vertex**, denoted by V in Figure 6.1.2.

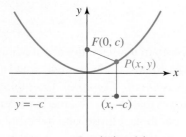

FIGURE 6.1.3 Parabola with vertex (0, 0) and focus on the y-axis

Parabola with Vertex (0, 0) To describe a parabola analytically, we use a rectangular coordinate system where the directrix is a horizontal line $y = -c$, where $c > 0$, and the focus is the point $F(0, c)$. Then we see that the axis of the parabola is along the y-axis, as FIGURE 6.1.3 shows. The origin is necessarily the vertex, since it lies on the axis c units from both the focus and the directrix. The distance from a point $P(x, y)$ to the directrix is

$$y - (-c) = y + c.$$

Using the distance formula, the distance from P to the focus F is

$$d(P, F) = \sqrt{(x - 0)^2 + (y - c)^2}.$$

From the definition of the parabola it follows that $d(P, F) = y + c$, or

$$\sqrt{(x - 0)^2 + (y - c)^2} = y + c.$$

By squaring both sides and simplifying, we obtain

$$x^2 + (y - c)^2 = (y + c)^2$$

$$x^2 + y^2 - 2cy + c^2 = y^2 + 2cy + c^2$$

or
$$x^2 = 4cy. \tag{1}$$

Equation (1) is referred to as the **standard form** of the equation of a parabola with focus $(0, c)$, directrix $y = -c$, $c > 0$, and vertex $(0, 0)$. The graph of any parabola with standard form (1) is symmetric with respect to the y-axis.

Equation (1) does not depend on the assumption that $c > 0$. However, the direction in which the parabola opens does depend on the sign of c. Specifically, if $c > 0$ the parabola opens *upward* as in Figure 6.1.3; if $c < 0$, the parabola opens *downward*.

If the focus of a parabola is assumed to lie on the x-axis at $F(c, 0)$ and the directrix is $x = -c$, then the x-axis is the axis of the parabola and the vertex is $(0, 0)$. If $c > 0$ the parabola opens to the right; if $c < 0$, it opens to the left. In either case, the **standard form** of the equation is

$$y^2 = 4cx. \tag{2}$$

The graph of any parabola with standard form (2) is symmetric with respect to the x-axis.

A summary of all this information for equations (1) and (2) is given in FIGURES 6.1.4 and 6.1.5, respectively. You may be surprised to see in Figure 6.1.4(b) that the directrix above the x-axis is labeled $y = -c$ and the focus on the negative y-axis has coordinates $F(0, c)$. Bear in mind that in this case the assumption is that $c < 0$ and so $-c > 0$. A similar remark holds for Figure 6.1.5(b).

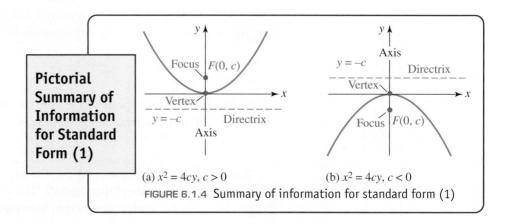

Pictorial Summary of Information for Standard Form (1)

(a) $x^2 = 4cy$, $c > 0$

(b) $x^2 = 4cy$, $c < 0$

FIGURE 6.1.4 Summary of information for standard form (1)

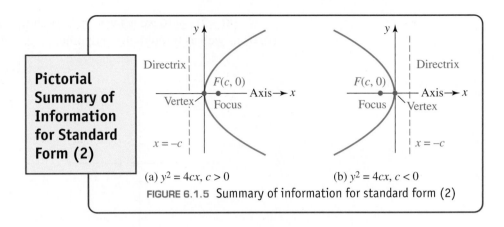

(a) $y^2 = 4cx$, $c > 0$ (b) $y^2 = 4cx$, $c < 0$

FIGURE 6.1.5 Summary of information for standard form (2)

Pictorial Summary of Information for Standard Form (2)

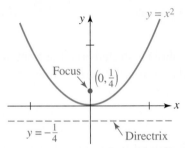

FIGURE 6.1.6 Graph of equation in Example 1

Graphing tip for equations (1) and (2). ▶

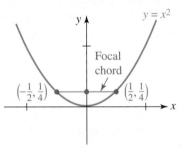

FIGURE 6.1.7 Focal chord for $y = x^2$

EXAMPLE 1   The Simplest Parabola

We first encountered the graph of $y = x^2$ in Section 2.2. By comparing this equation with (1) we see

$$\underset{\substack{\uparrow \\ 4c}}{x^2} = 1 \cdot y$$

and so $4c = 1$ or $c = \frac{1}{4}$. Therefore the graph of $y = x^2$ is a parabola with vertex at the origin, focus at $\left(0, \frac{1}{4}\right)$, and directrix $y = -\frac{1}{4}$. These details are indicated in the graph in **FIGURE 6.1.6**.

☐ **Focal Cord** Knowing the basic parabolic shape, all we need to know to sketch a *rough* graph of either equations (1) and (2) is the fact that the graph passes through its vertex $(0, 0)$ and the direction in which the parabola opens. To add more accuracy to the graph it is convenient to use the number c determined by the standard form equation to plot two additional points. Note that if we choose $y = c$ in (1), then $x^2 = 4c^2$ implies $x = \pm 2c$. Thus $(-2c, c)$ and $(2c, c)$ lie on the graph of $x^2 = 4cy$. Similarly, the choice $x = c$ in (2) implies $y = \pm 2c$, and so $(c, -2c)$ and $(c, 2c)$ are points on the graph of $y^2 = 4cx$. The *line segment* through the focus with endpoints $(-2c, c)$, $(2c, c)$ for equations with standard form (1), and $(c, -2c)$, $(c, 2c)$ for equations with standard form (2) is called the **focal chord**. In Example 1 we saw that $c = \frac{1}{4}$ for the equation $y = x^2$. Therefore, if we choose $y = \frac{1}{4}$, then $x^2 = \frac{1}{4}$ gives $x = \pm\frac{1}{2}$. Endpoints of the horizontal focal chord for $y = x^2$ are $\left(-\frac{1}{2}, \frac{1}{4}\right)$ and $\left(\frac{1}{2}, \frac{1}{4}\right)$. See **FIGURE 6.1.7**.

EXAMPLE 2   Equation of a Parabola

Find the equation in standard form of the parabola with directrix $x = 2$ and focus $(-2, 0)$. Graph.

Solution In **FIGURE 6.1.8** we have graphed the directrix and the focus. We see from their placement that the equation we seek is of the form $y^2 = 4cx$. Since $c = -2$, the parabola opens to the left and so

$$y^2 = 4(-2)x \qquad \text{or} \qquad y^2 = -8x.$$

As mentioned in the discussion preceding this example, if we substitute $x = c$, or in this case $x = -2$, into the equation $y^2 = -8x$ we can find two points on its graph. From

$y^2 = -8(-2) = 16$ we get $y = \pm 4$. As shown in FIGURE 6.1.9, the graph passes through $(0, 0)$ as well as through the endpoints $(-2, -4)$ and $(-2, 4)$ of the focal chord.

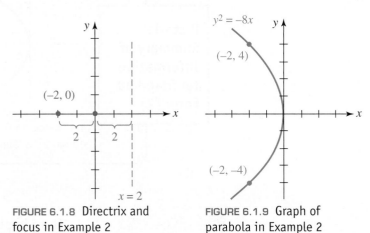

FIGURE 6.1.8 Directrix and
focus in Example 2

FIGURE 6.1.9 Graph of
parabola in Example 2

☐ **Parabola with Vertex (h, k)** Suppose that a parabola is shifted both horizontally and vertically so that its vertex is at the point (h, k) and its axis is the vertical line $x = h$. The **standard form** of the equation of the parabola is then

$$(x - h)^2 = 4c(y - k). \tag{3}$$

Similarly, if its axis is the horizontal line $y = k$, the **standard form** of the equation of the parabola with vertex (h, k) is

$$(y - k)^2 = 4c(x - h). \tag{4}$$

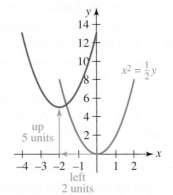

FIGURE 6.1.10 Graphs of (3) and (4) are shifts of the graphs of (1) and (2)

☐ **Shifts** The parabolas defined by the foregoing equations are identical in shape to the parabolas defined by equations (1) and (2) because (3) and (4) represent rigid transformations (shifts up, down, left, and right) of the graphs of (1) and (2). For example, the parabola

$$(x + 2)^2 = \tfrac{1}{2}(y - 5)$$

has vertex $(-2, 5)$. With the identifications $h = -2 < 0$ and $k = 5 > 0$, the graph of $(x + 2)^2 = \tfrac{1}{2}(y - 5)$ is the graph of $x^2 = \tfrac{1}{2}y$ shifted horizontally 2 units to the left followed by an upward vertical shift of 5 units. See the red graph in FIGURE 6.1.10.

For each of the equations, (1) and (2) or (3) and (4), the *distance* from the vertex to the focus, as well as the distance from the vertex to the directrix, is $|c|$.

FIGURE 6.1.11 Vertex and directrix in
Example 3

■ EXAMPLE 3   **Equation of a Parabola**

Find the equation in standard form of the parabola with vertex $(-3, -1)$ and directrix $y = 3$.

Solution We begin by graphing the vertex at $(-3, -1)$ and the directrix $y = 3$. From FIGURE 6.1.11 we can see that the parabola must open downward, and so its standard form is (3). This fact, plus the observation that the vertex lies 4 units below the directrix, indicates that the appropriate solution of $|c| = 4$ is $c = -4$. Substituting $h = -3$, $k = -1$, and $c = -4$ into (3) gives

$$[x - (-3)]^2 = 4(-4)[y - (-1)] \qquad \text{or} \qquad (x + 3)^2 = -16(y + 1). \quad ■$$

EXAMPLE 4   Find Everything

Find the vertex, focus, directrix, intercepts, and graph of the parabola

$$y^2 - 4y - 8x - 28 = 0. \tag{5}$$

Solution In order to write the equation in one of the standard forms we complete the square in y:

$$y^2 - 4y + 4 = 8x + 28 + 4 \quad \leftarrow \text{add 4 to both sides}$$
$$(y - 2)^2 = 8x + 32.$$

Thus the standard form of equation (5) is $(y - 2)^2 = 8(x + 4)$. Comparing this equation with (4) we conclude that the vertex is $(-4, 2)$ and that $4c = 8$ or $c = 2$. Thus the parabola opens to the right. From $c = 2 > 0$, the focus is 2 units to the right of the vertex at $(-4 + 2, 2)$ or $(-2, 2)$. The directrix is the vertical line 2 units to the left of the vertex, $x = -4 - 2$ or $x = -6$. Knowing the parabola opens to the right from the point $(-4, 2)$ also tells us that the graph has intercepts. To find the x-intercept we set $y = 0$ in (5) and find immediately that $x = -\frac{28}{8} = -\frac{7}{2}$. The x-intercept is $\left(-\frac{7}{2}, 0\right)$. To find the y-intercepts we set $x = 0$ in (5) and find from the quadratic formula that $y = 2 \pm 4\sqrt{2}$ or $y \approx 7.66$ and $y \approx -3.66$. The y-intercepts are $\left(0, 2 - 4\sqrt{2}\right)$ and $\left(0, 2 + 4\sqrt{2}\right)$. Putting all this information together we get the graph in FIGURE 6.1.12. ■

FIGURE 6.1.12 Graph of equation in Example 4

Review Section 2.7. ▶

☐ **Functions** Equations of the form given in (1), (2), (3) and (4) define functions implicitly.

EXAMPLE 5   Functions Defined Implicitly

(a) In Example 3, the final equation $(x + 3)^2 = -16(y + 1)$ is easily solved for y:

$$y + 1 = -\frac{1}{16}(x + 3)^2$$

$$y = -\frac{1}{16}(x + 3)^2 - 1 \quad \leftarrow \left\{ \begin{array}{l} \text{standard form of a} \\ \text{quadratic function,} \\ \text{(2) in Section 2.4} \end{array} \right.$$

$$= -\frac{1}{16}(x^2 + 6x + 9) - 1$$

$$y = -\frac{1}{16}x^2 - \frac{3}{8}x - \frac{25}{16} \quad \leftarrow \left\{ \begin{array}{l} \text{quadratic function} \\ f(x) = ax^2 + bx + c \end{array} \right.$$

(b) In Example 4, solving $(y - 2)^2 = 8x + 32$ for y gives

$$y = 2 \pm \sqrt{8x + 32}.$$

Thus two explicit functions defined implicitly by the original equation are

$$y = f(x) = 2 + 2\sqrt{2x + 8} \quad \text{and} \quad y = g(x) = 2 - 2\sqrt{2x + 8}.$$

Solving the inequality $2x + 8 \geq 0$ indicates that the domain of f and g is the interval $[-4, \infty)$. The graphs of f and g can be interpreted as rigid and nonrigid transformations of the graph of the square root function $y = \sqrt{x}$, and are, in turn, the top half and bottom half of the parabola in Figure 6.1.12. See FIGURE 6.1.13.

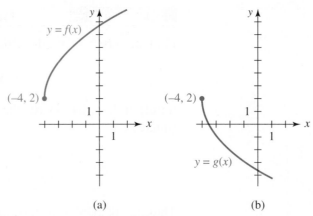

FIGURE 6.1.13 Graphs of functions in (b) of Example 5

Searchlight tribute to
World Trade Center

TV satellite dish

☐ **Applications of the Parabola** The parabola has many interesting properties that make it suitable for certain applications. Reflecting surfaces are often designed to take advantage of a reflection property of parabolas. Such surfaces, called **paraboloids**, are three-dimensional and are formed by rotating a parabola about its axis. As illustrated in FIGURE 6.1.14(a), rays of light (or electronic signals) from a point source located at the focus of a parabolic reflecting surface will be reflected along lines parallel to the axis. This is the idea behind the design of searchlights, some flashlights, and on-location satellite dishes. Conversely, if the incoming rays of light are parallel to the axis of a parabola, they will be reflected off the surface along lines passing through the focus. See Figure 6.1.14(b). Beams of light from a distant object such as a galaxy are essentially parallel, and so when these beams enter a reflecting telescope they are reflected by the parabolic mirror to the focus, where a camera is usually placed to capture the image over time. A parabolic home satellite dish operates on the same principle as the reflecting telescope; the digital signal from a TV satellite is captured at the focus of the dish antenna by a receiver.

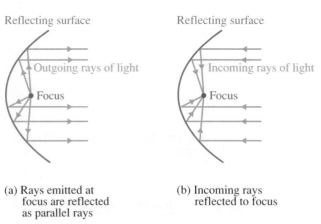

(a) Rays emitted at
focus are reflected
as parallel rays

(b) Incoming rays
reflected to focus

FIGURE 6.1.14 Parabolic reflecting surface

The Brooklyn Bridge is a suspension
bridge

Parabolas are also important in the design of suspension bridges. It can be shown that if the weight of the bridge is distributed uniformly along its length, then a support cable in the shape of a parabola will bear the load evenly.

The ball travels in a parabolic arc

The trajectory of an obliquely launched projectile—say, a basketball thrown from the free throw line—will travel in a parabolic arc.

Tuna, which prey on smaller fish, have been observed swimming in schools of 10 to 20 fish arrayed approximately in a parabolic shape. One possible explanation for this is that the smaller fish caught in the school of tuna will try to escape by "reflecting" off the parabola. As a result, they are concentrated at the focus and become easy prey for the tuna. See FIGURE 6.1.15.

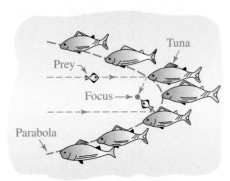

FIGURE 6.1.15 Tuna hunting in a parabolic arc

Exercises 6.1   Answers to selected odd-numbered problems begin on page ANS–21.

In Problems 1–24, find the vertex, focus, directrix, and axis of the given parabola. Graph the parabola.

1. $y^2 = 4x$

2. $y^2 = \frac{7}{2}x$

3. $y^2 = -\frac{4}{3}x$

4. $y^2 = -10x$

5. $x^2 = -16y$

6. $x^2 = \frac{1}{10}y$

7. $x^2 = 28y$

8. $x^2 = -64y$

9. $(y - 1)^2 = 16x$

10. $(y + 3)^2 = -8(x + 2)$

11. $(x + 5)^2 = -4(y + 1)$

12. $(x - 2)^2 + y = 0$

13. $y^2 + 12y - 4x + 16 = 0$

14. $x^2 + 6x + y + 11 = 0$

15. $x^2 + 5x - \frac{1}{4}y + 6 = 0$

16. $x^2 - 2x - 4y + 17 = 0$

17. $y^2 - 8y + 2x + 10 = 0$

18. $y^2 - 4y - 4x + 3 = 0$

19. $4x^2 = 2y$

20. $3(y - 1)^2 = 9x$

21. $-2x^2 + 12x - 8y - 18 = 0$

22. $4y^2 + 16y - 6x - 2 = 0$

23. $6y^2 - 12y - 24x - 42 = 0$

24. $3x^2 + 30x - 8y + 75 = 0$

In Problems 25–44, find an equation of the parabola that satisfies the given conditions.

25. Focus $(0, 7)$, directrix $y = -7$

26. Focus $(0, -5)$, directrix $y = 5$

27. Focus $(-4, 0)$, directrix $x = 4$

28. Focus $\left(\frac{3}{2}, 0\right)$, directrix $x = -\frac{3}{2}$

29. Focus $\left(\frac{5}{2}, 0\right)$, vertex $(0, 0)$

30. Focus $(0, -10)$, vertex $(0, 0)$

31. Focus $(2, 3)$, directrix $y = -3$

32. Focus $(1, -7)$, directrix $x = -5$

33. Focus $(-1, 4)$, directrix $x = 5$

34. Focus $(-2, 0)$, directrix $y = \frac{3}{2}$

35. Focus $(1, 5)$, vertex $(1, -3)$

36. Focus $(-2, 3)$, vertex $(-2, 5)$

37. Focus $(8, -3)$, vertex $(0, -3)$

38. Focus $(1, 2)$, vertex $(7, 2)$

39. Vertex $(0, 0)$, directrix $y = -\frac{7}{4}$

40. Vertex $(0, 0)$, directrix $x = 6$

41. Vertex $(5, 1)$, directrix $y = 7$

42. Vertex $(-1, 4)$, directrix $x = 0$

43. Vertex $(0, 0)$, through $(-2, 8)$, axis along the y-axis

44. Vertex $(0, 0)$, through $\left(1, \frac{1}{4}\right)$, axis along the x-axis

In Problems 45–48, find the x- and y-intercepts of the given parabola.

45. $(y + 4)^2 = 4(x + 1)$ **46.** $(x - 1)^2 = -2(y - 1)$

47. $x^2 + 2y - 18 = 0$ **48.** $y^2 - 8y - x + 15 = 0$

In Problems 49–52, find at least one function defined implicitly by the given equation. Give the domain of each function.

49. $(x - 1)^2 = \frac{1}{2}(y + 2)$ **50.** $(x + 3)^2 = 2(y - 4)$

51. $(y + 1)^2 = -5(x - 2)$ **52.** $y^2 - 12y = x - 11$

Applications

53. Spotlight A large spotlight is designed so that a cross section through its axis is a parabola and the light source is at the focus. Find the position of the light source if the spotlight is 4 ft across at the opening and 2 ft deep.

54. Reflecting Telescope A reflecting telescope has a parabolic mirror that is 20 ft across at the top and 4 ft deep at the center. Where should the eyepiece be located?

55. Light Ray Suppose that a light ray emanating from the focus of the parabola $y^2 = 4x$ strikes the parabola at $(1, -2)$. What is the equation of the reflected ray?

56. Suspension Bridge Suppose that two towers of a suspension bridge are 350 ft apart and the vertex of the parabolic cable is tangent to the road midway between the towers. If the cable is 1 ft above the road at a point 20 ft from the vertex, find the height of the towers above the road.

57. Another Suspension Bridge Two towers of a suspension bridge are 75 ft high (measured from the roadway) and are 250 ft apart. The vertex of the parabolic support cable is tangent to the road midway between the towers. Find the height of the cable above the roadway at a point 50 ft from one of the towers. See FIGURE 6.1.16.

FIGURE 6.1.16 Suspension bridge in Problem 57

58. Drainpipe Assume that the water gushing from the end of a horizontal pipe follows a parabolic arc with vertex at the end of the pipe. The pipe is 20 m above the ground. At a point 2 m below the end of the pipe, the horizontal distance from the water to a vertical line through the end of the pipe is 4 m. See FIGURE 6.1.17. Where does the water strike the ground?

FIGURE 6.1.17 Pipe in Problem 58

59. A Bullseye A dart thrower releases a dart 5 ft above the ground. The dart is thrown horizontally and follows a parabolic path. It hits the ground $10\sqrt{10}$ ft from the dart thrower. At a distance of 10 ft from the dart thrower, how high should a bullseye be placed in order for the dart to hit it?

60. Path of a Projectile The vertical position of a projectile is given by the equation $y = -16t^2$ and the horizontal position by $x = 40t$ for $t \geq 0$. By eliminating t between the two equations, show that the path of the projectile is a parabolic arc. Graph the path of the projectile.

61. Focal Width The focal width of a parabola is the length of the focal chord, that is, the line segment through the focus perpendicular to the axis, with endpoints on the parabola. See FIGURE 6.1.18.
(a) Find the focal width of the parabola $x^2 = 8y$.
(b) Show that the focal width of the parabola $x^2 = 4cy$ and $y^2 = 4cx$ is $4|c|$.

62. Parabolic Orbit The orbit of a comet is a parabola with the Sun at the focus. When the comet is 50,000,000 km from the Sun, the line from the comet to the Sun is perpendicular to the axis of the parabola. Use the result of Problem 61(b) to write an equation of the comet's path. (A comet with a parabolic path will not return to the Solar System.)

FIGURE 6.1.18 Focal width in Problem 61

Calculator/Computer Problems

In Problems 63 and 64, the given equation defines a parabola that has been rotated about the origin. Use the quadratic formula to find two functions f and g defined implicitly by the equation. Give the domain of each function. Use a graphing utility to obtain the graphs of f and g on the same coordinate axes.

63. $x^2 - 2xy + y^2 - 3x + 2y = 1$ **64.** $x^2 + 2xy + y^2 + 2x - 4y = 5$

For Discussion

65. Suppose that two parabolic reflecting surfaces face one another (with foci on a common axis). Any sound emitted at one focus will be reflected off the parabolas and concentrated at the other focus. FIGURE 6.1.19 shows the paths of two typical sound waves. Using the definition of a parabola on page 379, show that all waves will travel the same distance. [*Note:* This result is important for the following reason: If the sound waves traveled paths of different lengths, then the waves would arrive at the second focus at different times. The result would be interference rather than clear sound.]

66. The point on a parabola closest to the focus is the vertex. How would you go about proving this? Carry out your ideas.

67. For the comet in Problem 62, use the result of Problem 66 to determine the shortest distance between the Sun and the comet.

68. Tangent Lines In this problem it is necessary that you have studied Section 2.10.

(a) Any two distinct tangent lines to a parabola must intersect. Show that the tangent lines to a parabola at the endpoints of its focal chord are perpendicular. [*Hint:* Without loss of generality we may assume that an equation of the parabola is given by $x^2 = 4cy$.]

(b) Show that the tangent lines at the endpoints of the focal chord of a parabola intersect on its directrix.

Directrix · Path of sound wave · Directrix · Focus · Focus · Common axis

FIGURE 6.1.19 Parabolic reflecting surfaces in Problem 65

6.2 The Ellipse

INTRODUCTION The ellipse occurs frequently in astronomy. For example, the paths of the planets around the Sun are elliptical with the Sun located at one focus. Similarly, communication satellites, the Hubble Space Telescope, and the International Space Station revolve around the Earth in elliptical orbits with the Earth at one focus. In this section we define the ellipse and study some of its properties and applications.

DEFINITION 6.2.1 Ellipse

An **ellipse** is the set of points $P(x, y)$ in the plane such that the sum of the distances between P and two fixed points F_1 and F_2 is constant. The fixed points F_1 and F_2 are called **foci** (plural for **focus**). The midpoint of the line segment joining points F_1 and F_2 is called the **center** of the ellipse.

As shown in FIGURE 6.2.1, if P is a point on the ellipse and if $d_1 = d(F_1, P)$ and $d_2 = d(F_2, P)$ are the distances from the foci to P, then the preceding definition asserts that

$$d_1 + d_2 = k, \tag{1}$$

where $k > 0$ is some constant.

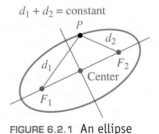

$d_1 + d_2 = $ constant

FIGURE 6.2.1 An ellipse

FIGURE 6.2.2 A way to draw an ellipse

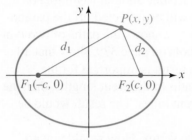

FIGURE 6.2.3 Ellipse with center (0, 0) and foci on the *x*-axis

On a practical level, equation (1) suggests a way of generating an ellipse. FIGURE 6.2.2 shows that if a string of length *k* is attached to a piece of paper by two tacks, then an ellipse can be traced out by inserting a pencil against the string and moving it in such a manner that the string remains taut.

☐ **Ellipse with Center (0, 0)** We now derive an equation of the ellipse. For algebraic convenience, let us choose $k = 2a > 0$ and put the foci on the *x*-axis with coordinates $F_1(-c, 0)$ and $F_2(c, 0)$ as shown in FIGURE 6.2.3. It follows from (1) that

$$\sqrt{(x + c)^2 + y^2} + \sqrt{(x - c)^2 + y^2} = 2a$$

or

$$\sqrt{(x + c)^2 + y^2} = 2a - \sqrt{(x - c)^2 + y^2}. \tag{2}$$

We square both sides of the second equation in (2) and simplify,

$$(x + c)^2 + y^2 = 4a^2 - 4a\sqrt{(x - c)^2 + y^2} + (x - c)^2 + y^2$$

$$a\sqrt{(x - c)^2 + y^2} = a^2 - cx.$$

Squaring a second time gives,

$$a^2[(x - c)^2 + y^2] = a^4 - 2a^2cx + c^2x^2$$

or

$$(a^2 - c^2)x^2 + a^2y^2 = a^2(a^2 - c^2). \tag{3}$$

Referring to Figure 6.2.3, we see that the points F_1, F_2, and P form a triangle. Because the sum of the lengths of any two sides of a triangle is greater than the remaining side, we must have $2a > 2c$ or $a > c$. Hence, $a^2 - c^2 > 0$. When we let $b^2 = a^2 - c^2$, then (3) becomes $b^2x^2 + a^2y^2 = a^2b^2$. Dividing this last equation by a^2b^2 gives

$$\frac{x^2}{a^2} + \frac{y^2}{b^2} = 1. \tag{4}$$

Equation (4) is called the **standard form** of the equation of an ellipse centered at $(0, 0)$ with foci $(-c, 0)$ and $(c, 0)$, where *c* is defined by $b^2 = a^2 - c^2$, and $a > b > 0$.

If the foci are placed on the *y*-axis, then a repetition of the above analysis leads to

$$\frac{x^2}{b^2} + \frac{y^2}{a^2} = 1. \tag{5}$$

Equation (5) is called the **standard form** of the equation of an ellipse centered at $(0, 0)$ with foci $(0, -c)$ and $(0, c)$, where *c* is defined by $b^2 = a^2 - c^2$ and $a > b > 0$.

☐ **Major and Minor Axes** The **major axis** of an ellipse is the line segment through its center, containing the foci, and with endpoints on the ellipse. For an ellipse with standard form equation (4) the major axis is horizontal, whereas for (5) the major axis is vertical. The line segment through the center, perpendicular to the major axis, and with endpoints on the ellipse, is called the **minor axis**. The two endpoints of the major axis are called the **vertices** of the ellipse. For (4) the vertices are the *x*-intercepts. Setting $y = 0$ in (4) gives $x = \pm a$. The vertices are then $(-a, 0)$ and $(a, 0)$. For (5) the vertices are the y-intercepts $(0, -a)$ and $(0, a)$. For equation (4), the endpoints of the minor axis are $(0, -b)$ and $(0, b)$; for (5) the endpoints are $(-b, 0)$ and $(b, 0)$. For either (4) or (5), the **length of the major axis** is $a - (-a) = 2a$; the **length of the minor axis** is $2b$. Since $a > b$, the major axis of an ellipse is always longer than its minor axis.

A summary of all this information for equations (4) and (5) is given in FIGURE 6.2.4.

Pictorial Summary of Information for Standard Forms (4) and (5)

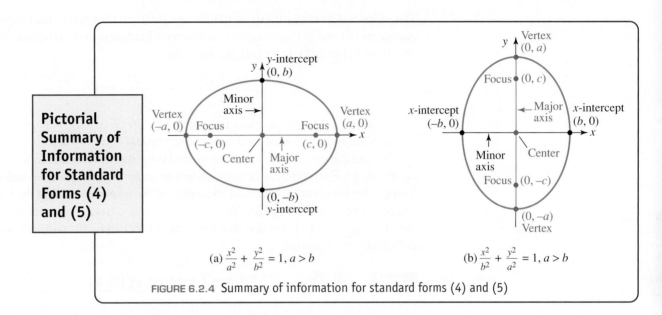

(a) $\dfrac{x^2}{a^2} + \dfrac{y^2}{b^2} = 1, a > b$ (b) $\dfrac{x^2}{b^2} + \dfrac{y^2}{a^2} = 1, a > b$

FIGURE 6.2.4 Summary of information for standard forms (4) and (5)

▮ EXAMPLE 1   Vertices and Foci

Find the vertices and foci of the ellipse whose equation is $3x^2 + y^2 = 9$. Graph.

Solution By dividing both sides of the equality by 9 the standard form of the equation is

$$\frac{x^2}{3} + \frac{y^2}{9} = 1.$$

We see that $9 > 3$ and so we identify the equation with (5). From $a^2 = 9$ and $b^2 = 3$, we see that $a = 3$ and $b = \sqrt{3}$. The major axis is vertical with vertices $(0, -3)$ and $(0, 3)$. The minor axis is horizontal with endpoints $(-\sqrt{3}, 0)$ and $(\sqrt{3}, 0)$. Of course, the vertices are also the y-intercepts and the endpoints of the minor axis are the x-intercepts. Now, to find the foci we use $b^2 = a^2 - c^2$ or $c^2 = a^2 - b^2$ to write $c = \sqrt{a^2 - b^2}$. With $a = 3, b = \sqrt{3}$, we get $c = \sqrt{9 - 3} = \sqrt{6}$. Hence, the foci are on the y-axis at $(0, -\sqrt{6})$ and $(0, \sqrt{6})$. The graph is given in FIGURE 6.2.5. ▮

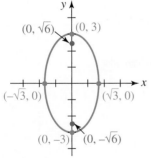

FIGURE 6.2.5 Ellipse in Example 1

▮ EXAMPLE 2   Equation of an Ellipse

Find an equation of an ellipse with a focus $(2, 0)$ and an x-intercept $(5, 0)$.

Solution Since the given focus is on the x-axis, we can find an equation in standard form (4). Consequently, $c = 2, a = 5, a^2 = 25$, and $b^2 = a^2 - c^2$ or $b^2 = 5^2 - 2^2 = 21$. The desired equation is

$$\frac{x^2}{25} + \frac{y^2}{21} = 1.$$ ▮

□ **Ellipse with Center (h, k)** When the center is at (h, k), the **standard form** of the equation of an ellipse is either

$$\frac{(x-h)^2}{a^2} + \frac{(y-k)^2}{b^2} = 1 \tag{6}$$

or

$$\frac{(x-h)^2}{b^2} + \frac{(y-k)^2}{a^2} = 1. \tag{7}$$

The ellipses defined by these equations are identical in shape to the ellipses defined by equations (4) and (5) since equations (6) and (7) represent rigid transformations of the graphs of (4) and (5). For example, the ellipse

$$\frac{(x-1)^2}{9} + \frac{(y+3)^2}{16} = 1$$

has center $(1, -3)$. Its graph is the graph of $x^2/9 + y^2/16 = 1$ shifted horizontally 1 unit to the right followed by a downward vertical shift of 3 units.

It is not a good idea to memorize formulas for the vertices and foci of an ellipse with center (h, k). Everything is the same as before: a, b, and c are positive and $a > b, a > c$. You can locate vertices, foci, and endpoints of the minor axis using the fact that a is the distance from the center to a vertex, b is the distance from the center to an endpoint on the minor axis, and c is the distance from the center to a focus. Also, the number c is still defined by the equation $b^2 = a^2 - c^2$.

EXAMPLE 3 Ellipse Centered at (h, k)

Find the center, vertices, and foci of the ellipse $4x^2 + 16y^2 - 8x - 96y + 84 = 0$. Graph.

Solution To write the given equation in one of the standard forms (6) or (7) we must complete the square in x and in y. Recall that in order to complete the square we want the coefficients of the quadratic terms x^2 and y^2 to be 1. To do this we factor 4 from both x^2 and x and factor 16 from both y^2 and y:

$$4(x^2 - 2x \quad) + 16(y^2 - 6y \quad) = -84.$$

Then from

$$\overbrace{4(x^2 - 2x + 1) + 16(y^2 - 6y + 9)}^{4 \cdot 1 \text{ and } 16 \cdot 9 \text{ are added to both sides}} = -84 + 4 \cdot 1 + 16 \cdot 9$$

we obtain

$$4(x-1)^2 + 16(y-3)^2 = 64$$

or

$$\frac{(x-1)^2}{16} + \frac{(y-3)^2}{4} = 1. \tag{8}$$

From (8) we see that the center of the ellipse is $(1, 3)$. Since the last equation has the standard form (6), we identify $a^2 = 16$ or $a = 4$ and $b^2 = 4$ or $b = 2$. The major axis is horizontal and lies on the horizontal line $y = 3$ passing through $(1, 3)$. This is the red horizontal dashed line segment in FIGURE 6.2.6. By measuring $a = 4$ units to the left and then to the right of the center along the line $y = 3$, we arrive at the vertices $(-3, 3)$ and $(5, 3)$. By measuring $b = 2$ units both down and up the vertical line $x = 1$ through the center, we arrive at the endpoints of the minor axis $(1, 1)$ and $(1, 5)$. The minor axis is the black dashed vertical line segment in Figure 6.2.6. Because $c^2 = a^2 - b^2 = 16 - 4 = 12, c = 2\sqrt{3}$. Finally, by measuring $c = 2\sqrt{3}$ units to the left and right of the center along $y = 3$, we obtain the foci $\left(1 - 2\sqrt{3}, 3\right)$ and $\left(1 + 2\sqrt{3}, 3\right)$.

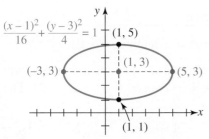

$$\frac{(x-1)^2}{16} + \frac{(y-3)^2}{4} = 1$$

FIGURE 6.2.6 Ellipse in Example 3

CHAPTER 6 CONIC SECTIONS

EXAMPLE 4   Functions Defined Implicitly

The equation $3x^2 + y^2 = 9$ of the ellipse in Example 1 defines at least two functions implicitly. Solving this equation for the variable y in terms of x yields

$$y^2 = 9 - 3x^2$$
$$y = \pm\sqrt{9 - 3x^2}.$$

Thus two explicit functions are

$$y = f(x) = \sqrt{9 - 3x^2} \qquad \text{and} \qquad y = g(x) = -\sqrt{9 - 3x^2}.$$

The x-intercepts of the graphs of f and g are the same, that is, $(-\sqrt{3}, 0)$ and $(\sqrt{3}, 0)$. The domain of each of these functions is the interval $[-\sqrt{3}, \sqrt{3}]$. The graphs of f and g are, in turn, the upper half-ellipse given in FIGURE 6.2.7(a) and the lower half-ellipse in Figure 6.2.7(b).

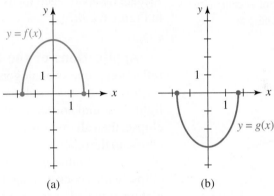

(a)   (b)

FIGURE 6.2.7 Graphs of functions in Example 4

To graph an ellipse on a calculator you may have to resort to superimposing the graphs of two half-ellipses defined by two functions in order to obtain the graph of the complete ellipse.

EXAMPLE 5   Equation of an Ellipse

Find an equation of the ellipse with center $(2, -1)$, vertical major axis of length 6, and minor axis of length 3.

Solution The length of the major axis is $2a = 6$; hence $a = 3$. Similarly, the length of the minor axis is $2b = 3$, so $b = \frac{3}{2}$. By sketching the center and the axes, we see from FIGURE 6.2.8 that the vertices are $(2, 2)$ and $(2, -4)$ and the endpoints of the minor axis are $(\frac{1}{2}, -1)$ and $(\frac{7}{2}, -1)$. Because the major axis is vertical, the standard form of the equation of this ellipse is

$$\frac{(x - 2)^2}{\left(\frac{3}{2}\right)^2} + \frac{(y - (-1))^2}{3^2} = 1 \qquad \text{or} \qquad \frac{(x - 2)^2}{\frac{9}{4}} + \frac{(y + 1)^2}{9} = 1.$$

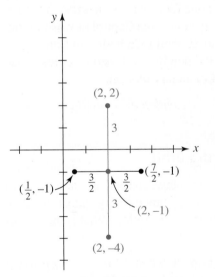

FIGURE 6.2.8 Graphical interpretation of data in Example 5

☐ **Eccentricity** Associated with each conic section is a number e called its **eccentricity**. The eccentricity of an ellipse is defined to be

$$e = \frac{c}{a},$$

where $c = \sqrt{a^2 - b^2}$. Since $0 < \sqrt{a^2 - b^2} < a$ implies $0 < \dfrac{\sqrt{a^2 - b^2}}{a} < 1$, the eccentricity of an ellipse satisfies $0 < e < 1$.

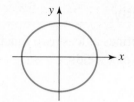

(a) *e* close to zero

(b) *e* close to 1

FIGURE 6.2.9 Effect of eccentricity on the shape of an ellipse

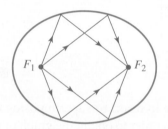

FIGURE 6.2.10 Reflection property of an ellipse

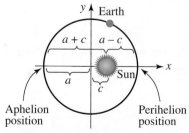

Aphelion position   Perihelion position

FIGURE 6.2.11 Graphical interpretation of data in Example 7

EXAMPLE 6   Example 3 Revisited

Determine the eccentricity of the ellipse in Example 3.

Solution In the solution of Example 3 we found that $a = 4$ and $c = 2\sqrt{3}$. Hence, the eccentricity of the ellipse is $e = (2\sqrt{3})/4 = \sqrt{3}/2 \approx 0.87$. ∎

Eccentricity is an indicator of the shape of an ellipse. When $e \approx 0$, that is, e is close to zero, the ellipse is nearly circular, and when $e \approx 1$ the ellipse is flattened or elongated. To see this, observe that if e is close to 0, it follows from $e = \sqrt{a^2 - b^2}/a$ that $c = \sqrt{a^2 - b^2} \approx 0$ and consequently $a \approx b$. As you can see from the standard equations in (4) and (5), this means that the shape of the ellipse is close to circular. Also, because c is the distance from the center of the ellipse to a focus, the two foci are close together near the center. See FIGURE 6.2.9(a). On the other hand, if $e \approx 1$ or $\sqrt{a^2 - b^2}/a \approx 1$, then $c = \sqrt{a^2 - b^2} \approx a$ and so $b \approx 0$. Also, $c \approx a$ means that the foci are far apart; each focus is close to a vertex. Thus, the ellipse is elongated as shown in Figure 6.2.9(b).

☐ **Applications of the Ellipse** Ellipses have a reflection property analogous to the one discussed in Section 6.1 for the parabola. It can be shown that if a light or sound source is placed at one focus of an ellipse, then all rays or waves will be reflected off the ellipse to the other focus. See FIGURE 6.2.10. For example, if a pool table is constructed in the form of an ellipse with a pocket at one focus, then any shot originating at the other focus will never miss the pocket.

Statuary Hall in Washington, DC

Similarly, if a ceiling is elliptical with two foci on (or near) the floor but considerably distant from each other, then anyone whispering at one focus will be heard at the other. Some famous "whispering galleries" are the Statuary Hall at the Capitol in Washington, DC, the Mormon Tabernacle in Salt Lake City, and St. Paul's Cathedral in London.

Using his Law of Universal Gravitation and the newly developed calculus, Isaac Newton was the first to prove Kepler's first law of planetary motion:

The orbit of each planet about the Sun is an ellipse with the Sun at one focus.

EXAMPLE 7   Eccentricity of Earth's Orbit

The perihelion distance of the Earth (the least distance between the Earth and the Sun) is approximately 9.16×10^7 miles, and its aphelion distance (the greatest distance between the Earth and the Sun) is approximately 9.46×10^7 miles. What is the eccentricity of Earth's orbit?

Solution Let us assume that the orbit of the Earth is as shown in FIGURE 6.2.11. From the figure we see that

$$a - c = 9.16 \times 10^7$$
$$a + c = 9.46 \times 10^7.$$

Solving this system of equations gives $a = 9.31 \times 10^7$ and $c = 0.15 \times 10^7$. Thus the eccentricity $e = c/a$ is

$$e = \frac{0.15 \times 10^7}{9.31 \times 10^7} \approx 0.016.$$

∎

The orbits of seven of the planets have eccentricities less than 0.1 and, hence, the orbits are not far from circular. The planet Mercury is the exception. The orbit of the dwarf planet Pluto has the eccentricity 0.25. Many of the asteroids and comets have highly eccentric orbits. The orbit of the asteroid Hildago is one of the most eccentric, with $e = 0.66$. Another notable case is the orbit of Comet Halley. See Problem 47 in Exercises 6.2.

NOTES FROM THE CLASSROOM

(a) Circle

(b) Ellipse

FIGURE 6.2.12 Lines and numbers

If you look up the definition of a **radius** of a circle you will likely find two definitions:

- A *line segment* from the center C of a circle to a point P on its circumference.
- The *distance* from the center C of a circle to a point P on its circumference.

Thus the word "radius" refers to either a *line* or a *number*. See FIGURE 6.2.12(a). So too, for an ellipse the terms **semimajor axis** and **semiminor axis** refer either to the *line segments* shown in Figure 6.2.12(b) or to the respective *lengths a* and *b* of these line segments. For example using the length interpretation of these terms, **Kepler's third law** of planetary motion is usually stated in the following manner:

The square of the orbital period of a planet is proportional to the cube of the semimajor axis of its orbit.

Clearly it is not the line segment that is cubed. So if T denotes the period or time it takes a planet (or any celestial body such as an asteroid) to make one complete orbit around the Sun and if its semimajor axis is a, then Kepler's third law becomes $T^2 = ka^3$, where k is a constant of proportionality.

Exercises 6.2   Answers to selected odd-numbered problems begin on page ANS–22.

In Problems 1–20, find the center, foci, vertices, endpoints of the minor axis, and eccentricity of the given ellipse. Graph the ellipse.

1. $\dfrac{x^2}{25} + \dfrac{y^2}{9} = 1$

2. $\dfrac{x^2}{16} + \dfrac{y^2}{4} = 1$

3. $x^2 + \dfrac{y^2}{16} = 1$

4. $\dfrac{x^2}{4} + \dfrac{y^2}{10} = 1$

5. $9x^2 + 16y^2 = 144$

6. $2x^2 + y^2 = 4$

7. $9x^2 + 4y^2 = 36$

8. $x^2 + 4y^2 = 4$

9. $\dfrac{(x-1)^2}{49} + \dfrac{(y-3)^2}{36} = 1$

10. $\dfrac{(x+1)^2}{25} + \dfrac{(y-2)^2}{36} = 1$

11. $(x+5)^2 + \dfrac{(y+2)^2}{16} = 1$

12. $\dfrac{(x-3)^2}{64} + \dfrac{(y+4)^2}{81} = 1$

13. $4x^2 + (y + \tfrac{1}{2})^2 = 4$

14. $36(x+2)^2 + (y-4)^2 = 72$

15. $5(x-1)^2 + 3(y+2)^2 = 45$

16. $6(x-2)^2 + 8y^2 = 48$

17. $25x^2 + 9y^2 - 100x + 18y - 116 = 0$

18. $9x^2 + 5y^2 + 18x - 10y - 31 = 0$

19. $x^2 + 3y^2 + 18y + 18 = 0$

20. $12x^2 + 4y^2 - 24x - 4y + 1 = 0$

In Problems 21–40, find an equation of the ellipse that satisfies the given conditions.

21. Vertices $(\pm 5, 0)$, foci $(\pm 3, 0)$

22. Vertices $(\pm 9, 0)$, foci $(\pm 2, 0)$

23. Vertices $(0, \pm 3)$, foci $(0, \pm 1)$

24. Vertices $(0, \pm 7)$, foci $(0, \pm 3)$

25. Vertices $(0, \pm 3)$, endpoints of minor axis $(\pm 1, 0)$

26. Vertices $(\pm 4, 0)$, endpoints of minor axis $(0, \pm 2)$

27. Vertices $(-3, -3)$, $(5, -3)$, endpoints of minor axis $(1, -1)$, $(1, -5)$

28. Vertices $(1, -6)$, $(1, 2)$, endpoints of minor axis $(-2, -2)$, $(4, -2)$

29. One focus $(0, -2)$, center at origin, $b = 3$

30. One focus $(1, 0)$, center at origin, $a = 3$

31. Foci $\left(\pm \sqrt{2}, 0\right)$, length of minor axis 6

32. Foci $\left(0, \pm \sqrt{5}\right)$, length of major axis 16

33. Foci $(0, \pm 3)$, passing through $\left(-1, 2\sqrt{2}\right)$

34. Vertices $(\pm 5, 0)$, passing through $\left(\sqrt{5}, 4\right)$

35. Vertices $(\pm 4, 1)$, passing through $\left(2\sqrt{3}, 2\right)$

36. Center $(1, -1)$, one focus $(1, 1)$, $a = 5$

37. Center $(1, 3)$, one focus $(1, 0)$, one vertex $(1, -1)$

38. Center $(5, -7)$, length of vertical major axis 8, length of minor axis 6

39. Endpoints of minor axis $(0, 5)$, $(0, -1)$, one focus $(6, 2)$

40. Endpoints of major axis $(2, 4)$, $(13, 4)$, one focus $(4, 4)$

In Problems 41–44, find a function f that defines the indicated half-ellipse. Give the domain of each function. The equations are from Problems 1, 3, 9, and 12.

41. $\dfrac{x^2}{25} + \dfrac{y^2}{9} = 1$; lower half-ellipse

42. $x^2 + \dfrac{y^2}{16} = 1$; upper half-ellipse

43. $\dfrac{(x-1)^2}{49} + \dfrac{(y-3)^2}{36} = 1$; upper half-ellipse

44. $\dfrac{(x-3)^2}{64} + \dfrac{(y+4)^2}{81} = 1$; lower half-ellipse

Applications

45. Mercury The orbit of the planet Mercury is an ellipse with the Sun at one focus. The length of the major axis of this orbit is 72 million miles and the length of the minor axis is 70.4 million miles. What is the least distance (perihelion) between Mercury and the Sun? What is the greatest distance (aphelion)?

46. What is the eccentricity of the orbit of Mercury in Problem 45?

47. Comet Halley The orbit of Comet Halley is an ellipse whose major axis is 3.34×10^9 miles long, and whose minor axis is 8.5×10^8 miles long. What is the eccentricity of the comet's orbit?

48. Earth Satellite A satellite orbits the Earth in an elliptical path with the center of the Earth at one focus. It has a minimum altitude of 200 miles and a maximum altitude of 1000 miles above the surface of the Earth. If the radius of the Earth is 4000 miles, what is an equation of the satellite's orbit?

FIGURE 6.2.13 Elliptical and circular gears in Problem 50

The Ellipse Park in Washington, DC

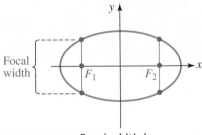

FIGURE 6.2.14 Focal width in Problem 54

49. Archway A semielliptical archway has a vertical major axis. The base of the arch is 10 ft across and the highest part of the arch is 15 ft. Find the height of the arch above the point on the base of the arch 3 ft from the center.

50. Gear Design An elliptical gear rotates about its center and is always kept in mesh with a circular gear that is free to move horizontally. See FIGURE 6.2.13. If the origin of the xy-coordinate system is placed at the center of the ellipse, then the equation of the ellipse in its present position is $x^2 + 3y^2 = 8$. The diameter of the circular gear equals the length of the minor axis of the elliptical gear. Given that the units are centimeters, how far does the center of the circular gear move horizontally during the rotation from one vertex of the elliptical gear to the next?

51. Carpentry A carpenter wishes to cut an elliptical top for a coffee table from a rectangular piece of wood that is 4 ft by 3 ft utilizing the entire length and width available. If the ellipse is to be drawn using the string and tack method illustrated in Figure 6.2.2, how long should the piece of string be and where should the tacks be placed?

52. Park Design The Ellipse is a park in Washington, DC. It is bounded by an elliptical path with a major axis of length 458 m and a minor axis of length 390 m. Find the distance between the foci of this ellipse.

53. Whispering Gallery Suppose that a room is constructed on a flat elliptical base by rotating a semiellipse $180°$ about its major axis. Then, by the reflection property of the ellipse, anything whispered at one focus will be distinctly heard at the other focus. If the height of the room is 16 ft and the length is 40 ft, find the location of the whispering and listening posts.

54. Focal Width The focal width of the ellipse is the length of a focal chord, that is, a line segment, perpendicular to the major axis, through a focus with endpoints on the ellipse. See FIGURE 6.2.14.
   (a) Find the focal width of the ellipse $x^2/9 + y^2/4 = 1$.
   (b) Show that, in general, the focal width of the ellipse $x^2/a^2 + y^2/b^2 = 1$ is $2b^2/a$.

55. Find an equation of the ellipse with foci $(0, 2)$ and $(8, 6)$ and fixed distance sum $2a = 12$. [*Hint*: Here the major axis is neither horizontal nor vertical; thus none of the standard forms from this section apply. Use the definition of the ellipse.]

56. Proceed as in Problem 55, and find an equation of the ellipse with foci $(-1, -3)$ and $(-5, 7)$ and fixed distance sum $2a = 20$.

For Discussion

57. The graph of the ellipse $x^2/4 + (y - 1)^2/9 = 1$ is shifted 4 units to the right. What are the center, foci, vertices, and endpoints of the minor axis for the shifted graph?

58. The graph of the ellipse $(x - 1)^2/9 + (y - 4)^2 = 1$ is shifted 5 units to the left and 3 units up. What are the center, foci, vertices, and endpoints of the minor axis for the shifted graph?

59. In engineering the eccentricity of an ellipse is often expressed only in terms of a and b. Show that $e = \sqrt{1 - b^2/a^2}$.

60. Look up the definition of **semi-ellipse**. Are the half-ellipses in Figure 6.2.7 semi-ellipses?

6.3 The Hyperbola

INTRODUCTION The definition of a hyperbola is basically the same as the definition of the ellipse with only one exception: the word *sum* is replaced by the word *difference*.

DEFINITION 6.3.1 Hyperbola

A **hyperbola** is the set of points $P(x, y)$ in the plane such that the difference of the distances between P and two fixed points F_1 and F_2 is constant. The fixed points F_1 and F_2 are called **foci** (plural for **focus**). The midpoint of the line segment joining points F_1 and F_2 is called the **center** of the hyperbola.

As shown in FIGURE 6.3.1, a hyperbola consists of two **branches**. If P is a point on the hyperbola, then

$$|d_1 - d_2| = k, \tag{1}$$

where $d_1 = d(F_1, P)$ and $d_2 = d(F_2, P)$.

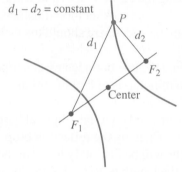

$d_1 - d_2 = $ constant

FIGURE 6.3.1 A hyperbola

☐ **Hyperbola with Center (0, 0)** Proceeding as for the ellipse, we place the foci on the x-axis at $F_1(-c, 0)$ and $F_2(c, 0)$ as shown in FIGURE 6.3.2 and choose the constant k to be $2a$ for algebraic convenience. It follows from (1) that

$$d_1 - d_2 = \pm 2a. \tag{2}$$

As drawn in Figure 6.3.2, P is on the right branch of the hyperbola and so $d_1 - d_2 = 2a > 0$. If P is on the left branch then the difference is $-2a$. Writing (2) as

$$\sqrt{(x + c)^2 + y^2} - \sqrt{(x - c)^2 + y^2} = \pm 2a$$

or

$$\sqrt{(x + c)^2 + y^2} = \pm 2a + \sqrt{(x - c)^2 + y^2}$$

we square, simplify, and square again:

$$(x + c)^2 + y^2 = 4a^2 \pm 4a\sqrt{(x - c)^2 + y^2} + (x - c)^2 + y^2$$

$$\pm a\sqrt{(x - c)^2 + y^2} = cx - a^2$$

$$a^2[(x - c)^2 + y^2] = c^2x^2 - 2a^2cx + a^4 \tag{3}$$

$$(c^2 - a^2)x^2 - a^2y^2 = a^2(c^2 - a^2).$$

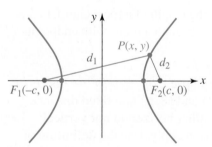

FIGURE 6.3.2 Hyperbola with center (0, 0) and foci on the x-axis

From Figure 6.3.2, we see that the triangle inequality gives

$$d_1 < d_2 + 2c \qquad \text{and} \qquad d_2 < d_1 + 2c,$$

or

$$d_1 - d_2 < 2c \qquad \text{and} \qquad d_2 - d_1 < 2c.$$

Using $d_1 - d_2 = \pm 2a$, the last two inequalities imply that $2a < 2c$ or $a < c$. Since $c > a > 0$, $c^2 - a^2$ is a positive constant. If we let $b^2 = c^2 - a^2$, (3) becomes $b^2x^2 - a^2y^2 = a^2b^2$ or, after dividing by a^2b^2,

$$\frac{x^2}{a^2} - \frac{y^2}{b^2} = 1. \tag{4}$$

Equation (4) is called the **standard form** of the equation of a hyperbola centered at (0, 0) with foci $(-c, 0)$ and $(c, 0)$, where c is defined by $b^2 = c^2 - a^2$.

When the foci lie on the y-axis, a repetition of the foregoing algebra leads to

$$\frac{y^2}{a^2} - \frac{x^2}{b^2} = 1. \tag{5}$$

Equation (5) is the **standard form** of the equation of a hyperbola centered at $(0, 0)$ with foci $(0, -c)$ and $(0, c)$. Here again, $c > a$ and $b^2 = c^2 - a^2$.

Note of Caution ▶

For the hyperbola (unlike the ellipse), bear in mind that in (4) and (5) there is no relationship between the relative sizes of a and b; rather, a^2 is always the denominator of the *positive term* and the intercepts *always* have $\pm a$ as a coordinate.

☐ **Transverse and Conjugate Axes** The line segment with endpoints on the hyperbola and lying on the line through the foci is called the **transverse axis**; its endpoints are called the **vertices** of the hyperbola. For the hyperbola described by equation (4), the transverse axis lies on the x-axis. Therefore, the coordinates of the vertices are the x-intercepts. Setting $y = 0$ gives $x^2/a^2 = 1$, or $x = \pm a$. Thus, as shown in FIGURE 6.3.3 the vertices are $(-a, 0)$ and $(a, 0)$; the **length of the transverse axis** is $2a$. Notice that by setting $y = 0$ in (4), we get $-y^2/b^2 = 1$ or $y^2 = -b^2$, which has no real solutions. Hence the graph of any equation in that form has no y-intercepts. Nonetheless, the numbers $\pm b$ are important. The line segment through the center of the hyperbola perpendicular to the transverse axis and with endpoints $(0, -b)$ and $(0, b)$ is called the **conjugate axis**. Similarly, the graph of an equation in standard form (5) has no x-intercepts. The conjugate axis for (5) is the line segment with endpoints $(-b, 0)$ and $(b, 0)$.

This information for equations (4) and (5) is summarized in Figure 6.3.3.

Pictorial Summary of Information for Standard Forms (4) and (5)

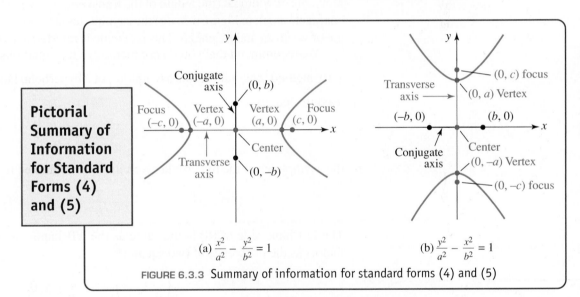

(a) $\dfrac{x^2}{a^2} - \dfrac{y^2}{b^2} = 1$

(b) $\dfrac{y^2}{a^2} - \dfrac{x^2}{b^2} = 1$

FIGURE 6.3.3 Summary of information for standard forms (4) and (5)

☐ **Asymptotes** Every hyperbola possesses a pair of slant asymptotes that pass through its center. These asymptotes are indicative of end behavior, and as such are an invaluable aid in sketching the graph of a hyperbola. Solving (4) for y in terms of x gives

$$y = \pm \frac{b}{a} x \sqrt{1 - \frac{a^2}{x^2}}.$$

As $x \to -\infty$ or as $x \to \infty$, $a^2/x^2 \to 0$, and thus $\sqrt{1 - a^2/x^2} \to 1$. Therefore, for large values of $|x|$, points on the graph of the hyperbola are close to the points on the lines

$$y = \frac{b}{a} x \qquad \text{and} \qquad y = -\frac{b}{a} x. \tag{6}$$

By a similar analysis we find that the slant asymptotes for (5) are

$$y = \frac{a}{b} x \qquad \text{and} \qquad y = -\frac{a}{b} x. \tag{7}$$

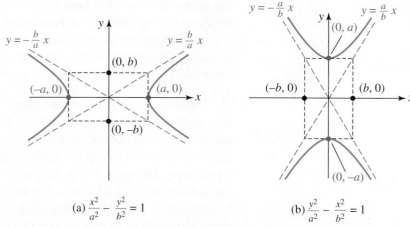

(a) $\dfrac{x^2}{a^2} - \dfrac{y^2}{b^2} = 1$ (b) $\dfrac{y^2}{a^2} - \dfrac{x^2}{b^2} = 1$

FIGURE 6.3.4 Hyperbolas (4) and (5) with slant asymptotes (in red) as the extended diagonals of the auxiliary rectangles (in black)

Each pair of asymptotes intersect at the origin, which is the center of the hyperbola. Note, too, in **FIGURE 6.3.4(a)** that the asymptotes are simply the *extended diagonals* of a rectangle of width $2a$ (the length of the transverse axis) and height $2b$ (the length of the conjugate axis); in Figure 6.3.4(b) the asymptotes are the extended diagonals of a rectangle of width $2b$ and height $2a$. This rectangle is referred to as the **auxiliary rectangle**.

We recommend that you *do not* memorize the equations in (6) and (7). There is an easy method for obtaining the asymptotes of a hyperbola. For example, since $y = \pm\dfrac{b}{a}x$ is equivalent to

$$\frac{x^2}{a^2} = \frac{y^2}{b^2}$$

the asymptotes of the hyperbola given in (4) are obtained from a single equation

$$\frac{x^2}{a^2} - \frac{y^2}{b^2} = 0. \tag{8}$$

The left-hand side of (8) is the same as the left-hand side of (4). Also, note that (8) factors as the difference of two squares:

$$\left(\frac{x}{a} - \frac{y}{b}\right)\left(\frac{x}{a} + \frac{y}{b}\right) = 0.$$

This is a mnemonic, or memory device. It ▶ has no geometric significance.

Setting each factor equal to zero and solving for y gives an equation of an asymptote. You do not even have to memorize (8) because it is simply the left-hand side of the standard form of the equation of a hyperbola given in (4). In like manner, to obtain the asymptotes for (5) just replace 1 by 0 in the standard form, factor $y^2/a^2 - x^2/b^2 = 0$, and solve for y.

EXAMPLE 1   Hyperbola Centered at $(0, 0)$

Find the vertices, foci, and asymptotes of the hyperbola $9x^2 - 25y^2 = 225$. Graph.

Solution We first put the equation into standard form by dividing the left-hand side by 225:

$$\frac{x^2}{25} - \frac{y^2}{9} = 1. \tag{9}$$

From this equation we see that $a^2 = 25$ and $b^2 = 9$, and so $a = 5$ and $b = 3$. Therefore the vertices are $(-5, 0)$ and $(5, 0)$. Since $b^2 = c^2 - a^2$ implies $c^2 = a^2 + b^2$, we have

FIGURE 6.3.5 Hyperbola in Example 1

$c^2 = 34$, and so the foci are $\left(-\sqrt{34}, 0\right)$ and $\left(\sqrt{34}, 0\right)$. To find the slant asymptotes we use the standard form (9) with 1 replaced by 0:

$$\frac{x^2}{25} - \frac{y^2}{9} = 0 \qquad \text{factors as} \qquad \left(\frac{x}{5} - \frac{y}{3}\right)\left(\frac{x}{5} + \frac{y}{3}\right) = 0.$$

Setting each factor equal to zero and solving for y gives the asymptotes $y = \pm 3x/5$. We plot the vertices and graph the two lines through the origin. Both branches of the hyperbola must become arbitrarily close to the asymptotes as $x \to \pm\infty$. See FIGURE 6.3.5. ■

■ EXAMPLE 2　　　Equation of a Hyperbola

Find an equation of the hyperbola with vertices $(0, -4)$, $(0, 4)$ and asymptotes $y = -\frac{1}{2}x$, $y = \frac{1}{2}x$.

Solution The center of the hyperbola is $(0, 0)$. This is revealed by the fact that the asymptotes intersect at the origin. Moreover, the vertices are on the y-axis and are 4 units on either side of the origin. Thus the equation we seek is of form (5). From (7) or Figure 6.3.4(b), the asymptotes must be of the form $y = \pm\dfrac{a}{b}x$ so that $a/b = 1/2$. From the given vertices we identify $a = 4$, and so

$$\frac{4}{b} = \frac{1}{2} \qquad \text{implies} \qquad b = 8.$$

The standard form of the equation of the hyperbola is then

$$\frac{y^2}{4^2} - \frac{x^2}{8^2} = 1 \qquad \text{or} \qquad \frac{y^2}{16} - \frac{x^2}{64} = 1. \qquad ■$$

☐ Hyperbola with Center (h, k)

When the center of the hyperbola is (h, k) the **standard form** analogues of equations (4) and (5) are, in turn,

$$\frac{(x - h)^2}{a^2} - \frac{(y - k)^2}{b^2} = 1 \tag{10}$$

and

$$\frac{(y - k)^2}{a^2} - \frac{(x - h)^2}{b^2} = 1. \tag{11}$$

As in (4) and (5), the numbers a^2, b^2, and c^2 are related by $b^2 = c^2 - a^2$.

You can locate vertices and foci using the fact that a is the distance from the center to a vertex and c is the distance from the center to a focus. The slant asymptotes for (10) can be obtained by factoring

$$\frac{(x - h)^2}{a^2} - \frac{(y - k)^2}{b^2} = 0$$

as

$$\left(\frac{x - h}{a} - \frac{y - k}{b}\right)\left(\frac{x - h}{a} + \frac{y - k}{b}\right) = 0.$$

Similarly, the asymptotes for (11) can be obtained from factoring

$$\frac{(y - k)^2}{a^2} - \frac{(x - h)^2}{b^2} = 0,$$

setting each factor equal to zero and solving for y in terms of x. As a check on your work, remember that (h, k) must be a point that lies on each asymptote.

6.3 The Hyperbola　　　**399**

EXAMPLE 3   **Hyperbola Centered at (*h, k*)**

Find the center, vertices, foci, and asymptotes of the hyperbola

$$4x^2 - y^2 - 8x - 4y - 4 = 0.$$

Graph.

Solution Before completing the square in *x* and *y*, we factor 4 from the two *x*-terms and factor -1 from the two *y*-terms so that the leading coefficient in each expression is 1. Then we have

$$4(x^2 - 2x \quad) + (-1)(y^2 + 4y \quad) = 4$$
$$4(x^2 - 2x + 1) - (y^2 + 4y + 4) = 4 + 4 \cdot 1 + (-1) \cdot 4$$
$$4(x - 1)^2 - (y + 2)^2 = 4$$
$$\frac{(x - 1)^2}{1} - \frac{(y + 2)^2}{4} = 1.$$

We see now that the center is $(1, -2)$. Since the term in the standard form involving *x* has the positive coefficient, the transverse axis is horizontal along the line $y = -2$, and we identify $a = 1$ and $b = 2$. The vertices are 1 unit to the left and to the right of the center at $(0, -2)$ and $(2, -2)$, respectively. From $b^2 = c^2 - a^2$, we have

$$c^2 = a^2 + b^2 = 1 + 4 = 5,$$

and so $c = \sqrt{5}$. Hence the foci are $\sqrt{5}$ units to the left and the right of the center $(1, -2)$ at $\left(1 - \sqrt{5}, -2\right)$ and $\left(1 + \sqrt{5}, -2\right)$.

To find the asymptotes, we solve

$$\frac{(x - 1)^2}{1} - \frac{(y + 2)^2}{4} = 0 \qquad \text{or} \qquad \left(x - 1 - \frac{y + 2}{2}\right)\left(x - 1 + \frac{y + 2}{2}\right) = 0$$

for *y*. From $y + 2 = \pm 2(x - 1)$ we find that the asymptotes are $y = -2x$ and $y = 2x - 4$. Observe that by substituting $x = 1$, both equations give $y = -2$, which means that both lines pass through the center. We then locate the center, plot the vertices, and graph the asymptotes. As shown in **FIGURE 6.3.6**, the graph of the hyperbola passes through the vertices and becomes closer and closer to the asymptotes as $x \to \pm\infty$. ∎

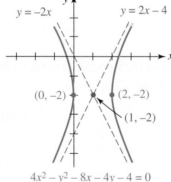

FIGURE 6.3.6 Hyperbola in Example 3

EXAMPLE 4   **Equation of a Hyperbola**

Find an equation of the hyperbola with center $(2, -3)$, passing through the point $(4, 1)$, and having one vertex $(2, 0)$.

Solution Since the distance from the center to one vertex is *a*, we have $a = 3$. From the location of the center and the vertex, it follows that the transverse axis is vertical and lies along the line $x = 2$. Therefore, the equation of the hyperbola must be of form (11):

$$\frac{(y + 3)^2}{3^2} - \frac{(x - 2)^2}{b^2} = 1, \tag{12}$$

(a) e close to 1

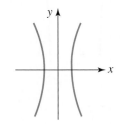

(b) e much greater than 1

FIGURE 6.3.7 Effect of eccentricity on the shape of a hyperbola

where b^2 is yet to be determined. Since the point $(4, 1)$ is on the graph on the hyperbola, its coordinates must satisfy equation (12). From

$$\frac{(1 + 3)^2}{3^2} - \frac{(4 - 2)^2}{b^2} = 1$$
$$\frac{16}{9} - \frac{4}{b^2} = 1$$
$$\frac{7}{9} = \frac{4}{b^2}$$

we find $b^2 = \frac{36}{7}$. We conclude that the desired equation is

$$\frac{(y + 3)^2}{3^2} - \frac{(x - 2)^2}{\frac{36}{7}} = 1.$$ ∎

☐ **Eccentricity** Like the ellipse, the equation that defines the **eccentricity** of a hyperbola is $e = c/a$. Except in this case the number c is given by $c = \sqrt{a^2 + b^2}$. Since $0 < a < \sqrt{a^2 + b^2}$, the eccentricity of a hyperbola satisfies $e > 1$. As with the ellipse, the magnitude of the eccentricity of a hyperbola is an indicator of its shape. FIGURE 6.3.7 shows examples of two extreme cases: $e \approx 1$ and e much greater than 1.

■ EXAMPLE 5   Eccentricity of a Hyperbola

Find the eccentricity of the hyperbola $\dfrac{(x - 2)^2}{2} - \dfrac{y^2}{36} = 1.$

Solution With the identifications $a^2 = 2$ and $b^2 = 36$, we get $c^2 = 2 + 36 = 38$. Thus the eccentricity of the given hyperbola is

$$e = \frac{c}{a} = \frac{\sqrt{38}}{\sqrt{2}} \approx 4.4.$$

We conclude that the hyperbola is one whose branches open widely as in Figure 6.3.7(b). ∎

■ EXAMPLE 6   Functions Defined Implicitly

Equations of the form given in (4), (5), (10), and (11) define at least two functions implicitly. Solving the equation $\dfrac{(x - 2)^2}{2} - \dfrac{y^2}{36} = 1$ of the hyperbola in Example 5 for y gives

$$\frac{y^2}{36} = \frac{(x - 2)^2}{2} - 1$$
$$y^2 = 18(x - 2)^2 - 36$$
$$y = \pm 3\sqrt{2}\sqrt{(x - 2)^2 - 2}.$$

Two functions are then

$$y = f(x) = 3\sqrt{2}\sqrt{(x - 2)^2 - 2} \quad \text{and} \quad y = g(x) = -3\sqrt{2}\sqrt{(x - 2)^2 - 2}.$$

The two x-intercepts of the graphs of f and g are the same, that is, $(2 - \sqrt{2}, 0)$ and $(2 + \sqrt{2}, 0)$. Also, the domain of f and of g is the union of intervals $(-\infty, 2 - \sqrt{2}] \cup [2 + \sqrt{2}, \infty)$. The graphs of f and g are, respectively, the two upper half-branches ($y \geq 0$) and the two lower half-branches ($y \leq 0$) of the hyperbola.

See FIGURES 6.3.8(a) and (b). When the graphs of f and g are plotted on the same rectangular coordinate system, we obtain the complete hyperbola with center $(2, 0)$ shown in Figure 6.3.8(c). The green dashed lines in the last figure are the asymptotes $y = 3\sqrt{2}(x - 2)$ and $y = -3\sqrt{2}(x - 2)$. As predicted from the eccentricity computed in Example 5, the branches of the hyperbola open widely so that they almost appear to be vertical lines.

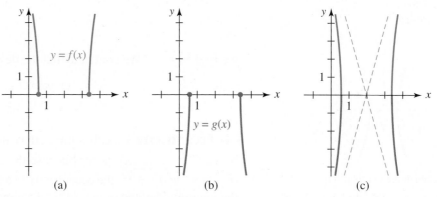

FIGURE 6.3.8 Graphs of functions in Example 6

☐ **Applications of the Hyperbola** The hyperbola has several important applications involving sounding techniques. In particular, several navigational systems utilize hyperbolas as follows. Two fixed radio transmitters at a known distance from each other transmit synchronized signals. The difference in reception times by a navigator determines the difference $2a$ of the distances from the navigator to the two transmitters. This information locates the navigator somewhere on the hyperbola with foci at the transmitters and fixed difference in distances from the foci equal to $2a$. By using two sets of signals obtained from a single master station paired with each of two secondary stations, the long-range navigation system LORAN locates a ship or plane at the intersection of two hyperbolas. See FIGURE 6.3.9.

FIGURE 6.3.9 The idea behind LORAN

The next example illustrates the use of a hyperbola in another situation involving sounding techniques.

EXAMPLE 7　　Locating a Big Blast

The sound of a dynamite blast is heard at different times by two observers at points A and B. Knowing that the speed of sound is approximately 1100 ft/s or 335 m/s, it is determined that the blast occurred 1000 meters closer to point A than to point B. If A and B are 2600 meters apart, show that the location of the blast lies on a branch of a hyperbola. Find an equation of the hyperbola.

Solution In FIGURE 6.3.10, we have placed the points A and B on the x-axis at $(1300, 0)$ and $(-1300, 0)$, respectively. If $P(x, y)$ denotes the location of the blast, then

$$d(P, B) - d(P, A) = 1000.$$

From the definition of the hyperbola on page 396 and the derivation following it, we see that this is the equation for the right branch of a hyperbola with fixed distance difference $2a = 1000$ and $c = 1300$. Thus the equation has the form

$$\frac{x^2}{a^2} - \frac{y^2}{b^2} = 1, \text{ where } x \geq 0,$$

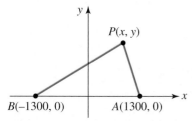

FIGURE 6.3.10 Graph for Example 7

or after solving for x,

$$x = a\sqrt{1 + \frac{y^2}{b^2}}.$$

With $a = 500$ and $c = 1300$, $b^2 = (1300)^2 - (500)^2 = (1200)^2$. Substituting in the foregoing equation gives

$$x = 500\sqrt{1 + \frac{y^2}{(1200)^2}} \qquad \text{or} \qquad x = \frac{5}{12}\sqrt{(1200)^2 + y^2}. \qquad \blacksquare$$

To find the exact location of the blast in Example 7 we would need another observer hearing the blast at a third point C. Knowing the time between when this observer hears the blast and when the observer at A hears the blast, we find a second hyperbola. The actual point of detonation is a point of intersection of the two hyperbolas.

There are many other applications of the hyperbola. As shown in FIGURE 6.3.11(a), a plane flying at a supersonic speed parallel to level ground leaves a hyperbolic sonic "footprint" on the ground. Like the parabola and ellipse, a hyperbola also possesses a reflection property. The Cassegrain reflecting telescope shown in Figure 6.3.11(b) utilizes a convex hyperbolic secondary mirror to reflect a ray of light back through a hole to an eyepiece (or camera) behind the parabolic primary mirror. This telescope construction makes use of the fact that a beam of light directed along a line through one focus of a hyperbolic mirror will be reflected on a line through the other focus. The most famous telescope in the world (or out of the world), the Hubble Space Telescope, is an example of a Cassegrain telescope. See the website:

http://hubblesite.org/the_telescope/hubble_essentials/

Orbits of objects in the universe can be parabolic, elliptic, or hyperbolic. When an object passes close to the Sun (or a planet), it is not necessarily captured by the gravitational field of the larger body. Under certain conditions, the object picks up a fractional amount of orbital energy of this much larger body and the resulting "slingshot-effect" orbit of the object as it passes the Sun is hyperbolic. See Figure 6.3.11(c).

Courtesy of NASA

Hubble Space Telescope

(a) Sonic footprint

(b) Cassegrain telescope

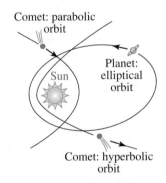

(c) Orbits around the Sun

FIGURE 6.3.11 Applications of hyperbolas

Exercises 6.3
Answers to selected odd-numbered problems begin on page ANS–23.

In Problems 1–20, find the center, foci, vertices, asymptotes, and eccentricity of the given hyperbola. Graph the hyperbola.

1. $\dfrac{x^2}{16} - \dfrac{y^2}{25} = 1$

2. $\dfrac{x^2}{4} - \dfrac{y^2}{4} = 1$

3. $\dfrac{y^2}{64} - \dfrac{x^2}{9} = 1$

4. $\dfrac{y^2}{6} - 4x^2 = 1$

5. $4x^2 - 16y^2 = 64$

6. $5x^2 - 5y^2 = 25$

7. $y^2 - 5x^2 = 20$

8. $9x^2 - 16y^2 + 144 = 0$

9. $\dfrac{(x-5)^2}{4} - \dfrac{(y+1)^2}{49} = 1$

10. $\dfrac{(x+2)^2}{10} - \dfrac{(y+4)^2}{25} = 1$

11. $\dfrac{(y-4)^2}{36} - x^2 = 1$

12. $\dfrac{\left(y-\frac{1}{4}\right)^2}{4} - \dfrac{(x+3)^2}{9} = 1$

13. $25(x-3)^2 - 5(y-1)^2 = 125$

14. $10(x+1)^2 - 2\left(y-\frac{1}{2}\right)^2 = 100$

15. $8(x+4)^2 - 5(y-7)^2 + 40 = 0$

16. $9(x-1)^2 - 81(y-2)^2 = 9$

17. $5x^2 - 6y^2 - 20x + 12y - 16 = 0$

18. $16x^2 - 25y^2 - 256x - 150y + 399 = 0$

19. $4x^2 - y^2 - 8x + 6y - 4 = 0$

20. $2y^2 - 9x^2 - 18x + 20y + 5 = 0$

In Problems 21–44, find an equation of the hyperbola that satisfies the given conditions.

21. Foci $(\pm5, 0)$, $a = 3$

22. Foci $(\pm10, 0)$, $b = 2$

23. Foci $(0, \pm4)$, one vertex $(0, -2)$

24. Foci $(0, \pm3)$, one vertex $\left(0, -\frac{3}{2}\right)$

25. Foci $(\pm4, 0)$, length of transverse axis 6

26. Foci $(0, \pm7)$, length of transverse axis 10

27. Center $(0, 0)$, one vertex $\left(0, \frac{5}{2}\right)$, one focus $(0, -3)$

28. Center $(0, 0)$, one vertex $(7, 0)$, one focus $(9, 0)$

29. Center $(0, 0)$, one vertex $(-2, 0)$, one focus $(-3, 0)$

30. Center $(0, 0)$, one vertex $(1, 0)$, one focus $(5, 0)$

31. Vertices $(0, \pm8)$, asymptotes $y = \pm2x$

32. Foci $(0, \pm3)$, asymptotes $y = \pm\frac{3}{2}x$

33. Vertices $(\pm2, 0)$, asymptotes $y = \pm\frac{4}{3}x$

34. Foci $(\pm5, 0)$, asymptotes $y = \pm\frac{3}{5}x$

35. Center $(1, -3)$, one focus $(1, -6)$, one vertex $(1, -5)$

36. Center $(2, 3)$, one focus $(0, 3)$, one vertex $(3, 3)$

37. Foci $(-4, 2)$, $(2, 2)$, one vertex $(-3, 2)$

38. Vertices $(2, 5)$, $(2, -1)$, one focus $(2, 7)$

39. Vertices $(\pm2, 0)$, passing through $\left(2\sqrt{3}, 4\right)$

40. Vertices $(0, \pm3)$, passing through $\left(\frac{16}{5}, 5\right)$

41. Center $(-1, 3)$, one vertex $(-1, 4)$, passing through $\left(-5, 3 + \sqrt{5}\right)$

42. Center $(3, -5)$, one vertex $(3, -2)$, passing through $(1, -1)$

43. Center $(2, 4)$, one vertex $(2, 5)$, one asymptote $2y - x - 6 = 0$

44. Eccentricity $\sqrt{10}$, endpoints of conjugate axis $(-5, 4)$, $(-5, 10)$

In Problems 45–48, find two functions f and g defined implicitly by the given equation. Give the domain of each function. Describe the graph of each function in words. The equations are from Problems 1, 3, 11, and 16.

45. $\dfrac{x^2}{16} - \dfrac{y^2}{25} = 1$

46. $\dfrac{y^2}{64} - \dfrac{x^2}{9} = 1$

47. $\dfrac{(y-4)^2}{36} - x^2 = 1$

48. $9(x-1)^2 - 81(y-2)^2 = 9$

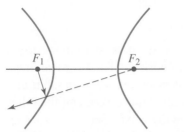

FIGURE 6.3.12 Reflecting property in Problem 50

FIGURE 6.3.13 Focal width in Problem 52

FIGURE 6.3.14 Sonic detectors in Problem 53

49. Three points are located at $A(-10, 16)$, $B(-2, 0)$, and $C(2, 0)$, where the units are kilometers. An artillery gun is known to lie on the line segment between A and C, and using sounding techniques it is determined that the gun is 2 km closer to B than to C. Find the point where the gun is located.

50. It can be shown that a ray of light emanating from one focus of a hyperbola will be reflected back along the line from the opposite focus. See FIGURE 6.3.12. A light ray from the left focus of the hyperbola $x^2/16 - y^2/20 = 1$ strikes the hyperbola at $(-6, -5)$. Find an equation of the reflected ray.

51. Find an equation of the hyperbola with foci $(0, -2)$ and $(8, 4)$ and fixed distance difference $2a = 8$. [*Hint*: See Problem 55 in Exercises 6.2.]

52. Focal Width The focal width of a hyperbola is the length of a focal chord, that is, a line segment, perpendicular to the line containing the transverse axis and through a focus, with endpoints on the hyperbola. See FIGURE 6.3.13.
(a) Find the focal width of the hyperbola $x^2/4 - y^2/9 = 1$.
(b) Show that, in general, the focal width of the hyperbola $x^2/a^2 - y^2/b^2 = 1$ is $2b^2/a$.

For Discussion

53. Sub Hunting Two sonar detectors are located at a distance d from one another. Suppose that a sound (such as a sneeze aboard a submarine) is heard at the two detectors with a time delay h between them. See FIGURE 6.3.14. Assume that sound travels in straight lines to the two detectors with speed v.
(a) Explain why h cannot be larger than d/v.
(b) Explain why, for given values of d, v, and h, the source of the sound can be determined to lie on one branch of a hyperbola. [*Hint*: Where do you suppose that the foci might be?]
(c) Find an equation of the hyperbola in part (b), assuming that the detectors are at the points $(0, d/2)$ and $(0, -d/2)$. Express the answer in the standard form $y^2/a^2 - x^2/b^2 = 1$.

54. Conjugate Hyperbolas The hyperbolas

$$\frac{x^2}{a^2} - \frac{y^2}{b^2} = 1 \qquad \text{and} \qquad \frac{y^2}{b^2} - \frac{x^2}{a^2} = 1$$

are said to be conjugates of each other.
(a) Find an equation of the hyperbola that is conjugate to

$$\frac{x^2}{25} - \frac{y^2}{144} = 1.$$

(b) Discuss how the graphs of conjugate hyperbolas are related.

55. Rectangular Hyperbola A rectangular hyperbola is one for which the asymptotes are perpendicular.
(a) Show that $y^2 - x^2 + 5y + 3x = 1$ is a rectangular hyperbola.
(b) Which of the hyperbolas given in Problems 1–20 are rectangular?

56. Suppose a hyperbola is rectangular. See Problem 55.
(a) How are the constants a and b related?
(b) Show that all rectangular hyperbolas have the same eccentricity.

6.4 The Polar Coordinate System

INTRODUCTION So far we have used the rectangular coordinate system to specify a point P in the plane. We can regard this system as a grid of horizontal and vertical lines. The coordinates (a, b) of a point P are determined by the intersection of two lines: one line $x = a$ is perpendicular to the horizontal reference line called the x-axis, and the other $y = b$ is perpendicular to the vertical reference line called the y-axis. See FIGURE 6.4.1(a). Another system for locating points in the plane is the **polar coordinate system**.

☐ **Terminology** To set up a **polar coordinate system**, we use a system of circles centered at a point O, called the **pole**, and straight lines or rays emanating from O. We take as a reference axis a horizontal half-line directed to the right of the pole and call it the **polar axis**. By specifying a directed (signed) distance r from O and an angle θ whose initial side is the polar axis and whose terminal side is the ray OP, we label the point P by (r, θ). We say that the ordered pair (r, θ) are the **polar coordinates** of P. See Figures 6.4.1(b) and 6.4.1(c) where we have assumed $r > 0$.

(a) Rectangular coordinate system

(b) Polar coordinate system

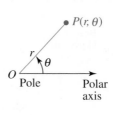

(c) Polar coordinates of P

FIGURE 6.4.1 Comparison of rectangular and polar coordinates of a point P

Although the measure of the angle θ can be either in degrees or radians, in calculus radian measure is used almost exclusively. Consequently, we shall use only radian measure in this discussion.

In the polar coordinate system we adopt the following conventions.

DEFINITION 6.4.1 Conventions in Polar Coordinates

(*i*) Angles $\theta > 0$ are measured counterclockwise from the polar axis, whereas angles $\theta < 0$ are measured clockwise.

(*ii*) To graph a point (r, θ), where $r < 0$, measure $|r|$ units along the ray $\theta + \pi$.

(*iii*) The coordinates of the pole O are $(0, \theta)$, where θ is any angle.

■ EXAMPLE 1   **Plotting Polar Points**

Plot the points whose polar coordinates are given.
(a) $(4, \pi/6)$ **(b)** $(2, -\pi/4)$ **(c)** $(-3, 3\pi/4)$

Solution (a) Measure 4 units along the ray $\pi/6$ as shown in FIGURE 6.4.2(a).

(b) Measure 2 units along the ray $-\pi/4$. See Figure 6.4.2(b).

(c) Measure 3 units along the ray $3\pi/4 + \pi = 7\pi/4$. Equivalently, we can measure 3 units along the ray $3\pi/4$ extended *backward* through the pole. Note carefully in Figure 6.4.2(c) that the point $(-3, 3\pi/4)$ is not in the same quadrant as the terminal side of the given angle.

(a)   (b)   (c)

FIGURE 6.4.2 Points in polar coordinates in Example 1

In contrast to the rectangular coordinate system, the description of a point in polar coordinates is not unique. This is an immediate consequence of the fact that

$$(r, \theta) \quad \text{and} \quad (r, \theta + 2n\pi), \quad n \text{ an integer},$$

are equivalent. To compound the problem, negative values of r can be used.

EXAMPLE 2   Equivalent Polar Points

The following polar coordinates are some alternative representations of the point $(2, \pi/6)$:

$$(2, 13\pi/6), \quad (2, -11\pi/6), \quad (-2, 7\pi/6), \quad (-2, -5\pi/6).$$

FIGURE 6.4.3 Relating polar and rectangular coordinates

☐ **Conversion of Polar Coordinates to Rectangular** By superimposing a rectangular coordinate system on a polar coordinate system, as shown in FIGURE 6.4.3, we can convert a polar description of a point to rectangular coordinates by using

$$x = r\cos\theta, \quad y = r\sin\theta. \tag{1}$$

These conversion formulas hold true for any values of r and θ in an equivalent polar representation of (r, θ).

EXAMPLE 3   Polar Coordinates to Rectangular

Convert $(2, \pi/6)$ in polar coordinates to rectangular coordinates.

Solution With $r = 2$, $\theta = \pi/6$, we have from (1),

$$x = 2\cos\frac{\pi}{6} = 2\left(\frac{\sqrt{3}}{2}\right) = \sqrt{3}$$

$$y = 2\sin\frac{\pi}{6} = 2\left(\frac{1}{2}\right) = 1.$$

Thus, $(2, \pi/6)$ is equivalent to $\left(\sqrt{3}, 1\right)$ in rectangular coordinates.

☐ **Conversion of Rectangular Coordinates to Polar** It should be evident from Figure 6.4.3 that x, y, r, and θ are also related by

$$r^2 = x^2 + y^2, \quad \tan\theta = \frac{y}{x}. \tag{2}$$

The equations in (2) are used to convert the rectangular coordinates (x, y) to the polar coordinates (r, θ).

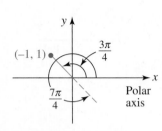

FIGURE 6.4.4 Point in Example 4

EXAMPLE 4   Rectangular Coordinates to Polar

Convert $(-1, 1)$ in rectangular coordinates to polar coordinates.

Solution With $x = -1$, $y = 1$, we have from (2)

$$r^2 = 2 \quad \text{and} \quad \tan\theta = -1.$$

Now, $r^2 = 2$ or $r = \pm\sqrt{2}$, and two of many angles that satisfy $\tan\theta = -1$ are $3\pi/4$ and $7\pi/4$. From FIGURE 6.4.4 we see that two polar representations for $(-1, 1)$ are $\left(\sqrt{2}, 3\pi/4\right)$ and $\left(-\sqrt{2}, 7\pi/4\right)$. ▄

In Example 4, observe that we cannot pair just *any* angle θ and *any* value r that satisfy (2); these solutions must also be consistent with (1). Because the points $\left(-\sqrt{2}, 3\pi/4\right)$ and $\left(\sqrt{2}, 7\pi/4\right)$ lie in the fourth quadrant, they are not polar representations of the second-quadrant point $(-1, 1)$.

There are instances in calculus when a rectangular equation must be expressed as a polar equation $r = f(\theta)$. The next example shows how to do this using the conversion formulas in (1).

EXAMPLE 5   Rectangular Equation to Polar Equation

Find a polar equation that has the same graph as the circle $x^2 + y^2 = 8x$.

Solution Substituting $x = r\cos\theta$, $y = r\sin\theta$, into the given equation we find

$$\begin{aligned} r^2\cos^2\theta + r^2\sin^2\theta &= 8r\cos\theta \\ r^2(\cos^2\theta + \sin^2\theta) &= 8r\cos\theta \quad \leftarrow \cos^2\theta + \sin^2\theta = 1 \\ r(r - 8\cos\theta) &= 0. \end{aligned}$$

The last equation implies that

$$r = 0 \quad \text{or} \quad r = 8\cos\theta.$$

Since $r = 0$ determines only the pole O, we conclude that a polar equation of the circle is $r = 8\cos\theta$. Note that the circle $x^2 + y^2 = 8x$ passes through the origin since $x = 0$ and $y = 0$ satisfy the equation. Relative to the polar equation $r = 8\cos\theta$ of the circle, the origin or pole corresponds to the polar coordinates $(0, \pi/2)$. ▄

EXAMPLE 6   Rectangular Equation to Polar Equation

Find a polar equation that has the same graph as the parabola $x^2 = 8(2 - y)$.

Solution We replace x and y in the given equation by $x = r\cos\theta$, $y = r\sin\theta$ and solve for r in terms of θ:

$$\begin{aligned} r^2\cos^2\theta &= 8(2 - r\sin\theta) \\ r^2(1 - \sin^2\theta) &= 16 - 8r\sin\theta \\ r^2 &= r^2\sin^2\theta - 8r\sin\theta + 16 \quad \leftarrow \left\{\begin{array}{l}\text{right side is a}\\ \text{perfect square}\end{array}\right. \\ r^2 &= (r\sin\theta - 4)^2 \\ r &= \pm(r\sin\theta - 4). \end{aligned}$$

Solving for r gives two equations,

$$r = \frac{4}{1 + \sin\theta} \quad \text{or} \quad r = \frac{-4}{1 - \sin\theta}.$$

Now recall that, by convention (*ii*) in Definition 6.4.1, (r, θ) and $(-r, \theta + \pi)$ represent the same point. You should verify that if (r, θ) is replaced by $(-r, \theta + \pi)$ in the second of these two equations, we obtain the first equation. In other words, the equations are equivalent and so we may simply take the polar equation of the parabola to be $r = 4/(1 + \sin\theta)$. ▄

EXAMPLE 7   Polar Equation to Rectangular Equation

Find a rectangular equation that has the same graph as the polar equation $r^2 = 9\cos 2\theta$.

Solution First, we use the trigonometric identity for the cosine of a double angle:

$$r^2 = 9(\cos^2\theta - \sin^2\theta). \quad \leftarrow \cos 2\theta = \cos^2\theta - \sin^2\theta$$

Then, from $r^2 = x^2 + y^2$, $\cos\theta = x/r$, $\sin\theta = y/r$, we have

$$x^2 + y^2 = 9\left(\frac{x^2}{x^2 + y^2} - \frac{y^2}{x^2 + y^2}\right) \quad \text{or} \quad (x^2 + y^2)^2 = 9(x^2 - y^2). \quad ▄$$

The next section will be devoted to graphing polar equations.

Exercises 6.4   Answers to selected odd-numbered problems begin on page ANS–24.

In Problems 1–6, plot the point with the given polar coordinates.

1. $(3, \pi)$
2. $(2, -\pi/2)$
3. $\left(-\frac{1}{2}, \pi/2\right)$
4. $(-1, \pi/6)$
5. $(-4, -\pi/6)$
6. $\left(\frac{2}{3}, 7\pi/4\right)$

In Problems 7–14, find alternative polar coordinates that satisfy

(a) $r > 0, \theta < 0$
(b) $r > 0, \theta > 2\pi$
(c) $r < 0, \theta > 0$
(d) $r < 0, \theta < 0$

for each point with the given polar coordinates.

7. $(2, 3\pi/4)$
8. $(5, \pi/2)$
9. $(4, \pi/3)$
10. $(3, \pi/4)$
11. $(1, \pi/6)$
12. $(3, 7\pi/6)$
13. $(9, 3\pi/2)$
14. $(5, \pi)$

In Problems 15–24, find the rectangular coordinates for each point with the given polar coordinates.

15. $\left(\frac{1}{2}, 2\pi/3\right)$
16. $(-1, 7\pi/4)$
17. $(-6, -\pi/3)$
18. $\left(\sqrt{2}, 11\pi/6\right)$
19. $(4, 5\pi/4)$
20. $(-5, \pi/2)$
21. $(-1, -5\pi/6)$
22. $(10, -4\pi/3)$
23. $(4, \pi/8)$
24. $(-8, 5\pi/12)$

In Problems 25–32, find polar coordinates that satisfy

(a) $r > 0$, $-\pi < \theta \le \pi$ (b) $r < 0$, $-\pi < \theta \le \pi$

for each point with the given rectangular coordinates.

25. $(-2, -2)$ 26. $(0, -4)$
27. $(1, -\sqrt{3})$ 28. $(\sqrt{6}, \sqrt{2})$
29. $(7, 0)$ 30. $(1, 2)$
31. $(-3, 4)$ 32. $(1, -1)$

In Problems 33–38, sketch the region on the plane that consists of points (r, θ) whose polar coordinates satisfy the given conditions.

33. $2 \le r < 4$, $0 \le \theta \le \pi$ 34. $2 < r \le 4$
35. $0 \le r \le 2$, $-\pi/2 \le \theta \le \pi/2$ 36. $r \ge 0$, $\pi/4 < \theta < 3\pi/4$
37. $-1 \le r \le 1$, $0 \le \theta \le \pi/2$ 38. $-2 \le r < 4$, $\pi/3 \le \theta \le \pi$

In Problems 39–50, find a polar equation that has the same graph as the given rectangular equation.

39. $y = 5$ 40. $x + 1 = 0$
41. $y = 7x$ 42. $3x + 8y + 6 = 0$
43. $y^2 = -4x + 4$ 44. $x^2 - 12y - 36 = 0$
45. $x^2 + y^2 = 36$ 46. $x^2 - y^2 = 1$
47. $x^2 + y^2 + x = \sqrt{x^2 + y^2}$ 48. $x^3 + y^3 - xy = 0$
49. $x^2 + y^2 = 5y$ 50. $2xy = 5$

In Problems 51–62, find a rectangular equation that has the same graph as the given polar equation.

51. $r = 2\sec\theta$ 52. $r\cos\theta = -4$
53. $r = 6\sin 2\theta$ 54. $2r = \tan\theta$
55. $r^2 = 4\sin 2\theta$ 56. $r^2\cos 2\theta = 16$
57. $r + 5\sin\theta = 0$ 58. $r = 2 + \cos\theta$

59. $r = \dfrac{2}{1 + 3\cos\theta}$ 60. $r(4 - \sin\theta) = 10$

61. $r = \dfrac{5}{3\cos\theta + 8\sin\theta}$ 62. $r = 3 + 3\sec\theta$

For Discussion

63. How would you express the distance d between two points (r_1, θ_1) and (r_2, θ_2) in terms of their polar coordinates?
64. You know how to find a rectangular equation of a line through two points with rectangular coordinates. How would you find a polar equation of a line through two points with polar coordinates (r_1, θ_1) and (r_2, θ_2)? Carry out your ideas by finding a polar equation of the line through $(3, 3\pi/4)$ and $(1, \pi/4)$. Find the polar coordinates of the x- and y-intercepts of the line.
65. In rectangular coordinates the x-intercepts of the graph of a function $y = f(x)$ are determined from the solutions of the equation $f(x) = 0$. In the next section we will graph polar equations $r = f(\theta)$. What is the significance of the solutions of the equation $f(\theta) = 0$?

6.5 Graphs of Polar Equations

INTRODUCTION The graph of a polar equation $r = f(\theta)$ is the set of points P with *at least* one set of polar coordinates that satisfies the equation. Since it is most likely that your classroom does not have a polar coordinate grid, to facilitate graphing and discussion of graphs of a polar equation $r = f(\theta)$, we will, as in the preceding section, superimpose a rectangular coordinate system over the polar coordinate system.

We begin with some simple polar graphs.

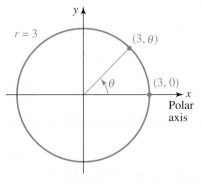

FIGURE 6.5.1 Circle in Example 1

EXAMPLE 1 A Circle Centered at the Origin

Graph $r = 3$.

Solution Since θ is not specified, the point $(3, \theta)$ lies on the graph of $r = 3$ for any value of θ and is 3 units from the origin. We see in **FIGURE 6.5.1** that the graph is the circle of radius 3 centered at the origin.

Alternatively, we know from (2) of Section 6.4 that $r = \pm\sqrt{x^2 + y^2}$ so that $r = 3$ yields the familiar rectangular equation $x^2 + y^2 = 3^2$ of a circle of radius 3 centered at the origin.

□ **Circles Centered at the Origin** In general, if a is any nonzero constant, the polar graph of

$$r = a \tag{1}$$

is a circle of radius $|a|$ with center at the origin.

EXAMPLE 2 A Ray Through the Origin

Graph $\theta = \pi/4$.

Solution Since r is not specified, the point $(r, \pi/4)$ lies on the graph for any value of r. If $r > 0$, then this point lies on the half-line in the first quadrant; if $r < 0$, then the point lies on the half-line in the third quadrant. For $r = 0$, the point $(0, \pi/4)$ is the pole or origin. Therefore, the polar graph of $\theta = \pi/4$ is the line through the origin that makes an angle of $\pi/4$ with the polar axis or positive x-axis. See **FIGURE 6.5.2**.

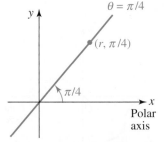

FIGURE 6.5.2 Line in Example 2

□ **Lines Through the Origin** In general, if α is any real constant, the polar graph of

$$\theta = \alpha \tag{2}$$

is a line through the origin that makes an angle of α radians with the polar axis. Lines described by (2) are called **radial lines**.

EXAMPLE 3 A Spiral

Graph $r = \theta$ where θ is measured in radians.

Solution As $\theta \geq 0$ increases, r increases and the points (r, θ) wind around the pole in a counterclockwise manner. This is illustrated by the blue portion of the graph in **FIGURE 6.5.3**. The red portion of the graph is obtained by plotting points for $\theta < 0$.

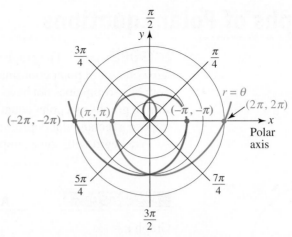

FIGURE 6.5.3 Graph of equation in Example 3

☐ **Spirals** Many graphs in polar coordinates are given special names. The graph in Example 3 is a special case of

$$r = a\theta, \tag{3}$$

where a is a constant. A graph of this equation is called a **spiral of Archimedes**. You are asked to graph other types of spiral curves in Problems 31 and 32 in Exercises 6.5.

In addition to basic point plotting, symmetry can often be utilized to graph a polar equation.

☐ **Symmetry** As shown in FIGURE 6.5.4, a polar graph can have three types of symmetry. A polar graph is **symmetric with respect to the y-axis** if whenever (r, θ) is a point on the graph, $(r, \pi - \theta)$ is also a point on the graph. A polar graph is **symmetric with respect to the x-axis** if whenever (r, θ) is a point on the graph, $(r, -\theta)$ is also a point on the graph. Finally, a polar graph is **symmetric with respect to the origin** if whenever (r, θ) is on the graph, $(-r, \theta)$ is also a point on the graph.

Symmetries of a snowflake

(a) Symmetry with respect to y-axis

(b) Symmetry with respect to x-axis

(c) Symmetry with respect to origin

FIGURE 6.5.4 Symmetries of a polar graph

We have the following tests for symmetries in polar coordinates.

THEOREM 6.5.1 Tests for Symmetry

The graph of a polar equation is:

   (*i*) **symmetric with respect to the y-axis** if replacing (r, θ) by $(r, \pi - \theta)$ results in an equivalent equation

   (*ii*) **symmetric with respect to the x-axis** if replacing (r, θ) by $(r, -\theta)$ results in an equivalent equation

   (*iii*) **symmetric with respect to the origin** if replacing (r, θ) by $(-r, \theta)$ results in an equivalent equation

---

In rectangular coordinates the description of a point is unique. Hence, in rectangular coordinates if a test for a particular type of symmetry fails, then we can definitely say that the graph does not possess that symmetry.

Because the polar description of a point is not unique, the graph of a polar equation may still have a particular type of symmetry even though the test for it fails. For example, observe that the graph of the $r = \theta$ in Figure 6.5.3 possesses symmetry with respect to the y-axis yet replacing (r, θ) by $(r, \pi - \theta)$ in the polar equation does not result in the same equation. Therefore, if one of the replacement tests in (*i*)–(*iii*) in Theorem 6.5.1 fails to yield the same polar equation, the best we can say is "no conclusion." Stated another way, the tests are *sufficient* for demonstrating a symmetry but are not *necessary*. See Problems 57–60 in Exercises 6.5.

EXAMPLE 4   Graphing a Polar Equation

Graph $r = 1 - \cos\theta$.

Solution One way of graphing this equation is to plot a few well-chosen points corresponding to $0 \le \theta \le 2\pi$. As the following table shows

θ	0	$\pi/4$	$\pi/2$	$3\pi/4$	π	$5\pi/4$	$3\pi/2$	$7\pi/4$	2π
r	0	0.29	1	1.71	2	1.71	1	0.29	0

as θ advances from $\theta = 0$ to $\theta = \pi/2$, r increases from $r = 0$ (the origin) to $r = 1$. See FIGURE 6.5.5(a). As θ advances from $\theta = \pi/2$ to $\theta = \pi$, r continues to increase from $r = 1$ to its maximum value of $r = 2$. See Figure 6.5.5(b). Then, for $\theta = \pi$ to $\theta = 3\pi/2$, r begins to decrease from $r = 2$ to $r = 1$. For $\theta = 3\pi/2$ to $\theta = 2\pi$, r continues to decrease and we end up again at the origin $r = 0$. See Figures 6.5.5(c) and 6.5.5(d).

By taking advantage of symmetry we could have simply plotted points for $0 \le \theta \le \pi$. From the trigonometric identity for the cosine function $\cos(-\theta) = \cos\theta$ it follows from (*ii*) of Theorem 6.5.1 that the graph of $r = 1 - \cos\theta$ is symmetric with respect to the x-axis. We can obtain the complete graph of $r = 1 - \cos\theta$ by reflecting in the x-axis that portion of the graph given in Figure 6.5.5(b).

(a)   (b)   (c)   (d)

FIGURE 6.5.5 Graph of equation in Example 4

☐ **Cardioids** The polar equation in Example 4 is a member of a family of equations that all have a "heart-shaped" graph that passes through the origin. A graph of any polar equation of the form

$$r = a \pm a\sin\theta \qquad \text{or} \qquad r = a \pm a\cos\theta \qquad (4)$$

is called a **cardioid**. The only difference in the graph of these four equations is their symmetry with respect to the y-axis ($r = a \pm a\sin\theta$) or symmetry with respect to the x-axis ($r = a \pm a\cos\theta$). See FIGURE 6.5.6 on page 414.

By knowing the basic shape and orientation of a cardioid, you can obtain a quick and accurate graph by plotting the four points corresponding to $\theta = 0$, $\theta = \pi/2$, $\theta = \pi$, and $\theta = 3\pi/2$.

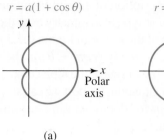

$r = a(1 + \cos\theta)$

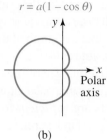

$r = a(1 - \cos\theta)$

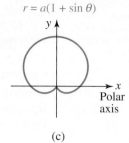

$r = a(1 + \sin\theta)$

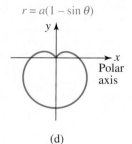

$r = a(1 - \sin\theta)$

(a)　　　　　(b)　　　　　(c)　　　　　(d)

FIGURE 6.5.6 Cardioids

☐ **Limaçons** Cardioids are special cases of polar curves known as **limaçons**:

$$r = a \pm b\sin\theta \quad \text{or} \quad r = a \pm b\cos\theta. \quad (5)$$

The shape of a limaçon depends on the relative magnitudes of a and b. Let us assume that $a > 0$ and $b > 0$. For $a/b < 1$, we get a **limaçon with an interior loop** as shown in FIGURE 6.5.7(a). When $a = b$ or equivalently $a/b = 1$ we get a **cardioid**. For $1 < a/b < 2$, we get a **dimpled limaçon** as shown in Figure 6.5.7(b). For $a/b \geq 2$, the curve is called a **convex limaçon**. See Figure 6.5.7(c).

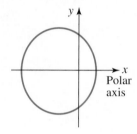

(a) Limaçon with interior loop ($a/b < 1$)　　(b) Dimpled limaçon ($1 < a/b < 2$)　　(c) Convex limaçon ($a/b \geq 2$)

FIGURE 6.5.7 Three kinds of limaçons

EXAMPLE 5　　A Limaçon

The graph of $r = 3 - \sin\theta$ is a convex limaçon, since $a = 3, b = 1$, and $a/b = 3 > 2$. ∎

EXAMPLE 6　　A Limaçon

The graph of $r = 1 + 2\cos\theta$ is a limaçon with an interior loop, since $a = 1, b = 2$, and $a/b = \frac{1}{2} < 1$. For $\theta \geq 0$, notice in FIGURE 6.5.8 the limaçon starts at $\theta = 0$ or $(3, 0)$. The graph passes through the y-axis at $(1, \pi/2)$ and then enters the origin $(r = 0)$ for the first angle for which $r = 0$ or $1 + 2\cos\theta = 0$ or $\cos\theta = -\frac{1}{2}$. This implies that $\theta = 2\pi/3$. At $\theta = \pi$, the curve passes through $(-1, \pi)$. The remainder of the graph can then be completed using the fact that it is symmetric with respect to the x-axis. ∎

FIGURE 6.5.8 Graph of equation in Example 6

$\theta = \dfrac{2\pi}{3}$

$r = 1 + 2\cos\theta$

$\theta = \dfrac{4\pi}{3}$

EXAMPLE 7　　A Rose Curve

Graph $r = 2\cos 2\theta$.

Solution Because $\cos 2\theta$ has period π and is an even function we have

$$\cos 2(\pi - \theta) = \cos(2\pi - 2\theta) = \cos 2\theta \quad \text{and} \quad \cos(-2\theta) = \cos 2\theta.$$

Therefore, by (*i*) and (*ii*) of Theorem 6.5.1 the graph is symmetric with respect to both the y- and x-axes. A moment of reflection should convince you that we need only

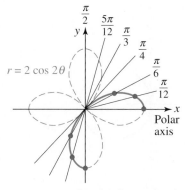

FIGURE 6.5.9 Graph of equation in Example 7

consider $0 \le \theta \le \pi/2$. Using the data in the following table, we see that the dashed portion of the graph given in FIGURE 6.5.9 is that completed by symmetry. The graph is called a **rose curve with four petals**.

θ	0	$\pi/12$	$\pi/6$	$\pi/4$	$\pi/3$	$5\pi/12$	$\pi/2$
r	2	1.7	1	0	-1	-1.7	-2

☐ **Rose Curves** In general, if n is a positive integer, the graphs of

$$r = a\sin n\theta \qquad \text{or} \qquad r = a\cos n\theta, \quad n \ge 2 \qquad (6)$$

are called **rose curves**, although as you can see in FIGURE 6.5.10 the curve looks more like a daisy. When n is odd, the number of **loops** or **petals** of the curve is n; if n is even, the curve has $2n$ petals. To graph a rose curve we can start by graphing one petal. To begin, we find an angle θ for which r is a maximum. This gives the center line of the petal. We then find corresponding values of θ for which the rose curve enters the origin ($r = 0$). To complete the graph we use the fact that the center lines of the petals are spaced $2\pi/n$ radians ($360/n$ degrees) apart if n is odd, and $2\pi/2n = \pi/n$ radians ($180/n$ degrees) apart if n is even. In Figure 6.5.10 we have drawn the graph of $r = a\sin 5\theta$, $a > 0$. The spacing between the center lines of the five petals is $2\pi/5$ radians ($72°$).

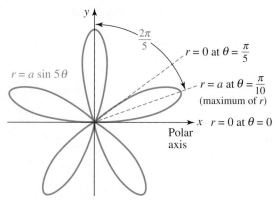

FIGURE 6.5.10 Rose curve with five petals

In Example 5 in Section 6.4 we saw that the polar equation $r = 8\cos\theta$ is equivalent to the rectangular equation $x^2 + y^2 = 8x$. By completing the square in x in the rectangular equation, we recognize

$$(x - 4)^2 + y^2 = 16$$

as a circle of radius 4 centered at $(4, 0)$ on the x-axis. Polar equations such as $r = 8\cos\theta$ or $r = 8\sin\theta$ are circles and are also special cases of rose curves. See FIGURE 6.5.11.

FIGURE 6.5.11 Graph of polar equation $r = 8\cos\theta$

☐ **Circles with Centers on an Axis** For $n = 1$ the equations in (6) become

$$r = a\sin\theta \qquad \text{or} \qquad r = a\cos\theta, \qquad (7)$$

and are polar equations of **circles** of diameter $|a|$ that pass through the origin and centers on a coordinate axis. The center of the graph of $r = a\sin\theta$ lies on the y-axis and its rectangular coordinates are $(0, a/2)$. The center of the graph of $r = a\cos\theta$ is on the x-axis and its rectangular coordinates are $(a/2, 0)$. FIGURE 6.5.12 illustrates the graphs of the equations in (7) in the two cases $a > 0$ and $a < 0$.

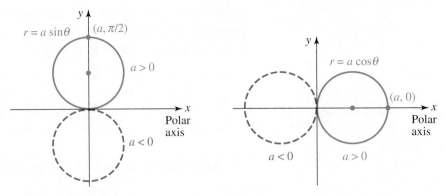

(a) Centers on the *y*-axis   (b) Centers on the *x*-axis

FIGURE 6.5.12 Circles through origin with centers on an axis

☐ **Lemniscates** The graphs of the polar equations

$$r^2 = a\cos 2\theta \qquad \text{or} \qquad r^2 = a\sin 2\theta, \tag{8}$$

where $a \neq 0$, are called **lemniscates**. Because $(-r)^2 = r^2$ it follows from (*iii*) of the tests for symmetry, Theorem 6.5.1, that the graphs of both of the equations in (8) are symmetric with respect to the origin. Moreover, from $\cos(-2\theta) = \cos 2\theta$ and (*ii*) of the tests for symmetry, we conclude that the graph of $r^2 = a\cos 2\theta$ is symmetric with respect to the *x*-axis. FIGURES 6.5.13(a) and 6.5.13(b) show typical graphs of the equations $r^2 = a\cos 2\theta$ and $r^2 = a\sin 2\theta$ for $a > 0$, respectively.

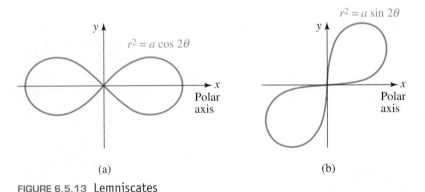

(a)   (b)

FIGURE 6.5.13 Lemniscates

EXAMPLE 8   A Lemniscate

Graph $r^2 = 4\cos 2\theta$.

Solution There are several ways of approaching this problem. But if we use symmetry we can keep point plotting to a minimum. From the discussion preceding this example we know that the graph of $r^2 = 4\cos 2\theta$ is symmetric with respect to the origin and to the *x*-axis. Solving for *r*, we will use $r = 2\sqrt{\cos 2\theta}$ to plot a few points in the first quadrant. Since $\cos 2\theta \geq 0$ for $0 \leq 2\theta \leq \pi/2$ we use $0 \leq \theta \leq \pi/4$ and a calculator to construct the following table.

θ	0	$\pi/12$	$\pi/8$	$\pi/6$	$\pi/4$
r	2	1.86	1.68	1.41	0

The five points obtained from the table are the red dots on the portion of the graph given in FIGURE 6.5.14(a). The graphs in Figures 6.5.14(b) and (c) are obtained

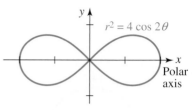

(a) Plotting points   (b) Reflection through origin   (c) Reflection in the *x*-axis

FIGURE 6.5.14 Graph of polar equation in Example 8

using, respectively, symmetry with respect to the origin followed by symmetry with respect to the *x*-axis.   ▮

☐ **Points of Intersection** In rectangular coordinates we can find the points (x, y) where the graphs of two functions $y = f(x)$ and $y = g(x)$ intersect by equating the *y*-values. The real solutions of the equation $f(x) = g(x)$ correspond to *all* the *x*-coordinates of the points where the graphs intersect. In contrast, problems may arise in polar coordinates when we try the same method to determine where the graphs of two polar equations $r = f(\theta)$ and $r = g(\theta)$ intersect.

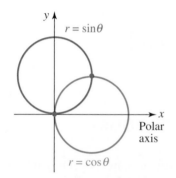

FIGURE 6.5.15 Intersecting circles in Example 9

▌EXAMPLE 9   Intersecting Circles

FIGURE 6.5.15 shows that the circles $r = \sin\theta$ and $r = \cos\theta$ have two points of intersection. By equating the *r* values, the equation $\sin\theta = \cos\theta$ leads to $\theta = \pi/4$. Substituting this value into either equation yields $r = \sqrt{2}/2$. Thus we have found only a single polar point $\left(\sqrt{2}/2, \pi/4\right)$ where the graphs intersect. From the figure, it is apparent that the graphs also intersect at the origin. But the problem here is that the origin or pole is $(0, \pi/2)$ on the graph of $r = \cos\theta$ but is $(0, 0)$ on the graph of $r = \sin\theta$. This situation is analogous to the curves reaching the same point at different times.   ▮

☐ **Rotation of Polar Graphs** In Section 2.2 we saw that if $y = f(x)$ is the rectangular equation of a function, then the graphs of $y = f(x - c)$ and $y = f(x + c), c > 0$, are obtained by *shifting* the graph of *f* horizontally *c* units to the right and to the left, respectively. In contrast, if $r = f(\theta)$ is a polar equation, then the graphs of $f(\theta - \gamma)$ and $f(\theta + \gamma)$, where $\gamma > 0$, can be obtained by *rotating* the graph of *f* by an amount γ. Specifically:

- The graph of $r = f(\theta - \gamma)$ is the graph of $r = f(\theta)$ rotated *counterclockwise* about the origin by an amount γ.
- The graph of $r = f(\theta + \gamma)$ is the graph of $r = f(\theta)$ rotated *clockwise* about the origin by an amount γ.

For example, the graph of the cardioid $r = a(1 + \cos\theta)$ is shown in Figure 6.5.6(a). The graph of $r = a(1 + \cos(\theta - \pi/2))$ is the graph of $r = a(1 + \cos\theta)$ rotated counterclockwise about the origin by an amount $\pi/2$. Its graph then must be that given in Figure 6.5.6(c). This makes sense, because the difference formula of the cosine gives

See the identity in (2) of Section 4.5. ▶ $r = a[1 + \cos(\theta - \pi/2)] = a[1 + \cos\theta\cos(\pi/2) + \sin\theta\sin(\pi/2)] = a(1 + \sin\theta).$

Similarly, rotating $r = a(1 + \cos\theta)$ counterclockwise about the origin by an amount π gives the equation

$$r = a[1 + \cos(\theta - \pi)] = a[1 + \cos\theta\cos\pi + \sin\theta\sin\pi] = a(1 - \cos\theta)$$

whose graph is given in Figure 6.5.6(b). As another example, take a look again at Figure 6.5.13. From

$$r^2 = a\cos 2\left(\theta - \frac{\pi}{4}\right) = a\cos\left(2\theta - \frac{\pi}{2}\right) = a\sin 2\theta$$

we see that the graph of the lemniscate in Figure 6.5.13(b) is the graph in Figure 6.5.13(a) rotated counterclockwise about the origin by an amount $\pi/4$.

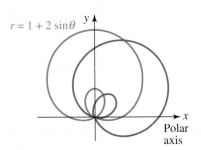

$r = 1 + 2\sin\theta$

FIGURE 6.5.16 Graphs of polar equations in Example 10

EXAMPLE 10　　Rotated Polar Graphs

Graph $r = 1 + 2\sin(\theta + \pi/4)$.

Solution The graph of the given equation is the graph of the limaçon $r = 1 + 2\sin\theta$ rotated clockwise about the origin by an amount $\pi/4$. In FIGURE 6.5.16 the blue graph is that of $r = 1 + 2\sin\theta$ and the red graph is the rotated graph. ∎

By using the sum formula for $\sin x$, (4) in Section 4.5, the polar equation of the rotated limaçon in Example 10 can be rewritten as

$$r = 1 + 2\sin\left(\theta + \frac{\pi}{4}\right)$$
$$= 1 + 2\sin\theta\,\cos\frac{\pi}{4} + 2\cos\theta\,\sin\frac{\pi}{4}$$
$$r = 1 + \sqrt{2}\sin\theta + \sqrt{2}\cos\theta.$$

NOTES FROM THE CLASSROOM

© CRM/Shutterstock, Inc.

(*i*) Although polar coordinates are extremely important in the study of calculus, Example 9 illustrates one of several frustrating difficulties of working in this coordinate system. Here is another:

A point can be on the graph of a polar equation even though its coordinates do not satisfy the equation.

You should verify that $(2, \pi/2)$ is an alternative polar description of the point $(-2, 3\pi/2)$. Moreover, verify that $(-2, 3\pi/2)$ is a point on the graph of $r = 1 + 3\sin\theta$ by showing that the coordinates satisfy the equation. However, note that the alternative coordinates $(2, \pi/2)$ do not satisfy the equation.

(*ii*) Many polar equations $r = f(\theta)$ are explicit functions where the independent variable is θ. For example,

$$r = 4,\ r = \theta,\ r = \sin 2\theta,\ r = 1 - \cos\theta,$$

are functions in that for an appropriate θ-value there is a single value of r defined by $r = f(\theta)$. But it is difficult to tell whether a polar graph is the graph of a function because:

The vertical-line test for a function is not applicable in the polar coordinate system.

In Example 1, $r = 3$ is a function whose graph is the circle in Figure 6.5.1. The same circle, whose rectangular equation is $x^2 + y^2 = 3^2$, is not the graph of a function of x. An equation such as $r^2 = 4\cos 2\theta$ in Example 8 is not a function of θ but defines implicitly the two functions $f(\theta) = 2\sqrt{\cos 2\theta}$ and $g(\theta) = -2\sqrt{\cos 2\theta}$.

(*iii*) The four-petal rose curve in Example 7 is obtained by plotting r for θ-values satisfying $0 \le \theta \le 2\pi$. See FIGURE 6.5.17. Do not assume this is true for every rose curve. Indeed, the five-petal rose curve discussed in Figure 6.5.10 is obtained using θ-values satisfying $0 \le \theta \le \pi$. In general, a rose curve $r = a\sin n\theta$ or $r = a\cos n\theta$ is traced out exactly once for $0 \le \theta \le 2\pi$ if n is even and once for $0 \le \theta \le \pi$ if n is odd.

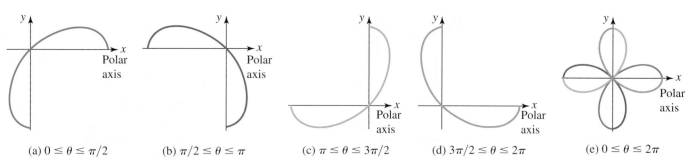

(a) $0 \le \theta \le \pi/2$　　(b) $\pi/2 \le \theta \le \pi$　　(c) $\pi \le \theta \le 3\pi/2$　　(d) $3\pi/2 \le \theta \le 2\pi$　　(e) $0 \le \theta \le 2\pi$

FIGURE 6.5.17 Plotting $r = 2\cos 2\theta$

Exercises 6.5　Answers to selected odd-numbered problems begin on page ANS–24.

In Problems 1–30, identify by name the graph of the given polar equation. Then sketch the graph of the equation.

1. $r = 6$
2. $r = -1$
3. $\theta = \pi/3$
4. $\theta = 5\pi/6$
5. $r = 2\theta, \theta \le 0$
6. $r = 3\theta, \theta \ge 0$
7. $r = 1 + \cos\theta$
8. $r = 5 - 5\sin\theta$
9. $r = 2(1 + \sin\theta)$
10. $2r = 1 - \cos\theta$
11. $r = 1 - 2\cos\theta$
12. $r = 2 + 4\sin\theta$
13. $r = 4 - 3\sin\theta$
14. $r = 3 + 2\cos\theta$
15. $r = 4 + \cos\theta$
16. $r = 4 - 2\sin\theta$
17. $r = \sin 2\theta$
18. $r = 3\sin 4\theta$
19. $r = 3\cos 3\theta$
20. $r = 2\sin 3\theta$
21. $r = \cos 5\theta$
22. $r = 2\sin 9\theta$
23. $r = 6\cos\theta$
24. $r = -2\cos\theta$
25. $r = -3\sin\theta$
26. $r = 5\sin\theta$
27. $r^2 = 4\sin 2\theta$
28. $r^2 = 4\cos 2\theta$
29. $r^2 = -25\cos 2\theta$
30. $r^2 = -9\sin 2\theta$

In Problems 31 and 32, the graph of the given equation is a spiral. Sketch its graph.

31. $r = 2^\theta, \theta \geq 0$ (logarithmic)   **32.** $r\theta = \pi, \theta > 0$ (hyperbolic)

In Problems 33–38, find an equation of the given polar graph.

33.

FIGURE 6.5.18 Graph for
Problem 33

34.

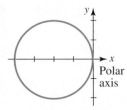

FIGURE 6.5.19 Graph for
Problem 34

35.

FIGURE 6.5.20 Graph for
Problem 35

36.

FIGURE 6.5.21 Graph for
Problem 36

37.

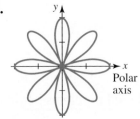

FIGURE 6.5.22 Graph for
Problem 37

38.

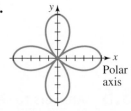

FIGURE 6.5.23 Graph for
Problem 38

In Problems 39–42, find the points of intersection of the graphs of the given pair of polar equations.

39. $r = 2, r = 4\sin\theta$ **40.** $r = \sin\theta, r = \sin 2\theta$
41. $r = 1 - \cos\theta, r = 1 + \cos\theta$ **42.** $r = 3 - 3\cos\theta, r = 3\cos\theta$

43. Suppose the red circle in Figure 6.5.15 is rotated clockwise about the origin by an amount $\pi/4$. Show that a polar equation of the rotated graph is given by

$$r = \frac{\sqrt{2}}{2}(\sin\theta + \cos\theta).$$

44. **(a)** Sketch the circle whose equation is given in Problem 43.
   (b) Find the polar coordinates of all intercepts.
   (c) Find the rectangular coordinates of the center of the circle.
   (d) Find the rectangular coordinates of the points at the end of the diameter that passes through the origin and the center.

In Problems 45 and 46, the given circle is rotated about the origin in the indicated direction and by the indicated amount. (a) Find a polar equation of the rotated circle (the red circle in the figure). (b) Use the polar equation of the rotated circle to find its standard-form rectangular equation. (c) Find polar and rectangular coordinates of the center of the rotated circle.

45. $r = 2\cos\theta$; clockwise, $\pi/6$

46. $r = -\sin\theta$; counterclockwise, $3\pi/4$

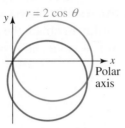

FIGURE 6.5.24 Graphs for Problem 45

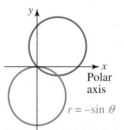

FIGURE 6.5.25 Graphs for Problem 46

Calculator/Computer Problems

47. Use a graphing utility to obtain the graph of the **bifolium** $r = 4\sin\theta\cos^2\theta$ and the circle $r = \sin\theta$ on the same axes. Find all points of intersection of the graphs.

48. Use a graphing utility to verify that the cardioid $r = 1 + \cos\theta$ and the lemniscate $r^2 = 4\cos\theta$ intersect at four points. Find these points of intersection of the graphs.

In Problems 49 and 50, the graphs of the equations (a)–(d) represent a rotation of the graph of the given equation. Try sketching these graphs by hand. If you have difficulties, then use a graphing utility.

49. $r = 1 + \sin\theta$
    (a) $r = 1 + \sin(\theta - \pi/2)$ **(b)** $r = 1 + \sin(\theta + \pi/2)$
    (c) $r = 1 + \sin(\theta - \pi/6)$ **(d)** $r = 1 + \sin(\theta + \pi/4)$

50. $r = 2 + 4\cos\theta$
    (a) $r = 2 + 4\cos(\theta + \pi/6)$ **(b)** $r = 2 + 4\cos(\theta - 3\pi/2)$
    (c) $r = 2 + 4\cos(\theta + \pi)$ **(d)** $r = 2 + 4\cos(\theta - \pi/8)$

51. Use a CAS to obtain graphs of the polar equation $r = a + \cos\theta$ for $a = 0, \frac{1}{4}, \frac{1}{2}, \frac{3}{4}, 1, \frac{5}{4}, \ldots, 3$.

52. Identify all the curves in Problem 51. What happens to the graphs as $a \to \infty$?

For Discussion

In Problems 53 and 54, suppose $r = f(\theta)$ is a polar equation. Graphically interpret the given property.

53. $f(-\theta) = f(\theta)$ (even function)   **54.** $f(-\theta) = -f(\theta)$ (odd function)

55. Show that $\cos(2\theta - \pi) = -\cos 2\theta$ and $\sin(2\theta - \pi) = -\sin 2\theta$. Use these identities to discuss how the graphs of the polar equations in Problems 29 and 30 are related to the graphs of $r^2 = 25\cos 2\theta$ and $r^2 = 9\sin 2\theta$.

56. Find the polar equation $r = f(\theta)$ of the red tangent line to the blue circle centered at the origin shown in FIGURE 6.5.26.

57. Suppose replacing (r, θ) by $(-r, -\theta)$ in a polar equation results in the same equation. What can be said about the graph of the equation?

58. Suppose replacing (r, θ) by $(-r, \pi - \theta)$ in a polar equation results in the same equation. What can be said about the graph of the equation?

FIGURE 6.5.26 Graphs for Problem 56

59. Suppose replacing (r, θ) by $(r, \pi + \theta)$ in a polar equation results in the same equation. What can be said about the graph of the equation?

60. **(a)** Use (*ii*) and (*iii*) of Theorem 6.5.1 to show that graph of the polar equation $r^2 = 1 + \cos\theta$ is symmetric with respect to x-axis and to the origin.

(b) As a consequence of the two symmetries demonstrated in part (a), the graph must also be symmetric with respect to the y-axis. Show, however, that (*i*) of Theorem 6.5.1 fails to prove this symmetry.

(c) Find an alternative test that proves y-axis symmetry.

(d) Use a graphing utility to obtain the graph the equation. [*Hint*: Use the functions $r = \sqrt{1 + \cos\theta}$ and $r = -\sqrt{1 + \cos\theta}$.]

6.6 Conic Sections in Polar Coordinates

INTRODUCTION In the first three sections of this chapter we derived equations for the parabola, ellipse, and hyperbola using the distance formula in rectangular coordinates. By using polar coordinates and the concept of eccentricity, we can now give one general definition of a conic section that encompasses all three curves.

DEFINITION 6.6.1 Conic Section

Let L be a fixed line in the plane called the **directrix**, and let F be a point not on the line called the **focus**. A **conic section** is the set of points P in the plane for which

$$\frac{d(P, F)}{d(P, Q)} = e \tag{1}$$

where $d(P, F)$ is the distance from P to F, $d(P, Q)$ the distance from P to L, and e is positive constant called the **eccentricity**.

A geometric interpretation of (1) is given in FIGURE 6.6.1. If

- $e = 1$, the conic is a **parabola**,
- $0 < e < 1$, the conic is an **ellipse**, and
- $e > 1$, the conic is a **hyperbola**.

Note that when $e = 1$ in (1) the resulting equation $d(P, F) = d(P, Q)$ is exactly the definition of a parabola given in Definition 6.1.1.

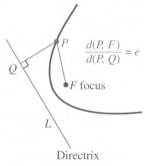

FIGURE 6.6.1 Geometric interpretation of (1)

☐ **Polar Equations of Conics** Equation (1) is readily interpreted using polar coordinates. Suppose F is placed at the pole and L is p units ($p > 0$) to the left of F perpendicular to the extended polar axis. We see from FIGURE 6.6.2 that (1) written as $d(P, F) = ed(P, Q)$ is the same as

$$r = e(p + r\cos\theta) \qquad \text{or} \qquad r - er\cos\theta = ep. \tag{2}$$

Solving for r yields

$$r = \frac{ep}{1 - e\cos\theta}. \tag{3}$$

FIGURE 6.6.2 Polar coordinate interpretation of (1)

To see that (3) yields the familiar equations of the conics, let us superimpose a rectangular coordinate system on the polar coordinate system with origin at the pole and the positive x-axis coinciding with the polar axis. We then express the first equation in (2) in rectangular coordinates and simplify:

$$\pm\sqrt{x^2 + y^2} = ex + ep$$
$$x^2 + y^2 = e^2x^2 + 2e^2px + e^2p^2$$
$$(1 - e^2)x^2 - 2e^2px + y^2 = e^2p^2. \tag{4}$$

Choosing $e = 1$, (4) becomes

$$-2px + y^2 = p^2 \qquad \text{or} \qquad y^2 = 2p\left(x + \tfrac{1}{2}p\right).$$

The last equation is the standard form of a parabola whose axis is the x-axis, vertex is at $\left(-\tfrac{1}{2}p, 0\right)$ and, consistent with the placement of F, whose focus is at the origin.

It is a good exercise in algebra to show that (3) yields standard form equations of an ellipse in the case $0 < e < 1$ and a hyperbola in the case $e > 1$. See Problem 49 in Exercises 6.6. Thus, depending on the value of e, the polar equation (3) can have three possible graphs as shown in FIGURE 6.6.3.

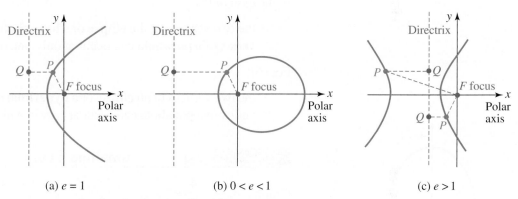

(a) $e = 1$ (b) $0 < e < 1$ (c) $e > 1$

FIGURE 6.6.3 Graphs of polar equation (3)

If we had placed the focus F to the *left* of the directrix in our derivation of the polar equation (3), then the equation $r = ep/(1 + e\cos\theta)$ would be obtained. When the directrix L is chosen parallel to the polar axis (that is, horizontal), then the equation of the conic is found to be either $r = ep/(1 - e\sin\theta)$ or $r = ep/(1 + e\sin\theta)$. A summary of the preceding discussion is given next.

THEOREM 6.6.1   Polar Equations of Conics

A polar equation of the form

$$r = \frac{ep}{1 \pm e\cos\theta} \tag{5}$$

or

$$r = \frac{ep}{1 \pm e\sin\theta} \tag{6}$$

is a conic section with focus at the origin and axis along a coordinate axis. The axis of the conic section is along the x-axis for equations of the form (5) and along the y-axis for equations of the form (6). The conic is a **parabola** if $e = 1$, an **ellipse** if $0 < e < 1$, and a **hyperbola** if $e > 1$.

EXAMPLE 1   **Identifying Conics**

Identify each of the following conics **(a)** $r = \dfrac{2}{1 - 2\sin\theta}$ **(b)** $r = \dfrac{3}{4 + \cos\theta}$.

Solution **(a)** A term-by-term comparison of the given equation with the polar form $r = ep/(1 - e\sin\theta)$ enables us to make the identification $e = 2$. Hence the conic is a hyperbola.

(b) In order to identify the conic section, we divide the numerator and the denominator of the given equation by 4. This puts the equation into the form

$$r = \dfrac{\frac{3}{4}}{1 + \frac{1}{4}\cos\theta}.$$

Then by comparison with $r = ep/(1 + e\cos\theta)$ we see that $e = \frac{1}{4}$. Hence the conic is an ellipse. ∎

☐ **Graphs** A rough graph of a conic defined by (5) or (6) can be obtained by knowing the orientation of its axis, finding the x- and y-intercepts, and finding the vertices. In the case of (5):

- the two vertices of the **ellipse** or a **hyperbola** occur at $\theta = 0$ and $\theta = \pi$; the vertex of a **parabola** can occur at only one of the values: $\theta = 0$ or $\theta = \pi$.

For (6):

- the two vertices of an **ellipse** or a **hyperbola** occur at $\theta = \pi/2$ and $\theta = 3\pi/2$; the vertex of a **parabola** can occur at only one of the values: $\theta = \pi/2$ or $\theta = 3\pi/2$.

EXAMPLE 2   **Graphing a Conic**

Graph $r = \dfrac{4}{3 - 2\sin\theta}$.

Solution By writing the equation as $r = \dfrac{\frac{4}{3}}{1 - \frac{2}{3}\sin\theta}$ we see that the eccentricity is $e = \frac{2}{3}$ and so the conic is an ellipse. Moreover, because the equation is of the form given in (6), we know that the axis of the ellipse is vertical along the y-axis. Now in view of the discussion preceding this example, we obtain:

$$\begin{aligned}
\textit{vertices: } & (4, \pi/2), \left(\tfrac{4}{5}, 3\pi/2\right) \\
\textit{x-intercepts: } & \left(\tfrac{4}{3}, 0\right), \left(\tfrac{4}{3}, \pi\right).
\end{aligned}$$

The graph of the equation is given in FIGURE 6.6.4. ∎

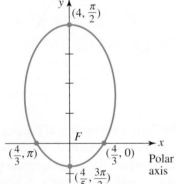

FIGURE 6.6.4 Graph of polar equation in Example 2

EXAMPLE 3   **Graphing a Conic**

Graph $r = \dfrac{1}{1 - \cos\theta}$.

Solution Inspection of the equation shows that it is of the form given in (5) with $e = 1$. Hence the conic section is a parabola whose axis is horizontal along the x-axis. Since r is undefined at $\theta = 0$, the vertex of the parabola occurs at $\theta = \pi$:

$$\begin{aligned}
\textit{vertex: } & \left(\tfrac{1}{2}, \pi\right) \\
\textit{y-intercepts: } & (1, \pi/2), (1, 3\pi/2).
\end{aligned}$$

The graph of the equation is given in FIGURE 6.6.5. ∎

FIGURE 6.6.5 Graph of polar equation in Example 3

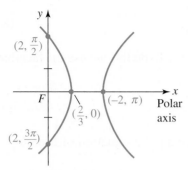

FIGURE 6.6.6 Graph of polar equation in Example 4

EXAMPLE 4　　Graphing a Conic

Graph $r = \dfrac{2}{1 + 2\cos\theta}$.

Solution From (5) we see that $e = 2$ and so the conic section is a hyperbola whose axis is horizontal along the x-axis. The vertices, the endpoints of the transverse axis of the hyperbola, occur at $\theta = 0$ and at $\theta = \pi$:

$$vertices:\ \left(\tfrac{2}{3}, 0\right), (-2, \pi)$$
$$y\text{-}intercepts:\ (2, \pi/2), (2, 3\pi/2).$$

The graph of the equation is given in FIGURE 6.6.6.

☐ **Rotated Conics** We saw in Section 6.5 that graphs of $r = f(\theta - \gamma)$ and $r = f(\theta + \gamma)$, $\gamma > 0$, are rotations of the graph of the polar equation $r = f(\theta)$ about the origin by an amount γ. Thus

$$\left.\begin{array}{l} r = \dfrac{ep}{1 \pm e\cos(\theta - \gamma)} \\[2mm] r = \dfrac{ep}{1 \pm e\sin(\theta - \gamma)} \end{array}\right\} \begin{array}{l}\text{conics rotated} \\ \text{counterclockwise} \\ \text{about the origin}\end{array} \qquad \left.\begin{array}{l} r = \dfrac{ep}{1 \pm e\cos(\theta + \gamma)} \\[2mm] r = \dfrac{ep}{1 \pm e\sin(\theta + \gamma)} \end{array}\right\} \begin{array}{l}\text{conics rotated} \\ \text{clockwise} \\ \text{about the origin}\end{array}$$

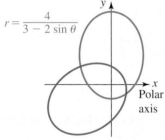

$r = \dfrac{4}{3 - 2\sin\theta}$

FIGURE 6.6.7 Graphs of polar equations in Example 5

EXAMPLE 5　　Rotated Conic

In Example 2 we saw that the graph of the polar equation $r = \dfrac{4}{3 - 2\sin\theta}$ is an ellipse with major axis along the y-axis. This is the blue graph in FIGURE 6.6.7. The graph of $r = \dfrac{4}{3 - 2\sin(\theta - 2\pi/3)}$ is the red graph in Figure 6.6.7 and is a counterclockwise rotation of the blue graph by the amount $2\pi/3$ radians (or 120°) about the origin. The major axis of the red graph lies along the radial line $\theta = 7\pi/6$.

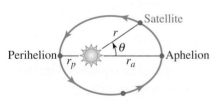

FIGURE 6.6.8 Orbit of satellite around the Sun

☐ **Applications** Equations of the type in (5) and (6) are well suited to describe a closed orbit of satellite around the Sun (Earth or Moon) since such an orbit is an ellipse with the Sun (Earth or Moon) at one focus. Suppose that an equation of the orbit is given by $r = ep/(1 - e\cos\theta)$, $0 < e < 1$, and r_p is the value of r at perihelion (perigee or perilune) and r_a is the value of r at aphelion (apogee or apolune). These are the points in the orbit, occurring on the x-axis, at which the satellite is closest and farthest, respectively, from the Sun (Earth or Moon). See FIGURE 6.6.8. It is left as an exercise to show that the eccentricity e of the orbit is related to r_p and r_a by

$$e = \frac{r_a - r_p}{r_a + r_p}. \tag{7}$$

Mercury is the closest planet to the Sun

EXAMPLE 6　　Polar Equation of an Orbit

Find a polar equation of the orbit of the planet Mercury around the Sun if $r_p = 2.85 \times 10^7$ miles and $r_a = 4.36 \times 10^7$ miles.

Solution From (7), the eccentricity of Mercury's orbit is

$$e = \frac{4.36 \times 10^7 - 2.85 \times 10^7}{4.36 \times 10^7 + 2.85 \times 10^7} = 0.21.$$

Hence
$$r = \frac{0.21p}{1 - 0.21\cos\theta}.$$ (8)

All we need to do now is to solve for the quantity $0.21p$. To do this we use the fact that aphelion occurs at $\theta = 0$:

$$4.36 \times 10^7 = \frac{0.21p}{1 - 0.21}.$$

The last equation yields $0.21p = 3.44 \times 10^7$. Hence a polar equation of Mercury's orbit is

$$r = \frac{3.44 \times 10^7}{1 - 0.21\cos\theta}.$$

Exercises 6.6 Answers to selected odd-numbered problems begin on page ANS-25.

In Problems 1–10, determine the eccentricity, identify the conic, and sketch its graph.

1. $r = \dfrac{2}{1 - \sin\theta}$

2. $r = \dfrac{2}{2 - \cos\theta}$

3. $r = \dfrac{16}{4 + \cos\theta}$

4. $r = \dfrac{5}{2 + 2\sin\theta}$

5. $r = \dfrac{4}{1 + 2\sin\theta}$

6. $r = \dfrac{-4}{\cos\theta - 1}$

7. $r = \dfrac{18}{3 - 6\cos\theta}$

8. $r = \dfrac{4\csc\theta}{3\csc\theta + 2}$

9. $r = \dfrac{6}{1 - \cos\theta}$

10. $r = \dfrac{2}{2 + 5\cos\theta}$

In Problems 11–14, determine the eccentricity e of the given conic. Then convert the polar equation to a rectangular equation and verify that $e = c/a$.

11. $r = \dfrac{6}{1 + 2\sin\theta}$

12. $r = \dfrac{10}{2 - 3\cos\theta}$

13. $r = \dfrac{12}{3 - 2\cos\theta}$

14. $r = \dfrac{2\sqrt{3}}{\sqrt{3} + \sin\theta}$

In Problems 15 and 16, convert the polar equation to a rectangular equation. Use the rectangular equation to verify that the focus of the conic is at the origin.

15. $r = \dfrac{2}{1 + \sin\theta}$

16. $r = \dfrac{1}{1 - \cos\theta}$

In Problems 17–22, find a polar equation of the conic with focus at the origin that satisfies the given conditions.

17. $e = 1$, directrix $x = 3$

18. $e = \frac{3}{2}$, directrix $y = 2$

19. $e = \frac{2}{3}$, directrix $y = -2$

20. $e = \frac{1}{2}$, directrix $x = 4$

21. $e = 2$, directrix $x = 6$

22. $e = 1$, directrix $y = -2$

23. Find a polar equation of the conic in Problem 17 if the graph is rotated clockwise about the origin by an amount $2\pi/3$.

24. Find a polar equation of the conic in Problem 18 if the graph is rotated counterclockwise about the origin by an amount $\pi/6$.

In Problems 25–30, find a polar equation of the parabola with focus at the origin and the given vertex.

25. $\left(\frac{3}{2}, 3\pi/2\right)$

26. $(2, \pi)$

27. $\left(\frac{1}{2}, \pi\right)$

28. $(2, 0)$

29. $\left(\frac{1}{4}, 3\pi/2\right)$

30. $\left(\frac{3}{2}, \pi/2\right)$

In Problems 31–36, identify the given rotated conic. Find the polar coordinates of its vertex or vertices.

31. $r = \dfrac{4}{1 + \cos(\theta - \pi/4)}$

32. $r = \dfrac{5}{3 + 2\cos(\theta - \pi/3)}$

33. $r = \dfrac{10}{2 - \sin(\theta + \pi/6)}$

34. $r = \dfrac{6}{1 + 2\sin(\theta + \pi/3)}$

35. $r = \dfrac{3}{2 - 3\cos(\theta + \pi/2)}$

36. $r = \dfrac{2}{3 - 3\sin(\theta - \pi)}$

Applications

37. Perigee Distance A communications satellite is 12,000 km above the Earth at its apogee. The eccentricity of its elliptical orbit is 0.2. Use (7) to find its perigee distance.

38. Orbit Find a polar equation $r = ep/(1 - e\cos\theta)$ of the orbit of the satellite in Problem 37.

39. Earth's Orbit Find a polar equation of the orbit of the Earth around the Sun if $r_p = 1.47 \times 10^8$ km and $r_a = 1.52 \times 10^8$ km.

40. Comet Halley (a) The eccentricity of the elliptical orbit of Comet Halley is 0.97 and the length of the major axis of its orbit is 3.34×10^9 mi. Find a polar equation of its orbit of the form $r = ep/(1 - e\cos\theta)$.
(b) Use the equation in part (a) to obtain r_p and r_a for the orbit of Comet Halley.

Calculator/Computer Problems

In Problems 41–44, use a graphing utility to graph the given two polar equations on the same coordinate axes.

41. $r = \dfrac{4}{4 + 3\cos\theta}$; $r = \dfrac{4}{4 + 3\cos(\theta - \pi/2)}$

42. $r = \dfrac{4}{6 - 3\sin\theta}$; $r = \dfrac{4}{6 - 3\sin(\theta - \pi)}$

43. $r = \dfrac{2}{1 - \sin\theta}$; $r = \dfrac{2}{1 - \sin(\theta + 3\pi/4)}$

44. $r = \dfrac{8}{3 + 5\cos\theta}$; $r = \dfrac{8}{3 + 5\cos(\theta - 2\pi/3)}$

Recent History The orbital characteristics (eccentricity, perigee, and major axis) of a satellite near the Earth gradually degrade over time due to many small forces acting on the satellite other than the gravitational force of the Earth. These forces include atmospheric drag, the gravitational attractions of the Sun and the Moon, and magnetic forces. Approximately once a month tiny rockets are activated for a few seconds in order to "boost" the orbital characteristics back into the desired range. These rocket engines are turned on longer to make a major change in the orbit of a satellite. The most fuel-efficient way to move from an inner orbit to an outer orbit, called a **Hohmann transfer orbit**, is

Next visit of Comet Halley to the Solar System will be in 2061

to add velocity in the direction of flight at the time the satellite reaches perigee on the inner orbit, follow the Hohmann transfer ellipse halfway around to its apogee, and add velocity again to achieve the outer orbit. A similar process (subtracting velocity at apogee on the outer orbit and subtracting velocity at perigee on the Hohmann transfer orbit) moves a satellite from an outer orbit to an inner orbit. This transfer orbit is named after the German engineer and astronomy/astronautics enthusiast **Walter Hohmann** (1880–1945) who first proposed it in his book *The Attainability of the Celestial Bodies* published in 1925.

In Problems 45–48, use a calculator or CAS to obtain the graphs of the given three polar equations on the same rectangular coordinate system. Use different colors for each graph.

45. Inner orbit $r = \dfrac{24}{1 + 0.2\cos\theta}$, Hohmann transfer $r = \dfrac{32}{1 + 0.6\cos\theta}$,

outer orbit $r = \dfrac{56}{1 + 0.3\cos\theta}$

46. Inner orbit $r = \dfrac{5.5}{1 + 0.1\cos\theta}$, Hohmann transfer $r = \dfrac{7.5}{1 + 0.5\cos\theta}$,

outer orbit $r = \dfrac{13.5}{1 + 0.1\cos\theta}$

47. Inner orbit $r = 9$, Hohmann transfer $r = \dfrac{15.3}{1 + 0.7\cos\theta}$,

outer orbit $r = 51$

48. Inner orbit $r = \dfrac{73.5}{1 + 0.05\cos\theta}$, Hohmann transfer $r = \dfrac{77}{1 + 0.1\cos\theta}$,

outer orbit $r = \dfrac{84.7}{1 + 0.01\cos\theta}$

For Discussion

49. Show that (2) yields standard form equations of an ellipse in the case $0 < e < 1$ and a hyperbola in the case $e > 1$.
50. Use the equation $r = ep/(1 - e\cos\theta)$ to derive the result in (7).

6.7 Parametric Equations

INTRODUCTION Rectangular equations and polar equations are not the only, and often not the most convenient, ways of describing curves in the coordinate plane. In this section we will consider a different way of representing a curve that is important in many applications. Let's consider one example. The motion of a particle along a curve, in contrast to a straight line, is called *curvilinear motion*. If it is assumed that a golf ball is hit off the ground, perfectly straight (no hook or slice), and that its path stays in a coordinate plane as shown in FIGURE 6.7.1, then it can be shown that its x- and y-coordinates at time t are given by

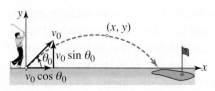

FIGURE 6.7.1 Path of golf ball

$$x = (v_0\cos\theta_0)t, \quad y = -\tfrac{1}{2}gt^2 + (v_0\sin\theta_0)t, \tag{1}$$

CHAPTER 6 CONIC SECTIONS

where θ_0 is the launch angle, v_0 is its initial velocity, and $g = 32 \text{ ft/s}^2$ is the acceleration due to gravity. These equations, which give the golf ball's position in the coordinate plane at time t, are said to be **parametric equations**. The third variable t in (1) is called a **parameter** and is restricted to some interval defined by $0 \le t \le T$, where $t = 0$ gives the origin $(0, 0)$, and $t = T$ is the time the ball hits the ground.

In general, a curve in a coordinate plane can be *defined* in terms of parametric equations.

DEFINITION 6.7.1 Plane Curve

A **plane curve** is a set C of ordered pairs $(f(t), g(t))$, where f and g are functions defined on a common interval I. The equations

$$x = f(t), \quad y = g(t), \quad \text{for } t \text{ in } I$$

are called **parametric equations** for C. The variable t is called a **parameter** and I is called the **parameter interval**.

It is also common practice to refer to $x = f(t), y = g(t)$, for t in I, as a **parameterization** for C.

The **graph** of a plane curve C is the set of all points (x, y) in the coordinate plane corresponding to the ordered pairs $(f(t), g(t))$. Hereafter, we will refer to a plane curve as a **curve** or as a **parameterized curve**.

◾ EXAMPLE 1   **Graph of a Parametric Curve**

Graph the curve C that has the parametric equations

$$x = t^2, \quad y = t^3, \quad -1 \le t \le 2.$$

Solution As shown in the accompanying table, for any choice of t in the interval $[-1, 2]$, we obtain a single ordered pair (x, y). By connecting the points with a curve, we obtain the graph in FIGURE 6.7.2.

FIGURE 6.7.2 Curve in Example 1

t	-1	$-\frac{1}{2}$	0	$\frac{1}{2}$	1	$\frac{3}{2}$	2
x	1	$\frac{1}{4}$	0	$\frac{1}{4}$	1	$\frac{9}{4}$	4
y	-1	$-\frac{1}{8}$	0	$\frac{1}{8}$	1	$\frac{27}{8}$	8

In Example 1, if we think in terms of motion and t as time, then as t increases from -1 to 2, a point P defined as (t^2, t^3) starts from $(1, -1)$, advances up the lower branch to the origin $(0, 0)$, passes to the upper branch, and finally stops at $(4, 8)$. In general, as we plot points corresponding to *increasing values* of the parameter, the curve C is traced out by $(f(t), g(t))$ in a certain *direction* indicated by the arrowheads on the curve in Figure 6.7.2. This direction is called the **orientation** of the curve C.

A parameter need have no relation to time. When the interval I over which f and g in (1) are defined is a closed interval $[a, b]$, we say that $(f(a), g(a))$ is the **initial point** of the curve C and that $(f(b), g(b))$ is its **terminal point**. In Example 1 the initial point is $(1, -1)$ and the terminal point is $(4, 8)$. If the terminal point is the same as the initial point, that is,

$$(f(a), g(a)) = (f(b), g(b)),$$

then C is a **closed curve**. If C is closed and does not cross itself, then it is called a **simple closed curve**. In FIGURE 6.7.3, A and B represent the initial and terminal points, respectively.

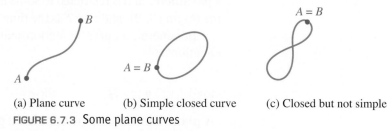

(a) Plane curve (b) Simple closed curve (c) Closed but not simple

FIGURE 6.7.3 Some plane curves

The next example illustrates a simple closed curve.

▌EXAMPLE 2 A Parameterization of a Circle

Find a parameterization for the circle $x^2 + y^2 = a^2$.

Solution The circle has center at the origin and radius a. If t represents the central angle, that is, an angle with vertex at the origin and initial side coinciding with the positive x-axis, then as shown in FIGURE 6.7.4 the equations

$$x = a\cos t, \quad y = a\sin t, \quad 0 \le t \le 2\pi \tag{2}$$

give every point P on the circle. For example, at $t = 0$ we get $x = a$ and $y = 0$; in other words, the initial point is $(a, 0)$. The terminal point corresponds to $t = 2\pi$ and is also $(a, 0)$. Since the initial and terminal points are the same, this proves the obvious: the curve C defined by the parametric equations in (2) is a closed curve. Note the orientation of C in Figure 6.7.4; as t increases from 0 to 2π, the point P traces out C in a counterclockwise direction. ▐

☐ **Changing the Parameter Interval** In Example 2, if we wish to describe *two* complete counterclockwise revolutions around the circle, we modify the parameter interval by writing

$$x = a\cos t, \quad y = a\sin t, \quad 0 \le t \le 4\pi.$$

On the other hand, we can describe *portions* of the circle by again modifying the parameter interval. For example, the upper semicircle $x^2 + y^2 = a^2$, $0 \le y \le a$, is defined parametrically by restricting the parameter interval to $[0, \pi]$,

$$x = a\cos t, \quad y = a\sin t, \quad 0 \le t \le \pi.$$

Observe that when $t = \pi$, the terminal point is now $(-a, 0)$. As seen in FIGURE 6.7.5(a) the orientation of the curve is counterclockwise. The same equations with a different parameter interval

$$x = a\cos t, \quad y = a\sin t, \quad -\pi/2 \le t \le \pi/2,$$

define the semicircle shown in Figure 6.7.5(b). The value $t = -\pi/2$ gives the initial point $(0, -a)$ and $t = \pi/2$ gives the terminal point $(0, a)$. Unlike the semicircle in Figure 6.7.5(a) the orientation of the curve in Figure 6.7.5(b) is seen to be clockwise.

☐ **Ellipse** It can also be seen from Figure 6.7.4 that a parameterization of an ellipse $x^2/a^2 + y^2/b^2 = 1$ is

$$x = a\cos t, \quad y = b\sin t, \quad 0 \le t \le 2\pi.$$

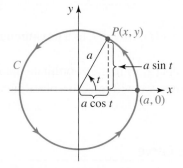

FIGURE 6.7.4 Circle in Example 2

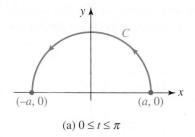

(a) $0 \le t \le \pi$

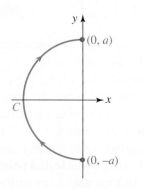

(b) $-\pi/2 \le t \le \pi/2$

FIGURE 6.7.5 Semicircles obtained from $x = a\cos t$, $y = a\sin t$ by restricting the parameter interval

Portions of the ellipse or changing the orientation of the parameterized curve can be done as illustrated in the preceding discussion.

☐ **Eliminating the Parameter** Given a set of parametric equations, we sometimes desire to eliminate or clear the parameter to obtain a rectangular equation for the curve. There is no well-defined method of eliminating the parameter; the method is dictated by the parametric equations. For example, to eliminate the parameter in (2), we simply square x and y and add the two equations:

$$x^2 + y^2 = a^2 \cos^2 t + a^2 \sin^2 t \qquad \text{implies} \qquad x^2 + y^2 = a^2$$

since $\sin^2 t + \cos^2 t = 1$. In the next example, we illustrate a substitution method.

■ EXAMPLE 3　　　Eliminating the Parameter

(a) In the first equation in (1) we can solve for t in terms of x and then substitute $t = x/(v_0 \cos \theta_0)$ into the second equation. This gives an equation involving only the variables x and y:

$$y = -\frac{g}{2(v_0 \cos \theta_0)^2}x^2 + (\tan \theta_0)x.$$

Since v_0, θ_0, and g are constants, the last equation has the form $y = ax^2 + bx$ and so the trajectory of any projectile launched at the angle $0 < \theta_0 < \pi/2$ is a parabolic arc.

(b) In Example 1, we can eliminate the parameter by solving the second equation for t in terms of y and then substituting in the first equation. We find

$$t = y^{1/3} \quad \text{and so} \quad x = (y^{1/3})^2 = y^{2/3}.$$

The curve shown in Figure 6.7.2 is only a portion of the graph of $x = y^{2/3}$. For $-1 \le t \le 2$, we have correspondingly $-1 \le y \le 8$. Thus, a rectangular equation for the curve in Example 1 is given by $x = y^{2/3}, -1 \le y \le 8$. ∎

☐ **Different Parameterizations** A curve C can have more than one parameterization. For example, an alternative parameterization for the circle in Example 2 is

$$x = a\cos 2t, \quad y = a\sin 2t, \quad 0 \le t \le \pi.$$

Note that the parameter interval is now $[0, \pi]$. We see that as t increases from 0 to π, the new angle $2t$ increases from 0 to 2π.

■ EXAMPLE 4　　　Alternative Parameterizations

(a) Consider the curve C that has the parametric equations $x = t, y = 2t^2$, $-\infty < t < \infty$. We can eliminate the parameter by using $t = x$ and substituting in $y = 2t^2$. This gives the rectangular equation $y = 2x^2$ that we recognize as a parabola. Moreover, since $-\infty < t < \infty$ is equivalent to $-\infty < x < \infty$, the point $(t, 2t^2)$ traces out the complete parabola $y = 2x^2$, $-\infty < x < \infty$.

(b) An alternative parameterization of the curve C in part (a) is given by $x = t^3/4$, $y = t^6/8$, $-\infty < t < \infty$. Using $t^3 = 4x$ and substituting in $y = t^6/8$ or $y = (t^3 \cdot t^3)/8$ gives $y = (4x)^2/8 = 2x^2$. Moreover, $-\infty < t < \infty$ implies $-\infty < t^3 < \infty$ and so $-\infty < x < \infty$. ∎

We note in Example 4 that a point on the curve C need not correspond to the same value of the parameter in each set of parametric equations for C. For example, $(1, 2)$ is obtained for $t = 1$ in $x = t, y = 2t^2$, but $t = \sqrt[3]{4}$ yields $(1, 2)$ in $x = t^3/4, y = t^6/8$.

EXAMPLE 5   **Example 4 Revisited**

One has to be careful when working with parametric equations. Eliminating the parameter in $x = t^2$, $y = 2t^4$, $-\infty < t < \infty$, would seem to yield the same parabola $y = 2x^2$ as in Example 4. However, this is *not* the case because for any value of t, satisfying $-\infty < t < \infty$, we have $t^2 \geq 0$ and so $x \geq 0$. In other words, the last set of equations $x = t^2$, $y = 2t^4$, $-\infty < t < \infty$ is a parametric representation of only the right-hand branch of the parabola, that is, $y = 2x^2$, $0 \leq x < \infty$.

Proceed with caution when eliminating the parameter. ▶

EXAMPLE 6   **Eliminating the Parameter**

Consider the curve C defined parametrically by
$$x = \sin t, \quad y = \cos 2t, \quad 0 \leq t \leq \pi/2.$$
Eliminate the parameter and obtain a rectangular equation for C.

Solution Using the double-angle formula $\cos 2t = \cos^2 t - \sin^2 t$, we can write
$$\begin{aligned} y &= \cos^2 t - \sin^2 t \\ &= (1 - \sin^2 t) - \sin^2 t \\ &= 1 - 2\sin^2 t \quad \leftarrow \text{substitute } \sin t = x \\ &= 1 - 2x^2. \end{aligned}$$

Now the curve C described by the parametric equations does not consist of the complete parabola, that is, $y = 1 - 2x^2$, $-\infty < x < \infty$. See FIGURE 6.7.6(a). For $0 \leq t \leq \pi/2$ we have $0 \leq \sin t \leq 1$ and $-1 \leq \cos 2t \leq 1$. This means that C is only that portion of the parabola for which the coordinates of a point $P(x, y)$ satisfy $0 \leq x \leq 1$ and $-1 \leq y \leq 1$. The curve C, along with its orientation, is shown in Figure 6.7.6(b). A rectangular equation for C is $y = 1 - 2x^2$ with the restricted domain $0 \leq x \leq 1$.

☐ **Intercepts** We can get intercepts of a curve C without finding its rectangular equation. For instance, in Example 6 we can find the x-intercept by finding the value of t in the parameter interval for which $y = 0$. The equation $\cos 2t = 0$ yields $2t = \pi/2$ so that $t = \pi/4$. The corresponding point at which C crosses the x-axis is $\left(\sqrt{2}/2, 0\right)$. Similarly, the y-intercept of C is found by solving $x = 0$. From $\sin t = 0$ we immediately conclude $t = 0$ and so the y-intercept is $(0, 1)$.

☐ **Applications of Parametric Equations** Cycloidal curves were a popular topic of study by mathematicians in the seventeenth century. Suppose a point $P(x, y)$, marked on a circle of radius a, is at the origin when its diameter lies along the y-axis. As the circle rolls along the x-axis, the point P traces out a curve C that is called a **cycloid**. See FIGURE 6.7.7.*

(a) $y = 1 - 2x^2$

(b) $x = \sin t$, $y = \cos 2t$,
    $0 \leq t \leq \pi/2$

FIGURE 6.7.6 Curve C in Example 6

(a) Circle rolling on x-axis

(b) Point P on the circle traces out this curve

FIGURE 6.7.7 Curve in red is a cycloid

*For an animation of the rolling circle go to http://mathworld.wolfram.com/Cycloid.html.

FIGURE 6.7.8 Sliding bead

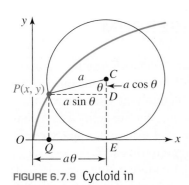

FIGURE 6.7.9 Cycloid in Example 7

Kimbell Art Museum

Two problems were extensively studied in the seventeenth century. Consider a flexible (frictionless) wire fixed at points A and B and a bead free to slide down the wire starting at P. See FIGURE 6.7.8. Is there a particular shape of the wire so that, regardless of where the bead starts, the time to slide down the wire to B will be the same? Also, what would the shape of the wire be so that the bead slides from P to B in the shortest time? The so-called **tautochrone** (same time) and **brachistochrone** (least time) were shown to be an inverted half-arch of a cycloid.

EXAMPLE 7   Parameterization of a Cycloid

Find a parameterization for the cycloid shown in Figure 6.7.7(b).

Solution A circle of radius a whose diameter initially lies along the y-axis rolls along the x-axis without slipping. We take as a parameter the angle θ (in radians) through which the circle has rotated. The point $P(x, y)$ starts at the origin, which corresponds to $\theta = 0$. As the circle rolls through an angle θ, its distance from the origin is the arc $PE = \overline{OE} = a\theta$. From FIGURE 6.7.9 we then see that the x-coordinate of P is

$$x = \overline{OE} - \overline{QE} = a\theta - a\sin\theta.$$

Now the y-coordinate of P is seen to be

$$y = \overline{CE} - \overline{CD} = a - a\cos\theta.$$

Hence parametric equations for the cycloid are

$$x = a\theta - a\sin\theta, \quad y = a - a\cos\theta. \tag{3}$$

As shown in Figure 6.7.7(a), one arch of a cycloid is generated by one rotation of the circle and corresponds to the parameter interval $0 \leq \theta \leq 2\pi$. ∎

The roofs of the parallel vaults of the Kimbell Art Museum in Fort Worth, Texas are cycloidal arches.

☐ **Parameterizations of Rectangular and Polar Curves** A curve C described by a continuous function $y = f(x)$ can always be parameterized by letting $x = t$. Parametric equations for C are then

$$x = t, \quad y = f(t). \tag{4}$$

Also, it is sometimes convenient to use parametric equations to plot the graphs of polar equations. This can be done using the conversion formulas $x = r\cos\theta$, $y = r\sin\theta$. Similarly, if $r = f(\theta)$, $\alpha \leq \theta \leq \beta$ describes a polar curve C, then a parameterization for C is given by

$$x = f(\theta)\cos\theta, \quad y = f(\theta)\sin\theta, \quad \alpha \leq \theta \leq \beta. \tag{5}$$

EXAMPLE 8   Parameterization of a Rectangular Curve

Find a parameterization of $f(x) = \sqrt{x - 1}$.

Solution If we let $x = t$, then from (4) a parameterization of the curve defined by f is

$$x = t, \quad y = \sqrt{t - 1}, \quad 1 \leq t < \infty.$$

Alternative Solution A curve can have many different parameterizations. If we let $x - 1 = t$, then another parameterization of the curve defined by f is

$$x = t + 1, \quad y = \sqrt{t}, \quad 0 \leq t < \infty. ∎$$

Note that the parameter ▶ interval is also changed.

EXAMPLE 9 **Parameterization of a Polar Curve**

Find a parameterization of the cardioid $r = 1 + \sin\theta$.

Solution With the identification $f(\theta) = 1 + \sin\theta$ and the knowledge that a complete cardioid is generated by allowing the values of θ range over the interval $[0, 2\pi]$, it follows from (5) that a parameterization of the polar curve defined by f is

$$x = (1 + \sin\theta)\cos\theta, \quad y = (1 + \sin\theta)\sin\theta, \quad 0 \le \theta \le 2\pi.$$

NOTES FROM THE CLASSROOM

In this section we have focused on **plane curves**, curves C defined parametrically in two dimensions. In the study of multivariable calculus you will see curves and surfaces in three dimensions that are defined by means of parametric equations. For example, a **space curve** C consists of a set of ordered triples $(f(t), g(t), h(t))$, where f, g, and h are defined on a common interval. Parametric equations for C are $x = f(t)$, $y = g(t)$, $z = h(t)$. For example, the **circular helix** such as shown in FIGURE 6.7.10 is a space curve whose parametric equations are

$$x = a\cos t, \quad y = a\cos t, \quad z = bt, \quad t \ge 0. \tag{6}$$

Surfaces in three dimensions can be represented by a set of parametric equations involving *two* parameters, $x = f(u, v)$, $y = g(u, v)$, $z = h(u, v)$. For example, the **circular helicoid** shown in FIGURE 6.7.11 arises from the study of minimal surfaces and is defined by the set of parametric equations similar to those in (6):

$$x = u\cos v, \quad y = u\sin v, \quad z = bv,$$

where b is a constant. The circular helicoid has a circular helix as its boundary. You might recognize the helicoid as the model for the rotating curved blade in machinery such as post hole diggers, ice augers, and snow blowers.

FIGURE 6.7.10
Circular helix

DNA is a
double helix

Helical antenna

FIGURE 6.7.11
Circular helicoid

Exercises 6.7   Answers to selected odd-numbered problems begin on page ANS–25.

In Problems 1 and 2, fill in the table for the given set of parametric equations. Find the x- and y-intercepts. Sketch the curve and indicate its orientation.

1. $x = t + 2, y = 3 + \frac{1}{2}t, \quad -\infty < t < \infty$

t	-3	-2	-1	0	1	2	3
x							
y							

2. $x = 2t + 1, y = t^2 + t, \quad -\infty < t < \infty$

t	-3	-2	-1	0	1	2	3
x							
y							

In Problems 3–10, sketch the curve that has the given set of parametric equations.

3. $x = t - 1, y = 2t - 1, \quad -1 \le t \le 5$
4. $x = t^2 - 1, y = 3t, \quad -2 \le t \le 3$
5. $x = \sqrt{t}, y = 5 - t, \quad t \ge 0$
6. $x = t^3 + 1, y = t^2 - 1, \quad -2 \le t \le 2$
7. $x = 3\cos t, y = 5\sin t, \quad 0 \le t \le 2\pi$
8. $x = 3 + 2\sin t, y = 4 + \sin t, \quad -\pi/2 \le t \le \pi/2$
9. $x = e^t, y = e^{3t}, \quad 0 \le t \le \ln 2$
10. $x = -e^t, y = e^{-t}, \quad t \ge 0$

In Problems 11–18, eliminate the parameter from the given set of parametric equations and obtain a rectangular equation that has the same graph.

11. $x = t^2, y = t^4 + 3t^2 - 1$
12. $x = t^3 + t + 4, y = -2(t^3 + t)$
13. $x = \cos 2t, y = \sin t, \quad -\pi/2 \le t \le \pi/2$
14. $x = e^t, y = \ln t, \quad t > 0$
15. $x = t^3, y = 3\ln t, \quad t > 0$
16. $x = \tan t, y = \sec t, \quad -\pi/2 < t < \pi/2$
17. $x = 4\cos t, y = 2\sin t, \quad 0 \le t \le 2\pi$
18. $x = -1 + \cos t, y = 2 + \sin t, \quad 0 \le t \le 2\pi$

In Problems 19–24, graphically show the difference between the given curves.

19. $y = x$ and $x = \sin t, y = \sin t$
20. $y = x^2$ and $x = -\sqrt{t}, y = t$
21. $y = \frac{1}{4}x^2 - 1$ and $x = 2t, y = t^2 - 1, \quad -1 \le t \le 2$
22. $y = -x^2$ and $x = e^t, y = -e^{2t}, \quad t \ge 0$
23. $x^2 - y^2 = 1$ and $x = \cosh t, y = \sinh t$ ← See (14) and (15) in Section 5.5.
24. $y = 2x - 2$ and $x = t^2 - 1, y = 2t^2 - 4$

In Problems 25–28 graphically show the difference between the given curves. Assume that $a > 0$ and $b > 0$,

25. $x = a\cos t, y = a\sin t, \quad 0 \le t \le \pi$
$x = a\sin t, y = a\cos t, \quad 0 \le t \le \pi$

26. $x = a\cos t, y = b\sin t, a > b, \quad \pi \le t \le 2\pi$
$x = a\sin t, y = b\cos t, a > b, \quad \pi \le t \le 2\pi$

27. $x = a\cos t, y = a\sin t, \quad -\pi/2 \le t \le \pi/2$
$x = a\cos 2t, y = a\sin 2t, \quad -\pi/2 \le t \le \pi/2$

28. $x = a\cos\dfrac{t}{2}, y = a\sin\dfrac{t}{2}, \quad 0 \le t \le \pi$

$x = a\cos\left(-\dfrac{t}{2}\right), y = a\sin\left(-\dfrac{t}{2}\right), \quad -\pi \le t \le 0$

In Problems 29–32, find the x- and y-intercepts of the given curves.

29. $x = t^2 - 2t, y = t + 1, \quad -2 \le t < 4$
30. $x = t^2 + t, y = t^2 + t - 6, \quad -5 \le t < 5$
31. $x = -1 + 2\cos t, y = 1 + 2\sin t, 0 \le t \le 2\pi$
32. $x = 1 + \sin t, y = \sin t - \cos t, 0 \le t \le 2\pi$

33. Show that parametric equations for a line through (x_1, y_1) and (x_2, y_2) are

$$x = x_1 + (x_2 - x_1)t, \quad y = y_1 + (y_2 - y_1)t, \quad -\infty < t < \infty.$$

What do these equations represent when $0 \le t \le 1$?

34. (a) Use the result of Problem 33 to find parametric equations of the line through $(-2, 5)$ to $(4, 8)$.
(b) Eliminate the parameter in part (a) to obtain a rectangular equation for the line.
(c) Find parametric equations for the line segment with $(-2, 5)$ as the initial point and $(4, 8)$ as the terminal point.

Applications

35. Fore! A famous golfer can generate a club head speed of approximately 130 mi/h or $v_0 = 190$ ft/s. If the golf ball leaves the ground at an angle $\theta_0 = 45°$, use (1) to find parametric equations for the path of the ball. What are the coordinates of the ball at $t = 2$ s?

36. Use the parametric equations obtained in Problem 35 to determine
(a) how long the golf ball is in the air,
(b) its maximum height, and
(c) the horizontal distance that the golf ball travels.

37. Piston Motion As shown in FIGURE 6.7.12, a piston is attached by means of a rod of length L to a circular crank mechanism of radius r. Parameterize the coordinates of the point P in terms of the angle ϕ shown in the figure.

38. Witch of Agnesi Consider a circle of radius a, which is tangent to the x-axis at the origin. Let B be a point on the horizontal line $y = 2a$ and let the line segment OB cut the circle at point A. As shown in FIGURE 6.7.13, the projection of AB on the vertical gives the line segment BP. Using the angle θ in the figure as a parameter, find parametric equations of the curve traced by the point P as A varies around the circle. The curve, more historically famous than useful, is called the witch of Agnesi.*

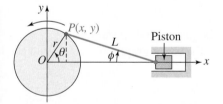

FIGURE 6.7.12 Crank mechanism in Problem 37

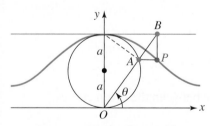

FIGURE 6.7.13 Witch of Agnesi in Problem 38

---

*No, the curve has nothing to do with witches and goblins. This curve, called *versoria*, which is Latin for a kind of rope, was included in a text on analytic geometry written in 1748 by the Italian mathematician **Maria Agnesi** (1718–1799). A translator of the text confused *versoria* with the Italian word *versiera*, which means *female goblin*. In English, *female goblin* became a *witch*.

Calculator/Computer Problems

In Problems 39–44, use a graphing utility to obtain the graph of the given set of parametric equations.

39. $x = 4\sin 2t, y = 2\sin t, \quad 0 \le t \le 2\pi$
40. $x = 6\cos 3t, y = 4\sin 2t, \quad 0 \le t \le 2\pi$
41. $x = 6\sin 4t, y = 4\sin t, \quad 0 \le t \le 2\pi$
42. $x = \cos t + t\sin t, y = \sin t - t\cos t, \quad 0 \le t \le 3\pi$
43. $x = 4\cos t - \cos 4t, y = 4\sin t - \sin 4t, \quad 0 \le t \le 2\pi$
44. $x = \cos^3 t, y = \sin^3 t, \quad 0 \le t \le 2\pi$

45. Trochoids In FIGURE 6.7.14 a blue circle of radius a rolls to the right without slipping on a horizontal line which we take to be the x-axis. The parametric equations

$$x = at - b\sin t, \quad y = a - b\cos t \qquad (7)$$

describe a family of curves, called trochoids, that are traced out by a point P that is located a distance b from the center of the blue circle. Note that the equations of a cycloid, (3) in Example 7, define a special trochoid when P lies on the circumference of the circle, that is, when $b = a$. If $b < a$, then the point P lies inside the circle and a trochoid is called a **curtate cycloid**. Use a graphing utility to obtain the graphs of the equations in (7) for the following cases. Use the parameter interval $[0, 8\pi]$.

(a) $a = 2, b = 1$ **(b)** $a = 1, b = 0.9$
(c) $a = 5, b = \frac{3}{2}$ **(d)** $a = 3, b = 2.3$

FIGURE 6.7.14 Rolling circle in Problem 45

As a bicycle tire rotates the reflector in its spokes travels in a curtate cycloid

Some violin makers create the arching of the back plate using a series of parallel arches in the shape of curtate cycloids

46. More Trochoids If $b > a$ in Problem 45, then the point P in Figure 6.7.14 lies outside the blue circle and the trochoid is called a **prolate cycloid**. Use a graphing utility to obtain the graphs of the equations in (7) for the following cases. Use the parameter interval $[-\pi, 9\pi]$.

(a) $a = 1, b = 2$ **(b)** $a = 1, b = 5$
(c) $a = 2, b = 2.5$ **(d)** $a = \frac{1}{2}, b = 4$

In Problems 47–50, use (5) to parameterize the curve whose polar equation is given. Use a graphing utility to obtain the graph of the resulting set of parametric equations.

47. $r = 2\sin\dfrac{\theta}{2}, \quad 0 \le \theta \le 4\pi$
 48. $r = 2\sin\dfrac{\theta}{4}, \quad 0 \le \theta \le 8\pi$

49. $r = 2\cos\dfrac{\theta}{5}, \quad 0 \le \theta \le 5\pi$
 50. $r = 2\cos\dfrac{3\theta}{2}, \quad 0 \le \theta \le 4\pi$

For Discussion

51. In FIGURE 6.7.15 the graphs of the parametric equations $x = f(t), y = g(t)$ are given, respectively, in the tx- and ty-planes. From these graphs, sketch the plane curve C defined by these parametric equations in the xy-plane.

52. The plane curve C defined by the parametric equations $x = f(t), y = g(t)$ is given in FIGURE 6.7.16. From this graph, sketch possible graphs of $x = f(t)$ and $y = g(t)$, respectively, in the tx- and ty-planes.

(a)   (b)

FIGURE 6.7.15 Graphs for Problem 51

FIGURE 6.7.16 Graph for Problem 52

6.8   3-Space

∫**Calculus** PREVIEW

FIGURE 6.8.1 Point in 2-space

In the plane, or **2-space**, one way of describing the position of a point P is to assign to it coordinates relative to two perpendicular coordinate axes called the x- and y-axes. The intersection of the two axes is called the origin and denoted by O. Recall,

- a vertical line $x = a$ consists of all points of the form (a, y), and
- a horizontal line $y = b$ consists of all points of the form (x, b).   (1)

If P is the point of intersection of the vertical line $x = a$ (perpendicular to the x-axis) and the horizontal line $y = b$ (perpendicular to the y-axis), then the ordered pair (a, b) is said to be the rectangular or Cartesian coordinates of the point. See FIGURE 6.8.1. In this section we extend this method of representation of a point to three dimensions.

☐ **Rectangular Coordinate System in 3-Space**   In three dimensions, or **3-space**, a rectangular coordinate system is constructed using three mutually perpendicular **coordinate axes**. The point at which these axes intersect is called the **origin** O. The axes drawn

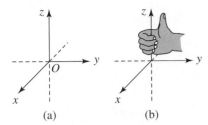

FIGURE 6.8.2 Three-dimensional coordinate axes

as solid lines in FIGURE 6.8.2(a) represent the positive axes, and are labeled in accordance with the so-called **right-hand rule** illustrated in Figure 6.8.2(b):

- If the fingers of the right hand, pointing in the direction of the positive x-axis, are curled toward the positive y-axis, then the thumb will point in the direction of a new axis perpendicular to the plane of the x- and y-axes. This new axis is labeled the z-axis.

The dashed lines in Figure 6.8.2(a) represent the negative axes. If the x- and y-axes are interchanged in Figure 6.8.2, the coordinate system is said to be **left-handed**. In 3-space, the graph of the equations $x = a$, $y = b$, and $z = c$ consist of all **ordered triples** or points of the form (a, y, z), (x, b, z), and (x, y, c), respectively. The graphs of the equations $x = a$, $y = b$, and $z = c$ are, in turn, planes perpendicular to the x-, y-, and z-axes. The point P at which these planes intersect can be represented by an **ordered triple** of numbers (a, b, c) said to be the **rectangular**, or **Cartesian**, **coordinates** of the point. The numbers a, b, and c are called the x-, y-, and z-coordinates of $P(a, b, c)$, respectively. See FIGURE 6.8.3.

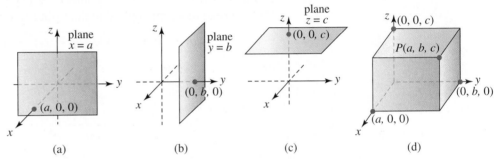

FIGURE 6.8.3 Three mutually perpendicular planes intersect in a point

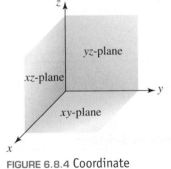

FIGURE 6.8.4 Coordinate planes

☐ **Octants** Each pair of coordinate axes determines a **coordinate plane**. As shown in green in FIGURE 6.8.4, the x- and y-axes determine the **xy-coordinate plane**, or simply, the **xy-plane**. Similarly, the y- and z-axes determine the **yz-plane**, and the x- and z-axes determine the **xz-plane**. The coordinate planes divide 3-space into eight regions known as **octants**. The octant in which all three coordinates of a point $P(a, b, c)$ are *positive* is called the **first octant**. There is no agreement for naming the other seven octants.

The following table summarizes the coordinates of a point either on a coordinate axis or in a coordinate plane. As seen in the table, we can also describe, say, the xy-plane by the simple equation $z = 0$. Similarly, the xz-plane is $y = 0$ and the yz-plane is $x = 0$. A point on a coordinate axes is not considered to be in any octant.

Axes	Coordinates	Plane	Coordinates
x	$(x, 0, 0)$	xy	$(x, y, 0)$
y	$(0, y, 0)$	xz	$(x, 0, z)$
z	$(0, 0, z)$	yz	$(0, y, z)$

■ EXAMPLE 1   **Graphing Points in 3-Space**

Graph the points $(4, 5, 6)$, $(3, -3, -1)$, and $(-2, -2, 0)$.

Solution Of the three points shown in FIGURE 6.8.5 only $(4, 5, 6)$ is in the first octant. The point $(-2, -2, 0)$ lies in the xy-plane.

FIGURE 6.8.5 Points in Example 1

☐ **Distance Formula** To find the **distance** between two points $P_1(x_1, y_1, z_1)$ and $P_2(x_2, y_2, z_2)$ in 3-space, let us first consider their projections onto the xy-plane. As seen in FIGURE 6.8.6, the distance between $(x_1, y_1, 0)$ and $(x_2, y_2, 0)$ follows from the usual distance formula in the plane and is $\sqrt{(x_2 - x_1)^2 + (y_2 - y_1)^2}$. Hence, from the Pythagorean theorem applied to the right triangle $P_1P_3P_2$, we have

$$[d(P_1, P_2)]^2 = [\sqrt{(x_2 - x_1)^2 + (y_2 - y_1)^2}]^2 + |z_2 - z_1|^2$$

or

$$d(P_1, P_2) = \sqrt{(x_2 - x_1)^2 + (y_2 - y_1)^2 + (z_2 - z_1)^2}. \tag{1}$$

FIGURE 6.8.6 Distance between two points in 3-space

EXAMPLE 2   **Distance Between Points in 3-Space**

Find the distance between $(2, -3, 6)$ and $(-1, -7, 4)$.

Solution From (1), the distance is

$$d = \sqrt{(2 - (-1))^2 + (-3 - (-7))^2 + (6 - 4)^2} = \sqrt{29}.$$

☐ **Midpoint Formula** The distance formula can be used to show that the coordinates of the **midpoint M** of the line segment in 3-space connecting the distinct points $P_1(x_1, y_1, z_1)$ and $P_2(x_2, y_2, z_2)$ are

$$M = \left(\frac{x_1 + x_2}{2}, \frac{y_1 + y_2}{2}, \frac{z_1 + z_2}{2} \right). \tag{2}$$

See Problem 68 in Exercises 6.8.

EXAMPLE 3   Midpoint in 3-Space

Find the coordinates of the midpoint of the line segment between the two points in Example 2.

Solution From (2) we obtain

$$\left(\frac{2 + (-1)}{2}, \frac{-3 + (-7)}{2}, \frac{6 + 4}{2}\right) \quad \text{or} \quad (\tfrac{1}{2}, -5, 5).$$ ▬

Review Section 1.4. ▶   ☐ **Sphere** Like a circle, a sphere can be defined in terms of the distance formula.

DEFINITION 6.8.1   Sphere

A **sphere** is the set of all points $P(x, y, z)$ in 3-space that are a given fixed distance r, called the **radius**, from a given fixed point C called the **center**.

If the center is $P_1(h, k, l)$, then a point $P(x, y, z)$ is on the sphere if and only if P_1 and P satisfy $[d(P_1, P)]^2 = r^2$, or

$$(x - h)^2 + (y - k)^2 + (z - l)^2 = r^2. \tag{3}$$

Equation (3) is called the **standard form** of the equation of a sphere.

FIGURE 6.8.7 Sphere in Example 4

EXAMPLE 4   **Graph of a Sphere**

Graph $x^2 + y^2 + z^2 = 25$.

Solution We identify $h = 0$, $k = 0$, $l = 0$, and $r^2 = 25 = 5^2$ in (3), and so the graph of $x^2 + y^2 + z^2 = 25$ is a sphere of radius 5 whose center is at the origin. The graph of the equation is given in FIGURE 6.8.7. ▬

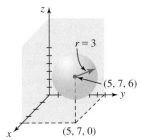

FIGURE 6.8.8 Sphere in Example 5

EXAMPLE 5   **Graph of a Sphere**

Graph $(x - 5)^2 + (y - 7)^2 + (z - 6)^2 = 9$.

Solution In this case we identify $h = 5$, $k = 7$, $l = 6$, and $r^2 = 9$. From (3) we see that the graph of $(x - 5)^2 + (y - 7)^2 + (z - 6)^2 = 3^2$ is a sphere with center $(5, 7, 6)$ and radius 3. Its graph lies entirely in the first octant and is shown in FIGURE 6.8.8. ▬

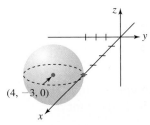

FIGURE 6.8.9 Sphere tangent to plane $y = 0$ in Example 6

EXAMPLE 6   **Equation of a Sphere**

Find an equation of the sphere whose center is $(4, -3, 0)$ that is tangent to the xz-plane.

Solution The perpendicular distance from the point $(4, -3, 0)$ to the xz-plane $(y = 0)$, and hence the radius of the sphere, is the absolute value of the y-coordinate, $|-3| = 3$. Thus, the standard form of the equation of the sphere is

$$(x - 4)^2 + (y + 3)^2 + z^2 = 3^2.$$

See FIGURE 6.8.9. ▬

To put an equation of a sphere in the standard form (3) it is necessary to complete the square in the three variables x, y, and z.

EXAMPLE 7 **Center and Radius**

Find the center and radius of the sphere whose equation is

$$2x^2 + 2y^2 + 2z^2 - 2x + y - 4z + 2 = 0.$$

Solution We first divide by 2, group like terms together, and then complete the square in x, y, and z:

$$(x^2 - x \quad) + (y^2 + \tfrac{1}{2}y \quad) + (z^2 - 2z \quad) = -1$$
$$[x^2 - x + (-\tfrac{1}{2})^2] + [y^2 + \tfrac{1}{2}y + (\tfrac{1}{4})^2] + [z^2 - 2z + (-1)^2]$$
$$= -1 + (-\tfrac{1}{2})^2 + (\tfrac{1}{4})^2 + (-1)^2$$
$$(x - \tfrac{1}{2})^2 + (y + \tfrac{1}{4})^2 + (z - 1)^2 = \tfrac{5}{16}.$$

From the last equation we see that the center and radius of the sphere are $(\tfrac{1}{2}, -\tfrac{1}{4}, 1)$ and $\tfrac{1}{4}\sqrt{5}$, respectively. ∎

☐ **Linear Equation in Three Variables** In the introduction to Section 2.3 we defined a linear equation in two variables to be $Ax + By + C = 0$. For various choices of the coefficients A, B, and C, the graph of a linear equation is a line in 2-space. The graph of a **linear equation in three variables**

$$Ax + By + Cz + D = 0, \tag{4}$$

A, B, C not all zero, is a **plane** in 3-space. We note that the simple equations $x = x_0$, $y = y_0$, and $z = z_0$, where x_0, y_0, and z_0 are constants, are special cases of (4). Here are two guidelines for graphing planes:

- The graphs of $x = x_0$, $y = y_0$, and $z = z_0$ are planes perpendicular to the x-, y-, and z-axes, respectively. See Figure 6.8.3.
- To graph a linear equation (4), find the x-, y-, and z-intercepts, or if necessary, find the trace of a plane in the coordinate planes.

A **trace** of a plane in a coordinate plane is the line of intersection of the plane with the coordinate plane. For example, by setting $z = 0$ we see that the trace of the plane $2x + 3y + 6z = 18$ in the xy-plane is the line $2x + 3y = 18$. To find the **intercepts** of a plane we use the fact that points on the x-, y-, and z-axes are of the form $(x, 0, 0)$, $(0, y, 0)$, and $(0, 0, z)$, respectively.

EXAMPLE 8 **Graph**

Graph the equation $2x + 3y + 6z = 18$.

Solution Setting:

$$\begin{aligned} y = 0, z = 0 \quad &\text{gives} \quad x = 9 \\ x = 0, z = 0 \quad &\text{gives} \quad y = 6 \\ x = 0, y = 0 \quad &\text{gives} \quad z = 3. \end{aligned}$$

As shown in FIGURE 6.8.10, we use the x-, y-, and z-intercepts $(9, 0, 0)$, $(0, 6, 0)$, and $(0, 0, 3)$ to draw the graph of the portion of the plane in the first octant. ∎

EXAMPLE 9 **Graph**

Graph the equation $x + y - z = 0$.

Solution First observe that the plane passes through the origin $(0, 0, 0)$. Now, the trace of the plane in the xz-plane ($y = 0$) is $z = x$, whereas its trace in the yz-plane ($x = 0$) is $z = y$. The traces are the two black lines in FIGURE 6.8.11. ∎

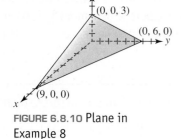

FIGURE 6.8.10 Plane in Example 8

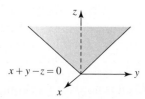

FIGURE 6.8.11 Plane in Example 9

(a) Trace in yz-plane

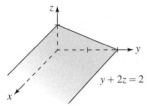

(b) Plane parallel to x-axis

FIGURE 6.8.12 Trace and plane in Example 10

☐ **Missing Variables** If *two* of the variables are missing in equation (4), then as we have already seen the plane is perpendicular to the coordinate axis corresponding to the variable present. For example, $z = 1$ is the equation of the plane perpendicular to the z-axis at the point $(0, 0, 1)$. Alternatively, we can interpret $z = 1$ as the equation of the plane through the point $(0, 0, 1)$ parallel to the coordinate plane corresponding to the missing two variables, in this case, parallel to the xy-plane. If *one* of the variables is missing in equation (4), the equation is then the same as the equation of the trace of the plane in the appropriate coordinate plane. The plane then is parallel to the coordinate axis corresponding to the missing variable. The following example illustrates this last idea.

EXAMPLE 10   **Graph**

Graph the equation $y + 2z = 2$.

Solution In 3-space the graph of the equation $y + 2z = 2$ is the graph of the set of ordered triples:

$$\{(x, y, z) \mid y + 2z = 2, x \text{ arbitrary}\}.$$

All we need do is draw the line $2z + y = 2$ in the yz-plane. Because the x variable is missing in the given equation the plane is drawn parallel to the x-axis. Necessarily the plane is perpendicular to the yz-plane. See **FIGURE 6.8.12**. ■

NOTES FROM THE CLASSROOM

(*i*) We have included this brief section on 3-space because in a typical three-semester course in calculus, the third semester deals primarily with calculus and vectors in three dimensions.

(*ii*) As we know in 2-space, two distinct points determine a line. Analogously, in 3-space, three noncollinear points (x_1, y_1, z_1), (x_2, y_2, z_2), (x_3, y_3, z_3) determine a plane. One way of finding an equation of the plane is by substituting the coordinates of the three points in (4) and solving the three simultaneous linear equations

$$\begin{aligned} Ax_1 + By_1 + Cz_1 + D &= 0 \\ Ax_2 + By_2 + Cz_2 + D &= 0 \\ Ax_3 + By_3 + Cz_3 + D &= 0 \end{aligned} \tag{5}$$

for A, B, and C in terms of D. We can then choose D to be any nonzero real number. See Problems 59–64 in Exercises 6.8.

(*iii*) The natural follow-up to this introduction to 3-space is to consider **multivariable functions**. The graph of a function f of one independent variable $y = f(x)$ is a *curve* in 2-space, whereas the graph of a function $z = f(x, y)$ of two independent variables is a *surface* in 3-space. For example, if we solve for z in Example 8 the resulting equation $z = -\frac{1}{3}x - \frac{1}{2}y + 3$ is a **linear function** whose graph, as we have seen, is a plane. If we solve the equation in Example 4 for z, then one result is a function $z = \sqrt{25 - x^2 - y^2}$ that describes the upper hemisphere of the sphere.

Graphing a function $z = f(x, y)$ can be difficult and may, at times, require a computer. If you have access to a computer lab, check to see whether the computers have 3D graphing software. The graphs of the polynomial function $z = 2x^2 - 2y^2 + 2$ and the trigonometric function $z = \sin xy$ shown in FIGURE 6.8.13 and FIGURE 6.8.14 were obtained using the computer algebra system (CAS) *Mathematica*.

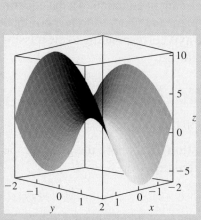

FIGURE 6.8.13 Graph of $z = 2x^2 - 2y^2 + 2$ for $-2 \le x \le 2, -2 \le y \le 2$

FIGURE 6.8.14 Graph of $z = \sin xy$ for $-2 \le x \le 2, -2 \le y \le 2$

Exercises 6.8 Answers to selected odd-numbered problems begin on page ANS-26.

In Problems 1–6, graph the given point.

1. $(1, 1, 5)$ **2.** $(0, 0, 4)$ **3.** $(3, 4, 0)$
4. $(6, 0, 0)$ **5.** $(6, -2, 0)$ **6.** $(5, -4, 3)$

In Problems 7–10, describe geometrically all points $P(x, y, z)$ whose coordinates satisfy the given conditions.

7. $z = 5$ **8.** $x = 1$
9. $x = 2, y = 3$ **10.** $x = 4, y = -1, z = 7$

11. Give the coordinates of the vertices of the rectangular parallelepiped whose sides are on the coordinate planes and the planes $x = 2, y = 5, z = 8$.

12. In FIGURE 6.8.15, two vertices are shown of a rectangular parallelepiped having sides parallel to the coordinate planes. Find the coordinates of the remaining six vertices.

13. Consider the point $P(-2, 5, 4)$.
  (a) If lines are drawn from P perpendicular to the coordinate planes, what are the coordinates of the point at the base of each perpendicular?
  (b) If a line is drawn from P to the plane $z = -2$, what are the coordinates of the point at the base of the perpendicular?
  (c) Find the point in the plane $x = 3$ that is closest to P.

14. Determine an equation of a plane parallel to a coordinate plane that contains the given pair of points.
  (a) $(3, 4, -5), (-2, 8, -5)$
  (b) $(1, -1, 1), (1, -1, -1)$
  (c) $(-2, 1, 2), (2, 4, 2)$

FIGURE 6.8.15 Parallelepiped in Problem 12

$(-1, 6, 7)$

$(3, 3, 4)$

In Problems 15–20, describe the set of points $P(x, y, z)$ in 3-space whose coordinates satisfy the given equation.

15. $xyz = 0$
17. $(x + 1)^2 + (y - 2)^2 + (z + 3)^2 = 0$
19. $z^2 - 25 = 0$

16. $x^2 + y^2 + z^2 = 0$
18. $(x - 2)(z - 8) = 0$
20. $x = y = z$

In Problems 21 and 22, find the distance between the given points.

21. $(3, -1, 2)$, $(6, 4, 8)$
22. $(-1, -3, 5)$, $(0, 4, 3)$

23. Find the distance from the point $(7, -3, -4)$
    (a) to the yz-plane and   **(b)** to the x-axis.
24. Find the distance from the point $(-6, 2, -3)$
    (a) to the xz-plane and   **(b)** to the origin.

In Problems 25–28, the given three points form a triangle. Determine which triangles are isosceles and which are right triangles.

25. $(0, 0, 0)$, $(3, 6, -6)$, $(2, 1, 2)$
27. $(1, 2, 3)$, $(4, 1, 3)$, $(4, 6, 4)$

26. $(0, 0, 0)$, $(1, 2, 4)$, $(3, 2, 2\sqrt{2})$
28. $(1, 1, -1)$, $(1, 1, 1)$, $(0, -1, 1)$

In Problems 29–32, use the distance formula to determine whether the given points are collinear.

29. $P_1(1, 2, 0)$, $P_2(-2, -2, -3)$, $P_3(7, 10, 6)$
30. $P_1(1, 2, -1)$, $P_2(0, 3, 2)$, $P_3(1, 1, -3)$
31. $P_1(1, 0, 4)$, $P_2(-4, -3, 5)$, $P_3(-7, -4, 8)$
32. $P_1(2, 3, 2)$, $P_2(1, 4, 4)$, $P_3(5, 0, -4)$

In Problems 33 and 34, solve for the unknown.

33. $P_1(x, 2, 3)$, $P_2(2, 1, 1)$; $d(P_1, P_2) = \sqrt{21}$
34. $P_1(x, x, 1)$, $P_2(0, 3, 5)$; $d(P_1, P_2) = 5$

In Problems 35 and 36, find the coordinates of the midpoint of the line segment between the given points.

35. $(1, 3, \frac{1}{2})$, $(7, -2, \frac{5}{2})$
36. $(0, 5, -8)$, $(4, 1, -6)$

37. The coordinates of the midpoint of the line segment between $P_1(x_1, y_1, z_1)$ and $P_2(2, 3, 6)$ are $(-1, -4, 8)$. Find the coordinates of P_1.
38. Let P_3 be the midpoint of the line segment between $P_1(-3, 4, 1)$ and $P_2(-5, 8, 3)$. Find the coordinates of the midpoint of the line segment
    (a) between P_1 and P_3 and   **(b)** between P_3 and P_2.

In Problems 39–42, sketch the graph of the given equation.

39. $x^2 + y^2 + z^2 = 9$
40. $x^2 + y^2 + (z - 3)^2 = 16$
41. $(x - 1)^2 + (y - 1)^2 + (z - 1)^2 = 1$
42. $(x + 3)^2 + (y + 4)^2 + (z - 5)^2 = 4$

In Problems 43–46, complete the square in x, y, and z to find the center and radius of the given sphere.

43. $x^2 + y^2 + z^2 + 8x - 6y - 4z - 7 = 0$
44. $4x^2 + 4y^2 + 4z^2 + 4x - 12z + 9 = 0$

45. $x^2 + y^2 + z^2 - 16z = 0$

46. $x^2 + y^2 + z^2 - x + y = 0$

In Problems 47–52, find an equation of a sphere that satisfies the given conditions.

47. Center $(-1, 4, 6)$; radius $\sqrt{3}$

48. Center $(0, -3, 0)$; diameter $\frac{5}{2}$

49. Center $(1, 1, 4)$; tangent to the xy-plane

50. Center $(5, 2, -2)$; tangent to the yz-plane

51. Center on the positive y-axis; radius 2; tangent to $x^2 + y^2 + z^2 = 36$

52. Center $(-3, 1, 2)$; passing through the origin

In Problems 53–58, graph the plane whose equation is given.

53. $5x + 2y + z = 10$

54. $3x + 2z = 9$

55. $3x + z - 6 = 0$

56. $3x + 4y - 2z - 12 = 0$

57. $-x + 2y + z = 4$

58. $3x - y - 6 = 0$

In Problems 59–64, use (4) and (5) to find an equation of a plane that contains the given points.

59. $(3, 5, 2), (2, 3, 1), (-1, -1, 4)$

60. $(0, 1, 0), (0, 1, 1), (1, 3, -1)$

61. $(-1, -1, 2), (1, 1, 1), (3, 2, -1)$

62. $(0, 0, 3), (0, -1, 0), (6, 0, 0)$

63. $(1, 2, -1), (4, 3, 1), (7, 4, 1)$

64. $(2, 1, 2), (4, 1, 0), (5, 2, -5)$

For Discussion

In Problems 65 and 66, discuss how the procedure used in Problems 59–64 can be used to find an equation of a plane that contains the given points. Carry out your ideas.

65. $(0, 0, 0), (1, 1, -1), (3, 2, 1)$

66. $(0, 0, 1), (0, 0, 5), (0, 2, 1)$

67. If you have ever sat at a four-legged table that rocks, you might consider replacing it with a three-legged table. Why?

68. Use the distance formula to prove that

$$M = \left(\frac{x_1 + x_2}{2}, \frac{y_1 + y_2}{2}, \frac{z_1 + z_2}{2} \right)$$

is the midpoint of the line segment between $P_1(x_1, y_1, z_1)$ and $P_2(x_2, y_2, z_2)$. [*Hint*: Show that $d(P_1, M) = d(M, P_2)$ and $d(P_1, P_2) = d(P_1, M) + d(M, P_2)$.]

In Problems 69–74, describe geometrically all points in 3-space whose coordinates satisfy the given condition(s).

69. $x^2 + y^2 + (z - 1)^2 = 4, 1 \le z \le 3$

70. $x^2 + y^2 + (z - 1)^2 = 4, z = 2$

71. $x^2 + y^2 + z^2 \ge 1$

72. $0 < (x - 1)^2 + (y - 2)^2 + (z - 3)^2 < 1$

73. $1 \le x^2 + y^2 + z^2 \le 9$

74. $1 \le x^2 + y^2 + z^2 \le 9, z \le 0$

In Problems 75 and 76, describe the surface in 3-space defined by the given set of points.

75. $\{(x, y, z) \,|\, x^2 + y^2 = 1\}$

76. $\{(x, y, z) \,|\, z = 1 - y^2\}$

Chapter 6 Review Exercises Answers to selected odd-numbered Problems begin on page ANS-27.

A. Fill in the Blanks

In Problems 1–22, fill in the blanks.

1. An equation in the standard form $y^2 = 4cx$ of a parabola with focus $(5, 0)$ is _____.

2. An equation in the standard form $x^2 = 4cy$ of a parabola through $(2, 6)$ is _____.

3. A rectangular equation of a parabola with focus $(1, -3)$ and directrix $y = -7$ is _____.

4. The directrix and vertex of a parabola are $x = -3$ and $(-1, -2)$, respectively. The focus of the parabola is _____.

5. The focus and directrix of a parabola are $(0, \frac{1}{4})$ and $y = -\frac{1}{4}$, respectively. The vertex of the parabola is _____.

6. The vertex and focus of the parabola $8(x + 4)^2 = y - 2$ are _____.

7. The eccentricity of a parabola is $e = $ _____.

8. The center and vertices of the ellipse $\dfrac{(x - 2)^2}{16} + \dfrac{(y + 5)^2}{4} = 1$ are _____.

9. The center and vertices of the hyperbola $y^2 - \dfrac{(x + 3)^2}{4} = 1$ are _____.

10. The oblique asymptotes of the hyperbola $y^2 - (x - 1)^2 = 1$ are _____.

11. The y-intercepts of the hyperbola $y^2 - (x - 1)^2 = 1$ are _____.

12. A horizontal line through the focus intersects the graph of the parabola $(x - 1)^2 = 16y$ at the points _____.

13. The center of a hyperbola with asymptotes $y = -\frac{5}{4}x + \frac{3}{2}$ and $y = \frac{5}{4}x + \frac{13}{2}$ is _____.

14. The eccentricity of the hyperbola $x^2 - y^2 = 1$ is _____.

15. If the graph of an ellipse is very elongated, then its eccentricity e is close to _____. (Fill in with 0 or 1.)

16. The rectangular coordinates of the point with polar coordinates $(-\sqrt{2}, 5\pi/4)$ are _____.

17. Polar coordinates of the point with rectangular coordinates $(0, -10)$ are _____.

18. Approximate polar coordinates of the point with rectangular coordinates $(-1, 3)$ are _____.

19. On the graph of the polar equation $r = 4\cos\theta$, two pairs of coordinates of the pole or origin are _____.

20. The equations $x = t + 2$, $y = 3 + \frac{1}{2}t$, $-\infty < t < \infty$, are a parametric representation of a _____.

21. The point on the curve defined by the parametric equations $x = -4 + 2\cos t$, $y = 2 + \sin t$, $-\infty < t < \infty$, corresponding to $t = 5\pi/2$ is _____.

22. The y-intercepts of the curve defined by the parametric equations $x = t^2 - 4$, $y = t^3 - 3t$, $-\infty < t < \infty$ are _____.

B. True/False

In Problems 1–22, answer true or false.

1. The axis of the parabola $x^2 = -4y$ is vertical. _____

2. The foci of an ellipse lie on its graph. _____

3. Rectangular coordinates of a point in the plane are unique. _____
4. For an ellipse, the length of the major axis is always greater than the length of the minor axis. _____
5. The vertex and focus are both on the axis of symmetry of a parabola. _____
6. The asymptotes for $(x - h)^2/a^2 - (y - k)^2/b^2 = 1$ must pass through (h, k). _____
7. An ellipse with eccentricity $e = 0.01$ is nearly circular. _____
8. The transverse axis of the hyperbola $x^2/9 - y^2/49 = 1$ is vertical. _____
9. The two hyperbolas $x^2 - y^2/25 = 1$ and $y^2/25 - x^2 = 1$ have the same pair of slant asymptotes. _____
10. The major axis of the ellipse $4(x + 1)^2 + 25(y - 3)^2 = 100$ lies on the line $y = 3$. _____
11. If P is a point on a parabola, then the perpendicular distance between P and the directrix equals the distance between P and the vertex. _____
12. If $y = 3x + 8$ is an asymptote of a hyperbola, then the slope of the other asymptote is $m = -3$. _____
13. The graph of $x^2 - 5y^2 = 5$ is symmetric about the x-axis, the y-axis, and the origin. _____
14. The graph of the polar equation $r = 5\sec\theta$ is a line. _____
15. $(3, \pi/6)$ and $(-3, -5\pi/6)$ are polar coordinates of the same point. _____
16. The graph of the ellipse $r = 90/(15 - \sin\theta)$ is nearly circular. _____
17. The graph of the curve $x = t^2$, $y = t^4 + 1$ is the same as the graph of $y = x^2 + 1$. _____
18. If P is a point on a parabola, then the perpendicular distance between P and the directrix equals the distance between P and the vertex. _____
19. The graph of $r = 2 + 4\sin\theta$ is a limaçon with an interior loop. _____
20. The graph of the rose curve $r = 5\sin 6\theta$ has 6 petals. _____
21. The polar graph $r^2 = 4\sin 2\theta$ is symmetric with respect to the origin. _____
22. The curve with parametric equations $x = 1 + \cos t$, $y = 1 + \sin t$, $0 \leq t \leq 2\pi$, is a circle of radius 1 centered at $(1, 1)$. _____

C. Review Exercises

In Problems 1–6, sketch the graph of the given conic section. Depending on the conic, find the center, vertex, vertices, endpoints of the minor axis, and asymptotes.

1. $(x - 1)^2 + 4(y - 1)^2 = 16$
2. $4y^2 - x^2 = 4$
3. $4x^2 + y^2 + 8x - 6y + 9 = 0$
4. $y^2 + 10y + 8x + 41 = 0$
5. $16y^2 - 9x^2 - 64y - 80 = 0$
6. $x^2 - 2x + 4y + 1 = 0$

In Problems 7 and 8, find a rectangular equation that has the same graph as the given polar equation.

7. $r = \cos\theta + \sin\theta$
8. $r(\cos\theta + \sin\theta) = 1$

In Problems 9 and 10, find a polar equation that has the same graph as the given rectangular equation.

9. $x^2 + y^2 - 4y = 0$
10. $(x^2 + y^2 - 2x)^2 = 9(x^2 + y^2)$.

11. Determine the rectangular coordinates of the vertices of the ellipse whose polar equation is $r = 2/(2 - \sin\theta)$.
12. Find a polar equation of the hyperbola with focus at the origin, vertices (in rectangular coordinates) $\left(0, -\frac{4}{3}\right)$ and $(0, -4)$, and eccentricity 2.

In Problems 13 and 14, find polar coordinates satisfying (a) $r > 0$, $-\pi < \theta \le \pi$, and (b) $r < 0$, $-\pi < \theta \le \pi$, for each point given in rectangular coordinates.

13. $(\sqrt{3}, -\sqrt{3})$ **14.** $(-\frac{1}{4}, \frac{1}{4})$

In Problems 15–26, identify and sketch the graph of the given polar equation.

15. $r = 5$ **16.** $\theta = -\pi/3$
17. $r = 5\sin\theta$ **18.** $r = -4\cos\theta$
19. $r = 4 - 4\cos\theta$ **20.** $r = 1 + \sin\theta$
21. $r = 2 + \sin\theta$ **22.** $r = 1 - 2\cos\theta$
23. $r = \sin 3\theta$ **24.** $r = 3\sin 4\theta$

25. $r = \dfrac{8}{3 - 2\cos\theta}$ **26.** $r = \dfrac{1}{1 + \cos\theta}$

27. Distance from a Satellite A satellite orbits the planet Neptune in an elliptical orbit with the center of the planet at one focus. If the length of the major axis of the orbit is 2×10^9 m and the length of the minor axis is 6×10^8 m, find the maximum distance between the satellite and the center of the planet.

28. Find an equation of the ellipse (in red) inscribed in the auxiliary rectangle (dashed black) of the hyperbola shown in FIGURE 6.R.1.

29. Mirror, Mirror . . . A parabolic mirror has a depth of 7 cm at its center and the distance across the top of the mirror is 20 cm. Find the distance from the vertex to the focus.

30. Identify the conic section that appears in the drawing of a wooden pencil in FIGURE 6.R.2.

In Problems 31 and 32, the graph of the given polar equation is rotated about the origin by the indicated amount.
    (a) Find a polar equation of the new graph.
    (b) Find a rectangular equation for the new graph.

31. $r = 2\cos\theta$; counterclockwise, $\pi/3$
32. $r = 1/(1 + \sin\theta)$; clockwise, $\pi/6$

In Problems 33 and 34, describe in words the graph of the given function.

33. $f(x) = \sqrt{36 - 9x^2}$ **34.** $f(x) = -\sqrt{36 + 9x^2}$

35. Find an equation of the ellipse when the center of $4x^2 + y^2 = 4$ is rigidly translated to the point $(-5, 2)$.

36. Find an equation of the hyperbola when the center of

$$9x^2 - y^2 - 54x - 2y + 71 = 0$$

is rigidly translated to the origin.

37. Find an equation of the parabola (blue) and of the ellipse (red) that have a common vertex $(6, 0)$ and common focus $(4, 0)$ as shown in FIGURE 6.R.3. Explain why this information does not determine a unique hyperbola (green). Find equations of two different hyperbolas that have a vertex $(6, 0)$ and a focus $(4, 0)$.

38. Describe in words the conic section and its general shape when its eccentricity is e is close to, but not equal to, 1.

39. Find an equation of a sphere that has a diameter with endpoints $(0, -4, 7)$ and $(2, 12, -3)$.

40. Find an equation of the plane such that the points (x, y, z) on the plane are equidistant from $(1, -2, 3)$ and $(2, 5, -1)$.

FIGURE 6.R.1 Ellipse in Problem 28

$$\frac{x^2}{a^2} - \frac{y^2}{b^2} = 1$$

Parabolic mirror in Problem 29

FIGURE 6.R.2 Pencil in Problem 30

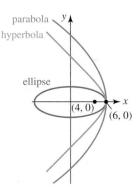

FIGURE 6.R.3 Graphs for Problem 37

© VIPDesignUSA/ShutterStock, Inc.

Final Examination

Try to answer the following questions without referring back to the text.

In Problems 1–20, fill in the blank.

1. Completing the square in x for $2x^2 + 6x + 5$ gives _____.
2. In the binomial expansion of $(1 - 2x)^3$ the coefficient of x^2 is _____.
3. In interval notation, the solution set of $\dfrac{x(x^2 - 9)}{x^2 - 25} \geq 0$ is _____.
4. If $a - 3$ is a negative number, then $|a - 3| =$ _____.
5. If $|5x| = 80$, then $x =$ _____.
6. If (a, b) is a point in the third quadrant, then $(-a, b)$ is a point in the _____ quadrant.
7. The point $(1, 7)$ is on a graph in the Cartesian plane. Give the coordinates of another point on the graph if the graph is:
   (a) symmetric with respect to the x-axis, _____
   (b) symmetric with respect to the y-axis, _____
   (c) symmetric with respect to the origin. _____
8. The lines $6x + 2y = 1$ and $kx - 9y = 5$ are parallel if $k =$ _____. The lines are perpendicular if $k =$ _____.
9. The complete factorization of the function $f(x) = x^3 - 2x^2 - 6x$ is _____.
10. The only potential rational zeros of $f(x) = x^3 + 4x + 2$ are _____.
11. The phase shift of the graph of $y = 5\sin(4x + \pi)$ is _____.
12. If $f(x) = x^4 \arctan(x/2)$, then the exact value of $f(-2)$ is _____.
13. If $\sin x = \frac{3}{5}$, $\pi/2 < x < \pi$, then $\sin 2x =$ _____.
14. $\arccos\left(-\frac{\sqrt{3}}{2}\right) =$ _____.
15. If $\left(\frac{1}{3}\right)^x = 81$, then $x =$ _____.
16. The domain of the function $y = \ln|x - 3|$ is _____.
17. $5\ln 2 - \ln\frac{2}{3} = \ln$ _____.
18. The graph of $y = \ln(2x + 5)$ has the vertical asymptote $x =$ _____.
19. The graph of the polar equation $r = \cos\theta$ is a _____.
20. The x- and y-intercepts of the curve C defined by the parametric equations $x = 6 + 3t$, $y = t^2 - 1$, $-\infty < t < \infty$ are _____.

In Problems 21–40, answer true or false.

21. The absolute value of any real number x is positive. _____
22. The inequality $|x| > -1$ has no solutions. _____
23. For any function f, if $f(a) = f(b)$, then $a = b$. _____
24. The graph of $y = f(x + c)$, $c > 0$, is the graph of $y = f(x)$ shifted c units to the right. _____
25. The points $(1, 3)$, $(3, 11)$, and $(5, 19)$ are collinear. _____
26. The function $f(x) = x^5 - 4x^3 + 2$ is an odd function. _____
27. $x + \frac{1}{4}$ is a factor of the function $f(x) = 64x^4 + 16x^3 + 48x^2 - 36x - 12$. _____
28. If $b^2 - 4ac < 0$, the graph of $f(x) = ax^2 + bx + c$, $a \neq 0$, does not cross the x-axis. _____

29. $f(x) = \dfrac{\sqrt{x}}{2x + 1}$ is a rational function. _____

30. If $f(x) = x^5 + 3x - 1$, then there exists a number c in $[-1, 1]$ such that $f(c) = 0$. _____

31. The graph of the function $f(x) = \dfrac{1}{x - 1} + \dfrac{1}{x - 2}$ has no x-intercepts. _____

32. $x = 0$ is a vertical asymptote for the graph of the rational function

$$f(x) = \dfrac{x^2 - 2x}{x}. \underline{\quad\quad}$$

33. The graph of $y = \cos(x/6)$ is the graph of $y = \cos x$ stretched horizontally. _____

34. $f(x) = \csc x$ is not defined at $x = \pi/2$. _____

35. The function $f(x) = e^{-4x^2}$ is not one-to-one. _____

36. The exponential function $f(x) = (\tfrac{3}{2})^x$ increases on the interval $(-\infty, \infty)$. _____

37. The domain of the function $f(x) = \ln x + \ln(x - 4)$ is $(4, \infty)$. _____

38. The solutions of the equation $\ln x^2 = \ln 3x$ are $x = 0$ and $x = 3$. _____

39. The major axis of the ellipse $x^2/9 + y^2/4 = 1$ is horizontal. _____

40. The focus of the parabola $y = 4x^2$ is $(0, 1)$. _____

41. Match the given interval with the appropriate inequality.

(*i*) $[2, 4]$	(*ii*) $[2, 4)$
(*iii*) $(2, 4)$	(*iv*) $(2, 4]$
(**a**) $|x - 3| \leq 1$	(**b**) $1 < x - 1 \leq 3$
(**c**) $-2 < 2 - x \leq 0$	(**d**) $|x - 3| < 1$

42. Write the solution of the absolute-value inequality $|3x - 1| > 7$ using interval notation.

43. The answer to a problem given in the back of a mathematics text is $1 + \sqrt{3}$, but your answer is $2/(\sqrt{3} - 1)$. Are you correct?

44. In which quadrants in the Cartesian plane is the quotient x/y negative?

45. Which one of the following equations best describes a circle that passes through the origin? The symbols $a, b, c, d,$ and e stand for different nonzero real constants.

(**a**) $ax^2 + by^2 + cx + dy + e = 0$	(**b**) $ax^2 + ay^2 + cx + dy + e = 0$
(**c**) $ax^2 + ay^2 + cx + dy = 0$	(**d**) $ax^2 + by^2 + cx + dy = 0$
(**e**) $ax^2 + ay^2 + e = 0$	(**f**) $ax^2 + ay^2 + cx + e = 0$

46. Match the given rational function f with the most appropriate phrase.

(*i*) $f(x) = \dfrac{x^4}{x^2 - 2}$	(*ii*) $f(x) = \dfrac{x^2}{x^2 + 2}$
(*iii*) $f(x) = \dfrac{x^5}{x^2 + 2}$	(*iv*) $f(x) = \dfrac{x^3}{x^2 + 2}$
(**a**) slant asymptote	(**b**) no asymptotes
(**c**) horizontal asymptote	(**d**) vertical asymptote

47. What is the range of the rational function $f(x) = \dfrac{10}{x^2 + 1}$?

48. What is the domain of the function $f(x) = \dfrac{\sqrt{x + 2}}{x^2}$?

49. Find an equation of the line that passes through the origin and through the point of intersection of the graphs of $x + y = 1$ and $2x - y = 7$.

50. Find a quadratic function f whose graph has the y-intercept $(0, -6)$ and the vertex of the graph is $(1, 4)$.

In calculus you are often required to rewrite a function either in a simpler form or in a form that is more helpful in solving the problem. In Problems 51–56, rewrite each function by following the given instruction. In calculus you would be expected to recognize what to do from the context of the actual problem.

51. $f(x) = \sqrt{x^6 + 4} - x^3$. Express f as a quotient using rationalization and simplification.

52. $f(x) = \dfrac{5x^3 - 4x^2\sqrt{x} + 8}{\sqrt[3]{x}}$. Carry out the indicated division and express each term as a power of x.

53. $f(x) = \dfrac{7x^2 - 7x - 6}{x^3 - x^2}$. Decompose f into partial fractions.

54. $f(x) = \dfrac{1}{1 + \sin x}$. Express f in terms of $\sec x$ and $\tan x$.

55. $f(x) = e^{3\ln x}$. Express f as a power of x.

56. $f(x) = |x^2 - 3x|$. Express f without absolute value signs.

In calculus you are often required to find zeros of a function. In Problems 57 and 58, solve the equation $f(x) = 0$ by following the given instruction.

57. $f(x) = x^2\frac{1}{2}(4 - x^2)^{-1/2}(-2x) + 2x\sqrt{4 - x^2}$. Rewrite f as a single expression without negative exponents.

58. $f(x) = 2\sin x\cos x - \sin x$. Find the zeros of f on the interval $[-\pi, \pi]$.

In Problems 59 and 60, compute and simplify the difference quotient
$$\frac{f(x + h) - f(x)}{h}$$ for the given function.

59. $f(x) = \dfrac{3x}{2x + 5}$

60. $f(x) = -x^3 + 10x^2$

61. Consider the trigonometric function $y = -8\sin(\pi x/3)$. What is the amplitude of the function? Give an interval over which one cycle of the graph is completed.

62. If $\tan\theta = \sqrt{5}$ and $\pi < \theta < 3\pi/2$, then what is the value of $\cos\theta$?

63. Suppose $f(x) = \sin x$ and $f(c) = 0.7$. What is the value of
$$2f(-c) + f(c + 2\pi) + f(c - 6\pi)?$$

64. Suppose $f(x) = \sin x$ and $g(x) = \ln x$. Solve $(f \circ g)(x) = 0$.

65. Find the x- and y-intercepts of the parabola whose equation is
$$(y + 4)^2 = 4(x + 1).$$

66. Find the center, foci, vertices, and endpoints of the minor axis of the ellipse whose equation is
$$x^2 + 2y^2 + 2x - 20y + 49 = 0.$$

67. The slant asymptotes of a hyperbola are $y = -5x + 2$ and $y = 5x - 8$. What is the center of the hyperbola?

68. From a point 220 ft from the base of a cell-phone antenna, a person measures a $30°$ angle of inclination from the ground to the top of the antenna. What is the angle of inclination to the top of the antenna if the person moves 100 ft closer to its base?

69. Iodine–131 is radioactive and is used in certain medical procedures. Assume that iodine–131 decays exponentially. If the half-life of I-131 is 8 days, then how much of a sample of 5 grams remains at the end of 15 days?

70. The polar coordinate equation $r = 3\cos 4\theta$ is a rose curve with eight petals. Find all radian-measure angles satisfying $0 \le \theta \le 2\pi$ for which $|r| = 3$.

71. Give the three trigonometric Pythagorean identities.

72. Without the aid of a calculator, find the exact value of

$$\cos 80° \cos 50° + \sin 80° \sin 50°.$$

73. Give the point that is common to the graphs of all exponential functions $f(x) = b^x, b > 0, b \ne 1$.

74. Give the y-intercept, the x-intercept, and horizontal asymptote for the graph of $f(x) = 4^x - 3$.

75. Describe how the graph of $y = \ln(-x)$ can be obtained from the graph of $y = \ln x$.

76. Find the asymptotes of the hyperbola

$$-x^2 + 10x + 9y^2 - 54y + 47 = 0.$$

77. Sketch the graph of the given function.

(a) $f(x) = \sqrt{4 - x^2}$ (b) $f(x) = -\frac{1}{2}\sqrt{4 - x^2}$

(c) $f(x) = \sqrt{x^2 - 4}$ (d) $f(x) = \sqrt{x^2 + 4}$

78. Without doing any work, describe in detail the graph of

$$r = \frac{10}{3 + 2\sin(\theta + 3\pi/4)}.$$

79. Without doing any work, describe in detail the graph of the polar equation $\theta = -\pi/6$.

80. Explain the difference between the graphs of the rectangular equation $y = x^2 - 1$ and the parametric equations $x = \sqrt{t}, y = t - 1, t \ge 0$.

© Sylvain Legare/ShutterStock, Inc.

Answers to Selected Odd-Numbered Problems

Exercises 1.1   Page 8

1. $a + 2 > 0$ **3.** $a + b \geq 0$ **5.** $2b + 4 \geq 100$

7. $(-\infty, 0)$

9. $[5, \infty)$

11. $(8, 10]$

13. $[-2, 4]$

15. $-7 \leq x \leq 9$ **17.** $x < 2$

19. $(-5, \infty)$

21. $(-\infty, 4]$

23. $\left(-\infty, \frac{3}{4}\right)$

25. $\left(-\infty, \frac{5}{2}\right]$

27. $(-10, 6)$

29. $(-5, 3)$

31. $[-10, -8)$

33. $[0, 6)$

35. $(-3, 3)$

37. $(-\infty, 0] \cup [5, \infty)$

39. $(2, 6)$

41. $\left[-\frac{3}{2}, 6\right]$

43. $(-\infty, -1) \cup (2, 4)$

45. $[-2, -1] \cup [1, 2]$

47. $(-\infty, -8)$

49. $(-\infty, -5] \cup (0, \infty)$

51. $\left(-\infty, \frac{1}{3}\right) \cup (1, \infty)$

53. $(-5, 0] \cup [1, \infty)$

55. $(-\infty, -1) \cup [1, 2]$

57. $(-\infty, -3) \cup (-1, 1)$

59. If x is the number, then $x < 8$.   **61.** $n > 10$

63. If x denotes the width, then $x > 7$.   **65.** $R > \frac{10}{3}$

67. $(12, 20)$

Exercises 1.2   Page 15

1. $4 - \pi$ **3.** $8 - \sqrt{63}$

5. 4 **7.** $-h$

9. $-x + 6$ **11.** 0

13. $-2x + 7$ **15.** 3

17. $2x - 2$ **19.** 4

21. $4; 5$ **23.** $3; 0$

25. $a = 2, b = 8$ **27.** $m = 4 + \pi, b = 4 + 2\pi$

29. $-\frac{1}{4}, \frac{3}{4}$ **31.** $-\frac{1}{2}, \frac{5}{6}$

33. $\frac{2}{3}, 2$

35. $\left(-\frac{4}{5}, \frac{4}{5}\right)$

37. $(-\infty, -10) \cup (4, \infty)$

39. $[3, 4]$

41. $\left(-\infty, -1 - \sqrt{2}\right] \cup \left[1 - \sqrt{2}, \infty\right)$

43. $\left(-\frac{7}{3}, 3\right)$

45. $(4.99, 5.01)$

47. $|x - 4| < 7$ **49.** $|x - 5| > 4$

51. $|x + 3| \geq 2, (-\infty, -5] \cup [-1, \infty)$

53. $|A_B - A_M| \leq 3$ **55.** $(11.95, 12.05)$

Exercises 1.3   Page 21

1.

3.

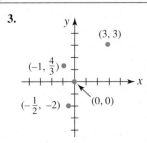

5. II   **7.** III
9. II   **11.** I
13. III   **15.** IV
17.

19. (3, 6)
21.

23.

25.   **27.** $2\sqrt{5}$ **29.** 10

31. 5   **33.** the triangle is neither isosceles nor a right triangle
35. the triangle is not isosceles but is a right triangle
37. the triangle is an isosceles right triangle
39. $(-3\sqrt{3}, 3)$ or $(3\sqrt{3}, 3)$
41. (a) $2x + y - 5 = 0$
    (b) The points (x, y) lie on the perpendicular bisector of the
         line segment joining A and B.
43. (6, 8) and (6, −4)   **45.** (−3, −3), (1, 1)   **47.** $(1, \frac{5}{2})$
49. $(-\frac{9}{2}, \frac{5}{2})$ **51.** $(3a, -\frac{3}{2}b)$ **53.** (5, −1)
55. (−7, −10)   **57.** 6   **59.** (2, −5)
61. $(\frac{7}{2}, \frac{13}{2}), (4, 7), (\frac{9}{2}, \frac{15}{2})$

Exercises 1.4   Page 30

1. center (0, 0), radius $\sqrt{5}$ **3.** center (0, 3), radius 7

5. center $(\frac{1}{2}, \frac{3}{2})$, radius 1

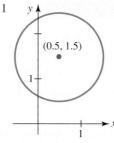

7. center (0, −4), radius 4   **9.** center (−1, 2), radius 3
11. center (10, −8), radius 6   **13.** center (−1, −4), radius $\sqrt{\frac{33}{2}}$
15. $x^2 + y^2 = 1$ **17.** $x^2 + (y − 2)^2 = 2$
19. $(x − 1)^2 + (y − 6)^2 = 8$ **21.** $x^2 + y^2 = 5$
23. $(x − 5)^2 + (y − 6)^2 = 36$
25.

27.

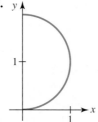

29. $y = 3 + \sqrt{4 − x^2}; x = \sqrt{4 − (y − 3)^2}$
31.

33.

35. $1 < (x + 2)^2 + (y − 2)^2 \le 9$
37. $(3 − \sqrt{13}, 0), (3 + \sqrt{13}, 0), (0, −6 − 2\sqrt{10}),$
     $(0, −6 + 2\sqrt{10})$
39. $(-1 − \sqrt{5}, 0), (−1 + \sqrt{5}, 0), (0, 2 − 2\sqrt{2}),$
     $(0, 2 + 2\sqrt{2})$
41. (0, 0), origin   **43.** $(−1, 0), (0, \frac{1}{2})$, no symmetry
45. (0, 0), x-axis   **47.** (−2, 0), (2, 0), (0, −4), y-axis
49. $(1 − \sqrt{3}, 0), (1 + \sqrt{3}, 0), (0, −2)$, no symmetry
51. $(0, 0), (−\sqrt{3}, 0), (\sqrt{3}, 0)$, origin
53. (0, −4), (0, 4), x-axis

55. $(0, -3), (0, 3)$, x-axis, y-axis, and origin
57. $(-\sqrt{7}, 0), (\sqrt{7}, 0)$, origin
59. $(-4, 0), (5, 0), (0, -\frac{10}{3})$, no symmetry
61. $(9, 0), (0, -3)$, no symmetry **63.** $(9, 0), (0, 9)$, no symmetry
65. $(-4, 0), (4, 0), (0, -4), (0, 4)$, x-axis, y-axis, and origin
67. x-axis, y-axis, and origin **69.** y-axis

71.

73.

75.
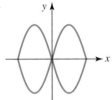

77. $(x - r)^2 + (y - r)^2 = r^2$
79. $32, 16\pi - 32$

---

17. $(-\infty, -3]$　　　　**19.** $(4, 12)$
21. $(-\infty, -10) \cup (10, \infty)$　　**23.** $\left(-\frac{1}{3}, 3\right)$
25. $\left[-1, \frac{5}{2}\right]$　　**27.** $(-1, 0) \cup (1, \infty)$　　**29.** $(0, 1) \cup (1, \infty)$
31. $(x - 1)^2 + (y - 1)^2 = 1,\quad (x - 2)^2 + (y - 2)^2 = 4$
33. $x^2 + (y + 1)^2 = 1,\quad x^2 + (y + 2)^2 = 4$　　**35.** $\frac{3}{10} < d_o < \frac{3}{4}$
37. (a) $\dfrac{1}{2x + 1}$　(b) $\frac{1}{2}$　　**39.** (a) $(\sqrt{x} + 2)(x + 4)$　(b) 32

Exercises 2.1　　Page 57

1. $24, 2, 8, 35$　　　　　　**3.** $0, 1, 2, \sqrt{6}$
5. $-\frac{3}{2}, 0, \frac{3}{2}, \sqrt{2}$
7. $-2x^2 + 3x, -8a^2 + 6a, -2a^4 + 3a^2, -50x^2 - 15x,$
$-8a^2 - 2a + 1, -2x^2 - 4xh - 2h^2 + 3x + 3h$
9. $-2, 2$　　　　　　　　**11.** $\left[\frac{1}{2}, \infty\right)$
13. $(-\infty, 1)$　　　　　　**15.** $\{x \mid x \neq 0, x \neq 3\}$
17. $\{x \mid x \neq 5\}$　　　　　**19.** $(-\infty, \infty)$
21. $[-5, 5]$　　　　　　　**23.** $(-\infty, 0] \cup [5, \infty)$
25. $(-2, 3]$　　　　　　　**27.** not a function
29. function　　　　　　　**31.** $[-4, 4], [0, 5]$
33. $[1, 9], [1, 6]$　　　　　**35.** $-\frac{6}{5}$
37. $2, 3$　　　　　　　　　**39.** $0, \frac{1}{3}, -9$
41. $-1, 1$　　　　　　　　**43.** $(8, 0), (0, -4)$
45. $\left(\frac{3}{2}, 0\right), \left(\frac{5}{2}, 0\right), (0, 15)$　　　**47.** $\left(0, -\frac{1}{4}\right)$
49. $(-2, 0), (2, 0), (0, 3)$
51. $f(x) = 4x - 11$　　　　**53.** $f(x) = -\frac{1}{4}x^3 - x + 1$
55. $f(x) = \dfrac{2x - 10}{x}$　　　　**57.** $0, -3.4, 0.3, 2, 3.8, 2.9; (0, 2)$
59. $3.6, 2, 3.3, 4.1, 2, -4.1; (-3.2, 0), (2.3, 0), (3.8, 0)$
61. (a) $2; 6; 120; 5040$　　　　(c) $(n + 1)(n + 2)$

Exercises 2.2　　Page 67

1. even　　**3.** neither even nor odd　　**5.** odd
7. even　　**9.** even　　**11.** odd　　**13.** neither even nor odd
15. (a)　　　　　　　　　(b)

17. (a)　　　　　　　　　(b)

19. $f(2) = 4, f(-3) = 7$　　**21.** $g(1) = 5, g(-4) = -8$
23. $(-2, 3), (3, -2)$　　　**25.** $(-8, 1), (-3, -4)$
27. $(-6, 2), (-1, -3)$　　**29.** $(2, 1), (-3, -4)$
31. $(-2, 15), (3, -60)$
33. (a)　　　　　　　　　(b)

Exercises 1.5　　Page 43

1. (a) $x + 5$　　(b) 10　　**3.** (a) $x - 6$　　(b) -5
5. (a) $\dfrac{x + 3}{x - 3}$　(b) -5　　**7.** (a) $x^2 + x + 1$　(b) 3
9. (a) $\dfrac{x^2 + x + 1}{x + 4}$　(b) $\frac{3}{5}$　　**11.** (a) $\dfrac{x + 1}{x^2 - x + 1}$　(b) 0
13. (a) $4 + h$　　(b) 4　　**15.** (a) $4(x + 2)$　　(b) 12
17. (a) $3 + 3x + x^2$　(b) 3　　**19.** (a) $2h^2 + h - 4$　(b) -4
21. (a) $\dfrac{1}{x + 4}$　(b) $\frac{1}{6}$　　**23.** (a) $\dfrac{1}{x + 10}$　(b) $\frac{1}{20}$
25. (a) $-\dfrac{4 + h}{4(2 + h)^2}$　(b) $-\frac{1}{4}$　　**27.** (a) $\dfrac{1}{\sqrt{x} + 3}$　(b) $\frac{1}{6}$
29. (a) $\sqrt{7 + x} + \sqrt{7}$ (b) $2\sqrt{7}$　**31.** (a) $5 + \sqrt{t}$　(b) 10
33. (a) $4\left(\sqrt{y^2 + y + 1} + \sqrt{y + 1}\right)$　(b) 8
35. $\dfrac{ax - 1}{ax}$　　　　**37.** $2(2x - 3)^3(2x - 1)(9x + 10)$
39. $\dfrac{6x(2 - x)}{(-4x + 6)^{3/2}}$　　**41.** $y' = \dfrac{x + y}{3y^2 - x}$
43. $y' = \dfrac{2x}{1 - 2y}$　　**45.** $y' = \dfrac{x^2 - 2xy + 2y + y^2}{2x}$

Chapter 1 Review Exercises　　Page 46

A. 1. $x \leq 9$　　　**3.** II　　**5.** $(2, -3)$
7. $(x + 2)^2 + (y + 5)^2 = 36$　**9.** $\sqrt{10}$
11. $\left(-\frac{5}{2}, 0\right), \left(\frac{5}{2}, 0\right), (0, -5)$　**13.** center $(8, 0)$, radius 8
15. $(-3, 4), (-3, -4)$　　**17.** $x^2 + y^2 > 36$
19. $x^2 + y^2 = 27$　　**21.** $|x - \sqrt{2}| > 3$

B. 1. false　　**3.** true　　**5.** false　　**7.** true
9. true　　**11.** true　　**13.** true　　**15.** false
17. true　　**19.** true　　**21.** true

C. 1. $a^2 < ab$　　　　**3.** $a < a + b$
5. 10　　　　　　**7.** $\leq$
9. $-4 \leq x \leq 3$　　**11.** $a = 4, b = 6$
13. (a) $-6 < x \leq 2$　　(b) $(-6, 2]$
15. (a) $x \geq -4$ or $-4 \leq x < \infty$　(b) $[-4, \infty)$

(c)

(d)

(e)

(e)

(f)

(f)

35. (a)

(b)

(g)

(c)

(d)

(h)

(e)

(f)

39. $y = (x - 1)^3 + 5$ **41.** $y = -(x + 7)^4$

43.   **45.**

37. (a)

(b)

(c)

Exercises 2.3   Page 75

1. $-\frac{7}{2}$;   **3.** 5;

(d)

5. -1;   **7.** $-\frac{5}{12}$

9. $\frac{3}{4}$; $(-4, 0)$, $(0, 3)$;

11. $\frac{2}{3}$; $\left(\frac{9}{2}, 0\right)$, $(0, -3)$;

13. $-\frac{2}{5}$; $(4, 0)$, $\left(0, \frac{8}{5}\right)$;

15. $-\frac{2}{3}$; $\left(\frac{3}{2}, 0\right)$, $(0, 1)$

17. $y = \frac{2}{3}x + \frac{4}{3}$

19. $y = 2$

21. $y = -x + 3$

23. $y = -2x + 7$

25. $y = 1$

27. $x = -2$

29. $y = -3x - 2$

31. $x = 5$

33. $y = -4x + 11$ **35.** $y = -\frac{1}{5}x - 5$ **37.** $\left(\frac{63}{16}, \frac{31}{8}\right)$

39. (a) and (c) are parallel, (b) and (e) are parallel; (a) and (c) are perpendicular to (b) and (e); (d) is perpendicular to (f)

41. (a) and (d) are perpendicular, (b) and (c) are perpendicular, (e) and (f) are perpendicular

43. $f(x) = \frac{1}{2}x + \frac{11}{2}$

45. $\left(-\frac{5}{6}, \frac{8}{3}\right)$;

47. $(-1, 3)$;

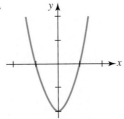

49. -9 **51.** $y = x + 3$ **53.** (a) $T_F = \frac{9}{5}T_C + 32$

55. 1,680; approximately 35.3 years

Exercises 2.4   Page 84

1.

3.

5.

7. (a) $(0, 0)$, $(-5, 0)$

   (b) $y = \left(x + \frac{5}{2}\right)^2 - \frac{25}{4}$

   (c) $\left(-\frac{5}{2}, -\frac{25}{4}\right)$, $x = -\frac{5}{2}$

   (d)

9. (a) $(-1, 0)$, $(3, 0)$, $(0, 3)$

   (b) $y = -(x - 1)^2 + 4$

   (c) $(1, 4)$, $x = 1$

(d)

11. (a) $(1, 0)$, $(2, 0)$, $(0, 2)$

   (b) $y = \left(x - \frac{3}{2}\right)^2 - \frac{1}{4}$

   (c) $\left(\frac{3}{2}, -\frac{1}{4}\right)$, $x = \frac{3}{2}$

   (d)

13. (a) $(0, 3)$

   (b) $y = 4\left(x - \frac{1}{2}\right)^2 + 2$

   (c) $\left(\frac{1}{2}, 2\right)$, $x = \frac{1}{2}$

   (d)

15. (a) $\left(1 - \sqrt{3}, 0\right)$, $\left(1 + \sqrt{3}, 0\right)$, $(0, 1)$

   (b) $y = -\frac{1}{2}(x - 1)^2 + \frac{3}{2}$

   (c) $\left(1, \frac{3}{2}\right)$, $x = 1$

   (d)

17. (a) $(5, 0)$, $(0, 25)$

   (b) $y = (x - 5)^2$

   (c) $(5, 0)$, $x = 5$

   (d)

19. Minimum functional value is $f\left(\frac{4}{3}\right) = -\frac{13}{3}$; $\left[-\frac{13}{3}, \infty\right)$

21. Increasing on $[0, \infty)$, decreasing on $(-\infty, 0]$

23. Increasing on $(-\infty, -3]$, decreasing on $[-3, \infty)$

25. The graph of $y = x^2$ is shifted horizontally 10 units to the right.

27. The graph of $y = x^2$ is compressed vertically, followed by a reflection in the x-axis, followed by a horizontal shift of 4 units to the left, followed by vertical shift of 9 units upward.

29. As the equation is given it can be interpreted as the graph of $y = x^2$ shifted horizontally 6 units to the right, followed by a reflection in the y-axis, followed by a vertical shift of 4 units downward.

31. $y = (x + 2)^2$ **33.** $y = -x^2 - 1$

35. $y = -(x - 1)^2 + 5$ **37.** $f(x) = 2x^2 + 3x + 5$

39. $f(x) = 4(x - 1)^2 + 2$

41. $(-4, 8), (1, 3),$ **43.** $(-2, 2), (0, 2),$

45. $(-\infty, 3 - \sqrt{2}) \cup (3 + \sqrt{2}, \infty)$ **47.** $\{-\sqrt{3}\}$

49. $(-2, 3), (4, 9); (-\infty, -2] \cup [4, \infty)$ **51.** 19

53. (a) $d^2 = 5x^2 - 10x + 25$ (b) $(1, 2)$

55. (a) $s(t) = -16t^2 + 64t + 6, v(t) = -32t + 64$
    (b) 70 ft, 0 ft/s   (c) 4 s, -64 ft/s

57. (a) 117.6 m, -9.8 m/s
    (b) in 5 seconds
    (c) -49 m/s

59. (a) The graph of $R(D) = -kD^2 + kPD$ is a parabola with vertex at $-b/2a = (-kP)/(-2k) = P/2$. Since k is positive, the graph opens downward, and so $R(D)$ is a maximum at this value. Since $R(D)$ measures the rate at which the disease spreads, we conclude that the disease spreads most rapidly when exactly one-half the population is infected.
    (b) 3×10^{-5}
    (c) approximately 48
    (d) approximately 62, 79, 102, and 130

Exercises 2.5   **Page 92**

1. $2, 4, -5$ **3.** $3, 0, 8, 2 + 2\sqrt{2}$

5. (a) 1   (b) 1   (c) 0   (d) 1
    (e) 1   (f) 0

7. (a) 3   (b) $-1, \sqrt{2}$ (c) $\sqrt[3]{-2}, 1$ (d) $\sqrt[3]{-3}, 0$
    (e) $\sqrt{3}$ (f) -2

9. $(0, 0),$ continuous,   **11.** $(0, 0),$ continuous,

13. x-intercepts are the points $(x, 0)$, where $-2 \le x < -1$, y-intercept is $(0, 2)$, the function is discontinuous at every integer value of x,

15. x-intercepts are the points $(x, 0)$, where $0 \le x < 1$, y-intercept is $(0, 0)$, the function is discontinuous at every integer value of x,

17. $(-3, 0), (0, 3),$ continuous,

19. $(-2, 0), (2, 0), (0, 2),$ continuous,

21. $(-3, 0), (1, 0), (0, -1),$ continuous,

23. $\left(\frac{5}{3}, 0\right), (0, -5),$ continuous,

25. $(-1, 0), (1, 0), (0, 1),$ continuous,

27. $(0, 0), (2, 0),$ continuous,

29. $(-2, 0), (2, 0), (0, 2),$ continuous,

31. $(1, 0), (0, 1),$ continuous,

ANS-6

33. (1, 0), (0, 1), continuous,

35.

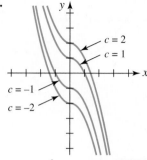

37. $\{-1, 1\}$

39. $f(x) = \begin{cases} x + 2, & x < 0 \\ -2x + 2, & 0 \le x < 2 \\ -2, & x \ge 2 \end{cases}$

41. $f(x) = \begin{cases} -x, & x < -3 \\ \sqrt{9 - x^2}, & -3 \le x < 3 \\ x, & x \ge 3 \end{cases}$

43.

45. $f(x) = \begin{cases} 1, & x > 0 \\ -1, & x < 0 \end{cases}$

47. $k = 1$

49. $g(x) = \lceil x \rceil = \begin{cases} -2, & -3 < x \le -2 \\ -1, & -2 < x \le -1 \\ 0, & -1 < x \le 0 \\ 1, & 0 < x \le 1 \\ 2, & 1 < x \le 2 \\ 3, & 2 < x \le 3 \end{cases}$

Exercises 2.6 **Page 100**

1. $(f + g)(x) = 3x^2 - x + 1$, domain: $(-\infty, \infty)$
$(f - g)(x) = -x^2 + x + 1$, domain: $(-\infty, \infty)$
$(fg)(x) = 2x^4 - x^3 + 2x^2 - x$, domain: $(-\infty, \infty)$
$(f/g)(x) = (x^2 + 1)/(2x^2 - x)$,
domain: real numbers except $x = 0$ and $x = \frac{1}{2}$

3. $(f + g)(x) = x + \sqrt{x - 1}$, domain: $[1, \infty)$,
$(f - g)(x) = x - \sqrt{x - 1}$, domain: $[1, \infty)$,
$(fg)(x) = x\sqrt{x - 1}$, domain: $[1, \infty)$,
$(f/g)(x) = x/\sqrt{x - 1}$, domain: $(1, \infty)$

5. $(f + g)(x) = 3x^3 - 3x^2 + 3x + 1$, domain: $(-\infty, \infty)$,
$(f - g)(x) = 3x^3 - 5x^2 + 7x - 1$, domain: $(-\infty, \infty)$,
$(fg)(x) = 3x^5 - 10x^4 + 16x^3 - 14x^2 + 5x$,
domain: $(-\infty, \infty)$,
$(f/g)(x) = (3x^3 - 4x^2 + 5x)/(1 - x)^2$,
domain: real numbers except $x = 1$

7. $(f + g)(x) = \sqrt{x + 2} + \sqrt{5 - 5x}$, domain: $[-2, 1]$,
$(f - g)(x) = \sqrt{x + 2} - \sqrt{5 - 5x}$, domain: $[-2, 1]$,
$(fg)(x) = \sqrt{5(x + 2)(1 - x)}$, domain: $[-2, 1]$,
$\left(\dfrac{f}{g}\right)(x) = \sqrt{\dfrac{x + 2}{5 - 5x}}$, domain: $[-2, 1)$

9. $(f + g)(x) = -x + 3$, domain: $(-\infty, \infty)$
$(f - g)(x) = 2x^2 - x - 15$, domain: $(-\infty, \infty)$
$(fg)(x) = -x^4 + x^3 + 15x^2 - 9x - 54$, domain: $(-\infty, \infty)$
$(f/g)(x) = -\dfrac{x + 2}{x + 3}$, domain: $(-\infty, -3) \cup (-3, 3) \cup (3, \infty)$

11. $10, 8, -1, 2, 0$

13. $(f \circ g)(x) = x$, domain: $[1, \infty)$,
$(g \circ f)(x) = \sqrt{x^2} = |x|$, domain: $(-\infty, \infty)$

15. $(f \circ g)(x) = \dfrac{1}{2x^2 + 1}$, domain: $(-\infty, \infty)$,
$(g \circ f)(x) = \dfrac{4x^2 - 4x + 2}{4x^2 - 4x + 1}$,
domain: real numbers except $x = \frac{1}{2}$

17. $(f \circ g)(x) = x$, $(g \circ f)(x) = x$

19. $(f \circ g)(x) = \dfrac{x^3 + 1}{x}$, $(g \circ f)(x) = \dfrac{x^2}{x^3 + 1}$

21. $(f \circ g)(x) = x + 1 + \sqrt{x - 1}$, $(g \circ f)(x) = x + 1 + \sqrt{x}$

23. $(f \circ f)(x) = 4x + 18$, $\left(f \circ \dfrac{1}{f}\right)(x) = \dfrac{6x + 19}{x + 3}$

25. $(f \circ f)(x) = x^4$, $\left(f \circ \dfrac{1}{f}\right)(x) = \dfrac{1}{x^4}$

27. $(f \circ g \circ h)(x) = |x - 1|$ **29.** $(f \circ g \circ g)(x) = 54x^4 + 7$

31. $(f \circ f \circ f)(x) = 8x - 35$ **33.** $f(x) = x^5, g(x) = x^2 - 4x$

35. $f(x) = x^2 + 4\sqrt{x}, g(x) = x - 3$

37.

39. $y = \begin{cases} x + 3, & -3 \le x < -2 \\ x + 2, & -2 \le x < -1 \\ x + 1, & -1 \le x < 0 \\ x, & 0 \le x < 1 \\ x - 1, & 1 \le x < 2 \\ x - 2, & 2 \le x < 3 \end{cases}$

41.

43.

45.

47. **(a)** $(-2, 3), (1, 0)$ **(b)** $d = -x^2 - x + 2$ **(c)** $\frac{9}{4}$

49. $d = \sqrt{10,000 + 250,000r^2}$; approximately 2,502 ft

1. $y = 2x$, $y = -2x$

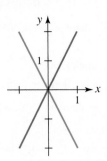

3. $y = \sqrt{5-x}$, $y = -\sqrt{5-x}$

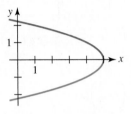

5. $y = \sqrt{1-(x+1)^2}$, $y = -\sqrt{1-(x+1)^2}$

7. $y = 1$, $y = x^3$

9. $y = -2 + \sqrt{4-x}$, $y = -2 - \sqrt{4-x}$

11. $y = \dfrac{x}{x-1}$, domain: $(-\infty, 1) \cup (1, \infty)$

13. $y = -\dfrac{3}{4}x + \dfrac{4}{x}$, domain: $(-\infty, 0) \cup (0, \infty)$

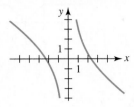

15. $y = \sqrt[3]{x+1}$, domain: $(-\infty, \infty)$

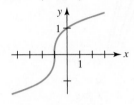

17. $y = (1 - \sqrt{x})^2$, domain: $[0, 1]$

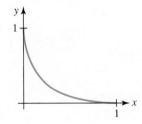

19. $y = x$, domain: $[0, \infty)$; $y = -x$, domain: $[0, \infty)$

21. $y = \sqrt{x^2 - 4}$, $y = -\sqrt{x^2 - 4}$,
domain of each: $(-\infty, -2] \cup [2, \infty)$

23. $y = \frac{1}{2}(-x + \sqrt{12 - 3x^2})$, $y = \frac{1}{2}(-x - \sqrt{12 - 3x^2})$,
domain of each is: $[-2, 2]$

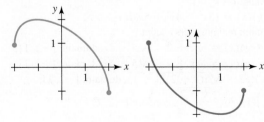

25. $y = \dfrac{x^2}{\sqrt{1 - x^2}}$, $y = -\dfrac{x^2}{\sqrt{1 - x^2}}$, domain of each is: $(-1, 1)$

27. $y = \sqrt{x^2 + \sqrt{x}}$, $y = -\sqrt{x^2 + \sqrt{x}}$,
$y = \sqrt{x^2 - \sqrt{x}}$, $y = -\sqrt{x^2 - \sqrt{x}}$,
domain of first and second functions is: $[0, \infty)$, domain of third and fourth functions is: $[1, \infty)$.

29. $y = \sqrt[4]{16 - x^4}$, $y = -\sqrt[4]{16 - x^4}$, domain of each is: $[-2, 2]$

31. $y = \sqrt{x(x^2 - 3)}$, $y = -\sqrt{x(x^2 - 3)}$,
domain of each is: $[-\sqrt{3}, 0] \cup [\sqrt{3}, \infty)$

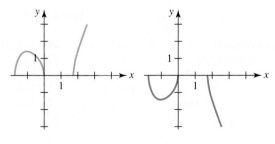

1. not one-to-one   **3.** not one-to-one
5. one-to-one

7. one-to-one,

9. not one-to-one,

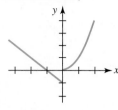

25. domain is $[4, \infty)$; range is $[0, \infty)$

27. $f^{-1}(x) = \dfrac{4}{x^2}$, $(0, \infty)$, $(0, \infty)$

29. $f^{-1}(x) = -\frac{1}{2}x + 3$,   **31.** $f^{-1}(x) = \sqrt[3]{x - 2}$,

33. $f^{-1}(x) = (2 - x)^2$, $(-\infty, 2]$

35. $f^{-1}(x) = \frac{1}{7}x^3$

37. $f^{-1}(x) = \dfrac{3}{x + 1}$, domain of f^{-1} is the set of real numbers except $x = -1$, range of f^{-1} is the set of real numbers except $y = 0$, range of f is the set of real numbers except $y = -1$

39. $f^{-1}(x) = \dfrac{x + 1}{2x}$, domain of f^{-1} is the set of real numbers except $x = 0$, range of f^{-1} is the set of real numbers except $y = \frac{1}{2}$, range of f is the set of real numbers except $y = 0$

41. $f^{-1}(x) = \dfrac{3x}{2x - 7}$, domain of f^{-1} is the set of real numbers except $x = \frac{7}{2}$, range of f^{-1} is the set of real numbers except $y = \frac{3}{2}$, range of f is the set of real numbers except $y = \frac{7}{2}$

43. $f(-1) = 19$ **45.** $f^{-1}(8) = 1$
47. $f^{-1}(-20) = 0$ **49.** $(20, 2)$
51. $(12, 9)$

53.

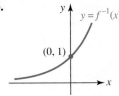

$y = f^{-1}(x)$

$(0, 1)$

55.

$(\frac{3}{2}, 0)$

$y = f(x)$

$(0, -1)$

57. $f^{-1}(x) = \frac{1}{2}\sqrt{x - 2}$, $[2, \infty)$

f

f^{-1}

59. $f^{-1}(x) = 2\sqrt{1 - x^2}$, $[0, 1]$

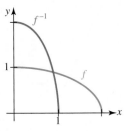

f^{-1}

f

Exercises 2.9   Page 121

1. $S(x) = x + \dfrac{50}{x}$; $(0, \infty)$ **3.** $S(x) = 3x^2 - 4x + 2$; $[0, 1]$

5. $A(x) = 100x - x^2$; $[0, 100]$ **7.** $A(x) = 2x - \frac{1}{2}x^2$; $[0, 4]$

9. $d(x) = \sqrt{2x^2 + 8}$; $(-\infty, \infty)$

11. $S(x) = 2x^4 - 14x^2 + 40$; $[-2, 2]$

13. $P(A) = 4\sqrt{A}$; $(0, \infty)$ **15.** $d(C) = C/\pi$; $(0, \infty)$

17. $A(h) = \dfrac{1}{\sqrt{3}}h^2$; $(0, \infty)$ **19.** $A(x) = \dfrac{1}{4\pi}x^2$; $(0, \infty)$

21. $s(h) = \dfrac{30h}{25 - h}$; $[0, 25)$; $s \to \infty$ as $h \to 25^-$

23. $S(w) = 3w^2 + \dfrac{1200}{w}$; $(0, \infty)$

25. $d(t) = 20\sqrt{13t^2 + 8t + 4}$; $(0, \infty)$

27. $V(h) = \begin{cases} 120h^2, & 0 \le h < 5 \\ 1200h - 3000, & 5 \le h \le 8 \end{cases}$; $[0, 8]$

29. $f(x) = x - x^2$; $(-\infty, \infty)$

31. $F(x) = 2x + \dfrac{16,000}{x}$; $(0, \infty)$

33. $C(x) = 4x + \dfrac{640,000}{x}$; $(0, \infty)$

35. $A(x) = \frac{1}{2}xp - x^2$; $[0, \frac{1}{2}p]$

37. **(a)** $A(x) = x^2 + \dfrac{128,000}{x}$; $(0, \infty)$

     (b) $A(x) = 2x^2 + \dfrac{128,000}{x}$; $(0, \infty)$

39. $V(x) = 20x - 40x^2$; $[0, \frac{1}{2}]$

41. $A(x) = 40 + 4x + \dfrac{64}{x}$; $(0, \infty)$

43. $L(x) = x + \dfrac{8x}{\sqrt{x^2 - 64}}$; $(8, \infty)$

45. $L(x) = \dfrac{1}{4\pi}(L^2x - x^3)$; $[0, L]$

47. $V(x) = 5x\sqrt{64 - x^2}$; $[0, 8]$

49. $T(x) = \frac{1}{3}\sqrt{x^2 + 1} + \frac{1}{2}\sqrt{x^2 - 8x + 17}$; $[0, 4]$

Exercises 2.10   Page 133

1. (a) $6 + h$ **(b)** 6 **(c)** $y = 6x - 15$

3. (a) $-1 + h$ **(b)** -1 **(c)** $y = -x - 1$

5. (a) $-23 - 12h - 2h^2$ **(b)** -23 **(c)** $y = -23x + 32$

7. (a) $\dfrac{1}{2(-1 + h)}$ **(b)** $-\frac{1}{2}$ **(c)** $y = -\frac{1}{2}x - 1$

9. (a) $\dfrac{1}{\sqrt{4 + h} + 2}$ **(b)** $\frac{1}{4}$ **(c)** $y = \frac{1}{4}x + 1$

11. (a) 0 **(b)** $f'(x) = 0$

13. (a) $-8x - 4h$ **(b)** $f'(x) = -8x$

15. (a) $6x + 3h - 1$ **(b)** $f'(x) = 6x - 1$

17. (a) $3x^2 + 3xh + h^2 + 5$ **(b)** $f'(x) = 3x^2 + 5$

19. (a) $\dfrac{1}{(4 - x)(4 - x - h)}$ **(b)** $f'(x) = \dfrac{1}{(4 - x)^2}$

21. (a) $\dfrac{-1}{(x - 1)(x + h - 1)}$ **(b)** $f'(x) = \dfrac{-1}{(x - 1)^2}$

23. (a) $1 - \dfrac{1}{x(x + h)}$ **(b)** $f'(x) = 1 - \dfrac{1}{x^2}$

25. (a) $\dfrac{2}{\sqrt{x + h} + \sqrt{x}}$ **(b)** $f'(x) = \dfrac{1}{\sqrt{x}}$

27. $(2, 17)$; 11; $y = 11x - 5$

29. $(1, 2)$; 8; $y = 8x - 6$ **31.** $(\frac{1}{2}, \frac{5}{2})$; -3; $y = -3x + 4$

33. (a) $3x + 3a$ **(b)** $f'(a) = 6a$

35. (a) $10(x^2 + ax + a^2)$ **(b)** $f'(a) = 30a^2$

37. (a) $\dfrac{-1}{ax}$ **(b)** $f'(a) = \dfrac{-1}{a^2}$

39. (a) $\dfrac{\sqrt{7}}{\sqrt{x} + \sqrt{a}}$ **(b)** $f'(a) = \dfrac{1}{2}\sqrt{\dfrac{7}{a}}$

Chapter 2 Review Exercises   Page 134

A. 1. $-\frac{1}{3}$ **3.** $(-\infty, 5)$

5. $x = 0, x = 2$ **7.** $f(x) = \dfrac{x}{x + 1}$ **9.** $k = -\frac{6}{5}$

11. $m = -\frac{3}{2}$ **13.** $(1 - \sqrt{2}, 0), (1 + \sqrt{2}, 0), (0, -1)$

15. $f(x) = \frac{7}{4}(x + 2)^2$ **17.** $(10, 2)$

19. $(0, 5)$ **21.** $a = -\frac{1}{4}, a = 0,$ and $a = 8$

23. second   **25.** $[0.5, 7]$

27. approximately $(1.4, 0), (2.7, 0),$ and $(6, 0)$ **29.** $[2, 4], [6, 7]$

31. approximately the interval $(1.4, 2.7)$ **33.** 1.2

B. 1. true   **3.** false   **5.** true   **7.** true

9. true   **11.** true   **13.** true   **15.** true

17. false   **19.** false   **21.** true   **23.** false

C. 1. $f(x) = x^2, g(x) = \dfrac{3x - 5}{x}$

3. (a) $y = (x + 3)^3 - 2$ **(b)** $y = x^3 - 7$

    (c) $y = (x - 1)^3$ **(d)** $y = -x^3 + 2$

    (e) $y = -x^3 - 2$ **(f)** $y = 3x^3 - 6$

5. domain: $(\pi/2, 3\pi/2)$, range: $(-\infty, \infty)$

7. $f(x) = \sqrt{x}$; $y = 5 - \sqrt{x + 3}$

9. $f(x) = x^2$; $y = 2 - (x - 3)^2$

11. $f(x) = \begin{cases} x + 1, & x < 0 \\ -x + 1, & 0 \le x < 1 \\ x - 1, & x \ge 1 \end{cases}$ **13.** $(-\infty, \infty)$

15. $\{x \,|\, x \ne -3, x \ne 2\}$ **17.** $f^{-1}(x) = -1 + \sqrt[3]{x}$

19. $A(h) = h^2(1 - \pi/4)$ **21.** $d(s) = \sqrt{3}s$

23. (a) $d(t) = 6t$ (b) $d(t) = \sqrt{90^2 + (90 - 6t)^2}$

25. $S(x) = 20x + \dfrac{5}{x}, x > 0$

27. $A(x) = 2x(1 - \pi x)$, where x is the radius of the semicircle

29. $f'(x) = -6x + 16, y = 4x + 24$

31. $f'(x) = \dfrac{1}{x^3}, y = 8x - 6$ **33.** $(0, 0), \left(\frac{2}{3}, -\frac{4}{27}\right)$

Exercises 3.1 Page 150

1.

3.

5.

7.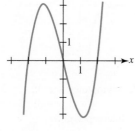

9. odd **11.** neither even nor odd

13. (f) **15.** (e) **17.** (b)

19. $f(x) = 2x^4 + 3x^2 - 6$ **21.** $f(x) = -7x(x - 1)(x + 3)^2$

23.

25.

27.

29.

31.

33.

35.

37.

39.

41.

43.

45. (b) $f(x) = (x - 1)^2(x + 2)$ **47.** $k = -\frac{7}{16}$ **49.** $k = -\frac{10}{3}$

51. the odd positive integers **53.** $[-2, -1], [-1, 0], [1, 2]$

55. $[0, 15]$;

57. $V(h) = \frac{1}{3}\pi(R^2 h - h^3)$

Exercises 3.2 Page 157

1. $f(x) = x^2 \cdot 8 + (4x - 7)$

3. $f(x) = (x^2 + x - 1) \cdot (5x - 12) + (21x - 11)$

5. $f(x) = (x + 2)^2 \cdot (2x - 4) + (5x + 21)$

7. $f(x) = (3x^2 - x) \cdot (9x + 3) + (4x - 2)$

9. $f(x) = (x^3 - 2) \cdot (6x^2 + 4x + 1) + (12x^2 + 8x + 2)$

11. $r = 6$ **13.** $r = \frac{29}{8}$

15. $r = 76$ **17.** $f(2) = 2$

19. $f(-5) = -74$ **21.** $f\left(\frac{1}{2}\right) = \frac{303}{16}$

23. $q(x) = 2x + 3, r = 11$
25. $q(x) = x^2 - 4x + 12, r = -34$
27. $q(x) = x^3 + 2x^2 + 4x + 8, r = 32$
29. $q(x) = x^4 - 4x^3 + 16x^2 - 8x + 32, r = -132$
31. $q(x) = x^2 - 2x + \sqrt{3}, r = 0$
33. $f(-3) = 51$ **35.** $f(\sqrt{2}) = -4\sqrt{2}$
37. $f(1) = 1$ **39.** $f(4) = 5369$
41. $k = -1$ **43.** $k = -\frac{1}{5}$
45. $k = -4$

Exercises 3.3 Page 165

1. $4 - 3i, 2 - i, 1 - 5i$ **3.** $9 + 15i, 5 + 13i, 35i$
5. $6 + 4i, 14 + 6i, -35 - 30i$
7. $-4 - 8i, -6 - 14i, 28 - 26i$
9. $\frac{9}{2} - \frac{11}{2}i, -\frac{7}{2} + \frac{5}{2}i, -4 - 8i$ **11.** $8 + 6i, 18 + 26i$
13. $-50i, -250 - 250i$
15. $f(x) = 4(x - \frac{1}{4})(x - 1)^2$ **17.** 5 is not a zero
19. $f(x) = 3(x + \frac{2}{3})(x - 2 + \sqrt{2})(x - 2 - \sqrt{2})$
21. $f(x) = 4(x + 3)(x - 5)(x - \frac{1}{2})(x + \frac{1}{2})$
23. $f(x) = 9(x - 1)(x + \frac{1}{3})^2(x + 8)$
25. $f(x) = (x - 1)^3(x + 5)(x - 5)$
27. $x - 5$ is not a factor
29. $f(x) = (x - 1)(x + \frac{1}{2} + \frac{1}{2}\sqrt{7}i)(x + \frac{1}{2} - \frac{1}{2}\sqrt{7}i)$
31. $x - \frac{1}{3}$ is not a factor
33. $f(x) = (x - 1)(x - 2)(x - 2i)(x + 2i)$
35. $f(x) = 2(x - 1)^2(x + 1)(x + \frac{3}{2})$
37. $f(x) = 3(x - \frac{5}{3})(x + 2i)(x - 2i)$
39. $f(x) = 5(x + \frac{2}{5})(x + 1 - i)(x + 1 + i)$
41. $f(x) = (x - 3)(x + 3)(x - 1 + 2i)(x - 1 - 2i)$
43. $f(x) = (x - 2)(x - 1)(x + 3)^2$
$= x^4 + 3x^3 - 7x^2 - 15x + 18$
45. $f(x) = x^5 - 6x^4 + 10x^3$ **47.** $f(x) = x^2 - 2x + 37$
49. $f(x) = 5x^2 - 10x + 10$
51. $f(x) = 84x^4 - 126x^3 - 126x - 84$
53. 0 is a simple zero, $\frac{5}{4}$ is a zero of multiplicity two, $\frac{1}{2}$ is a zero of multiplicity three
55. $-\frac{2}{3}$ is a zero of multiplicity two, $\frac{2}{3}$ is a zero of multiplicity two
57. $-1, -\frac{1}{2}, 3$
59. $k = -36, f(x) = 2(x - 3)(x + 1 - \sqrt{5}i)(x + 1 + \sqrt{5}i)$
61. $f(x) = -\frac{1}{16}(x - 4)(x + 2)^2$

Exercises 3.4 Page 173

1. $-\frac{2}{5}$ **3.** 3
5. $\frac{1}{2}$ (multiplicity 2) **7.** no rational zeros
9. $\frac{1}{3}, \frac{3}{2}$ **11.** 0, 1
13. $-3, 0, 2$ **15.** $\frac{3}{2}$
17. $-\frac{1}{5}$ **19.** $-\frac{1}{2}$ (multiplicity two), $\frac{1}{3}$ (multiplicity two)
21. $\frac{3}{8}, -\frac{1}{2} - \frac{1}{2}\sqrt{5}, -\frac{1}{2} + \frac{1}{2}\sqrt{5}$;
$f(x) = (8x - 3)(x + \frac{1}{2} + \frac{1}{2}\sqrt{5})(x + \frac{1}{2} - \frac{1}{2}\sqrt{5})$
23. $\frac{4}{5}, \frac{5}{2}, -\sqrt{2}, \sqrt{2}$;
$f(x) = (5x - 4)(2x - 5)(x + \sqrt{2})(x - \sqrt{2})$
25. $-4, -1, 1, -\sqrt{5}, \sqrt{5}$;
$f(x) = (x + 4)(x + 1)(x - 1)(x + \sqrt{5})(x - \sqrt{5})$
27. $0, 1, 3, -1 - \sqrt{2}, -1 + \sqrt{2}$;
$f(x) = 4x(x - 1)(x - 3)(x + 1 + \sqrt{2})(x + 1 - \sqrt{2})$
29. $-1, \frac{1}{4}$ (multiplicity 2); $f(x) = (x + 1)(4x - 1)^2(x^2 - 2x + 3)$

31. $-\frac{1}{2}$ **33.** $-\frac{3}{2}, 2, -2 - \sqrt{3}, -2 + \sqrt{3}$
35. 1 (multiplicity 3)
37. $f(x) = 3x^4 - x^3 - 39x^2 + 49x - 12$
39. $\frac{3}{4}$ **41.** $f(x) = -\frac{1}{6}(x - 1)(x - 2)(x - 3)$
43. 3 inches or $\frac{1}{2}(7 - \sqrt{33}) \approx 0.63$ inches

Exercises 3.5 Page 185

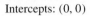

1.

x	3.1	3.01	3.001	3.0001	3.00001
$f(x)$	62	602	6,002	60,002	600,002
x	2.9	2.99	2.999	2.9999	2.99999
$f(x)$	-58	-598	-5,998	-59,998	-599,998

3. Asymptotes: $x = 2, y = 0$ **5.** Asymptotes: $x = -1, y = 1$
Intercepts: $(0, -\frac{1}{2})$ Intercepts: $(0, 0)$

7. Asymptotes: $x = -\frac{3}{2}, y = 2$ **9.** Asymptotes: $x = -1, y = -1$
Intercepts: $(\frac{9}{4}, 0), (0, -3)$ Intercepts: $(1, 0), (0, 1)$

11. Asymptotes: $x = 1, y = 0$ **13.** Asymptotes: $x = 0, y = 0$
Intercepts: $(0, 1)$ Intercepts: none

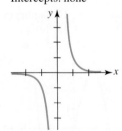

15. Asymptotes: $x = 1, x = -1,$ **17.** Asymptotes: $x = 0, x = 2,$
$y = 0$ $y = 0$
Intercepts: $(0, 0)$ Intercepts: none

19. Asymptotes: $x = 0, y = -1$
Intercepts: $(-1, 0), (1, 0)$

21. Asymptotes: $x = 1, y = -2$
Intercepts: $(-2, 0), (2, 0),$
$(0, 8)$

41. Hole in the graph at $x = 1$
Intercepts: $(-1, 0), (0, 1)$

43. Hole in the graph at $x = -1$
Intercepts: none

23. Asymptotes: $x = 0, y = x$
Intercepts: $(-3, 0), (3, 0)$

25. Asymptotes: $x = -2,$
$y = x - 2$
Intercepts: $(0, 0)$

45. $R \to 5$ as $r \to \infty$;

27. Asymptotes: $x = 1, y = x - 1$
Intercepts: $(3, 0), (-1, 0), (0, 3)$

47. $I(x) \to \infty$ as $x \to 0^+$; $I(x) \to \infty$ as $x \to 100^-$

Exercises 3.6   Page 193

1. $\dfrac{A}{x} + \dfrac{B}{x + 1}$

3. $\dfrac{A}{x} + \dfrac{B}{x^2} + \dfrac{C}{x + 1}$

5. $\dfrac{A}{x} + \dfrac{B}{x^2} + \dfrac{C}{x + 1} + \dfrac{D}{(x + 1)^2}$

7. $\dfrac{Ax + B}{x^2 + 1} + \dfrac{Cx + D}{(x^2 + 1)^2}$

9. $\dfrac{\frac{1}{2}}{x} - \dfrac{\frac{1}{2}}{x + 2}$

11. $-\dfrac{6}{x + 1} - \dfrac{3}{x - 5}$

13. $\dfrac{\frac{3}{2}}{x + 1} - \dfrac{10}{x + 2} + \dfrac{\frac{21}{2}}{x + 3}$

15. $\dfrac{\frac{3}{2}}{x + 4} + \dfrac{\frac{3}{2}}{x - 4}$

17. $\dfrac{5}{x - 3} + \dfrac{9}{(x - 3)^2}$

19. $-\dfrac{\frac{1}{4}}{x} + \dfrac{\frac{1}{4}}{x^2} + \dfrac{\frac{1}{4}}{x + 2} + \dfrac{\frac{1}{4}}{(x + 2)^2}$

21. $-\dfrac{\frac{11}{27}}{x} - \dfrac{\frac{7}{9}}{x^2} + \dfrac{\frac{1}{3}}{x^3} + \dfrac{\frac{1}{2}}{x - 1} - \dfrac{\frac{5}{54}}{x + 3}$

23. $\dfrac{1}{x - 1} + \dfrac{5x - 2}{x^2 + 9}$

25. $\dfrac{\frac{36}{7}}{2x - 3} + \dfrac{-\frac{4}{7}x + \frac{26}{7}}{x^2 - x + 1}$

27. $-\dfrac{\frac{7}{4}}{t + 1} + \dfrac{\frac{9}{4}}{t - 1} + \dfrac{-\frac{1}{2}t - 4}{t^2 + 1}$

29. $\dfrac{2x}{x^2 + 2} - \dfrac{x}{x^2 + 1}$

31. $\dfrac{1}{x^2 + 1} + \dfrac{2x}{(x^2 + 1)^2}$

33. $-\dfrac{10}{x + 2} + \dfrac{2}{x - 2} + \dfrac{8}{x + 3}$

35. $x^3 + x + \dfrac{\frac{1}{2}}{x - 1} + \dfrac{\frac{1}{2}}{x + 1}$

37. $\dfrac{1}{2} - \dfrac{\frac{13}{3}}{x + 2} + \dfrac{\frac{13}{6}}{2x + 1}$

39. $x^3 + 2x^2 + 3x + 6 + \dfrac{\frac{64}{5}}{x - 2} + \dfrac{\frac{1}{5}x + \frac{2}{5}}{x^2 + 1}$

Exercises 3.7   Page 199

1. $\frac{7}{16}$

3. $\frac{1}{2}$

5. (a) $\frac{27}{4}$ **(b)** $\frac{33}{4}$

7. (a) 20   **(b)** 20

---

29. Asymptotes: $x = 1, x = 0, y = x + 1$
Intercepts: $(2, 0)$

31. $(3, 0)$;

33. $(4, 4)$;

35. $(-3, -6)$

37. $y = \dfrac{x - 5}{x - 2}$

39. $y = \dfrac{3x(x - 3)}{(x + 1)(x - 2)}$

9. $\frac{85}{4}$

11. 6.85; 7.15

13. 9.32 acres; 8.48 acres

Chapter 3 Review Exercises   Page 201

A. 1. $(1, 0)$; $(0, 0)$, $(5, 0)$
3. $f(x) = x^4$
5. $k = \frac{2}{3}$
7. $x = 1, x = 4$
9. $y = -\frac{1}{2}$
11. $n = 0, n = 1, n = 2$
13. $-i, -1 + i, -1 - i$
15. 4
17. $y = 2x + 4, x = 0, x = 1$
19. the origin   **21.** even

B. 1. true   **3.** true   **5.** true   **7.** true
9. true   **11.** false   **13.** false   **15.** false
17. true   **19.** true   **21.** true   **23.** true

C. 1. $3x^3 - \frac{1}{2}x + 1 + \dfrac{-\frac{1}{2}x + 5}{2x^2 - 1}$

3. $7x^3 + 14x^2 + 22x + 53 + \dfrac{109}{x - 2}$

5. $r = f(-3) = -198$ **7.** vertical asymptote

9. $\pm 1, \pm 3, \pm 5, \pm 15, \pm\frac{1}{2}, \pm\frac{3}{2}, \pm\frac{5}{2}, \pm\frac{15}{2}, \pm\frac{1}{4}, \pm\frac{3}{4}, \pm\frac{5}{4}, \pm\frac{15}{4}, \pm\frac{1}{8}, \pm\frac{3}{8}, \pm\frac{5}{8}, \pm\frac{15}{8}$

11. $f(x) = (x - 2)\left(x - \frac{7}{2} + \frac{1}{2}\sqrt{3}i\right)\left(x - \frac{7}{2} - \frac{1}{2}\sqrt{3}i\right)$

13. $k = -\frac{21}{2}$ **15.** $k = \frac{3}{2}$

17. $f(x) = 3x^2(x + 2)^2(x - 1)$ **19.** $f(x) = \dfrac{4(x + 2)}{(x - 2)(x + 4)}$

21. (f)   **23.** (d)   **25.** (h)   **27.** (c)   **29.** (b)

31. $y = 0, x = -4, x = 2, (-2, 0), \left(0, -\frac{1}{4}\right)$,

33. $\dfrac{\frac{1}{3}}{x} + \dfrac{\frac{1}{4}}{x - 1} - \dfrac{\frac{7}{12}}{x + 3}$ **35.** $\dfrac{1}{x^2 + 9} - \dfrac{9}{(x^2 + 9)^2}$

Exercises 4.1   Page 213

1.

3.

5.

7.

9.

11.

13.

15.

17. 10.6547°   **19.** 5.17°
21. 210° 46′48″   **23.** 30°48′36″
25. $\pi/18$ **27.** $5\pi/12$
29. $3\pi/2$ **31.** $-23\pi/18$
33. 40°   **35.** 120°
37. 225°   **39.** 177.62°
41. 155°   **43.** 110°
45. −205°   **47.** $7\pi/4$
49. 1.3π **51.** $2\pi - 4 \approx 2.28$
53. $-\pi/4$
55. (a) 41.75°   (b) 131.75°
57. (a) The given angle is greater than 90°.   (b) 81.6°
59. (a) $\pi/4$ (b) $3\pi/4$
61. (a) The given angle is greater than $\pi/2$.   (b) $\pi/3$
63. 216°, $6\pi/5$ **65.** (a) 1.5 radians   (b) 85.94°
67. (a) $\frac{5}{3}$ radians   (b) 95.49°   **69.** 15 in
71. 1.047 m   **73.** 32.4 ft²
75. 9.42 m²

77. (a) $\frac{1}{2}(R^2 - r^2)\theta$ (b) $(R^2 - r^2)\dfrac{\theta\pi}{360}$

79. 30° or $\pi/6$ radian
81. (a) 120° or $2\pi/3$ radians   (b) 60° or $\pi/3$ radians
    (c) 45° or $\pi/4$ radian
83. (a) 16 h   (b) 2 h
85. (a) 0.000072921 rad/s   (b) 3.074641 km/s
87. 1.15 statute miles
89. (a) 3π rad/s   (b) 300π cm/s
91. 11.5 mi²
93. (a) 711.1 rev/min   (b) 4468 rad/min

Exercises 4.2   Page 224

1. $\sqrt{21}/5$ **3.** $-\sqrt{5}/3$
5. $\pm 3\sqrt{5}/7$ **7.** $\pm 2\sqrt{6}/5 \approx \pm 0.98$
9. $\sin t = \pm 1/\sqrt{5}$, $\cos t = \pm 2/\sqrt{5}$
11. (a) −1   (b) 0   **13.** (a) 0   (b) 1
15. $\pi/3, \sqrt{3}/2, -\frac{1}{2}$ **17.** $\pi/4, -\sqrt{2}/2, -\sqrt{2}/2$
19. $\pi/6, -\frac{1}{2}, \sqrt{3}/2$ **21.** $\pi/4, -\sqrt{2}/2, \sqrt{2}/2$
23. $\pi/6, -\frac{1}{2}, -\sqrt{3}/2$ **25.** $\pi/3, \sqrt{3}/2, \frac{1}{2}$
27. $\sqrt{3}/2$ **29.** $\sqrt{2}/2$
31. −1
33. $\sin t = \sin(t + 2\pi)$ for $t = \pi$
35. $\sin(-t) = -\sin t$ for $t = 3 + \pi$
37. $\cos t = \cos(-t)$ for $t = 0.43$
39. $\sqrt{2}/2$ **41.** $-\sqrt{3}/2$
43. $\sqrt{3}/2$ **45.** $-\sqrt{3}/2$
47. 0, π **49.** $\pi/4, 7\pi/4$
51. 30°, 330°   **53.** 225°, 315°
55. 4.81 m
57. (a) 978.6019, 979.6517, 980.2304, 982.2751
    (b) minimum at $\phi = 0°$, maximum at $\phi = \pm 90°$

1.

3.

5.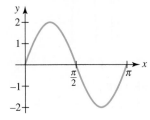

7. $y = -3\sin x$

9. $y = 1 - 3\cos x$

11. $(n, 0)$, where n is an integer

13. $((2n + 1)\pi, 0)$, where n is an integer

15. $(\pi/4 + n\pi, 0)$, where n is an integer

17. $(\pi/2, 0)$; $(\pi/2 + 2n\pi, 0)$, where n is an integer

19. $y = 3\sin 2x$ **21.** $y = \frac{1}{2}\cos \pi x$

23. $y = -\sin \pi x$

25. amplitude: 4; period: 2   **27.** amplitude: 3; period: 1

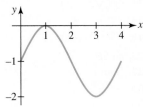

29. amplitude: 2; period: π **31. (a)**

(b) amplitude: 4
(c) $M = 7$, $m = -1$
(e) $[-1, 7]$

33. (a)

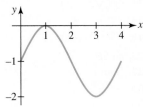

(b) amplitude: 1
(c) $M = 0$, $m = -2$
(e) $[-2, 0]$

35. (a)

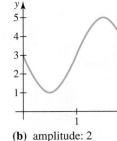

(b) amplitude: 2
(c) $M = 5$, $m = 1$
(e) $[1, 5]$

37. $y = -1 + \frac{1}{2}\sin x$

39. amplitude: 1; period: 2π
phase shift: $-\pi/6$

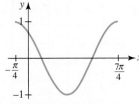

41. amplitude: 1; period: 2π
phase shift: $\pi/4$

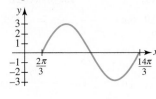

43. amplitude: 4; period: π
phase shift: $-3\pi/4$

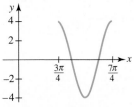

45. amplitude: 3; period: 4π
phase shift: $-2\pi/3$

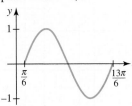

47. amplitude: 4; period: 6
phase shift: -1

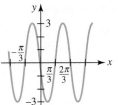

49. $y = -5 + 3\cos\left(6x + \dfrac{3\pi}{2}\right)$

51.

$y = 3\sin(3x - \pi)$ $y = 3\cos(3x - \pi)$

53.

 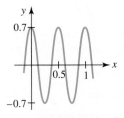

$y = 0.7\sin[4\pi(x - 4)]$ $y = 0.7\cos[4\pi(x - 4)]$

57.

59.

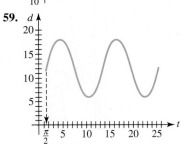

1.

x	$\frac{2\pi}{3}$	$\frac{3\pi}{4}$	$\frac{5\pi}{6}$	π	$\frac{7\pi}{6}$	$\frac{5\pi}{4}$	$\frac{4\pi}{3}$	$\frac{3\pi}{2}$	$\frac{5\pi}{3}$	$\frac{7\pi}{4}$	$\frac{11\pi}{6}$	2π
$\tan x$	$-\sqrt{3}$	-1	$-\frac{1}{\sqrt{3}}$	0	$\frac{1}{\sqrt{3}}$	1	$\sqrt{3}$	—	$-\sqrt{3}$	-1	$-\frac{1}{\sqrt{3}}$	0
$\cot x$	$-\frac{1}{\sqrt{3}}$	-1	$-\sqrt{3}$	—	$\sqrt{3}$	1	$\frac{1}{\sqrt{3}}$	0	$-\frac{1}{\sqrt{3}}$	-1	$-\sqrt{3}$	—

3. $\sqrt{3}$ **5.** undefined   **7.** $-2/\sqrt{3}$ **9.** -1

11. -2 **13.** undefined   **15.** -2 **17.** $\sqrt{2}$

19. $\cot x = -\dfrac{1}{2}$, $\sec x = -\sqrt{5}$, $\cos x = -\dfrac{1}{\sqrt{5}}$, $\sin x = \dfrac{2}{\sqrt{5}}$, $\csc x = \dfrac{\sqrt{5}}{2}$

21. $\sin x = \frac{3}{4}$, $\cos x = \dfrac{\sqrt{7}}{4}$, $\tan x = \dfrac{3}{\sqrt{7}}$, $\cot x = \dfrac{\sqrt{7}}{3}$, $\sec x = \dfrac{4}{\sqrt{7}}$

23. $\csc x = 3$, $\cos x = -\dfrac{2\sqrt{2}}{3}$, $\sec x = -\dfrac{3}{2\sqrt{2}}$, $\tan x = -\dfrac{1}{2\sqrt{2}}$, $\cot x = -2\sqrt{2}$

25. $\sec x = \frac{13}{12}$, $\sin x = -\frac{5}{13}$, $\csc x = -\frac{13}{5}$, $\tan x = -\frac{5}{12}$, $\cot x = -\frac{12}{5}$

27. $\tan x = 3$, $\cot x = \frac{1}{3}$, $\sec x = \pm\sqrt{10}$, $\csc x = \pm\dfrac{\sqrt{10}}{3}$

29. period: 1; x-intercepts: $(n, 0)$, where n is an integer; asymptotes: $x = \dfrac{2n+1}{2}$, n an integer;

31. period: $\pi/2$; x-intercepts: $\left(\dfrac{2n+1}{4}\pi, 0\right)$, where n is an integer; asymptotes: $x = n\pi/2$, n an integer;

33. period: 2π; x-intercepts: $\left(\dfrac{\pi}{2} + 2n\pi, 0\right)$, where n is an integer; asymptotes: $x = \dfrac{3\pi}{2} + 2n\pi$, n an integer;

35. period: 1; x-intercepts: $\left(\frac{1}{4} + n, 0\right)$, where n is an integer; asymptotes: $x = n$, n an integer;

37. period: 2π; asymptotes: $x = \dfrac{2n+1}{2}\pi$, n an integer;

39. period: 2; asymptotes: $x = n$, n an integer;

41. period: $2\pi/3$; asymptotes: $x = \dfrac{n\pi}{3}$, n an integer;

43. period: π; asymptotes: $x = \dfrac{2n-1}{4}\pi$, n an integer;

45. $\cot x = -\tan\left(x - \dfrac{\pi}{2}\right)$ **47.** $a\sin\theta$

49. $a\tan\theta$ **51.** $\tan\theta$

53. $\frac{\sqrt{3}}{3}\sin\theta\cos\theta$ **55.** $\frac{\sqrt{7}}{7}\cos\theta$

1. $\frac{\sqrt{2}}{4}(1 + \sqrt{3})$ **3.** $\frac{\sqrt{2}}{4}(1 + \sqrt{3})$

5. $-\frac{\sqrt{2}}{4}(1 + \sqrt{3})$ **7.** $2 + \sqrt{3}$

9. $\frac{\sqrt{2}}{4}(1 - \sqrt{3})$ **11.** $\frac{\sqrt{2}}{4}(\sqrt{3} - 1)$

13. $-\frac{\sqrt{2}}{4}(1 + \sqrt{3})$ **15.** $-2 + \sqrt{3}$

17. $\frac{\sqrt{2}}{4}(1 - \sqrt{3})$ **19.** $\frac{\sqrt{2}}{4}(\sqrt{3} + 1)$

21. $-\frac{\sqrt{2}}{4}(1 + \sqrt{3})$ **23.** $\sin 2\beta$

25. $\cos(2\pi/5)$ **27.** $\frac{1}{2}\tan 6t$

29. (a) $\frac{5}{9}$ (b) $-\frac{2\sqrt{14}}{9}$ (c) $-\frac{2\sqrt{14}}{5}$

31. (a) $\frac{3}{5}$ (b) $\frac{4}{5}$ (c) $\frac{4}{3}$

33. (a) $-\frac{119}{169}$ (b) $-\frac{120}{169}$ (c) $\frac{120}{119}$

35. $\frac{1}{2}\sqrt{2 + \sqrt{3}}$ **37.** $\frac{1}{2}\sqrt{2 + \sqrt{2}}$

39. $\frac{1}{2}\sqrt{2 - \sqrt{2}}$ **41.** $-2 - \sqrt{3}$

43. $-2\sqrt{2 + \sqrt{3}}$

45. (a) $\frac{2\sqrt{13}}{13}$ (b) $\frac{3\sqrt{13}}{13}$ (c) $\frac{3}{2}$

47. (a) $-\sqrt{(5 - \sqrt{5})/10}$ **(b)** $\sqrt{(5 + \sqrt{5})/10}$
(c) $-\frac{1}{2}(1 + \sqrt{5})$
49. (a) $\frac{\sqrt{30}}{6}$ **(b)** $\frac{\sqrt{6}}{6}$ **(c)** $\frac{\sqrt{5}}{5}$
61. $\frac{1}{2} - \frac{1}{2}\cos 10x$ **63.** $\frac{3}{8} + \frac{1}{2}\cos 2x + \frac{1}{8}\cos 4x$
65. $\frac{5}{16} - \frac{15}{32}\cos 2x + \frac{3}{16}\cos 4x - \frac{1}{32}\cos 6x$
67. $y = 2\cos 2x$, amplitude 2, period π
69. $y = \sin 4x$, amplitude 1, period $\pi/2$
71. (a) $-\frac{2}{9}(\sqrt{10} + 1)$ **(b)** $\frac{1}{9}(\sqrt{5} - 4\sqrt{2})$
(c) $\frac{2}{9}(1 - \sqrt{10})$ **(d)** $\frac{1}{9}(\sqrt{5} + 4\sqrt{2})$
73. $2\sqrt{2 + \sqrt{3}} \approx 3.86$

Exercises 4.6 **Page 260**

1. $\frac{1}{2}(\cos\theta + \cos 7\theta)$ **3.** $\frac{1}{2}(\cos 3x - \cos 7x)$
5. $\frac{1}{2}\left(\cos x + \cos\dfrac{5x}{3}\right)$ **7.** $\frac{1}{2}(\sin 20x - \sin 2x)$
9. $\sin 4\beta - \sin 2\beta$ **11.** $-\cos 2x$
13. $\frac{1}{2} - \frac{\sqrt{3}}{4}$ **15.** $\frac{1}{4}$
17. $\frac{\sqrt{2}}{4} + \frac{\sqrt{3}}{4}$ **19.** $-2\cos 3y\sin 2y$
21. $-2\sin\dfrac{5x}{2}\sin 2x$ **23.** $2\cos 4x\cos 2x$
25. $2\sin\dfrac{\omega_1 + \omega_2}{2}t\cos\dfrac{\omega_1 - \omega_2}{2}t$ **27.** $-\sin 3t\sin 2t$
29. 0 **31.** 1
33. $-\frac{\sqrt{6}}{2}$ **35.** $\frac{\sqrt{2}}{2}$
37. $f(t) = 0.06\sin 750\pi t\cos 250\pi t$

Exercises 4.7 **Page 268**

1. 0 **3.** π
5. $\pi/3$ **7.** $-\pi/3$
9. $\pi/4$ **11.** $-\pi/6$
13. $-\pi/4$ **15.** $\frac{4}{5}$
17. $-\frac{\sqrt{5}}{2}$ **19.** $\frac{\sqrt{5}}{5}$
21. $\frac{5}{3}$ **23.** $\frac{1}{5}$
25. 1.2 **27.** $\pi/16$
29. 0 **31.** $\pi/4$
33. $\dfrac{x}{\sqrt{1 + x^2}}$ **35.** $\dfrac{x}{\sqrt{1 - x^2}}$
37. $\dfrac{\sqrt{1 - x^2}}{x}$ **39.** $\dfrac{\sqrt{1 + x^2}}{x}$

41.

43.
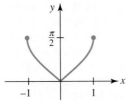

45.

47.

---

49. domain: $(-\infty, \infty)$,
range: $(0, \pi)$

51. domain: $(-\infty, -1] \cup [1, \infty)$,
range: $[0, \pi/2) \cup (\pi/2, \pi]$

53. (a) x in $(-\infty, \infty)$ **(b)** x in $(0, \pi)$
55. (a) x in $(-\infty, -1] \cup [1, \infty)$ **(b)** x in $[0, \pi/2) \cup (\pi/2, \pi]$
59. 0.9273 **61.** -0.7297
63. 2.5559 **65.** 19.9°, 70.1°
67. 5.76°
69. (a) $\pi/4$ **(b)** 0.942 radians or 53.97°

Exercises 4.8 **Page 277**

1. $x = \dfrac{\pi}{3} + 2n\pi$ or $x = \dfrac{2\pi}{3} + 2n\pi$, where n is an integer
3. $x = \dfrac{\pi}{4} + 2n\pi$ or $x = \dfrac{7\pi}{4} + 2n\pi$, where n is an integer
5. $x = \dfrac{5\pi}{6} + n\pi$, where n is an integer
7. $x = \pi + 2n\pi = (2n + 1)\pi$, where n is an integer
9. $x = n\pi$, where n is an integer
11. $x = \dfrac{3\pi}{2} + 2n\pi$, where n is an integer
13. $\theta = 60° + 360°n$, $\theta = 120° + 360°n$, where n is an integer
15. $\theta = 135° + 180°n$, where n is an integer
17. $\theta = 120° + 360°n$, $\theta = 240° + 360°n$, where n is an integer
19. $x = n\pi$, where n is an integer
21. no solutions
23. $\theta = 120° + 360°n$, $\theta = 240° + 360°n$, where n is an integer
25. $\theta = 90° + 180°n$, $\theta = 135° + 180°n$, where n is an integer
27. $x = \dfrac{\pi}{2} + n\pi$, where n is an integer
29. $\theta = 10° + 120°n$, $\theta = 50° + 120°n$, where n is an integer
31. $x = \dfrac{\pi}{2} + 2n\pi$, where n is an integer
33. $x = n\pi, x = \dfrac{2\pi}{3} + 2n\pi, x = \dfrac{4\pi}{3} + 2n\pi$, where n is an integer
35. $\theta = 30° + 360°n$, $\theta = 150° + 360°n$, $\theta = 270° + 360°n$, where n is an integer
37. $x = \dfrac{\pi}{2} + n\pi$, where n is an integer
39. $x = n\pi$, where n is an integer
41. $\theta = 2n\pi$, $\theta = \dfrac{\pi}{2} + 2n\pi$, where n is an integer
43. $x = \dfrac{\pi}{6} + 2n\pi$, $x = \dfrac{5\pi}{6} + 2n\pi$, where n is an integer
45. $\theta = 90° + 180°n$, $\theta = 360°n$, where n is an integer
47. $\{0, \pi/10, 3\pi/10, \pi/2, 7\pi/10, 9\pi/10, \pi, 11\pi/10, 13\pi/10, 3\pi/2, 17\pi/10, 19\pi/10\}$
49. $\{0, \pm 2\pi/5, \pm 2\pi/3, \pm 4\pi/5\}$
51. $\{0, \pm\pi/3, \pm 2\pi/3, \pm\pi/2\}$
53. $(\pi/3, 0)$, $(2\pi/3, 0)$, $(\pi, 0)$ **55.** $\left(\frac{2}{3}, 0\right)$, $\left(\frac{10}{3}, 0\right)$, $\left(\frac{14}{3}, 0\right)$
57. $(\pi, 0)$, $(2\pi, 0)$, $(3\pi, 0)$ **59.** $(\pi/3, 0)$, $(\pi, 0)$, $(5\pi/3, 0)$

61. The equation has infinitely many solutions.

63. The equation has infinitely many solutions.

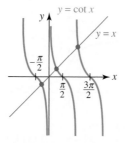

65. 1.37, 1.82
67. −1.02, 0.55
69. 0.58, 1.57, 1.81
71. 30°, 150°
73. 60°
75. $t = \frac{1}{120}\left(\frac{1}{6} + n\right)$, where n is an integer.
77. **(a)** 36.93 million square kilometers **(b)** $w = 31$ weeks
(c) August

Exercises 4.9   Page 283

1. $y = \sqrt{2}\sin(\pi x + 3\pi/4)$;
amplitude: $\sqrt{2}$; period: 2;
phase shift: $\frac{3}{4}$; one cycle
of the graph is

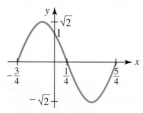

3. $y = 2\sin(2x + \pi/6)$;
amplitude: 2; period: π;
phase shift: $\pi/12$; one
cycle of the graph is

5. $y = \sin(x + 5\pi/4)$;
amplitude: 1; period: 2π;
phase shift: $5\pi/4$; one cycle
of the graph is

7. $(\frac{1}{8}, 0)$, $(\frac{5}{8}, 0)$
9. $x = \pi/4, x = \pi/2$
11. $y = \frac{\sqrt{13}}{4}\sin(2t + 0.5880)$; amplitude: $\frac{\sqrt{13}}{4}$ feet;
period: π seconds; frequency: $1/\pi$ cycles per second
13. $y = \frac{\sqrt{5}}{2}\sin(4t + 4.2487)$; amplitude: $\frac{\sqrt{5}}{2}$ feet;
period: $\pi/2$ seconds; frequency: $2/\pi$ cycles per second
15. $y_0 = -\frac{5\sqrt{3}}{4}, v_0 = \frac{5}{2}$ **17.** $I = I_0\sin(\omega t + \theta + \phi)$
19. **(a)** $\theta(t) = \theta_0\cos\omega t$ **21.** $x \approx 0.4636, x \approx 3.6052$

Exercises 4.10   Page 288

1. $\sin\theta = \frac{4}{5}$, $\cos\theta = \frac{3}{5}$, $\tan\theta = \frac{4}{3}$, $\csc\theta = \frac{5}{4}$, $\sec\theta = \frac{5}{3}$,
$\cot\theta = \frac{3}{4}$
3. $\sin\theta = 3\sqrt{10}/10$, $\cos\theta = \sqrt{10}/10$, $\tan\theta = 3$,
$\csc\theta = \sqrt{10}/3$, $\sec\theta = \sqrt{10}$, $\cot\theta = \frac{1}{3}$
5. $\sin\theta = \frac{2}{5}$, $\cos\theta = \sqrt{21}/5$, $\tan\theta = 2\sqrt{21}/21$, $\csc\theta = \frac{5}{2}$,
$\sec\theta = 5\sqrt{21}/21$, $\cot\theta = \sqrt{21}/2$

7. $\sin\theta = \frac{1}{3}$, $\cos\theta = 2\sqrt{2}/3$, $\tan\theta = \sqrt{2}/4$, $\csc\theta = 3$,
$\sec\theta = 3\sqrt{2}/4$, $\cot\theta = 2\sqrt{2}$
9. $\sin\theta = y/\sqrt{x^2 + y^2}$, $\cos\theta = x/\sqrt{x^2 + y^2}$, $\tan\theta = y/x$,
$\csc\theta = \sqrt{x^2 + y^2}/y$, $\sec\theta = \sqrt{x^2 + y^2}/x$, $\cot\theta = x/y$
11. $b = 2.04, c = 4.49$ **13.** $a = 11.71, c = 14.19$
15. $\alpha = 60°, \beta = 30°, a = 2.6$
17. $\alpha = 21.8°, \beta = 68.2°, c = 10.8$
19. $\alpha = 48.6°, \beta = 41.4°, b = 7.9$ **21.** $a = 8.5, c = 21.7$
23. 36.53   **25.** 35.3°
27. approximately 0.951   **29.** $A = \frac{\sqrt{3}}{4}s^2$

Exercises 4.11   Page 293

1. 52.1 m   **3.** 66.4 ft
5. 409.7 ft   **7.** height: 15.5 ft; distance: 12.6 ft
9. 8.7°   **11.** 34,157 ft ≈ 6.5 mi
13. 20.2 ft   **15.** 6617 ft
17. Yes, since the altitude of the storm is 6.3 km.
19. approximately 1052 ft   **21.** 227,100 mi
23. $h = c/(\cot\alpha + \cot\beta)$ **25.** height is 1.37, area is 4.81
27. 80.87°   **29.** **(a)** 51.8°   **(b)** 13.8°   **(c)** 1258 ft
31. $h(\theta) = 1.25\tan\theta$
33. $\theta(x) = \arctan\left(\dfrac{1}{x}\right) - \arctan\left(\dfrac{1}{2x}\right)$, x measured in meters

Exercises 4.12   Page 303

1. 241.21   **3.** 65.53
5. $\gamma = 80°, a = 20.16, c = 20.16$
7. $\alpha = 92°, b = 3.01, c = 3.89$
9. $\alpha = 79.61°, \gamma = 28.39°, a = 12.41$
11. no solution
13. $\alpha = 24.46°, \beta = 140.54°, b = 12.28$;
$\alpha = 155.54°, \beta = 9.46°, b = 3.18$
15. no solution
17. $\alpha = 45.58°, \gamma = 104.42°, c = 13.56$;
$\alpha = 134.42°, \gamma = 15.58°, c = 3.76$
19. no solution   **21.** 9.76   **23.** 15.80 ft
25. 9.07 m   **27.** 10.35 ft
29. 8.81°   **31.** the vessel sailing at 5 knots

Exercises 4.13   Page 308

1. $\alpha = 37.59°, \beta = 77.41°, c = 7.43$
3. $\alpha = 52.62°, \beta = 83.33°, \gamma = 44.05°$
5. $\alpha = 25°, \beta = 57.67°, c = 7.04$
7. $\alpha = 76.45°, \beta = 57.10°, \gamma = 46.45°$
9. $\alpha = 27.66°, \beta = 40.54°, \gamma = 111.80°$
11. $\alpha = 36.87°, \beta = 53.13°, \gamma = 90°$
13. $\alpha = 26.38°, \beta = 36.34°, \gamma = 117.28°$
15. $\beta = 10.24°, \gamma = 147.76°, a = 6.32$
17. 35.94 nautical miles
19. **(a)** S33.66° W   **(b)** S2.82° E
21. $\alpha = 119.45°, \beta = 67.98°$
23. 91.77, 176.18   **25.** **(a)** 63.72 ft   **(b)** 42.43 ft

Exercises 4.14   Page 315

1. $\frac{1}{2}$ **3.** −1   **5.** $-\sqrt{3}/2$ **7.** 5
9. 0   **11.** −2   **13.** 0   **15.** 1
17. 1   **19.** $y = x$ **21.** $y = \frac{1}{2} + \frac{\sqrt{3}}{2}(x - \frac{\pi}{6})$
23. $f'(x) = -\sin x$ **25.** $f'(x) = 5\cos 5x$

A. 1. $36°$ **3.** $(-\sqrt{3}/2, 1/2)$ **5.** $\sqrt{3}$
   7. $3/(2\sqrt{2})$ **9.** $(0, -2)$ **11.** $2/\sqrt{5}$
 13. $6\sin(\pi x/2)$ **15.** $\sqrt{2}$ **17.** 10
 19. $\pi/5$

B. 1. false   **3.** true   **5.** true
   7. true   **9.** true   **11.** true
 13. true   **15.** false   **17.** false
 19. false

C. 1. $\pi/2, \pi$ **3.** $\pi/6, 5\pi/6, 7\pi/6, 11\pi/6$
   5. $0, \pi/2, 2\pi$
   7. $\gamma = 80°, a = 5.32, c = 10.48$
   9. $\alpha = 15.76, \beta = 99.44°, \gamma = 29.56°$
 11. $2\pi/3$ **13.** $3/\sqrt{7}$
 15. 0 **17.** $\frac{12}{13}$
 19. $\sqrt{1 - x^2}$
 21. $y = -\sin x; y = \cos(x + \pi/2)$
 23. $y = 1 + \frac{1}{2}\sin(x + \pi/2); y = 1 + \frac{1}{2}\cos x$
 25. 42.61 m   **27.** 118 ft
 29. (a) $42.35°$ **31.** $16{,}927.6$ km
 33. front: $18.88°$; back: $35.12°$ **35.** $\theta \approx 109.62°$
 37. (b) approximately $63°$ and $27°$
 39. 162.3 ft   **41.** approximately $30{,}000$ ft/min
 45. approximately 114 in   **47.** $V(\theta) = 160\sin 2\theta$
 49. $L(\theta) = 3\csc\theta + 4\sec\theta$ **51.** $V(\theta) = 360 + 75\cot\theta$
 53. $A(\phi) = 100\cos\phi + 50\sin 2\phi$
 55. (a) $r(\phi) = 3959\cos\phi$ **(b)** 1574.71 mi   **(c)** 613 mi   **(d)** 69 mi

 57. (a) $\theta(s) = 2\tan^{-1}\left(\dfrac{15}{s - 15}\right)$ **(b)** π

 59. $A(s) = s^2(\frac{\pi}{3} - \frac{\sqrt{3}}{4})$

1.

3.

5.
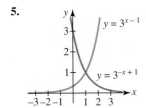

7. $(0, -4); y = -5$; increasing
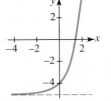

9. $(0, 2); y = 3$; increasing

11. $(0, -1 + e^{-3}); y = -1$; increasing

13. $f(x) = 6^x$
17. $f(x) = 3^{-x}$
21. $(-2, \infty)$
25. $(-10, 0), (0, 10)$
29. $x > 4$
33.

37.

41.

47. approximately 1.175
49.

51. $y = \frac{26}{9}x + \frac{29}{9}$

15. $f(x) = (e^{-2})^x = e^{-2x}$
19. $(5, \infty)$
23. $(2, 0), (0, -3)$
27. $(-2, 0), (-3, 0), (0, 0)$
31. $x < 2$
35.

39.

43.

57. approximately $-0.77, 2, 4$

1. $-\frac{1}{2} = \log_4\frac{1}{2}$
5. $-s = \log_t v$
9. $(\sqrt{3})^8 = 81$
13. -7
17. e
21. $\frac{1}{7}$
25. $(0, \infty); (1, 0), x = 0$

29. $(-3, \infty); (5, 0), x = -3$
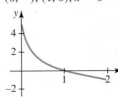

3. $4 = \log_{10}10{,}000$
7. $2^7 = 128$
11. $b^y = u$
15. 3
19. 36
23. $f(x) = \log_7 x$
27. $(-\infty, 0); (-1, 0), x = 0$

31. $(0, \infty); (e, 0), x = 0$

33. $-1 < x < 0$

35. $f(x) = \begin{cases} \ln x, & x > 0 \\ \ln(-x), & x < 0 \end{cases}$, $(-1, 0)$, $(1, 0)$, $x = 0$

37.

39. $\left(\frac{3}{2}, \infty\right)$

41. the interval $(-3, 3)$

43. $(1, \infty)$

45.

47. $f^{-1}(x) = \log_4(x - 2)$; $(2, \infty)$, $(-\infty, \infty)$

49. $f^{-1}(x) = 2 + e^{x-1}$; $(-\infty, \infty)$, $(2, \infty)$

51. $\log_{10} 50$

53. $\ln(x^2 - 2)$

55. $\ln 1 = 0$

57. 0.3011

59. 1.8063

61. 0.3495

63. 0.2007

65. -0.0969

67. 1.6609

69. $\ln y = 10 \ln x + \frac{1}{2}\ln(x^2 + 5) - \frac{1}{3}\ln(8x^3 + 2)$

71. $\ln y = 5\ln(x^3 - 3) + 8\ln(x^4 + 3x^2 + 1)$
$\qquad -\frac{1}{2}\ln x - 9\ln(7x + 5)$

77. $\{x \mid x \neq n\pi, n \text{ and integer}\}$

Exercises 5.3   Page 349

1. 2

3. 2

5. 1.0802

7. 2

9. 3

11. 0.3495

13. 2.7712

15. ± 4

17. ± 3

19. -0.8782

21. 32

23. 45

25. $\pm\frac{1}{10}$

27. 81

29. $\frac{3}{2}$

31. 100

33. 2, 8

35. 1

37. 4

39. $\frac{7}{2}$

41. 0, 2

43. 1, 16

45. $\log_5\left(1 + \sqrt{2}\right) = \dfrac{\ln\left(1 + \sqrt{2}\right)}{\ln 5} \approx 0.5476$

47. e^{-2}, e

49. 0

51. $(-3, 0)$

53. $\left(1 + \frac{\ln 3}{\ln 4}, 0\right) \approx (1.7925, 0)$

55. $(-1.0397, 0)$

57. $(2.3219, 0)$

59. $28, -36$

61. approximately -0.6606

63. approximately $-0.4386, 1.4386$

65. $\sqrt{10} \approx 3.1623$

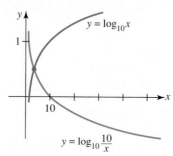

67. e^{-3}, e^3

69. 9

71. $(2.1944, 51)$

73. $(7, 2)$

75. $(0.8326, \frac{1}{2})$, $(-0.8326, \frac{1}{2})$

77. 1.2110

Exercises 5.4   Page 358

1. (a) $P(t) = P_0 e^{0.3466t}$ **(b)** $5.66 P_0$ **(c)** 8.64 h

3. 2,344 **5.** 201

7. (a) 82 **(b)** 8.53 days **(c)** 2000

(d)

9. $A(t) = 200 e^{-0.005077t}$; 177 mg

11. approximately 100 g, 50 g, 25 g

13. approximately $k = -0.08664$; 34.58 days

15. $0.6730 A_0$; $0.2264 A_0$

17. approximately 16,253 years old

19. approximately 92%

21. (a) $220.2°$ F **(b)** 9.2 minutes **(c)** $80°$ F

23. 8.74 minutes

25. approximately 1.6 hours

27. $\$4,851,651.95$; 2.35×10^{15}

29. $\$3080.37$ in interest

31. (a) 1.25×10^9 yr **(b)** 89.5% of P_0; 10.5% of P_0

33. (a) 6 days **(b)** 19.84%

35. $t = -\dfrac{L}{R}\ln\left(1 - \dfrac{IR}{E}\right)$

37. approximately 25 times more intense
39. 5.5   **41.** 6
43. 7.6   **45.** 5×10^{-4}
47. 2.5×10^{-7} **49.** 10 times as acidic
51. 163 times as acidic   **53.** (a) 5.62×10^{25} ergs
55. 10^{-2} watts/cm^2 **57.** 103.7 dB
59. (a) 2.46 mm (b) 0.79 mm, 0.19 mm (c) 7.7×10^{-6} mL
61. $S = 0.20247\,w^{0.425}\,h^{0.725}$

45. $f(x) = 5 + \left(\frac{1}{2}\right)^x$
47. After doubling, it will take another 9 hours to double again. In other words, it will take a total of 18 hours for the population to grow to 4 times the initial population.
49. $0.0625A_0$ or $6\frac{1}{4}\%$ of the initial quantity A_0
51. 4.3%   **53.** $t = \dfrac{1}{c}[\ln b - \ln(\ln a - \ln y)]$
55. \$51,955.78

Exercises 5.5   Page 368

3. $\dfrac{1}{x}\log_{10} e = \dfrac{1}{x\ln 10}$ **5.** $\dfrac{1}{x}$

7. $e^{5x}\left(\dfrac{e^{5h} - 1}{h}\right)$ **15.** (a) $\dfrac{\sqrt{13}}{2}$ (b) $\dfrac{-3}{\sqrt{13}}, -\dfrac{\sqrt{13}}{3}, \dfrac{2}{\sqrt{13}}, -\dfrac{2}{3}$

17. $\dfrac{17}{8}$ **19.** $\dfrac{7}{24}$

21. $\dfrac{x^2 - 1}{2x}$ **23.** $\dfrac{4x^2 + 1}{4x^2 - 1}$

29. -1.4436 **31.** 0.9624

35. (a) 625 ft   (b) (299.9, 0), 599.8 ft

37. $5e^{5x}$

Chapter 5 Review Exercises   Page 372

A. 1. $(0, 5), y = 6$ **3.** $(-3, 0), x = -4$
5. -1 **7.** 1000
9. 6   **11.** $\frac{1}{9}$
13. 3   **15.** $3 + e$
17. -7.8577 **19.** 64
21. 8   **23.** 0
25. $f^{-1}(x) = e^{2-x}$

B. 1. true   **3.** true
5. true   **7.** false
9. true   **11.** false
13. true   **15.** true
17. true   **19.** true
21. false   **23.** true
25. true

C. 1. $\log_5 0.2 = -1$ **3.** $9^{1.5} = 27$
5. -2 **7.** $-\frac{1}{2}$
9. $1 - \log_2 7 = 1 - \dfrac{\ln 7}{\ln 2}$ **11.** $-2 + \ln 6$
13. $m = -\dfrac{1}{r}\ln(P/S)$ **15.**

17. C, D, A, B   **19.** $\dfrac{3^{1-h} - 3}{h}$

21. (ii) **23.** (iv)
25. (iii) **27.** $y = x - 3$
29. $y = 8x$ **31.** $y = 4x^2 - 4x + 6$
33. $y = 3(x^2 + 1)$ **35.** $y = 5e^{-3x}$

37. $f^{-1}(x) = -\ln(x - 5)$ **39.** $f(x) = \begin{cases} \ln(x - 2), & x > 2 \\ \ln(2 - x), & x < 2 \end{cases}$

41. upward shift of 1 unit   **43.** $f(x) = 5e^{(-\frac{1}{6}\ln 5)x} = 5e^{-0.2682x}$

Exercises 6.1   Page 385

1. Vertex: $(0, 0)$
Focus: $(1, 0)$
Directrix: $x = -1$
Axis: $y = 0$

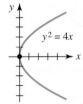

3. Vertex: $(0, 0)$
Focus: $\left(-\frac{1}{3}, 0\right)$
Directrix: $x = \frac{1}{3}$
Axis: $y = 0$

5. Vertex: $(0, 0)$
Focus: $(0, -4)$
Directrix: $y = 4$
Axis: $x = 0$

7. Vertex: $(0, 0)$
Focus: $(0, 7)$
Directrix: $y = -7$
Axis: $x = 0$

9. Vertex: $(0, 1)$
Focus: $(4, 1)$
Directrix: $x = -4$
Axis: $y = 1$

11. Vertex: $(-5, -1)$
Focus: $(-5, -2)$
Directrix: $y = 0$
Axis: $x = -5$

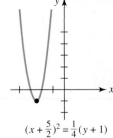

13. Vertex: $(-5, -6)$
Focus: $(-4, -6)$
Directrix: $x = -6$
Axis: $y = -6$

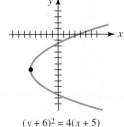

15. Vertex: $\left(-\frac{5}{2}, -1\right)$
Focus: $\left(-\frac{5}{2}, -\frac{15}{16}\right)$
Directrix: $y = -\frac{17}{16}$
Axis: $x = -\frac{5}{2}$

17. Vertex: $(3, 4)$
Focus: $\left(\frac{5}{2}, 4\right)$
Directrix: $x = \frac{7}{2}$
Axis: $y = 4$

$(y - 4)^2 = -2(x - 3)$

19. Vertex: $(0, 0)$
Focus: $\left(0, \frac{1}{8}\right)$
Directrix: $y = -\frac{1}{8}$
Axis: $x = 0$

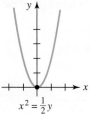

$x^2 = \frac{1}{2}y$

21. Vertex: $(3, 0)$
Focus: $(3, -1)$
Directrix: $y = 1$
Axis: $x = 3$

$(x - 3)^2 = -4y$

23. Vertex: $(-2, 1)$
Focus: $(-1, 1)$
Directrix: $x = -3$
Axis: $y = 1$

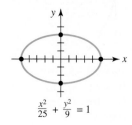

$(y - 1)^2 = 4(x + 2)$

25. $x^2 = 28y$
27. $y^2 = -16x$
29. $y^2 = 10x$
31. $(x - 2)^2 = 12y$
33. $(y - 4)^2 = -12(x - 2)$
35. $(x - 1)^2 = 32(y + 3)$
37. $(y + 3)^2 = 32x$
39. $x^2 = 7y$
41. $(x - 5)^2 = -24(y - 1)$
43. $x^2 = \frac{1}{2}y$
45. $(3, 0), (0, -2), (0, -6)$
47. $\left(-3\sqrt{2}, 0\right), \left(3\sqrt{2}, 0\right), (0, 9)$
49. $f(x) = 2x^2 - 4x$; domain of f is $(-\infty, \infty)$
51. $f(x) = -1 + \sqrt{10 - 5x}, g(x) = -1 - \sqrt{10 - 5x}$;
domain of f and g is $(-\infty, 2]$
53. At the focus 6 in. from the vertex
55. $y = -2$
57. 27 ft
59. 4.5 ft
61. (a) 8
63. $f(x) = x - 1 + \sqrt{x + 2}, g(x) = x - 1 - \sqrt{x + 2}$;
domain of f and g is $[-2, \infty)$

Exercises 6.2   Page 393

1. Center: $(0, 0)$
Foci: $(\pm 4, 0)$
Vertices: $(\pm 5, 0)$
Minor axis endpoints: $(0, \pm 3)$
Eccentricity: $\frac{4}{5}$

$\frac{x^2}{25} + \frac{y^2}{9} = 1$

3. Center: $(0, 0)$
Foci: $\left(0, \pm \sqrt{15}\right)$
Vertices: $(0, \pm 4)$
Minor axis endpoints: $(\pm 1, 0)$
Eccentricity: $\frac{\sqrt{15}}{4}$

$\frac{x^2}{1} + \frac{y^2}{16} = 1$

5. Center: $(0, 0)$
Foci: $\left(\pm \sqrt{7}, 0\right)$
Vertices: $(\pm 4, 0)$
Minor axis endpoints: $(0, \pm 3)$
Eccentricity: $\frac{\sqrt{7}}{4}$

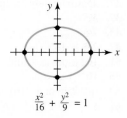

$\frac{x^2}{16} + \frac{y^2}{9} = 1$

7. Center: $(0, 0)$
Foci: $\left(0, \pm \sqrt{5}\right)$
Vertices: $(0, \pm 3)$
Minor axis endpoints: $(\pm 2, 0)$
Eccentricity: $\frac{\sqrt{5}}{3}$

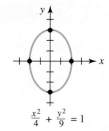

$\frac{x^2}{4} + \frac{y^2}{9} = 1$

9. Center: $(1, 3)$
Foci: $\left(1 \pm \sqrt{13}, 3\right)$
Vertices: $(-6, 3), (8, 3)$
Minor axis endpoints: $(1, -3), (1, 9)$
Eccentricity: $\frac{\sqrt{13}}{7}$

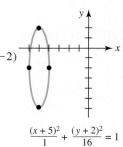

$\frac{(x - 1)^2}{49} + \frac{(y - 3)^2}{36} = 1$

11. Center: $(-5, -2)$
Foci: $\left(-5, -2 \pm \sqrt{15}\right)$
Vertices: $(-5, -6), (-5, 2)$
Minor axis endpoints: $(-6, -2), (-4, -2)$
Eccentricity: $\frac{\sqrt{15}}{4}$

$\frac{(x + 5)^2}{1} + \frac{(y + 2)^2}{16} = 1$

13. Center: $\left(0, -\frac{1}{2}\right)$
Foci: $\left(0, -\frac{1}{2} \pm \sqrt{3}\right)$
Vertices: $\left(0, -\frac{5}{2}\right), \left(0, \frac{3}{2}\right)$
Minor axis endpoints: $\left(-1, -\frac{1}{2}\right), \left(1, -\frac{1}{2}\right)$
Eccentricity: $\frac{\sqrt{3}}{2}$

$\frac{x^2}{1} + \frac{\left(y + \frac{1}{2}\right)^2}{4} = 1$

15. Center: $(1, -2)$
Foci: $\left(1, -2 \pm \sqrt{6}\right)$
Vertices: $\left(1, -2 \pm \sqrt{15}\right)$
Minor axis endpoints: $(-2, -2), (4, -2)$
Eccentricity: $\sqrt{\frac{2}{5}}$

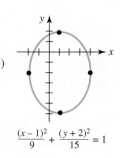

$\frac{(x - 1)^2}{9} + \frac{(y + 2)^2}{15} = 1$

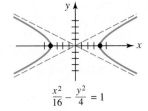

17. Center: $(2, -1)$
Foci: $(2, -5), (2, 3)$
Vertices: $(2, -6), (2, 4)$
Minor axis endpoints: $(-1, -1), (5, -1)$
Eccentricity: $\frac{4}{5}$

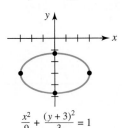

$$\frac{(x-2)^2}{9} + \frac{(y+1)^2}{25} = 1$$

19. Center: $(0, -3)$
Foci: $(\pm\sqrt{6}, -3)$
Vertices: $(-3, -3), (3, -3)$
Minor axis endpoints: $(0, -3 \pm \sqrt{3})$
Eccentricity: $\frac{\sqrt{6}}{3}$

$$\frac{x^2}{9} + \frac{(y+3)^2}{3} = 1$$

21. $\dfrac{x^2}{25} + \dfrac{y^2}{16} = 1$ **23.** $\dfrac{x^2}{8} + \dfrac{y^2}{9} = 1$

25. $\dfrac{x^2}{1} + \dfrac{y^2}{9} = 1$ **27.** $\dfrac{(x-1)^2}{16} + \dfrac{(y+3)^2}{4} = 1$

29. $\dfrac{x^2}{9} + \dfrac{y^2}{13} = 1$ **31.** $\dfrac{x^2}{11} + \dfrac{y^2}{9} = 1$

33. $\dfrac{x^2}{3} + \dfrac{y^2}{12} = 1$ **35.** $\dfrac{x^2}{16} + \dfrac{(y-1)^2}{4} = 1$

37. $\dfrac{(x-1)^2}{7} + \dfrac{(y-3)^2}{16} = 1$ **39.** $\dfrac{x^2}{45} + \dfrac{(y-2)^2}{9} = 1$

41. $f(x) = -\frac{3}{5}\sqrt{25 - x^2}$; domain is $[-5, 5]$
43. $f(x) = 3 + \frac{6}{7}\sqrt{49 - (x-1)^2}$; domain is $[-6, 8]$
45. greatest distance is 43.5 millions miles; least distance is 28.5 million miles
47. approximately 0.97 **49.** 12 ft
51. The piece of string should be 4 ft long. The tacks should be placed $\sqrt{7}/2$ ft from the center of the rectangle on the major axis of the ellipse.
53. on the major axis, 12 ft to either side from the center of the room
55. $5x^2 - 4xy + 8y^2 - 24x - 48y = 0$

Exercises 6.3 **Page 403**

1. Center: $(0, 0)$
Foci: $(\pm\sqrt{41}, 0)$
Vertices: $(\pm 4, 0)$
Asymptotes: $y = \pm\frac{5}{4}x$
Eccentricity: $\frac{\sqrt{41}}{4}$

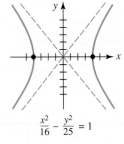

$$\frac{x^2}{16} - \frac{y^2}{25} = 1$$

3. Center: $(0, 0)$
Foci: $(0, \pm\sqrt{73})$
Vertices: $(0, \pm 8)$
Asymptotes: $y = \pm\frac{8}{3}x$
Eccentricity: $\frac{\sqrt{73}}{8}$

$$\frac{y^2}{64} - \frac{x^2}{9} = 1$$

5. Center: $(0, 0)$
Foci: $(\pm 2\sqrt{5}, 0)$
Vertices: $(\pm 4, 0)$
Asymptotes: $y = \pm\frac{1}{2}x$
Eccentricity: $\frac{\sqrt{5}}{2}$

$$\frac{x^2}{16} - \frac{y^2}{4} = 1$$

7. Center: $(0, 0)$
Foci: $(0, \pm 2\sqrt{6})$
Vertices: $(0, \pm 2\sqrt{5})$
Asymptotes: $y = \pm\sqrt{5}x$
Eccentricity: $\sqrt{\frac{6}{5}}$

$$\frac{y^2}{20} - \frac{x^2}{4} = 1$$

9. Center: $(5, -1)$
Foci: $(5 \pm \sqrt{53}, -1)$
Vertices: $(3, -1), (7, -1)$
Asymptotes: $y = -1 \pm \frac{7}{2}(x - 5)$
Eccentricity: $\frac{\sqrt{53}}{2}$

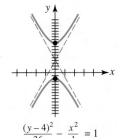

$$\frac{(x-5)^2}{4} - \frac{(y+1)^2}{49} = 1$$

11. Center: $(0, 4)$
Foci: $(0, 4 \pm \sqrt{37})$
Vertices: $(0, -2), (0, 10)$
Asymptotes: $y = 4 \pm 6x$
Eccentricity: $\frac{\sqrt{37}}{6}$

$$\frac{(y-4)^2}{36} - \frac{x^2}{1} = 1$$

13. Center: $(3, 1)$
Foci: $(3 \pm \sqrt{30}, 1)$
Vertices: $(3 \pm \sqrt{5}, 1)$
Asymptotes: $y = 1 \pm \sqrt{5}(x - 3)$
Eccentricity: $\sqrt{6}$

$$\frac{(x-3)^2}{5} - \frac{(y-1)^2}{25} = 1$$

15. Center: $(-4, 7)$
Foci: $(-4, 7 \pm \sqrt{13})$
Vertices: $(-4, 7 \pm 2\sqrt{2})$
Asymptotes: $y = 7 \pm \sqrt{\frac{8}{5}}(x + 4)$
Eccentricity: $\sqrt{\frac{13}{8}}$

$$\frac{(y-7)^2}{8} - \frac{(x+4)^2}{5} = 1$$

17. Center: $(2, 1)$
Foci: $\left(2 \pm \sqrt{11}, 1\right)$
Vertices: $\left(2 \pm \sqrt{6}, 1\right)$
Asymptotes: $y = 1 \pm \sqrt{\frac{5}{6}}(x - 2)$
Eccentricity: $\sqrt{\frac{11}{6}}$

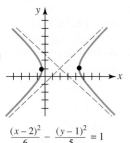

$$\frac{(x-2)^2}{6} - \frac{(y-1)^2}{5} = 1$$

19. Center: $(1, 3)$
Foci: $\left(1, 3 \pm \frac{1}{2}\sqrt{5}\right)$
Vertices: $(1, 2), (1, 4)$
Asymptotes: $y = 3 \pm 2(x - 1)$
Eccentricity: $\frac{\sqrt{5}}{2}$

$$\frac{(y-3)^2}{1} - \frac{(x-1)^2}{\frac{1}{4}} = 1$$

21. $\dfrac{x^2}{9} - \dfrac{y^2}{16} = 1$ **23.** $\dfrac{y^2}{4} - \dfrac{x^2}{12} = 1$

25. $\dfrac{x^2}{9} - \dfrac{y^2}{7} = 1$ **27.** $\dfrac{y^2}{\frac{25}{4}} - \dfrac{x^2}{\frac{11}{4}} = 1$

29. $\dfrac{x^2}{4} - \dfrac{y^2}{5} = 1$ **31.** $\dfrac{y^2}{64} - \dfrac{x^2}{16} = 1$

33. $\dfrac{x^2}{4} - \dfrac{y^2}{\frac{64}{9}} = 1$ **35.** $\dfrac{(y+3)^2}{4} - \dfrac{(x-1)^2}{5} = 1$

37. $\dfrac{(x+1)^2}{4} - \dfrac{(y-2)^2}{5} = 1$ **39.** $\dfrac{x^2}{4} - \dfrac{y^2}{8} = 1$

41. $\dfrac{(y-3)^2}{1} - \dfrac{(x+1)^2}{4} = 1$ **43.** $\dfrac{(y-4)^2}{1} - \dfrac{(x-2)^2}{4} = 1$

45. $f(x) = \frac{5}{4}\sqrt{x^2 - 16}, g(x) = -\frac{5}{4}\sqrt{x^2 - 16}$;
domain of f and g is $(-\infty, -4] \cup [4, \infty)$

47. $f(x) = 4 + 6\sqrt{x^2 + 1}, g(x) = 4 - 6\sqrt{x^2 + 1}$;
domain of f and g is $(-\infty, \infty)$

49. $(-7, 12)$

51. $7y^2 - 24xy + 24x + 82y + 55 = 0$

Exercises 6.4   Page 409

1.

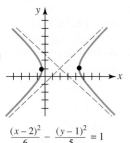

$(3, \pi)$ O Polar axis

3.

O $\frac{\pi}{2}$ Polar axis
$\left(-\frac{1}{2}, \frac{\pi}{2}\right)$

5. $\left(-4, -\frac{\pi}{6}\right)$

O $-\frac{\pi}{6}$ Polar axis

7. (a) $(2, -5\pi/4)$ (b) $(2, 11\pi/4)$
     (c) $(-2, 7\pi/4)$ (d) $(-2, -\pi/4)$

9. (a) $(4, -5\pi/3)$ (b) $(4, 7\pi/3)$
     (c) $(-4, 4\pi/3)$ (d) $(-4, -2\pi/3)$

11. (a) $(1, -11\pi/6)$ (b) $(1, 13\pi/6)$
     (c) $(-1, 7\pi/6)$ (d) $(-1, -5\pi/6)$

---

13. (a) $(9, -\pi/2)$ (b) $(9, 7\pi/2)$
     (c) $(-9, \pi/2)$ (d) $(-9, -3\pi/2)$

15. $\left(-\frac{1}{4}, \frac{\sqrt{3}}{4}\right)$ **17.** $\left(-3, 3\sqrt{3}\right)$

19. $\left(-2\sqrt{2}, -2\sqrt{2}\right)$ **21.** $\left(\frac{\sqrt{3}}{2}, \frac{1}{2}\right)$

23. $(3.696, 1.531)$

25. (a) $\left(2\sqrt{2}, -3\pi/4\right)$ (b) $\left(-2\sqrt{2}, \pi/4\right)$

27. (a) $(2, -\pi/3)$ (b) $(-2, 2\pi/3)$

29. (a) $(7, 0)$ (b) $(-7, \pi)$

31. (a) $(5, 2.214)$ (b) $(-5, -0.927)$

33.

35.

37.

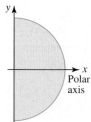

39. $r = 5 \csc \theta$
41. $\theta = \tan^{-1} 7$
43. $r = 2/(1 + \cos \theta)$
45. $r = 6$
47. $r = 1 - \cos \theta$
49. $r = 5\sin\theta$
51. $x = 2$
53. $(x^2 + y^2)^3 = 144x^2y^2$

55. $(x^2 + y^2)^2 = 8xy$ **57.** $x^2 + y^2 + 5y = 0$
59. $8x^2 - 12x - y^2 + 4 = 0$ **61.** $3x + 8y = 5$

Exercises 6.5   Page 419

1. circle   **3.** line through the pole

$r = 6$

$\theta = \dfrac{\pi}{3}$

5. spiral   **7.** cardioid

$r = 2\theta$ $r = 1 + \cos \theta$

9. cardioid   **11.** limaçon with interior loop

$r = 2(1 + \sin \theta)$ $r = 1 - 2 \cos \theta$

13. dimpled limaçon

15. convex limaçon

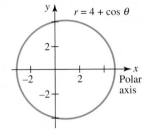

1. $e = 1$, parabola

$$r = \frac{2}{1 - \sin\theta}$$

3. $e = \frac{1}{4}$, ellipse

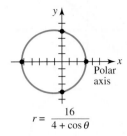

$$r = \frac{16}{4 + \cos\theta}$$

17. rose curve

19. rose curve

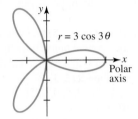

5. $e = 2$, hyperbola

$$r = \frac{4}{1 + 2\sin\theta}$$

7. $e = 2$, hyperbola

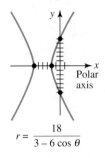

$$r = \frac{18}{3 - 6\cos\theta}$$

21. rose curve

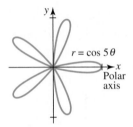

23. circle with center on x-axis

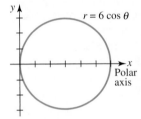

9. $e = 1$, parabola

$$r = \frac{6}{1 - \cos\theta}$$

11. $e = 2$, $\dfrac{(y-4)^2}{4} - \dfrac{x^2}{12} = 1$

25. circle with center on y-axis

27. lemniscate

13. $e = \frac{2}{3}$, $\dfrac{\left(x - \frac{24}{5}\right)^2}{\frac{1296}{25}} + \dfrac{y^2}{\frac{144}{5}} = 1$ **15.** $x^2 = 4(1 - y)$

17. $r = \dfrac{3}{1 + \cos\theta}$ **19.** $r = \dfrac{4}{3 - 2\sin\theta}$

21. $r = \dfrac{12}{1 + 2\cos\theta}$ **23.** $r = \dfrac{3}{1 + \cos(\theta + 2\pi/3)}$

25. $r = \dfrac{3}{1 - \sin\theta}$ **27.** $r = \dfrac{1}{1 - \cos\theta}$

29. lemniscate

31.

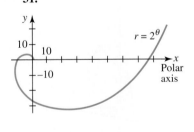

29. $r = \dfrac{1}{2 - 2\sin\theta}$

31. parabola, vertex: $(2, \pi/4)$

33. ellipse, vertices: $(10, \pi/3)$ and $\left(\frac{10}{3}, 4\pi/3\right)$

35. hyperbola, vertices: $(-3, -\pi/2)$, $(\frac{3}{5}, \pi/2)$

37. $r_p = 8000$ km   **39.** $r = \dfrac{1.495 \times 10^8}{1 - 0.0167\cos\theta}$

33. $r = \frac{5}{2}$

35. $r = 4 - 3\cos\theta$

37. $r = 2\cos 4\theta$

39. $(2, \pi/6)$, $(2, 5\pi/6)$

41. $(1, \pi/2)$, $(1, 3\pi/2)$, origin

45. (a) $r = 2\cos(\theta + \pi/6) = \sqrt{3}\cos\theta - \sin\theta$

   (b) $(x - \frac{\sqrt{3}}{2})^2 + (y + \frac{1}{2})^2 = 1$

   (c) $(1, -\pi/6)$, $(\frac{\sqrt{3}}{2}, -\frac{1}{2})$

47. $(\sqrt{3}/2, \pi/3)$, $(\sqrt{3}/2, 2\pi/3)$, origin

1.

t	-3	-2	-1	0	1	2	3
x	-1	0	1	2	3	4	5
y	$\frac{3}{2}$	2	$\frac{5}{2}$	3	$\frac{7}{2}$	4	$\frac{9}{2}$

intercepts: $(-4, 0)$, $(0, 2)$

3.

5.

1, 3, 5.

7.

9.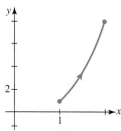

7. The set $\{(x, y, 5) \,|\, x, y$ real numbers$\}$ is a plane perpendicular to the z-axis, 5 units above the xy-plane.

9. The set $\{(2, 3, z) \,|\, z$ a real number$\}$ is a line perpendicular to the xy-plane at $(2, 3, 0)$.

11. $(2, 0, 0), (2, 5, 0,) (2, 0, 8), (2, 5, 8), (0, 5, 0), (0, 5, 8), (0, 0, 8), (0, 0, 0)$

13. (a) $(-2, 5, 0), (-2, 0, 4), (0, 5, 4)$ **(b)** $(-2, 5, -2)$ **(c)** $(3, 5, 4)$

11. $y = x^2 + 3x - 1, x \geq 0$ **13.** $x = 1 - 2y^2, -1 \leq x \leq 1$

15. The union of the three coordinate planes

17. The point $(-1, 2, -3)$

15. $y = \ln x, x > 0$ **17.** $\dfrac{x^2}{16} + \dfrac{y^2}{4} = 1$

19. The union of the planes $z = 5$ and $z = -5$

21. $\sqrt{70}$ **23. (a)** 7   **(b)** 5

19.

25. right triangle   **27.** isosceles triangle

29. collinear   **31.** not collinear

33. 6 or -2 **35.** $\left(4, \tfrac{1}{2}, \tfrac{3}{2}\right)$ **37.** $(-4, -11, 10)$

21.

39.   **41.**

23.

43. center $(-4, 3, 2)$, radius 6   **45.** center $(0, 0, 8)$, radius 8

47. $(x + 1)^2 + (y - 4)^2 + (z - 6)^2 = 3$

49. $(x - 1)^2 + (y - 1)^2 + (z - 4)^2 = 16$

51. $x^2 + (y - 4)^2 + z^2 = 4$ or $x^2 + (y - 8)^2 + z^2 = 4$

25.

53.   **55.**

27.

57.   **59.** $5x - 3y + z = 2$

29. $(3, 0); (0, 1), (0, 3)$

31. $(-1 - \sqrt{3}, 0), (-1 + \sqrt{3}, 0); (0, 1 + \sqrt{3}), (0, 1 - \sqrt{3})$

33. the line segment between (x_1, y_1) and (x_2, y_2)

35. $x = 95\sqrt{2}\,t, y = -16t^2 + 95\sqrt{2}\,t, t \geq 0;$
$\left(190\sqrt{2}, 190\sqrt{2} - 64\right) \approx (268.70, 204.70)$

37. $x = \pm\sqrt{r^2 - L^2 \sin^2\phi}, y = L\sin\phi$

61. $3x - 2y + 2z = 3$ **63.** $-x + 3y = 5$

A. **1.** $y^2 = 20x$

  3. $(x-1)^2 = 8(y+5)$

  5. $(0,0)$

  7. 1

  9. $(-3,0);\ (-3,-1),\ (-3,1)$

  11. $(0,-\sqrt{2}),\ (0,\sqrt{2})$

  13. $(-2,4)$

  15. 1

  17. $(10, 3\pi/2)$

  19. $(0, \pi/2),\ (0, 3\pi/2)$

  21. $(-4,3)$

B. **1.** true

  3. true

  5. true

  7. true

  9. true

  11. false

  13. true

  15. true

  17. false

  19. true

  21. true

C. **1.** Center $(1,1)$,
Vertices $(-3,1),\ (5,1)$,
Minor axis endpoints
$(1,3),\ (1,-1)$

  3. Center $(-1,3)$,
Vertices $(-1,5),\ (-1,1)$,
Minor axis endpoints
$(-2,3),\ (0,3)$

  5. Center $(0,2)$,
Vertices $(0,-1),\ (0,5)$,
Asymptotes $y = 2 - \frac{3}{4}x$

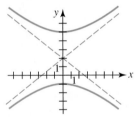

  7. $(x-\frac{1}{2})^2 + (y-\frac{1}{2})^2 = \frac{1}{2}$

  9. $r = 4\sin\theta$

  11. $(0,2),\ (0,-\frac{2}{3})$

  13. (a) $(\sqrt{6}, -\pi/4)$
     (b) $(-\sqrt{6}, 3\pi/4)$

15. circle of radius 5 centered at the origin

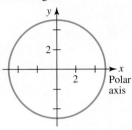

17. circle with center on y-axis

19. cardioid

21. convex limaçon

23. rose curve

25. ellipse

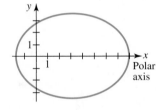

27. 1.95×10^9 m

29. $\frac{25}{7}$ cm

31. (a) $r = 2\cos(\theta - \pi/3)$ or $r = \cos\theta + \sqrt{3}\sin\theta$
   (b) $(x-\frac{1}{2})^2 + (y - \frac{\sqrt{3}}{2})^2 = 1$

33. upper half-ellipse

35. $(x+5)^2 + \dfrac{(y-2)^2}{4} = 1$

37. parabola: $y^2 = -8(x-6)$, ellipse: $x^2/6^2 + y^2/(2\sqrt{5})^2 = 1$

39. $(x-1)^2 + (y-4)^2 + (z-2)^2 = 90$

Index

A

Abel, Niels Henrik, 169
Absolute value:
    definition of ($|\ |$), 10
    as distance on the number line, 12
    equations involving, 12
    of a function, 91
    inequalities involving, 13
    properties of, 11
Absolute-value function, 90
Acceleration due to gravity:
    on the Earth, 84
    as a function of latitude, 225–226
    on the Moon, 87
Acute angle, 211
Addition of functions:
    definition of, 96
    domain of, 97
Addition of two sinusoidal functions, 280
Adjacent side (adj), 285
Agnesi, Maria, 436
Algebra, Fundamental Theorem of, 161
Algebraic expression, 34–35
Algebraic function, 199
Ambient temperature, 354
Ambiguous case (Law of Sines), 302
Amplitude:
    of cosine graphs, 228–230, 232
    definition of, 228
    of sine graphs, 228–230, 232
Angels flight (Los Angeles), 322
Angle(s):
    acute, 211
    between two intersecting lines, 271
    central, 209
    complementary, 211
    conversions between degrees and
        radians, 210–211
    coterminal, 207
    definition of, 207
    of depression, 291
    of elevation, 291
    of inclination of a line, 271
    initial side of, 207
    obtuse, 211
    quadrantal, 211
    reference, 222
    right, 211
    standard position of, 207
    straight, 211
    supplementary, 211
    terminal side of, 207
    vertex of, 207
Angular measure:
    degrees, 207, 208
    minutes, 208
    negative, 207
    positive, 207
    radians, 209–210
    seconds, 208
Angular speed, 215
Annuity, 377
Aphelion, 392, 394, 425
Apogee, 425
Apollonius, 379
Apolune, 425
Apothem, 299
Approaching infinity:
    from the negative direction ($x \to -\infty$), 54, 177
    from the positive direction ($x \to \infty$), 54, 177
Approaching a number:
    from the left ($x \to a^-$), 39, 176
    from the left and the right ($x \to a$), 39, 177
    from the right ($x \to a^+$), 39, 177
Approximating the zeros of a function, 56
Arc length on a circle, 212
Arccosecant function, 268, 269
Arccosine function, 265
Arccotangent function, 268, 269
Arcsecant function, 268, 269
Arcsine function, 263
Arctangent function, 266
Area:
    of a circular sector, 213
    of a circular segment, 327
    of an equilateral triangle, 290
    Heron's formula for, 309
    inscribed right triangle, 289
    of the intersection of two circles, 217, 299, 310
    of an isosceles triangle, 289
    of an oblique triangle, 299–300
    of a regular polygon, 317
    of a rhombus, 310
    of a slice of pizza, 213
    of a trapezoid, 201
    under a graph, 60, 194–198
Arithmetic combinations of functions:
    definition of, 96
    domain of, 97
Arithmetic mean, 10
Arrow notation, 39, 176–177
Asteroid Hildago, 393
Asymptotes:
    of exponential functions, 331
    finding, 179–180, 182
    horizontal, 178
    of a hyperbola, 397–398, 399
    of logarithmic functions, 337–338
    oblique, 178
    of rational functions, 178–179, 182
    slant, 178
    of trigonometric functions, 242–243
    vertical, 177

Index

of orbit of dwarf planet Pluto, 393
of orbit of planet Earth, 392
of orbit of planet Mercury, 425
of a parabola, 422, 423
E. Coli bacteria, 352
Effective half-life, 361
Eiffel tower, 297
El Castillo, 324–325
Elevation, angle of, 291
Eliminating the parameter, 431
Ellipse:
    center of, 387, 388, 390
    definition of, 387
    eccentricity of, 391, 422, 423
    focal width of, 395
    foci of, 387
    length of major axis, 389
    major axis of, 388–389
    minor axis of, 388–389
    polar equation of, 423
    reflection property of, 392
    standard form of, 388, 337
    vertices of, 388–389
Ellipse Park in Washington DC, 395
End behavior of a function:
    definition of, 53, 143–144
    types of, 144
Endpoints of an interval, 4
Envelope curves, 335
Equation of motion, 282
Equation in two variables:
    definition of, 23
    equivalent, 24
    graph of, 24
    satisfying, 24
    solution set of, 24
Equilateral arch, 327
Equilateral triangle, 290
Equilibrium position, 282
Equivalent equations, 24
Equivalent inequalities, 2
Eratosthenes, 215
Euler, Leonhard, 365
Even function, 61, 220
Existence of a limit, 39–40
Expansion of a binomial, 36
Explicitly defined function, 104
Exponent, 329
Exponential equations, 344–345
Exponential function:
    asymptote of, 331
    base of, 329
    definition of, 329
    domain of, 329
    graphs of, 330–334
    natural, 290–291

    one-to-one property of, 345
    properties of, 331
Exponential mathematical models, 351
Exponents, laws of, 330
Extraneous solutions, 276, 346
Extremum, relative, 145
Eye pupil diameter, 364

F

Factor Theorem, 160
Factored expression, 35, 160, 162
Factorial function, 59
Factoring, 35
Factorization:
    of the difference of two cubes, 35
    of the difference of two squares, 35
    of a polynomial, 157, 160, 161–162
    of the sum of two cubes, 35
Fahrenheit to Celsius conversion, 77
Family of lines, 73
Ferrari, Lodovico, 169
First octant, 439
First quadrant, 17
Flatiron building, 322
Floor function ($\lfloor\ \rfloor$), 89
Focal chord, 381, 395
Focal width, 386, 395
Focus:
    of a conic section, 379, 387, 396, 422
    of an ellipse, 387, 422
    of a hyperbola, 396, 422
    of a parabola, 379, 422
Folium of Descartes, 106
Fontana, Niccolò, 169
Frac function, 96
Fractional expression:
    addition of, 37
    definition of, 34–35
Freely falling object, 84
Frequency of simple harmonic motion, 282
Function(s):
    absolute-value, 90–91
    absolute value of a, 91
    algebraic, 199
    arccosecant, 268, 269
    arccosine, 265
    arccotangent, 268, 269
    arcsecant, 268, 269
    arcsine, 263
    arctangent, 266
    arithmetic combinations of, 96
    ceiling, 95
    compositions of, 97–98
    constant, 70
    continuous, 90, 148

L

Lascaux cave drawings, 359
Latitude, 226, 326–327
Latitude, circle of, 326–327
Law of Cosines, 305
Law of Sines, 300
Law of Universal Gravitation, 392
Laws of exponents, 329–330
Laws of logarithms, 339–340
LAX, 364
Leading coefficient of a polynomial, 141
Leaning Tower of Pisa, 297
Left-handed coordinate system, 439
Legs of a right triangle, 285
Leibniz, Wilhelm Gottfried, 198
Lemniscate, 416
Length of major axis, 393
Length of minor axis, 393
Length of transverse axis, 393
Lens equation, 48
Less than ($<$), 2
Less than or equal to ($\leq$), 1–2
Libby, Willard, 354
Lighthouse of Alexandria, 296
Limaçon:
   convex, 414
   definition, 414
   dimpled, 414
   with an interior loop, 414
Limit(s):
   of a difference quotient, 128, 129
   existence of, 39
   evaluating, 40
   indeterminate form of, 40, 312
   nonexistence of, 43
   notation for, 39
   one-sided, 43
   trigonometric, 311–314
Line of latitude, 326
Line of longitude, 326
Line(s) in the plane:
   family of, 73, 78
   graphs of, 71, 74
   horizontal, 73
   intercepts of, 75
   parallel, 73–74
   parametric equations of, 436
   perpendicular, 73–74
   point-slope form of, 72
   points of intersection of, 75
   slope-intercept form of, 73
   vertical, 73
   with no slope, 71
   with slope, 72
Linear depreciation, 9, 77
Linear equation:
   in three variables, 442
   in two variables, 70
Linear function:
   definition of, 71, 141, 443
   domain of, 71
Linear inequality in one variable, 5
Linear speed, 215
Local behavior, 142, 145
Local extremum:
   definition of, 145
   maximum, 145
   minimum, 145
Location of a cell phone, 216
Logarithmic equations, 344–345
Logarithmic function:
   asymptote of, 338
   common, 339
   definition of, 337
   domain of, 337
   equations involving, 344
   graph of, 337
   as an inverse, 336
   natural, 339
   one-to-one property of, 345
   properties of, 338
Logarithmic mathematical models, 356–358
Logarithms:
   base e, 339
   base 10, 339
   change of base for, 348
   laws of, 340
Logistic function, 352
Longitude, 326
LORAN, 402
Losing solutions, 273

M

Mach number, 256, 279
Major axis of an ellipse:
   definition of, 388
   length of, 389
Malthus, Thomas R., 351
Mathematica, 96, 106, 172, 341, 444
Mathematical model, 351
Maximum, 141
Mercury's orbit, polar equation of, 425–426
Meridian, 326
Midpoint of a line segment:
   on the real number line, 12
   in 3-space, 440
   in 2-space, 21
Minimum, 141
Minor axis of an ellipse:
   definition of, 388
   length of, 389
Minus infinity, 4

Index

negative direction on, 1
origin on, 1
positive direction on, 1
Real numbers:
  rational, 1
  irrational, 1
Real part of a complex number, 159
Reciprocal function, 52, 176
Reciprocal trigonometric functions, 241
Rectangular coordinate system:
  axes of, 17
  coordinates of a point in, 18
  definition of, 17
  distance formula in, 20
  origin of, 17
  quadrants in, 17
  translation in, 63
  in 3-space, 438–439
  in 2-space, 17
Rectangular hyperbola, 405
Reducing powers of sine and cosine, 254
Reed, Lowell, 358
Reference angle, 222
Reflecting surfaces, 384, 387, 392, 403
Reflection of a graph:
  in x-axis, 29, 64–65, 79
  in y-axis, 29, 64–65
  in the line $y = x$, 112
Reflection property:
  of an ellipse, 392
  of a hyperbola, 403
  of a parabola, 384, 387
Refraction of light, 279
Regular n-gon, 317
Regular partition, 195
Regular polygon, 299
Relative extremum:
  definition of, 145
  maximum, 145
  minimum, 145
Remainder, 154
Remainder Theorem, 155
Repeated zero, 146
Repeating decimals, 1
Restricted domains, 113, 263, 265, 266, 269
Rhombus, 310
Richter, Charles F., 356
Richter scale, 356
Right angle, 211
Right cylinder, 126
Right-hand rule, 439
Right triangle:
  definition of, 211
  inscribed in a circle, 289
  solving a, 287
  trigonometry, 285

Rigid transformation of a graph:
  definition of, 63
  horizontal shifts, 63, 233
  vertical shifts, 63, 231
  reflections, 64
Rise, 71
Robotic arm, 308
Root of an equation, 55
Rose curves:
  definition of, 414–415
  number of loops or petals, 415
Rotated conics, 425
Rotation of polar graphs, 417–418, 425
Run, 71

S

Saarinen, Eero, 370
Sample points, 194
Satisfying an equation, 24
Secant function:
  algebraic signs of, 239
  asymptotes of, 243
  cycle of, 243
  definition of, 218, 239, 286
  domain of, 239
  even function, 242
  graph of, 243
  inverse of, 268, 269
  period of, 241–242, 244
  Pythagorean identity for, 240, 241
  range of, 239
  symmetry of graph of, 243
  values of, 240
Secant, hyperbolic:
  definition of, 267
  graph of, 268
Secant line, 127
Seconds (angular measure), 208
Sector of a circle, 212–213
Segment of a circle, 327
Semicircles, 27, 104
Semi-ellipse, 395
Semimajor axis, 393
Semiminor axis, 393
Set-builder notation, 3, 53
Set(s):
  disjoint, 6
  intersection of, 4
  interval notation for, 3–4
  union of, 6
SFO, 292
Shift, phase, 232–233
Shifted graphs:
  of a cosine function, 232, 233
  of an ellipse, 390

Index

Review of Formulas from Geometry

Area A, Circumference C, Volume V, Surface Area S

RECTANGLE	PARALLELOGRAM	TRAPEZOID
$A = lw, \; C = 2l + 2w$	$A = bh$	$A = \frac{1}{2}(a + b)h$

RIGHT TRIANGLE	TRIANGLE	EQUILATERAL TRIANGLE
Pythagorean Theorem: $c^2 = a^2 + b^2$	$A = \frac{1}{2}bh, \; C = a + b + c$	$h = \frac{\sqrt{3}}{2}s, \; A = \frac{\sqrt{3}}{4}s^2$

CIRCLE	CIRCULAR RING	CIRCULAR SECTOR
$A = \pi r^2, \; C = 2\pi r$	$A = \pi(R^2 - r^2)$	$A = \frac{1}{2}r^2\theta, \; s = r\theta$

ELLIPSE	ELLIPSOID	SPHERE
		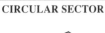
$A = \pi ab$	$V = \frac{4}{3}\pi abc$	$V = \frac{4}{3}\pi r^3, \; S = 4\pi r^2$

RIGHT CYLINDER	RIGHT CIRCULAR CYLINDER	RECTANGULAR PARALLELEPIPED
$V = Bh, \; B$ area of base	$V = \pi r^2 h, \; S = 2\pi rh$ (lateral side)	$V = lwh, \; S = 2(hl + lw + hw)$

CONE	RIGHT CIRCULAR CONE	FRUSTUM OF A CONE
$V = \frac{1}{3}Bh, \; B$ area of base	$V = \frac{1}{3}\pi r^2 h, \; S = \pi r \sqrt{r^2 + h^2}$	$V = \frac{1}{3}\pi h(r_1^2 + r_1 r_2 + r_2^2)$

Review of Trigonometry

Unit Circle Definition of Sine and Cosine

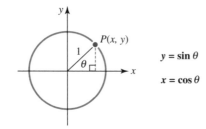

$$y = \sin\theta$$
$$x = \cos\theta$$

Other Trigonometric Functions

$$\tan\theta = \frac{y}{x} = \frac{\sin\theta}{\cos\theta}, \qquad \cot\theta = \frac{x}{y} = \frac{\cos\theta}{\sin\theta}$$

$$\sec\theta = \frac{1}{x} = \frac{1}{\cos\theta}, \qquad \csc\theta = \frac{1}{y} = \frac{1}{\sin\theta}$$

Right Triangle Definition of Sine and Cosine

$$\sin\theta = \frac{\text{opp}}{\text{hyp}}$$

$$\cos\theta = \frac{\text{adj}}{\text{hyp}}$$

Other Trigonometric Functions

$$\tan\theta = \frac{\text{opp}}{\text{adj}}, \qquad \cot\theta = \frac{\text{adj}}{\text{opp}}$$

$$\sec\theta = \frac{\text{hyp}}{\text{adj}}, \qquad \csc\theta = \frac{\text{hyp}}{\text{opp}}$$

Signs of Sine and Cosine

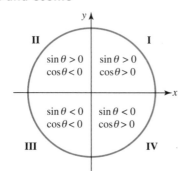

Values of Sine and Cosine for Special Angles

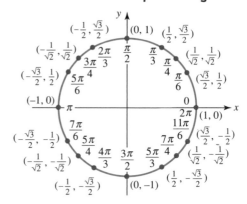

Bounds for Sine and Cosine

$$-1 \le \sin\theta \le 1 \quad \text{and} \quad -1 \le \cos\theta \le 1$$

Periodicity of Trigonometric Functions

$$\sin(\theta + 2\pi) = \sin\theta, \qquad \cos(\theta + 2\pi) = \cos\theta$$

$$\sec(\theta + 2\pi) = \sec\theta, \qquad \csc(\theta + 2\pi) = \csc\theta$$

$$\tan(\theta + \pi) = \tan\theta, \qquad \cot(\theta + \pi) = \cot\theta$$

Cofunction Identities

$$\sin\left(\frac{\pi}{2} - \theta\right) = \cos\theta$$

$$\cos\left(\frac{\pi}{2} - \theta\right) = \sin\theta$$

$$\tan\left(\frac{\pi}{2} - \theta\right) = \cot\theta$$

Pythagorean Identities

$$\sin^2\theta + \cos^2\theta = 1$$

$$1 + \tan^2\theta = \sec^2\theta$$

$$1 + \cot^2\theta = \csc^2\theta$$

Even/Odd Identities

Even	Odd
$\cos(-\theta) = \cos\theta$	$\sin(-\theta) = -\sin\theta$
$\sec(-\theta) = \sec\theta$	$\csc(-\theta) = -\csc\theta$
	$\tan(-\theta) = -\tan\theta$
	$\cot(-\theta) = -\cot\theta$